Electrochemistry and Photo-Electrochemistry of Nanomaterials

Electrochemistry and Photo-Electrochemistry of Nanomaterials
Fundamentals and Applications

Edited by

GHULAM YASIN
School of Environment and Civil Engineering,
Dongguan University of Technology, Guangdong,
P.R. China

SHUMAILA IBRAHEEM
Institute for Advanced Study, College of Physics and
Optoelectronic Engineering, Shenzhen University,
Shenzhen, Guangdong, P.R. China

ANUJ KUMAR
Department of Building Energy Efficiency, CSIR-Central
Building Research Institute, Roorkee, Uttarakhand, India

TUAN ANH NGUYEN
Institute for Tropical Technology, Vietnam Academy of
Science and Technology, Hanoi, Vietnam

THANDAVARAYAN MAIYALAGAN
Department of Chemistry, SRM Institute of Science and
Technology, Chennai, Tamil Nadu, India

ELSEVIER

Elsevier
Radarweg 29, PO Box 211, 1000 AE Amsterdam, Netherlands
125 London Wall, London EC2Y 5AS, United Kingdom
50 Hampshire Street, 5th Floor, Cambridge, MA 02139, United States

Copyright © 2025 Elsevier Inc. All rights are reserved, including those for text and data mining, AI training, and similar technologies.

Publisher's note: Elsevier takes a neutral position with respect to territorial disputes or jurisdictional claims in its published content, including in maps and institutional affiliations.

No part of this publication may be reproduced or transmitted in any form or by any means, electronic or mechanical, including photocopying, recording, or any information storage and retrieval system, without permission in writing from the publisher. Details on how to seek permission, further information about the Publisher's permissions policies and our arrangements with organizations such as the Copyright Clearance Center and the Copyright Licensing Agency, can be found at our website: www.elsevier.com/permissions.

This book and the individual contributions contained in it are protected under copyright by the Publisher (other than as may be noted herein).

Notices
Knowledge and best practice in this field are constantly changing. As new research and experience broaden our understanding, changes in research methods, professional practices, or medical treatment may become necessary.

Practitioners and researchers must always rely on their own experience and knowledge in evaluating and using any information, methods, compounds, or experiments described herein. In using such information or methods they should be mindful of their own safety and the safety of others, including parties for whom they have a professional responsibility.

To the fullest extent of the law, neither the Publisher nor the authors, contributors, or editors, assume any liability for any injury and/or damage to persons or property as a matter of products liability, negligence or otherwise, or from any use or operation of any methods, products, instructions, or ideas contained in the material herein.

ISBN: 978-0-443-18600-4

For Information on all Elsevier publications
visit our website at https://www.elsevier.com/books-and-journals

Publisher: Matthew Deans
Acquisitions Editor: Stephen Jones
Editorial Project Manager: Sara Greco
Production Project Manager: Anitha Sivaraj
Cover Designer: Matthew Limbert

Typeset by MPS Limited, Chennai, India

Contents

List of contributors		ix
1.	**Nanomaterials for electrochromic application**	**1**
	M.S. Sekhavat and F.E. Ghodsi	
	Key points	1
	1.1 Introduction	1
	1.2 Essential parameters for conventional electrochromic device performance	4
	1.3 Electrochromic nanomaterials based on transition metal oxide	7
	1.4 Electrochromic nanomaterials based on Prussian blue	14
	1.5 Electrochromic based on organic nanomaterials	25
	1.6 Conclusion	28
	References	29
2.	**Electrochemical biosensing**	**35**
	Lamees Abbas, Maria Hany, Mariam Alnaqbi, Amani Al-Othman and Muhammad Tawalbeh	
	2.1 Introduction	35
	2.2 Types of biosensors	38
	2.3 Applications	44
	2.4 Advantages	51
	2.5 Limitations	52
	2.6 Conclusions	53
	References	54
3.	**Solar cells at the nanoscale**	**59**
	Leanne Shahin, Aya ElGazar, Taima Al Hazaimeh, Abdullah Ali, Amani Al-Othman and Muhammad Tawalbeh	
	3.1 Introduction	59
	3.2 Solar cells	60
	3.3 Nanotechnology materials in solar cells	66
	3.4 Advantages of nanoscale solar cells	70
	3.5 Challenges and limitations of nanotechnology in solar cells	80
	3.6 Conclusions	81
	References	82

vi   Contents

4. Nanomaterials for photo-electrochemical solar cells   89
Bahadur Ali

4.1	Introduction	89
4.2	Fundamentals of photo-electrochemical solar cells	99
4.3	Types of nanomaterials in photo-electrochemical solar cells	102
4.4	Fabrication methods for nanomaterial-based photo-electrochemical solar cells	120
4.5	Characterization of nanomaterials for photo-electrochemical solar cells	123
4.6	Role of nanomaterials in enhancing the performance of photo-electrochemical solar cells	126
4.7	Challenges and opportunities in the use of nanomaterials for photo-electrochemical solar cells	127
4.8	Summary and concluding remarks	129
	References	131

5. Nanomaterials for fuel cells' electrodes   133
Hafsah Azfar Khan, Bana Al Kurdi, Hind Alqassem, Abdullah Ali, Amani Al-Othman and Muhammad Tawalbeh

5.1	Introduction	133
5.2	Types and applications of fuel cells	134
5.3	Applications of fuel cells	139
5.4	State-of-the-art nanomaterials for fuel cell electrodes	140
5.5	Novel materials for nanoelectrodes in fuel cells	144
5.6	Conclusions	151
	References	152

6. Nanoelectrochemistry in microbial fuel cells   159
Yunfeng Qiu, Yanxia Wang, Xusen Cheng, Yanping Wang, Qingwen Zheng, Zheng Zhang, Zhuo Ma and Shaoqin Liu

6.1	Introduction	159
6.2	Overview of microbial fuel cells	161
6.3	Factors influencing anodic nanoelectrochemistry	170
6.4	Anodic modification materials	177
6.5	Reconstructing EET pathways in microorganisms	191
6.6	Conclusions and perspectives	197
	References	200

Contents vii

7. Nanoelectrochemistry in next generation lithium batteries   211

Moon San, Do Youb Kim, Myeong Hwan Lee, Jungdon Suk and
Yongku Kang

7.1 Introduction	211
7.2 Nanoelectrochemistry in the cathode materials	212
7.3 Nanoelectrochemistry in the anode materials	222
7.4 Nanoelectrochemistry in the electrolytes	232
7.5 Nanoelectrochemistry in the separators	237
References	243

8. Coordination materials for supercapacitors   251

Diab Khalafallah, Mohamed S. Abdel-Latif, Mohamed A Ibrahim
and Qinfang Zhang

8.1 Introduction	251
8.2 An overview of electrochemical SCs	254
8.3 Cost-efficient coordination materials for electrochemical SCs	263
8.4 Manipulating the energy storage capability of electrodes	265
8.5 Conclusions and perspectives	282
References	283

9. Coordination materials for metal−sulfur batteries   287

Dominika Capková and Miroslav Almáši

9.1 Introduction	287
9.2 Mechanism of metal−sulfur batteries	289
9.3 Challenges of metal−sulfur batteries	291
9.4 Requirements for host materials	291
9.5 5 Lithium−sulfur batteries	293
9.6 Other metal−sulfur battery systems	308
9.7 Conclusion	319
Acknowledgments	320
References	320

10. Nano/photoelectrochemistry for environmental applications   333

Zahraa Alqallaf, Hamda Bukhatir, Fayne D'Souza, Abdullah Ali,
Amani Al-Othman and Muhammad Tawalbeh

10.1 Introduction	333
10.2 Nanomaterials used in environmental applications	335
10.3 Environmental applications of nanoelectrochemistry	338
10.4 Conclusions	351
References	351

viii Contents

11. Nanomaterials for electrochemical chlorine evolution reaction in the Chlor-alkali process 359

Waseem Ahmad, Kaidi Zhang, Yu Zou, Liang Wang, Mengyang Dong, Huai Qin Fu, Huajie Yin, Yonggang Jin, Porun Liu and Huijun Zhao

11.1 Introduction	359
11.2 Chlor-alkali process	361
11.3 Electrocatalysts for CER	370
11.4 Conclusion and future perspectives	385
References	389

12. Photothermal properties of metallic nanostructures for biomedical application 395

Dorothy Bardhan and Sujit Kumar Ghosh

12.1 Introduction	395
12.2 Nonradiative properties	398
12.3 Theory of plasmonic heating	400
12.4 Laser heating of nanostructures	403
12.5 Biological tissue transparency window	405
12.6 Photothermal nanostructures	407
12.7 Applications	408
12.8 Challenges that lie ahead	414
12.9 Conclusions and outlook	416
References	417

13. Nanophoto/electrochemistry for green energy production 427

Rana Ahmed Aly, Abdulwahab Alaamer, Tala Ashira, Saeed Najib Alkhajeh, Abdullah Ali, Amani Al-Othman and Muhammad Tawalbeh

13.1 Introduction	427
13.2 Characteristic properties of nanomaterials and conventional nanomaterials	428
13.3 Electrochemistry and photoelectrochemistry of nanomaterials technologies	431
13.4 Industrial applications, challenges, and limitations	445
13.5 Conclusions	448
References	449

Index 453

List of contributors

Lamees Abbas
Department of Chemical and Biological Engineering, American University of Sharjah, Sharjah, United Arab Emirates

Mohamed S. Abdel-Latif
Engineering Physics and Mathematics Department, Faculty of Engineering, Tanta University, Tanta, Egypt

Waseem Ahmad
Centre for Catalysis and Clean Energy, Gold Coast Campus, Griffith University, Gold Coast, QLD, Australia

Abdulwahab Alaamer
Department of Chemical and Biological Engineering, American University of Sharjah, Sharjah, United Arab Emirates

Abdullah Ali
Department of Chemical and Biological Engineering, American University of Sharjah, Sharjah, United Arab Emirates

Bahadur Ali
Department of Chemistry, College of Science, Mathematics, and Technology, Wenzhou-Kean University, Wenzhou, China

Saeed Najib Alkhajeh
Department of Chemical and Biological Engineering, American University of Sharjah, Sharjah, United Arab Emirates

Miroslav Almáši
Department of Inorganic Chemistry, Faculty of Sciences, Pavol Jozef Šafárik University in Košice, Košice, Slovak Republic

Mariam Alnaqbi
Department of Chemical and Biological Engineering, American University of Sharjah, Sharjah, United Arab Emirates

Amani Al-Othman
Department of Chemical and Biological Engineering, American University of Sharjah, Sharjah, United Arab Emirates

Zahraa Alqallaf
Department of Chemical and Biological Engineering, American University of Sharjah, Sharjah, United Arab Emirates

Hind Alqassem
Department of Chemical and Biological Engineering, American University of Sharjah, Sharjah, United Arab Emirates

x   List of contributors

Rana Ahmed Aly
Department of Chemical and Biological Engineering, American University of Sharjah, Sharjah, United Arab Emirates

Tala Ashira
Department of Chemical and Biological Engineering, American University of Sharjah, Sharjah, United Arab Emirates

Dorothy Bardhan
Department of Chemistry, Assam University, Silchar, Assam, India

Hamda Bukhatir
Department of Chemical and Biological Engineering, American University of Sharjah, Sharjah, United Arab Emirates

Dominika Capková
Department of Physical Chemistry, Faculty of Sciences, Pavol Jozef Šafárik University in Košice, Košice, Slovak Republic; Department of Chemical Sciences, Bernal Institute, University of Limerick, Limerick, Ireland; Department of Electrical and Electronic Technology, Faculty of Electrical Engineering and Communication, Brno University of Technology, Brno, Czech Republic

Xusen Cheng
Faculty of Life Science and Medicine, School of Medicine and Health, Harbin Institute of Technology, Harbin, Longjiang Hei, P.R. China

Mengyang Dong
Centre for Catalysis and Clean Energy, Gold Coast Campus, Griffith University, Gold Coast, QLD, Australia

Fayne D'Souza
Department of Chemical and Biological Engineering, American University of Sharjah, Sharjah, United Arab Emirates

Aya ElGazar
Department of Chemical and Biological Engineering, American University of Sharjah, Sharjah, United Arab Emirates

Huai Qin Fu
Centre for Catalysis and Clean Energy, Gold Coast Campus, Griffith University, Gold Coast, QLD, Australia

F.E. Ghodsi
Department of Physics, Faculty of Science, University of Guilan, Namjoo Avenue, Rasht, Guilan, Iran

Sujit Kumar Ghosh
Physical Chemistry Section, Department of Chemistry, Jadavpur University, Kolkata, West Bengal, India

Maria Hany
Department of Chemical and Biological Engineering, American University of Sharjah, Sharjah, United Arab Emirates

Taima Al Hazaimeh
Department of Chemical and Biological Engineering, American University of Sharjah, Sharjah, United Arab Emirates

Mohamed A Ibrahim
Faculty of Engineering, Aswan University, Aswan, Egypt

Yonggang Jin
Commonwealth Scientific and Industrial Research Organization (CSIRO) Mineral Resources, Pullenvale, QLD, Australia

Yongku Kang
Advanced Energy Materials Research Center, Korea Research Institute of Chemical Technology (KRICT), Daejeon, Republic of Korea; Department of Chemical Convergence Materials, University of Science & Technology (UST), Daejeon, Republic of Korea

Diab Khalafallah
School of Materials Science and Engineering, Yancheng Institute of Technology, Yancheng, P.R. China; Mechanical Design and Materials Department, Faculty of Energy Engineering, Aswan University, Aswan, Egypt

Hafsah Azfar Khan
Department of Chemical and Biological Engineering, American University of Sharjah, Sharjah, United Arab Emirates

Do Youb Kim
Advanced Energy Materials Research Center, Korea Research Institute of Chemical Technology (KRICT), Daejeon, Republic of Korea

Bana Al Kurdi
Department of Chemical and Biological Engineering, American University of Sharjah, Sharjah, United Arab Emirates

Myeong Hwan Lee
Advanced Energy Materials Research Center, Korea Research Institute of Chemical Technology (KRICT), Daejeon, Republic of Korea

Porun Liu
Centre for Catalysis and Clean Energy, Gold Coast Campus, Griffith University, Gold Coast, QLD, Australia

Shaoqin Liu
Faculty of Life Science and Medicine, School of Medicine and Health, Harbin Institute of Technology, Harbin, Longjiang Hei, P.R. China

Zhuo Ma
Faculty of Life Science and Medicine, School of Medicine and Health, Harbin Institute of Technology, Harbin, Longjiang Hei, P.R. China

Yunfeng Qiu
Faculty of Life Science and Medicine, School of Medicine and Health, Harbin Institute of Technology, Harbin, Longjiang Hei, P.R. China

xii List of contributors

Moon San
Advanced Energy Materials Research Center, Korea Research Institute of Chemical
Technology (KRICT), Daejeon, Republic of Korea

M.S. Sekhavat
Department of Physics, Faculty of Science, University of Guilan, Namjoo Avenue, Rasht,
Guilan, Iran

Leanne Shahin
Department of Chemical and Biological Engineering, American University of Sharjah,
Sharjah, United Arab Emirates

Jungdon Suk
Advanced Energy Materials Research Center, Korea Research Institute of Chemical
Technology (KRICT), Daejeon, Republic of Korea; Department of Chemical
Convergence Materials, University of Science & Technology (UST), Daejeon,
Republic of Korea

Muhammad Tawalbeh
Sustainable and Renewable Energy Engineering Department, University of Sharjah,
Sharjah, United Arab Emirates; Sustainable Energy & Power Systems Research Centre,
RISE, University of Sharjah, Sharjah, United Arab Emirates

Liang Wang
Centre for Catalysis and Clean Energy, Gold Coast Campus, Griffith University,
Gold Coast, QLD, Australia

Yanping Wang
Faculty of Life Science and Medicine, School of Medicine and Health, Harbin Institute of
Technology, Harbin, Longjiang Hei, P.R. China

Yanxia Wang
Faculty of Life Science and Medicine, School of Medicine and Health, Harbin Institute of
Technology, Harbin, Longjiang Hei, P.R. China

Huajie Yin
Key Laboratory of Materials Physics, Centre for Environmental and Energy
Nanomaterials, Anhui Key Laboratory of Nanomaterials and Nanotechnology, CAS
Center for Excellence in Nanoscience, Institute of Solid State Physics, Hefei Institutes of
Physical Science, Chinese Academy of Sciences, Hefei, Anhui, China

Kaidi Zhang
Centre for Catalysis and Clean Energy, Gold Coast Campus, Griffith University,
Gold Coast, QLD, Australia

Qinfang Zhang
School of Materials Science and Engineering, Yancheng Institute of Technology,
Yancheng, P.R. China

Zheng Zhang
Faculty of Life Science and Medicine, School of Medicine and Health, Harbin Institute of
Technology, Harbin, Longjiang Hei, P.R. China

Huijun Zhao
Centre for Catalysis and Clean Energy, Gold Coast Campus, Griffith University, Gold Coast, QLD, Australia

Qingwen Zheng
Faculty of Life Science and Medicine, School of Medicine and Health, Harbin Institute of Technology, Harbin, Longjiang Hei, P.R. China

Yu Zou
Centre for Catalysis and Clean Energy, Gold Coast Campus, Griffith University, Gold Coast, QLD, Australia

CHAPTER 1

Nanomaterials for electrochromic application

M.S. Sekhavat and F.E. Ghodsi
Department of Physics, Faculty of Science, University of Guilan, Namjoo Avenue, Rasht, Guilan, Iran

Key points

- Nanomaterials with active reaction surface play important role in improving heterogeneous electron transport and homogeneous ion transport for ECD.
- EC nanomaterials can be in the form of nanoparticles, nanodots, nanorods, nanotubes, nanowires, nanofibers, nanofilms, nanosheets, nanoclusters, nanocrystal architectures, and other special structures.
- Electrochromic nanomaterials divided into two groups of inorganic (WO_3, V_2O_5, TiO_2, PB, etc.) and organic (viologens, polyaniline, polypyrrole, etc.) compounds, which are examples of common polymeric and inorganic EC materials based on transition metal oxide.
- The color change in EC nanomaterials originates from the redox processes by applying a potential difference between EC and IS layers.
- Several substrates including glass, textiles, plastic, metal and fibers can be well deposited by present EC nanomaterials.

1.1 Introduction

Smart materials can significantly change their properties under external stimulations. Temperature, humidity, electric or magnetic field, heat, or light are common external stimulations. These materials include electrochromic, mechanochromic, chemochromic, thermochromic, and photochromic. There are many applications for these materials, among which we can mention their use in making windows, cars and airplanes, screens, sunglasses, electrochromic mirrors, etc. (Lampert, 2004). In recent years, due to population growth and global warming, the need for energy in the world has increased. Since most of this consumed energy in buildings, in heating, cooling, and ventilation devices is wasted through the windows

Electrochemistry and Photo-Electrochemistry of Nanomaterials
DOI: https://doi.org/10.1016/B978-0-443-18600-4.00001-6
© 2025 Elsevier Inc. All rights reserved, including those for text and data mining,
AI training, and similar technologies.

of building, researchers tried to make smart windows using electrochromic materials to prevent energy loss (Granqvist et al., 2009).

The term electrochromism was first proposed by Platt (1961) to show a color change by an electric field. Chang (1976), Faughnan and Crandall (1980), Lampert (1984), Byker (1994), and Granqvist (1995) have developed the electrochromism phenomenon. Electrochromism is an occurrence that includes light and electrochemical reactions at the same time. Electrochromic material is a type of material which its optical properties (transmittance, reflectance, absorbance, or color) can be changed reversibly under the application of an external voltage, depending on the insertion and extraction of ions and electrons. The color change of EC materials is due to oxidation—reduction reaction involves transfer of electron. Many electrochromic materials perform electron transfer in the visible/ultraviolet spectral region of the electromagnetic wave spectrum. The changes of colors are between the transparent state (colorless) and the absorption state (colorful) in the visible region, where the change of color occurs in one color (Assis et al., 2016; Yuan et al., 2011). In some electrochromic materials, the possibility of oxidation–reduction reaction exists more than once, and in each oxidation—reduction reaction, a specific color is produced (Argun et al., 2004; Delongchamp & Hammond 2004). In this case, these materials are referred to polyelectrochromic materials with multicolor coloring (Mortimer, 1991).

The most important electrochromic devices are smart windows and are used in modern and energy efficient architecture. Nowadays, on sunny summer days, smart windows with smoked glass are used to control the heat and light of the room. Smart windows allow for the control of various forms of light (visible, IR, UV) by applying electric field. The electric field changes the properties of materials and causes them to absorb or reflect light by creating a chemical reaction. The electric field causes to move the ions in the structure of the electrochromic device from the ion storage layer to the ion conduction layer, return to the electrochromic layer, and darken the glass. When the electric field is cut off, the process reverses and the glass becomes transparent again. One of the characteristics of electrochromic materials is the ability to adjust their color, so that their turbidity can be adjusted by changing the amount of current (Georg et al., 2008). Of course, in recent years, windows called photoelectrochromic smart windows have been proposed, which are very popular among researchers today (Kolay et al., 2022; Lavagna et al., 2021). These windows not only have all the advantages of photochromic ones (such as the ability to control the color condition) but also they are fed by the light of their descent. In these windows, the light absorption process is physically separate from the coloring process, allowing for the optimization of each separately. On the other hand, the limitations in which a photochromic material must have all the characteristics (color change, switching speed, light stability, etc.) alone are greatly reduced) Bechinger et al., 1996).

In principle, the electrochromic device has the same behavior as a rechargeable battery, in which the electrochromic material is separated from a counter electrode by a suitable solid or liquid electrolyte, and color changes occur by charging and discharging the electrochemical cell with an applied potential of several volts give (Somani & Radhakrishnan, 2002). A complete electrochromic device (ECD) is a combination of five layers stacked on top of each other, and these five layers are sandwiched between two substrates in a layered configuration. Two conductive electrodes (semiconductors) are generally made of glass or transparent flexible sheets (such as PET) so that the color changes of the electrochromic material can be observed. There are different layers between the two electrodes (Fig. 1.1). The ion conducting layer is an electrolyte that can be an organic (viscous polymer) or inorganic (most often oxide films are used) substance and it allows the movement of ions from the ion storage layer to the electrochromic layer, which leads to oxidation. The material becomes electrochromic and changes its color (Pawlicka, 2009). Glass or transparent plastic is also used to protect the inner layers. Of course, by covering the outer surface of the transparent layer, this device attains

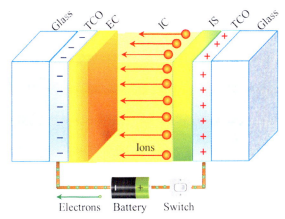

Figure 1.1 A schematic illustration of an electrochromic device. Transparent conductor oxide (TCO), electrochromic (EC), ion conductor (IC), and ion storage (IS) layers.

reflective properties and turns into electrochromic mirrors. The performance of the electrochromic device due to the application of a small voltage is equivalent to 1.5 to 2 V.

Nanomaterials with active reaction surface can play an important role in optimizing heterogeneous electron transport and homogeneous ion transport for ECD and other optoelectronic devices. In recent years, the use of a variety of new methods to introduce nanomaterials to make ECD with much better performance is rapidly developing. We hope that this review will be a way to open a special look at more and better use of nanomaterials to improve ECD properties and optoelectronic technologies. Here, the investigations are reviewed with an emphasis on the EC layer and optimization methods that include increasing the level of active reactions between the electrodes. By reducing the dimensions to the nanoscale, we especially encounter an increase in the porosity and roughness of the surface, which due to physical and chemical changes, causes the electrical and optical characteristics to change (Pham et al., 2004; Zhao et al., 2017). In addition, some conductive nanomaterials can be used as conductive substrates to replace traditional indium tin oxide (ITO) or fluorine-doped tin dioxide (FTO), which suffer from brittleness defects (Liu et al., 2017; Lu et al. 2015), which are a way to produce flexible and soft ECDs of the next generation. We also have hybrid nanostructured EC materials that include two categories: the first category is nanostructured EC materials that are combined with nonnanostructured EC materials, and the second category includes chemical/physical combinations of various nanostructured EC materials. The principle of using hybrid materials is to eliminate the defects of EC materials and increase their efficiency.

From the point of view of morphology and structure, nanomaterials can be divided into four categories: zero-dimensional (0D) materials, which mainly refer to one-dimensional nanoparticles and nanodots; one-dimensional (1D) materials, which include nanorods, nanotubes, nanowires, and nanofibers; two-dimensional (2D) materials, which are nanofilms, nanosheets; and three-dimensional (3D) materials, which include nanoclusters, nanocrystal architectures, and other special structures (Yang et al., 2020).

1.2 Essential parameters for conventional electrochromic device performance

The performance of an electrochromic device depends on different factors. The type of electrochromic material which is used in the device, the method of preparing electrochromic film, the amount of applied voltage, the type of electrodes and electrolyte are some important factors among these. Electrochromic material can have various color efficiency, switching time, color storage memory, color contrast percentage, contrast ratio, and stability of different color cycles (Argun, 2004). Switching time is the time that takes for the electrochromic material to change their color after applying voltage. Therefore, the coloration time (τ_c) describes the time it takes for the material to change from transparent to absorbent mode, whereas the bleaching time (τ_b) is defined as the time that the reverse process is undertaken. The ionic conductivity of electrolytes, ion diffusion process inside the device, and electronic conductivity of electrodes can affect the switching time of coloring or bleaching. The coloration efficiency (CE (λ)) at a certain wavelength is the color change density (ΔOD) per electric charge density (Q_s) introduced into the electrochromic material. The unit of coloration efficiency is square centimeter per Coulomb. The color change density is the logarithm of the transmission percentage in the bleaching (colorless) state (T_b) to the transmission percentage in the colored state (T_c) at the maximum wavelength. In other words, the color change density is the amount of surface color change given in an electrochromic layer, by applying a unit of electric charge per area. The coloration efficiency is given by the Eq. 1.1a (Sekhavat & Ghodsi 2023):

$$CE(\lambda) = \frac{\Delta OD}{Q_s} \tag{1.1a}$$

where

$$\Delta OD = \log \frac{T_b}{T_c} \tag{1.1b}$$

The electric charge density (Q_s) as the stored charge in ECDs is dependent on the applied voltage, which is related to the areal capacitance (C_s) according to following equation (Falahatgar et al., 2014):

$$Q_s = C_s(V_f - V_i) = \frac{1}{v} \int_{V_i}^{V_f} I(v)dv, \tag{1.2}$$

where V_i and V_f are initial and final voltage, respectively, I the current density (A/cm^2) and v is sweep rate (mV/s), ΔV (V_f-V_i) is the applied potential window (V), and the integral term is the area under the CV curve. Color storage memory is the duration of the remaining color state of the electrochromic material after the applied voltage is cut off. For instance, when the glass turns from colorless to blue and suddenly the electricity is cut off, it takes about 2 days for the color to disappear. Then, to change the color of the glass again (darker or more transparent), the required voltage must be applied again in the original direction (or vice versa).

The color contrast percentage is the highest percentage of difference that occurs between the colorless (T_b) and colored (T_c) states in one wavelength is called the color contrast percentage and it can be defined as the difference between the most colorful and the most colorless part of an electrochromic material at a certain wavelength:

$$\Delta\%T = \%T_b - \%T_c. \tag{1.3}$$

The contrast ratio is a functional parameter for intensity change between color and noncolor modes in the systems such as ECD, which is obtained from the following relationship:

$$CR(\lambda) = \frac{T_b(\lambda)}{T_c(\lambda)}. \tag{1.4}$$

In electrochromic materials, the ratio of inserted charge (Q_{in}) to extracted charge (Q_{ex}) during the coloration and bleaching states can be defined as a measure of reversibility which is given by (Hopmann et al., 2023):

$$\Gamma = \frac{Q_{in}}{Q_{ex}}. \tag{1.5}$$

If the value of Γ is close to unity, it indicates good reversibility in the oxidation and reduction process. Irreversible insertion of charges into the EC (IS) layer or extraction of the charges into IC material indicates the redox process is unbalanced. The stability of the color cycle is the maintaining color state of the electrochromic material as a result of repeating successive cycles of coloring and bleaching states.

The aforementioned cases were the criteria for comparing and checking the electrochromic properties of different materials. Here, we classify electrochromic nanomaterials into three categories, inorganic, Prussian blue, and organic, and cover some significant studies in the field of electrochromism that have improved electrochromic properties focus on EC layer.

1.3 Electrochromic nanomaterials based on transition metal oxide

The films of a number of transition metal oxides, after being placed as an electrode in suitable electrochemical cells, show electrochromic properties with the reversible ion insertion/extraction mechanism. What causes the reversible color change is the unpaired single electron in the d orbital of metal elements in transition metal oxides/hydroxides, which cause the instability of the electron layer structure. These oxides are divided into two categories, anodic (the oxide of V, Cr, Mn, Fe, Co, Ni, Rh, and Ir metals) and cathodic (the oxide of Ti, Nb, Mo, Ta, and W metals). Anodic oxides change color in oxidation processes and cathodic oxides in reduction processes. The combination of these two groups leads to the creation of a complete electrochromic device. Many transition metal oxides that are prepared and used in the form of films have electrochromic properties, for example, iridium (Buckley et al., 1976), rhodium (Burke & O'Sakan, 1978), ruthenium (Burke & Whelan, 1979), tungsten (Burke et al., 1980), manganese, and cobalt (Burke & Murphy, 1980). The properties of this group of electrochromic materials are classified as EC minerals. One of the most studied materials is tungsten oxide, whose electrochromic properties were first reported in 1969 by Deb (1969).

In Table 1.1, the colors resulting from oxidation−reduction of the most important metal oxides are reported. Preparation of electrochromic minerals is done by different methods such as thermal evaporation, electrochemical deposition, sol−gel, etc. When the electrical conductivity of the electrochromic film is important, the use of metal oxides is preferred due to the superiority of metal oxides over polymeric materials in

Table 1.1 Color change of electrochromic metal oxides (Somani & Radhakrishnan, 2002).

Electrochromic material	natural color	Color in reduced state	Color in oxidation state
MoO_3	Transparent	Blue	—
V_2O_5	Yellow	Pale blue	—
Nb_2O_5	Yellow	Blue	—
WO_3	Transparent	Blue	—
$Ir (OH)_3$	Transparent	—	Blue black
$Ni (OH)_2$	Transparent or pale green	—	Reddish brown

conductivity. These materials are solid in normal state and are used as solid in the electrochromic device.

Tungsten oxide (WO_3) is the most suitable material for energy saving applications due to its better color performance. WO_3 thin films can be prepared by various techniques such as physical vapor deposition, chemical vapor deposition, sol–gel , and magnetron sputtering methods. Tungsten oxide has a nearly cubic structure that can be simply described as a type of hollow perovskite formed by WO_6 octahedra. The empty space inside the cube made it possible for guest ions to enter. Injection and extraction of electrons and metal cations (Li^+, H^+, etc.) play an important role in changing its color. WO_3 is a cathode electrochromic material whose layer color is blue, which can be made colorless by electrochemical oxidation. In the case of Li^+ cations, the electrochemical reaction can be written as the following equation (Somani & Radhakrishnan, 2002):

$$WO_3 + xM^+ + xe^- \rightarrow M_xWO_3,$$
$$\text{Transparent} \qquad\qquad \text{Blue} \tag{1.6}$$

where M^+ represents cations, for example, H^+, Li^+, K^+, and Na^+. When EC films of tungsten trioxide are prepared by traditional methods, usually by magnetron sputtering to make a compact film (Deb, 2008), in addition to the disadvantages such as high cost, complex preparation processes, etc., it was observed that the switching performance of EC films was not fast enough and great enough to be used in ECD construction. Therefore, the researchers, in their studies, synthesized WO_x nanostructures in different sizes and dimensions and investigated them in ECD, where they encountered improved switching and increased electrochromic performances.

Liu et al. rapidly fabricated PVP (vinyl pyrrolidone)-coated $W_{18}O_{49}$ nanowire thin films with a diameter of less than 5 nm and a length of tens of micrometers by modified Langmuir–Blodgett (LB) techniques. The LB technique is a good method for ordering a large number of nanostructures including nanoparticles, nanorods, and nanowires. PVP was used in the work as a surfactant to improve nanowire assembly. Single-layer $W_{18}O_{49}$ nanowires with periodic structures can easily exhibit reversible electrochromic properties between negative and positive voltages. The staining/bleaching time for the $W_{18}O_{49}$ nanowire monolayer was about 2 seconds, which is much faster than traditional tungsten oxide nanostructures. In addition, nanowire devices exhibit excellent stability in color change cycles, which makes it a promising alternative for fabricating a variety of electrochromic devices (Liu et al., 2013). Also, Ma et al. showed that WO_3 nanostructured layers with vertical alignment can be fabricated on FTO-coated glass layers using moldless hydrothermal technique. Their studies showed that a variety of WO_3 nanostructures including 1D nanobricks, 1D nanorods and nanowires, and 3D flower nanorods can be obtained by adjusting the composition of the precursor solution. So that, the shape and size of nanostructures can be controlled by urea content and solvent composition. These nanostructured films exhibited enhanced electrochromic performance due to the large tunnels in the hexagonally structured WO_3 and the large active surface area available for electrochemical reactions. A large optical modulation of 66% and fast switching speeds of 6.7 and 3.4 seconds are achieved for coloration and bleaching, respectively, and a high coloration efficiency of 106.8 cm^2/C for the cylindrical nanorod array film(Ma et al., 2013).

Two-dimensional metal oxides (TMOs) are very attractive and remarkable materials due to their combined advantages of high active surface area, enhanced electrochemical properties, and stability. Among them, two-dimensional tungsten oxide (WO_3) nanosheets have high potential in electrochromic (EC) devices due to their high electrochemical characteristics. Azam et al. achieved a novel solution phase synthesis of 2D WO_3 nanosheets through simple oxidation of 2D tungsten disulfide (WS_2) nanosheets exfoliated from WS_2 bulk powder. The EC device showed a color modulation of 62.57% at 700 nm, which is 3.43 times higher than that of the conventional device using bulk WO_3 powder, while also improving the switching response time. This research shows the promising presence of two-dimensional WO_3 nanosheets for other applications such as sensors, catalysts, thermoelectrics and energy conversion(Azam et al., 2018).

For EC nanomaterials with metal oxide, doping/embedding other types of metals is a common and efficient way to reduce the size of the structure, which often happens by improving the active surface of EC (Wang et al., 2019a). In fact, the difference in the atomic radius of different metals causes a change in the lattice and creating empty space. Such defects cause electron mobility and ion transfer increase, and as a result improve EC properties. Zhan et al. also synthesized Ti-doped WO_3 nanocomposite materials using a facile wet bath method. Their results showed that the regular crystal structure of WO_3 nanosheets (length 100 nm at the beginning) changed to a long irregular crystal structure (size 10—20 nm). For the optimized Ti-doped films, the transmission contrast value and color efficiency were obtained as 67.6% and 87.7 cm^2/ C, respectively, which were greatly improved compared to pure WO_3 (about 43.6% and 48.3 cm^2/C). This performance optimization shows the good effects of Ti-doping in the WO_3 structure, which is the result of reducing the crystallite size and increasing the surface area along with increasing the proton diffusion coefficient for the EC reaction (Zhan et al., 2017).

Despite the features of high color contrast and easy synthesis, WO_3 has poor stability and low electrochromic efficiency in successive cycles. In a research, Li et al. (2023) proposed the combination of amorphous WO_3 (a-WO_3) on crystalline WO_3 (c-WO_3), which was done in order to form a composite film and improve electrochromic properties. Synergistic mechanism of crystalline and amorphous WO_3 and improving the electrochromic properties of the film were investigated. The Li^+ emission of the amorphous/crystalline film was more than three times that of the crystalline film. Detailed experimental data show that this film has high electrochromic properties because the crystalline WO_3 enhances the cyclic stability and the amorphous WO_3 enhances the ion storage and transport capabilities. The crystallinity control strategy increased the electrochromic properties, which can be a promising method to study electrochromism and improve its properties in the future.

Considering that the larger effective surface area or wider grain boundaries of tungsten oxide layers make them more suitable for use in electrochromic devices, so Giannouli and Leftheriotis (2011) proposed a simple method to increase the roughness and thus the effective surface area of WO_3 films. This method is based on the tendency of the peroxy-tungstate precursor to form large aggregates in its solution over time. To achieve this goal, a systematic study of the effect of precursor aging on the properties of the resulting WO_3 film was carried out. Fig. 1.2A shows the scanning electron microscopy (SEM) images of the film surface 6 hours after the preparation of the solution. Fig. 1.2B and C shows the films prepared 23 and 48 hours after the formation of the precursor, respectively, and Fig. 1.2D shows the surface of a film deposited using a precursor solution that was aged for 72 hours. It can be seen that the surface of the film has initially only a few large grains of relatively uniform WO_3 particles. As the solution aging time increases, more cracks and aggregates appear on the resulting films. Large aggregates have a grape-like structure and consist of spherical particles with a diameter of 50 to 200 nm, and large cracks are formed around the aggregates. It was found that as the aging

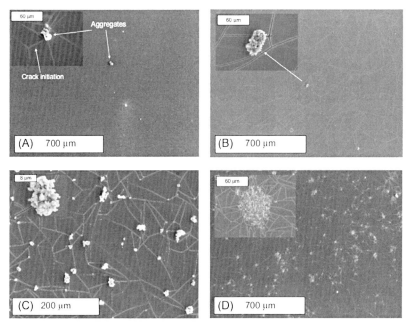

Figure 1.2 SEM images of WO$_3$ films prepared with a solution aged for (A) 6 h, (B) 23 h, (C) 48 h, and (D) 72 h (Giannouli & Leftheriotis, 2011).

time of the precursor solution increases, more and larger aggregates are formed. These aggregates caused large cracks on the surface of the film, thus increasing the effective surface area. The films prepared by this method had the highest voltammetric charge density per film thickness unit. It was also observed that the dyeing efficiency of the films prepared using this method was higher than other films throughout the visible spectrum and especially in the near infrared. This method improved the electrochromic performance of the layers, which is mainly due to the increase of their surface area.

One of the anodic electrochromic materials is V_2O_5 with a layered crystal structure whose color can be reversibly changed between yellow and pale blue (Hsiao et al., 2021). In its crystal structure, the environment of vanadium atoms can be considered as a disordered tetragonal pyramid, where vanadium atoms form V-O bonds with 5 surrounding oxygen atoms. Single crystal V_2O_5 is an oxygen-free semiconductor that has point defects in the form of V^{4+}. Consequently, the color change process of V_2O_5 is under the simultaneous incorporation of Li^+ and e^-, which reduces V^{5+} to V^{4+} according to the following equation and induces color change (Li et al., 2021):

$$V_2O_5 + xM^+ + xe^- \rightarrow M_xV_2O_5 \qquad (1.7)$$
$$\text{Yellow} \qquad\qquad\qquad \text{Pale Blue}$$

Vanadium pentoxide (V_2O_5) is an important material for making electrochromic devices due to its special properties among lithium ions/separators, but it has poor performance in long cycles. Rui et al. (2013) reported that V_2O_5 nanostructures exhibit high current density due to their larger specific surface area. However, the nanostructured films do not exhibit good cyclic stability. In order to improve the intercalation properties of ions, the doping of transition metal ions has been the focus of researchers. Studies have been reported in which the electrochemical properties of V_2O_5 can be gradually improved by doping metal ions in its matrix, such as Mn, Cu, Zn, and Ag (Coustier et al., 1997, 1999; Park 2005). Wei et al. (2015) reported that doping with titanium (Ti) significantly improves the cyclic stability of vanadium and exhibits good performance in electrochromic devices along with WO_3. X-ray diffraction studies show that the presence of Ti element reduces the crystallinity of the V_2O_5 film and changes the structure of the layer and reduced the degree of crystallinity and provided more free space for Li^+ intercalation and deintercalation. The SEM image shows the not-so-smooth surface of pure V_2O_5, which contains tens of nanometer-sized nanoparticles (Fig. 1.3). These nanoparticles create effective contact with electrolyte particles. However, as the Ti content increases, the grain boundaries blur and disappear. Ti-doped layers are softer and more compact than pure V_2O_5 films, and Li^+ diffusion inside the material is slower and more difficult than in liquid electrolyte or along grain boundaries. This reason could explain the lower charge capacity of higher doped V_2O_5 films compared to pure V_2O_5 films during charge/discharge cycles (Fig. 1.4A). The cyclic stability of V_2O_5 films increases with increasing Ti doping. The electrochromic device assembled with the optimized V_2O_5 electrode (V:Ti = 2:1) was observed after 200,000 cycles between the lowest (2%) and the highest (62%) passivity, without significant degradation in performance (Fig. 1.4B). A similar behavior has been investigated in the presence of Mo doped in V_2O_5 (Mjejri et al., 2019).

Wang et al. (2023) developed a photochemical route at room temperature to deposit macroporous a-V_2O_5 layers, which significantly enhances the multicolor transparent properties of EC. They proposed a bubble

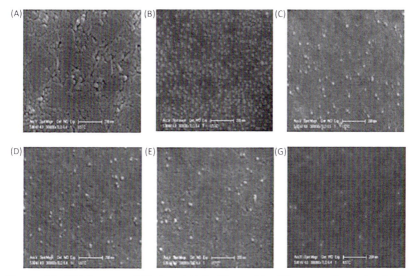

Figure 1.3 SEM images of Ti-doped V$_2$O$_5$ films with different molar ratios (V/Ti): (A) 1:0, (B) 5:1, (C) 4:1, (D) 3:1, (E) 2:1, and (F) 1:1 (Wei et al., 2015).

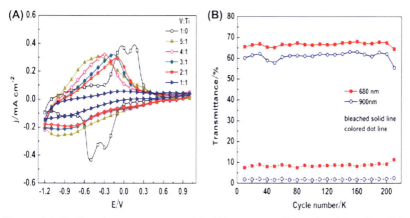

Figure 1.4 Cyclic voltammogram at a 20 mV/s scan rate (A) and stability (B) of Ti-doped V$_2$O$_5$ films with different Ti concentrations (Wei et al., 2015).

patterning mechanism to clarify the generation of large amount of macropores in vanadium oxide films. In this mechanism, UV-induced ozone reacts with chlorine in the vanadium precursor layers to produce gaseous products that lead to the formation of bubbles. Amorphous phase coupled with coarse porosity to V$_2$O$_5$ films with significantly enhanced multicolor

Figure 1.5 (A) Schematic illustration showing the synthesis, (B) SEM image of core (TiO$_2$)/ shell (Co$_3$O$_4$) nanowires array grown on FTO-coated glass substrate. (C) Stability test with response time analysis (Mishra et al., 2018).

EC performance, which can be attributed to the surface quasi-capacitive mechanism. The quasi-capacitive EC process exhibits faster electrochemical kinetics, higher charge density, and less structural degradation than conventional ion insertion/extraction electrochromism, which has been shown to result in increased response times, color palettes, and cyclic stability. This feature enables the fabrication of transparent multicolor EC devices, which show a promising future for the manufacture of sunglasses.

Combining different nanostructured materials is another common way to improve EC performance. For example, Mishra et al. created novel nanostructures by combining TiO$_2$ nanorods and Co$_3$O$_4$ nanoparticles through a facile two-step hydrothermal process following electropatterning techniques (Fig. 1.5A). To make core − shell nanostructures, Co$_3$O$_4$ deposited on TiO$_2$ nanorods (with a diameter size of about 200 nm and a length less than 2 μm) by electrochemical deposition method. The SEM images of Co$_3$O$_4$@ TiO$_2$ core − shell nanorods grown on the FTO substrate has been shown in Fig. 1.5B. It was observed that there was no change in the color of the sample (combined TiO$_2$ nanorods and Co$_3$O$_4$ nanoparticles) after 1500 cycles, which indicates the good performance of the nanostructure's composition, while only Co$_3$O$_4$ sample lost more than 80% of it after 300 complete cycles (Fig. 1.5C) (Mishra et al., 2018).

1.4 Electrochromic nanomaterials based on Prussian blue

One of the most popular nanomaterials used for anodic electrochromic materials is Prussian blue (PB), which is a blue nanoporous pigment and a prototype of a number of metallohexacyanoferrites, an important class of compounds, which are unstable and insoluble. Although Prussian blue is an important pigment and is used on a large scale for use in paints, varnishes, printing inks, and other dyes, the electrochemical properties of PB were not well known. Johann Konrad Dippel was discovered PB in Berlin in 1706 (Kraft 2008). The electrochromic properties of PB were first studied in 1978 by Neff (Itaya et al., 1986). Neff showed a method for the preparation of PB thin films on platinum (Pt) and gold (Au) and its oxidation/reduction behavior. This was the first report on the electrochemical and electrochromic properties of PB, which subsequently promoted numerous investigations on the properties of PB thin films. PB is used as a sensitizer to increase the sensitivity to light in order to improve the electrochromic response of conductive polymers (especially polypyrrole and polyaniline). Prussian blue, known as ferric hexacyanoferrite, is a typical mixed-valence compound, in which iron atoms exist in different oxidation states (Fe^{2+}/Fe^{3+}) in a face-centered cubic (fcc) lattice. The iron (III) ions (high spin) form the central part of the cube. They are connected by nitrogen atoms in an octahedral environment, while the iron (II) ions (low spin) are in the midpoints of each side (Fig. 1.6). They are cubic and surrounded by carbon atoms. These two types of iron are related to cyanide groups so that carbon atoms coordinate with other iron and nitrogen atoms in an octahedral manner (Wang et al., 2018).

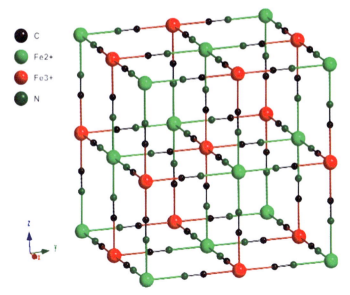

Figure 1.6 Prussian blue unit cell.

Prussian blue belongs to a group of materials called Metal Hexacyano, whose chemical formula is generally as follows:

$$M_x\left[M'(CN)_6\right]_y \tag{1.8}$$

which is placed in place of M and M' intermediate metals and in the case of Prussian blue both of them are iron. Also, M and M' can be different transition metals (Mortimer et al., 2015). Iron ions in Prussian blue can replace other divalent or trivalent ions of transition metals such as nickel, manganese, cadmium, copper, zinc, and their compounds, which are called Prussian blue analog compounds (PBA) (Yang et al., 2022). In the 200 years since the discovery of Prussian blue, this substance has been widely used in many scientific fields, including chemistry, medicine, nanomaterials, and various industrial applications. PB and PBA have excellent properties, such as high absorption and long-term electrochemical performance. Also, this material has attracted a lot of attention as electrocatalysis, biological fields, ion sensors, ion storage, and magnetic properties. Prussian blue, as an electrochromic material, can participate in oxidation and reduction process with several independent electrochromic states.

The color of Prussian blue is blue, and in the process of reduction, it turns into a colorless urate salt or Prussian white (PW) (Ricci & Palleschi 2005):

$$\underset{\text{Insoluble PB}}{Fe^{III}\left[Fe^{II}(CN)_6\right]_3} + 4K^+ + 4e^- \Leftrightarrow \underset{\text{Prussian White (Everitt salt)}}{K_4Fe_4^{II}\left[Fe^{II}(CN)_6\right]_3} \tag{1.9}$$

$$\underset{\text{Soluble PB}}{KFe^{III}Fe^{II}(CN)_6} + K^+ + e^- \Leftrightarrow \underset{\text{Prussian White (Everitt salt)}}{K_2Fe_4^{II}Fe^{II}(CN)_6} \tag{1.10}$$

where K is the cation produced by the electrolyte. In the oxidation process, PB is converted to BG according to the following equation:

$$\underset{\text{Insoluble PB}}{Fe^{III}\left[Fe^{II}(CN)_6\right]_3} + 3A^- \Leftrightarrow 3e^- + \underset{\text{Berlin Green-BG}}{Fe_4^{III}\left[Fe^{III}(CN)_6A\right]_3} \tag{1.11}$$

$$\underset{\text{Soluble PB}}{KFe^{III}Fe^{II}(CN)_6} \Leftrightarrow 2/3K^+ + 2/3e^- + \underset{\text{Berlin Green-BG}}{K_{1/3}\left(Fe^{III}(CN)_6\right)_{2/3}\left(Fe^{II}(CN)_6A\right)_{1/3}}$$

$$\tag{1.12}$$

where A is the anion produced by the electrolyte. Although in bulk form, BG has a constant composition with anion composition. There is a continuous composition range between PB and PY in the fully oxidized state for the films, which appears as golden yellow called PY (Somani & Radhakrishnan, 2002). PB is an anodic electrochromic material that can form a complementary ECD with a high transition change with a cathodic electrochromic material. PB is commonly used to improve the electrochromic properties of conducting polymers. Prussian blue film is often prepared by electrochemical deposition.

Prussian blue applications are very diverse and practical. For example, Zakaria et al. (2018) fabricated a photoelectrochemical solar cell using Prussian blue-sensitized TiO_2 layer as photoanode, in which graphite was used as counter electrode and KI/I as electrolyte. Prussian blue was added as a nanomaterial on TiO_2 using electrochemical deposition method. The maximum absorption of TiO_2 sensitive to Prussian blue was at 684 nm. It was observed that the absorption of light increased in the combination of TiO_2 and Prussian blue, which in turn increased the efficiency of the solar cell. This research showed that Prussian blue can be used as an effective sensitizer in solar cells. Konchi and Wolfbeis (1998) showed that Prussian blue films made with polypyrrole can be potential optical pH sensors for the pH range of 5−9. They investigated spectral changes as a function of pH using semiconductor light sources and detectors. These films seem to be an alternative to indicator-based pH sensors, which are renewable and reversible. Also, Prussian blue acts as an electrocatalytic agent for reducing or oxidizing compounds other than hydrogen peroxide. The first efforts in this direction have been reported by Ogura et al. (1994) who proposed the use of polyaniline electrodes modified with Prussian blue for the detection of carbon dioxide.

Reducing the material size from micro- to nanoscale is one of the effective ways to improve the functional properties of EC. It has been observed that nanoscale metal complexes can significantly reduce the kinetic barriers of the original crystal structure and optimize the switching performance during electrochemical cycles (Wang et al., 2019b). Assis et al. (2015) have used Prussian blue as the electrochromic layer and CeO_2-TiO_2 as the counter electrodes of solid-state electrochromic devices (ECDs) with hydroxypropyl cellulose (HPC) membrane as electrolyte. All the presented results show that it is possible to use PB as an electrochromic coating and HPC-based membrane as a solid electrolyte in the development of a new generation of ECD. The transmittance of this ECD after applied potentials of −2.6 and 2.0 V for 15, 30, and 60 seconds has been shown in Fig. 1.7A. The largest color contrast percentage occurs

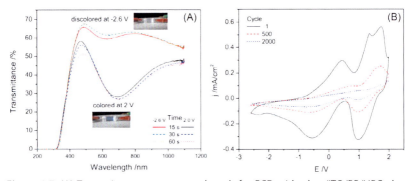

Figure 1.7 (A) Transmittance versus wavelength for ECD with glass/ITO/PB/HPC-electrolyte/CeO$_2$–TiO$_2$/ITO/glass configuration for applied potentials of −2.6 V to +2.0 V. (B) Cyclic voltammetry for cycle 1500 and 2000 (Assis et al., 2015).

between 600 and 800 nm wavelengths. The color changes (ΔT) are of 32% for 15 seconds and of 35% for 30 and 60 seconds of applied potentials at 686 nm wavelength. In the chronoamprometric (CA) measurements, the cyclic voltammetry was carried out for 1500 and 2000 cycles. Fig. 1.7B demonstrates the 1st cycle voltammograms where two cathodic and three anodic peaks are seen. After consecutive chronoamperometric cycling by applying −2.6 and 2.0 V for each 15 seconds, the voltammograms shapes modified. Although the two cathodic peaks remain at the same potentials (−0.5 and 1.0 V) and shapes after 500 CA cycles the current density considerably decrease when compared with the values of the 1st cycle. The shape and the positions of the anodic peaks alter to the cathodic ones, contrarily. The peak at 0.5 V shifts to 0.8 V and the peak allocated at 1.4 V joint with the peak at 1.7 V. After 200 cycles, the cathodic peaks keep their potentials at −0.5 and 1.0 V; nevertheless, the values of current density reduce. A rise of the anodic current density is only seen at 2.0 V; on the other hand, the anodic peaks nearly vanish, totally. During CA cycling of ECD, the observed cyclic voltammograms changes with the suggested configuration clearly shows that PB redox reactions are not reversible, entirely. Consequently, a reduction of the charge density causes a diminution of color contrast percentage.

The effects of pulsed electrical deposition parameters such as current density and frequency on the electrochemical properties of Prussian blue thin films were investigated by Najafisayar and Bahrololoom (2013). Their results showed that the use of pulsed deposition technique reduced the size of the Prussian blue particles accumulated in the Prussian blue thin film, nevertheless did not change the morphology of the film. Furthermore, increase in current density and the use of higher frequencies during electrical deposition reduce the charge transfer resistance of the films. In this work, the electrochromic properties of the material were not investigated and the effect of the pulse on the morphology of Prussian blue was mostly studied. Isfahani et al. (2019) investigated the effect of time on the preparation of Prussian blue thin film. By using a direct voltage of 0.4 V for the electrical deposition of Prussian blue on the surface of the ITO substrate, they prepared five different films with deposition times of 25, 50, 75, 100, and 150 seconds. From XRD analysis, they found that the structure of all samples is amorphous. The electrochemical results showed that the sample prepared at the deposition time of 75 seconds has better electrochromic properties. Controlling the deposition time resulted in a 38.9-fold increase in the contrast ratio and the corresponding values for the optical density (ΔOD) of the Prussian blue films.

In a study, Cheng et al. (2007) improved the stability of Prussian blue by using nanocomposite in lithium-based electrolyte. To prepare the NPB film, an alcoholic solution containing well-dispersed ITO nanoparticles was prepared by an ultrasonic tank and then sprayed onto the glass heated at 200°C by an airbrush with a pressure set at about 15 psi. The prepared nanoparticle-containing ITO layer was used as a conductive medium layer with nanoscale pores for the PB film. It has been seen from the SEM images (Fig. 1.8A and B) that there are inevitable cracks on the PB films as well as particle-like clusters, which become larger and the diameter of the clusters increases as the deposition time increases. On the other hand, it can be seen in the SEM micrograph that the nanostructured ITO film consists of almost spherical nanoparticles with an average size of about 40 nm. The results indicate that the Prussian blue nanocomposite film had an almost constant contrast ratio and did not decrease in charge density until about 2000 cycles, and almost no changes were observed in its morphology. In contrast, Prussian blue loses its charge density up to about 100 cycles (Fig. 1.8C and D). According to the results, the switching speed and the contrast ratio, considering that, the electrons have enough paths to move through the porous NPB film and the liquid electrolyte can penetrate the film to shorten the ion diffusion path, can be significantly improve attention.

Elshorbagy et al. (2017) used the spray pyrolysis technique and compared the results against the electrochemical deposition method. The EDS surface has unavoidable large cracks (Fig. 1.9A) that cause resistance within

20   Electrochemistry and Photo-Electrochemistry of Nanomaterials

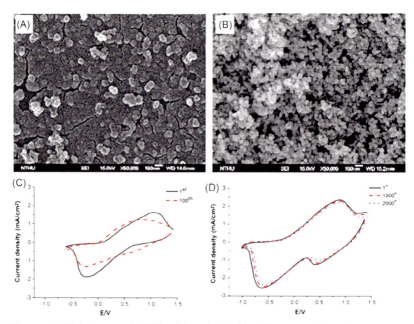

Figure 1.8 SEM images of PB film (A) and NPB film (B) on ITO glass substrates with the deposition current density of 20 A/cm^2 and deposition time of 200 s. Cyclic voltammograms of PB film with potential varying between -0.6 and $+1.4$ V (C) and NPB film with potential varying between -1 and $+1.4$ V at sweep rate of 200 mV/s (D) in 1 M LiClO4/PC electrolyte (Cheng et al., 2007).

Figure 1.9 SEM image for PB EC thin film deposited on FTO substrate (A) by electrodeposition (EDS), (B) by spray pyrolysis (SPS). Charge density versus time for (C) EDS and (D) SPS for 1st, 40th, and 70th cycles (Elshorbagy et al., 2017).

the layer, thus, may reduce stability and adhesion, but spray pyrolysis solves this to some extent and creates layers with much smaller cracks, which is due to low deposition temperature and long deposition time. The results showed that the use of their proposed technique eliminates the cracks on the surface of the Prussian blue film to a great extent, and as a result, it is more stable than the films obtained by the usual electrochemical deposition method (Fig. 1.9B). It was also found from the electrochemical analysis that the stability of the sample prepared by spray pyrolysis is higher than the electrochemical deposition method, and its color efficiency (η) was almost 3 times higher than the samples prepared by the electrochemical deposition method. The response time after 70 cycles for spray pyrolysis was 7 seconds, which was smaller than the time of 20 seconds obtained from electrochemical deposition method (Fig. 1.9C).

An innovative deposition technique, called electrothermophoresis (ETP), that involves the use of a temperature gradient and a pulsed potential, has developed by Sekhavat and Ghodsi (2023) for deposition of Prussian blue (PB) thin films. In the work, the PB thin films were deposited in five different conditions as direct current (DCEP), 100 Hz square pulse (PCEP1), 10 kHz square pulse (PCEP2), 10 kHz square pulse with applying of temperature gradient in the reverse direction (PCETP1, the substrate was taken at lower temperature side), and 10 kHz square pulse with applying of temperature gradient in the normal direction (PCETP2, the substrate was taken at higher temperature side). The temperature gradient was taken between 10°C and 70°C.

Fig. 1.10 shows the SEM and cross-section images of the deposited PB thin films prepared in different conditions. As can see from figure, the formation of three layers is obvious. A closely uniform layer of PB thin film is formed on the FTO substrate with nanoparticle-like grains over films. By applying the inverse temperature gradient, the grain size (with an average size of 45 nm) takes its smallest value. In this condition by increasing the surface roughness, the effective surface area of the layer rises, which facilitates surface ion exchange. When the pulse frequency increases from 100 Hz to 10 kHz, the size of grains has been reduced. More nuclei form on the surface of the cathode in the higher frequency, which causes to decrease the growth time. The structure is more tending to new nucleation and less toward growth with increasing frequency, which causes the growth of new and smaller nuclei. At room temperature, the nuclei growth is mostly dominant. The grain density rises and its size diminishes when the direct potential or lower frequency pulses are used. Thus, by increasing the pulse frequency, the surface of the films is more uniform, compact, and granular.

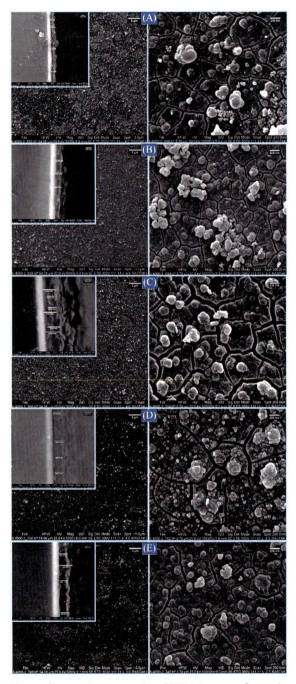

Figure 1.10 The surface and cross section FESEM images of samples in different mode. (A) DCEP, (B) PCEP1, (C) PCEP2, (D) PCETP1, and (E) PCETP2 samples (Sekhavat & Ghodsi, 2023).

Nanomaterials for electrochromic application 23

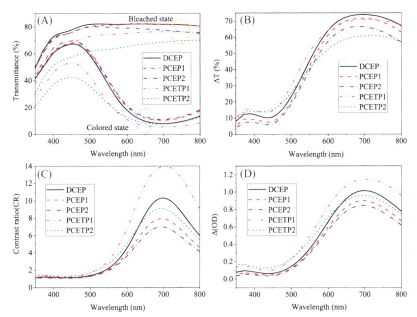

Figure 1.11 (A) Optical transmittance in the bleached and colored states, (B) the transmittance change (ΔT), (C) contrast ratio (CR), and (D) optical density versus wavelength for DCEP, PCEP1, PCEP2, PCETP1 and PCETP2 samples (Sekhavat & Ghodsi, 2023).

They found that the temperature gradient causes the grains to become smaller, and by changing the deposition frequency from Hz to kHz in the pulsed potential mode, the transition of colored and bleached states (ΔT), contrast ratio (CR), and optical density (ΔOD) decrease (Fig. 1.11). The contrast ratio and optical density increase by using the innovative method. Also, the sample indicates good reversibility to oxidation and reduction processes, since the charge ratio of insertion and extraction (Q_{in}/Q_{ex}) is almost close to unity.

Li et al. (2018) investigated the cyclic stability of Prussian blue films in electrolytes. They showed that acidification of the electrolyte by adding some acid to it will cause a significant increase in the cycles completed by the Prussian blue film. Acidic KCl solutions can significantly improve the durability of PB film, and KCl solutions showed better performance with the addition of hydrochloric acid and pH change. In addition, they investigated the cycling stability of PB films in $LiClO_4/PC$ electrolyte containing different acids (Fig. 1.12A). The CV curve of the Prussian blue (PB)

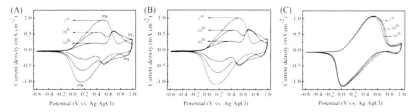

Figure 1.12 The cyclic voltammetry (CV) of the Prussian blue (PB) film in LiClO4/PC electrolytes with different acid: (A) LiClO4/PC, (B) LiClO4/PC/HCl, and (C) LiClO4/PC/HAc (Li et al., 2018).

film measured in LiClO$_4$/PC electrolyte containing hydrochloric acid (LiClO$_4$/PC/HCl) is shown Fig. 1.12B. The results show that the stability of the PB film is not significantly affected with the addition of hydrochloric acid. Fig. 1.12C displays the CV curves of the film tested in LiClO4/PC electrolyte containing acetic acid (LiClO4/PC/HAc). The addition of acetic acid to the electrolyte had a much longer cycle life than some previously demonstrated Prussian blue films, showing no obvious fading after 2000 cycles. This shows that the method of adding acetic acid to the electrolyte provides an effective way to improve the cyclic stability of the PB film in LiClO$_4$/PC.

In general, nonnanostructured organic EC materials can be combined with inorganic EC nanomaterials to create a new hybrid EC material that exhibits better properties. For example, Hu et al. (2016) synthesized a new nanocomposite, Prussian blue (PB) nanoparticles and polyaniline: polystyrene sulfonate (PANI:PSS). They prepared nanocomposite thin film by wet coating method on conductive ITO glass and synthesized PB−PANI:PSS nanocomposite by two-step polymerization. The CV curves of PANI: PSS, PB, and PB−PANI:PSS films at a scan rate of 10 mV/s and between −0.5 and +0.8 V potential window are shown in Fig. 1.13A. The films exhibit three optical states: highly transparent in the reduced state (−0.5 V vs. Ag/AgCl), green at +0.2 V by PANI, and blue-green at +0.5 V. Furthermore, it showed higher optical changes (52.3%) than the single component (about 40%). The photographs of as prepared PANI: PSS, PB, and PB−PANI: PSS thin films presented in Fig. 1.13B, where they exhibit grass green, cerulean blue, blue-green. As a result, this PB − PANI: PSS composite thin films can potentially be suitable for multicolor electrochromic applications.

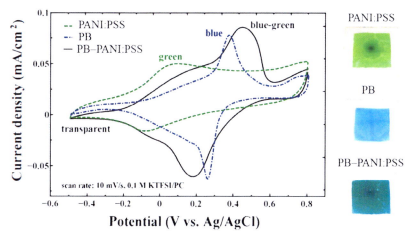

Figure 1.13 (A) CV curves for PANI: PSS, PB, and PB−PANI: PSS thin films scanned from −0.5 to +0.8 V in 0.1 M KTFSI/PC solution at a scan rate of 10 mV/s. (B) Photos of thin films (Hu et al., 2016).

1.5 Electrochromic based on organic nanomaterials

Organic electrochromic nanomaterials include conductive polymers (CPs), metallopolymers, metallophthalocyanines, and viologen. Today, organic electrochromic nanomaterials have attracted a lot of attention, and one of the reasons for its importance is the fast switching of these materials. In fact, what causes their high color efficiency and accelerates color change in them is due to the flexibility of polymer chains. This saves the energy used to change the color. Organic electrochromic nanomaterials are in the form of solution, and after applying voltage in the structure of the electrochromic device, they either turn into a solid or remain in the same form of solution (Mortimer, 1991).

In recent years, scientists discovered a special type of conjugated polymer, called polyethylene, which could be a good conductor of electricity after applying an impurity treatment that creates a structural modification in this polymer. The term coupled is due to the presence of alternating single and double bonds in the polymer chain structure. In general, conductive polymers are classified as semiconductors. Although some of them, like polystyrene, are more conductive than some metals. Due to the special continuity in the chain structure of such polymers, it is possible for electrons to be locally shared among chain atoms. These local electrons can move among the system and assume the role of a charge carrier and cause conductivity and current transfer in the polymer. It should be noted that coupled polymers do not have a valence band like pure metals, and there is a covalent bond between its components, so they are not intrinsically conductive. The impurity process allows electrons to flow due to the existence of the coupled structure and the conduction band in this type of polymers. When the pollution process happens, the electrons in the coupled systems that have lost their bonds find the ability to move and flow between the polymer chains (Shirakawa et al., 1977). The most famous conductive polymers and the date of their discovery can be seen in Table 1.2. All conductive polymers have the ability to show electro-chromic properties. A good conductive polymer is polypyrrole, which is blue/purple in oxidized state and yellow/green in natural state. In conductive polymers, polythiophene and polyaniline are receiving a lot of attention. The color change range of polythiophene is from red to blue and the color change of polyaniline is from transparent yellow to green.

Metallophthalocyanines are important industrial pigments that are mainly used for coloring plastics and metal surfaces and were used as active materials for molecular devices such as chemical sensors for a long

Table 1.2 Conductive polymers CPs.

Name	Discovery year	Structure
Polystyrene	1977	
Polypyrrole	1979	
Polythiophene	1981	
Polyphenylene	1979	
Poly (p-phenylene vinylene)	1979	
Polyaniline	1980	

time (Takamura et al., 2002). It has been observed that many metals (zinc, magnesium, aluminum, iron, cobalt, titanium, copper, etc.) can combine with phthalocyanine, and these metals can be placed in the center of a phthalocyanine ring or between two phthalocyanine rings (Boileau et al., 2019). Another organic electrochromic material is viologen, whose nitrogen can be attached to different side groups. Viologen is colorless in dication state, and as a result of recovery and acceptance of electron, it turns into radical cation state, which is strongly colored and its color is blue-violet. In the case of the second regeneration and the production of two radicals, the intensity of the produced color will be very low and almost colorless. The side groups attached to nitrogen, as a result of creating resonance, lead to the production of different energy levels for the molecule and produce different colors. The correct selection of side groups attached to nitrogen will allow the production of the desired color. For example, if alkyl groups are attached, the ability to produce violet/blue color is created in viologen, or the attachment of aryl groups such as 4-cyanophenyl leads to the production of green color (Małachowski, 2011). Although the synthesis of viologen and polyviologen derivatives is time-consuming, however due to their importance and application, they are produced in different ways (Liu et al., 2007).

Among the organic class of electrochromic nanomaterials, the electrochromic properties of viologen have been widely studied due to its lower driving voltage for color changes (showing three primary colors), high contrast, tunable electrochromic properties, and easy cell fabrication. However, some side processes during the cycle, such as comproportionation, dimerization, and recrystallization (Porter & Vaid, 2005), lead to the loss of viologens and its irreversibility, which is the reason for the poor stability of viologen-based ECDs. One of the ways is the synthesis of mineral-viologen nanocomposites as EC hybrid material, which provides better coloration and shows long-term stability. The used inorganic electrochromic nanomaterials are WO_3, NiO, TiO_2, V_2O_5, and Prussian blue (PB). Fan et al. (2016) used uniformly dispersed Prussian blue (wPB) in water as an electrochromic ink and spray-coated it on the ITO surface. This electrode was placed as an anodic electrode in front of a cathodic electrode composed of poly (butyl viologen) (PBV). Succinonitrile with 0.1 M potassium bis(trifluoromethanesulfonyl)imide and SiO_2 nanoparticles were used as the solid electrolyte. The resulting thin film yielded high color contrast from blue violet to transparent with high color effi ciency, short switching time, and long-term stability (up to 1000 cycles).

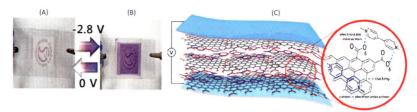

Figure 1.14 (A, B) Photographs for a flexible MV^{2+}—GQD ECD (containing 100 mM MV^{2+} with 8 mg/mL GQD in PVA with ITO-on-PET). (C) Illustration of an electrolyte-free flexible electrochromic device of MV^{2+}—GQDs (Hwang et al., 2014).

Li et al. (2017) reported an electrochromic device consisting of ethylviologen and Ti-doped V_2O_5 electrochromic materials with long stability. The results showed good electrochromic performance at the wavelength of 600 nm with a transmission percentage between 8% and 70%. The stability of the sample was obtained up to 250,000 cycles at 60°C. Graphene/graphene oxide is used as an active nanomaterial in EC applications due to its special physical properties such as optical transparency and high charge carrier mobility at room temperature. Hwang et al. (2014) reported a flexible, electrolyte-free electrochromic system in which they fabricated an electrochromic with viologen—graphene quantum dot (MV^{2+}—GQD) nanocomposites stabilized electrostatically through strong intermolecular forces. This new device showed very stable switching performance, high durability under repeated application of operating voltage, and thermal stability to 80°C. This device exhibits a continuous reversible color change from colorless to purple in the applied potential range of −2.8 to 0.0 V (Fig. 1.14A and B). A schematic of the device structure based on MV^{2+}—GQD is shown in Fig. 1.14C. The results herein can provide a guideline for stability improvements and design simplifications for future flexible nano-ECDs.

1.6 Conclusion

In this chapter, we investigated electrochromic nanomaterials and their performance in electrochromic devices. In general, electrochromic nanomaterials are divided into two categories: anodic materials that change color in oxidation processes and cathodic materials that change color during reduction processes. They are also divided into inorganic and organic categories in terms of their composition, bonding, and properties. Considering the important and diverse use of Prussian blue, which is a natural anodic pigment, we examined it separately. What distinguishes an electrochromic material from its counterparts is its high efficiency and long stability in oxidation and reduction cycles. Considering that nanomaterials with an active reaction surface can play an important role in optimizing heterogeneous electron transfer and homogeneous ion transfer, as a result, we briefly reviewed the studies based on the use of nanomaterials to improve the performance of electrochromic nanomaterials. In almost all of them, the presence of nanostructures has improved the electrochromic properties of the materials, which promise a bright future in making electrochromic parts with optimized capabilities and higher efficiency.

References

Argun, A. A. (2004). Multicolored electrochromism in polymers: Structures and devices. *Chemistry of Materials: A Publication of the American Chemical Society, 6,* 4401−4412.

Argun, A. A., Aubert, P., Thompson, B. C., Schwendeman, I., Gaupp, C. L., Hwang, J., Pinto, N. J., Tanner, D. B., MacDiarmid, A. G., & Reynolds, J. R. (2004). Multicolored electrochromism in polymers: Structures and devices. *Chemistry of Materials: a Publication of the American Chemical Society, 16,* 4401.

Assis, L. M. N., Leones, R., Kanicki, J., Pawlicka, A., & Silva, M. M. (2016). Prussian blue for electrochromic devices. *Journal of Electroanalytical Chemistry, 777,* 33−39.

Assis, L. M. N., Sabadini, R. C., Santos, L. P., Kanicki, J., èapkowski, M., & Pawlicka, A. (2015). Electrochromic device with Prussian blue and HPC-based electrolyte. *Electrochimica Acta, 182,* 878.

Azam, A., Kim, J., Park, J., Novak, T. G., Tiwari, A. P., Song, S. H., Kim, B., & Jeon, S. (2018). Two-dimensional WO_3 nanosheets chemically converted from layered WS_2 for high-performance electrochromic devices. *Nano Letters, 18*(9), 5646−5651.

Bechinger, C., Ferrere, S., Zaban, A., Sprague, J., & Gregg, B. A. (1996). Photoelectrochromic windows and displays. *Nature, 383,* 608−610.

Boileau, N. T., Cranston, R., Melville, B., Mirka, O. A., & Lessard, B. H. (2019). Metal phthalocyanine organic thin-film transistors: Changes in electrical performance and stability in response to temperature and environment. *RSC Advances, 9,* 21478−21485.

Buckley, D. N., Burke, L. D., & Mukahy, J. K. (1976). The oxygen electrode. Part 7— Influence of some electrical and electrolyte variables on the charge capacity of iridium in the anodic region. *Journal of the Chemical Society, Faraday Transactions, 72*(1), 1896−1902.

Burke, L. D., & Murphy, O. J. (1980). Electrochromic behaviour of oxide films grown on cobalt and manganese in base. *Journal of Electroanalytical Chemistry, 109,* 373−377.

Burke, L. D., & O'Sakan, E. J. M. (1978). Enhanced oxide growth at a rhodium surface in base under potential cycling conditions. *Journal of Electroanalytical Chemistry, 93*(1), 11−18.

Burke, L. D., Thomey, T. A. M., & Whelan, D. P. (1980). Growth of an electrochromic film on tungsten in acid under potential cycling conditions. *Journal of Electroanalytical Chemistry and Interfacial Electrochemistry., 107,* 201−204.

Burke, L. D., & Whelan, D. P. (1979). The behaviour of ruthenium anodes in base. *Journal of Electroanalytical Chemistry, 103,* 179−187.

Byker, H. J. (1994). Commercial developments in electrochromics. *Proceedings of the Electrochemical Society*, 94–2, 1–13.

Chang, I. F. (1976). Electrochromic and electrochemichromic materials and phenomena. In A. R. Kmetz, & F. K. Willisen (Eds.), *Non-emissive Electrooptical Displays* (pp. 155–196). New York: Plenum Press.

Cheng, K. C., Kai, J. J., & Chen, F. R. (2007). Improving the durability of Prussian blue based on nano-composite thin film in Li^+ based liquid electrolyte. *Electrochimica Acta*, 52, 6554–6560.

Coustier, F., Hil, J., Owens, B. B., Passerini, S., & Smyrl, W. H. (1999). Doped vanadium oxides as host materials for lithium intercalation. *Journal of the Electrochemical Society*, 146, 1355.

Coustier, F., Passerini, S., & Smyrl, W. H. (1997). Dip-coated silver-doped V_2O_5 xerogels as host materials for lithium. *Solid State Ionics*, 100, 247–258.

Deb, S. K. (1969). A novel electrophotographic system. *Applied Optics*, 3, 192–195.

Deb, S. K. (2008). Opportunities and challenges in science and technology of WO_3 for electrochromic and related applications. *Solar Energy Materials and Solar Cells*, 92, 245–258.

Delongchamp, D. M., & Hammond, P. T. (2004). Multiple-color electrochromism from layer-by-layer-assembled polyaniline/prussian blue nanocomposite thin films. *Chemistry of Materials: a Publication of the American Chemical Society*, 16, 4799–4805.

Elshorbagy, M. H., Ramadan, R., & Abdelhady, K. (2017). Preparation and characterization of spray-deposited efficient Prussian blue electrochromic thin film. *Optik*, 129, 130–139.

Falahatgar, S. S., Ghodsi, F. E., Tepehan, F. Z., Tepehan, G. G., & Turhan, I. (2014). Electrochromic performance, wettability and optical study of copper manganese oxide thin films: Effect of annealing temperature. *Applied Surface Science*, 289, 289–299.

Fan, M., Kao, S., Chang, T., Vittal, R., & Ho, K. (2016). A high contrast solid-state electrochromic device based on nano-structural Prussian blue and poly(butyl viologen) thin films. *Solar Energy Materials and Solar Cells*, 145, 35–41.

Faughnan, B. W., & Crandall, R. S. (1980). Electrochromic displays based on WO3. In J. I. Pankove (Ed.), *Display Devices* (pp. 181–211). Berlin: Springer Verlag.

Georg, A., Graf, W., & Wittwer, V. (2008). Switchable windows with tungsten oxide. *Vacuum*, 82, 730–735.

Giannouli, M., & Leftheriotis, G. (2011). The effect of precursor aging on the morphology and electrochromic performance of electrodeposited tungsten oxide films. *Solar Energy Materials and Solar Cells*, 95(7), 1932–1939.

Granqvist, G. (1995). C.Handbook of Inorganic Electrochromic Materials. Amsterdam: Elsevier.

Granqvist, C. G., Lansaker, P. C., Mlyuka, N. R., Niklasson, G. A., & Avendano, E. (2009). Progress in chromogenics: New results for electrochromic and thermochromic materials and devices. *Solar Energy Materials and Solar Cells*, 93, 2032–2039.

Hopmann, E., Zhang, W., Li, H., & Elezzabi, A. Y. (2023). Advances in electrochromic device technology through the exploitation of nanophotonic and nanoplasmonic effects. *Nanophotonics*, 12(4), 637–657.

Hsiao, Y. S., Chang-Jian, C. W., Syu, W. L., Yen, S. C., Huang, J. H., Weng, H. C., Lu, C. Z., & Hsu, S. C. (2021). Enhanced electrochromic performance of carbon-coated V_2O_5 derived from a metal−organic framework. *Applied Surface Science*, 542, 148498.

Hu, C., Kawamoto, T., Tanaka, H., Takahashi, A., Lee, K., Kao, S., Liao, Y., & Ho, K. (2016). Water processable Prussian blue−polyaniline:polystyrene sulfonate nanocomposite (PB−PANI:PSS) for multi-color electrochromic applications. *Journal of Materials Chemistry C*, 4, 10293–10300.

Hwang, E., Seo, S., Bak, S., Lee, H., Min, M., & Lee, H. (2014). An electrolyte-free flexible electrochromic device using electrostatically strong graphene quantum dot—viologen nanocomposites. *Advanced Materials, 26*(30), 5129—5136.

Isfahani, V. B., Memarian, N., Dizaji, H. G., Arab, A., & Silva, M. M. (2019). The Physical and electrochromic properties of Prussian blue thin films electrodeposited on ITO electrodes. *Electrochimica Acta, 304*, 282—291.

Itaya, K., Uchida, I., & Neff, V. D. (1986). Electrochemistry of polynuclear transition metal cyanides: Prussian blue and its analogues. *Accounts of Chemical Research, 19*(6), 162—168.

Kolay, A., Maity, D., Flint, H., Gibson, E. A., & Deep, M. (2022). Self-switching photoelectrochromic device with low cost, plasmonic and conducting Ag nanowires decorated V_2O_5 and PbS quantum dots. *Solar Energy Materials and Solar Cells., 239*, 11674.

Konchi, R., & Wolfbeis, O. S. (1998). Composite film of Prussian blue and N-substituted polypyrroles: Fabrication and application to optical determination of pH. *Analytical Chemistry, 70*, 2544—2550.

Kraft, A. (2008). On the discovery and history of Prussian blue. *Bulletin for the History of Chemistry, 33*, 61—67.

Lampert, C. M. (1984). Electrochromic materials and devices for energy efficient windows. *Solar Energy, Materials, 11*, 1—27.

Lampert, C. M. (2004). Chromogenic smart materials. *Materials Today, 7*, 28—35.

Lavagna, L., Syrrokostas, G., Fagiolari, L., Amici, J., Francia, C., Bodoardo, S., Leftheriotis, G., & Bella, F. (2021). Platinum-free photoelectrochromic devices working with copper-based electrolytes for ultra-stable smart windows. *Journal of Materials Chemistry A, 9*, 19687.

Li, J., Zhuang, Y., Chen, J., Li, B., Wang, L., Liu, S., & Zhao, Q. (2021). Two-dimensional materials for electrochromic applications. *EnergyChem, 3*(5), 100060.

Li, M., Weng, D., Wei, Y., Zheng, J., & Xu, C. (2017). Durability-reinforced electrochromic device Based on surface-confined Ti-doped V_2O_5 and solution-phase viologen. *Electrochimica Acta, 248*, 206—214.

Li, Z., Liu, Z., Zhao, L., Chen, Y., Li, J., & Yan, W. (2023). Efficient electrochromic efficiency and stability of amorphous/crystalline tungsten oxide film. *Journal of Alloys and Compounds, 930*, 167405.

Li, Z. T., Tang, Y. H., Zhou, K. L., Wang, H., & Yan, H. (2018). Improving electrochromic cycle life of prussian blue by acid addition to the electrolyte. *Materials (Basel), 12*(1), 28.

Liu, H., Pan, B., & Liou, G. (2017). Highly transparent AgNW/PDMS stretchable electrodes for elastomeric electrochromic devices. *Nanoscale, 9*, 2633—2639.

Liu, J., Zheng, J., Wang, J., Xu, J., Li, H., & Yu, S. (2013). Ultrathin $W_{18}O_{49}$ nanowire assemblies for electrochromic devices. *Nano Letters, 13*(8), 3589—3593.

Liu, M. O., Chen, I. M., & Lin, J. L. (2007). Microwave-assisted synthesis of viologens and polyviologens and their preliminary electrochromic effects. *Materials Letters, 61*, 5227—5231.

Lu, H., Chou, C., Wu, J., Lin, J., & Liou, G. (2015). Highly transparent and flexible polyimide—AgNW hybrid electrodes with excellent thermal stability for electrochromic applications and defogging devices. *Journal of Materials Chemistry C, 3*, 3629—3635.

Ma, D., Shi, G., Wang, H., Zhang, Q., & Li, Y. (2013). Morphology-tailored synthesis of vertically aligned 1D WO_3 nano-structure films for highly enhanced electrochromic performance. *Journal of Materials Chemistry A, 1*, 684—691.

Małachowski, M. J. (2011). New organic electrochromic materials and their applications. *Journal of Achievements in Materials and Manufacturing Engineering., 48*, 14 23.

Mishra, S., Yogi, P., Sagdeo, P. R., & Kumar, R. (2018). $TiO_2-Co_3O_4$ core–shell nanorods: Bifunctional role in better energy storage and electrochromism. *ACS Applied Energy Materials, 1*(2), 790–798.

Mjejri, I., Gaudon, M., & Rougier, A. (2019). Mo addition for improved electrochromic properties of V_2O_5 thick films. *Solar Energy Materials and Solar Cells, 198*, 19–25.

Mortimer, R. J. (1991). Organic electrochromic materials. *Electrochimica Acta, 44*, 2971–2981.

Mortimer, R. J., Rosseinsky, D. R., & Monk, P. M. S. (2015). *Electrochromic materials and devices* (pp. 91–106). Wiley- VCH.

Najafisayar, P., & Bahrololoom, M. E. (2013). Pulse electrodeposition of Prussian Blue thin films. *Thin Solid Films, 542*, 45–51.

Ogura, K., Higasa, M., & Yano, J. (1994). Electroreduction of CO_2 to C_2 and C_3 compounds on bis(4,5-dihydroxybenzene-1,3-disulphonato) ferrate (II)-fixed polyaniline/Prussian blue-modified electrode in aqueous solutions. *Journal of Electroanalytical Chemistry, 379*, 373–377.

Park, H. (2005). Manganese vanadium oxides as cathodes for lithium batteries. *Solid State Ionics, 176*, 307–312.

Pawlicka, A. (2009). Development of electrochromic devices. *Recent Patents on Nanotechnology, 3*, 177–181.

Pham, H. H., Gourevich, I., Oh, J. K., Jonkman, J. E. N., Kumacheva, E., & Multidye, A. (2004). Nanostructured material for optical data storage and security data encryption. *Advanced Materials, 16*(6), 516–520.

Platt, J. R. (1961). Electrochromism, a possible change of color producible in dyes by an electric field. *The Journal of Chemical Physics, 34*, 862.

Porter, W. W., & Vaid, T. P. (2005). Isolation and characterization of phenyl viologen as a radical cation and neutral molecule. *The Journal of Organic Chemistry, 70*(13), 5028–5035.

Ricci, F., & Palleschi, G. (2005). Sensor and biosensor preparation, optimization and applications of Prussian blue modified electrodes. *Biosensors & Bioelectronics, 21*, 389–407.

Rui, X., Lu, Z., Yin, Z., Sim, D. H., Xiao, N., Lim, T. M., Hng, H. H., Zhang, H., & Yan, Q. (2013). Oriented molecular attachments through sol-gel chemistry for synthesis of ultrathin hydrated vanadium pentoxide nanosheets and their applications. *Small (Weinheim an der Bergstrasse, Germany), 9*, 716–721.

Sekhavat, M. S., & Ghodsi, F. E. (2023). Improving the electrochromic performance of Prussian blue (PB) thin films by using an innovative electrothermophoresis method. *Journal of Materials Research, 38*, 2852–2862.

Shirakawa, H., Louis, E. J., MacDiarmid, A. G., Chiang, C. K., & Heeger, A. J. (1977). Synthesis of electrically conducting organic polymers: Halogen derivatives of polyacetylene, $(CH)_x$. *Journal of the Chemical Society. Chemical Communications, 16*, 578–580.

Somani, P. R., & Radhakrishnan, S. (2002). Electrochromic materials and devices: present and future. *Materials Chemistry and Physics, 77*, 117–133.

Takamura, T., Moriyama, M., Komatsu, T., & Shimoyama, Y. (2002). Molecular orientations in langmuir-blodgett and vacuum-deposited films of VO-phthalocyanine. *Japanese Journal of Applied Physics, 38*, 2928–2933.

Wang, B., Han, Y., Wang, X., Bahlawane, N., Pan, H., Yan, M., & Jiang, Y. (2018). Prussian blue analogs for rechargeable batteries. *iScience, 3*, 110–133.

Wang, J., Zhao, W., Tam, B., Zhang, H., Zhou, Y., Yong, L., & Cheng, W. (2023). Pseudocapacitive porous amorphous vanadium pentoxide with enhanced multicolored electrochromism. *Chemical Engineering Journal, 452*, 139655.

Wang, Y. C., Lu, H. C., Hsiao, L. Y., Lu, Y. A., & Ho, K. C. (2019b). A complementary electrochromic device composed of nanoparticulated ruthenium purple and Fe (II)-based metallo-supramolecular polymer. *Solar Energy Materials and Solar Cells, 200*, 109929.

Wang, Z., Wang, H., Gu, X., & Cui, H. (2019a). Hierarchical structure WO_3/TiO_2 complex film with enhanced electrochromic performance. *Solid State Ionics, 338,* 168—176.

Wei, Y., Zhou, J., Zheng, J., & Xu, C. (2015). Improved stability of electrochromic devices using Ti-doped V_2O_5 film. *Electrochimica Acta, 166,* 277—284.

Yang, G., Zhang, Y., Cai, Y., Yang, B., Gu, C., & Zhang, S. X. (2020). Advances in nanomaterials for electrochromic devices. *Chemical Society Reviews, 49,* 8687—8720.

Yang, Y., Zhou, J., Wang, L., Jiao, Z., Xiao, M., Huang, Q., Liu, M., Shao, Q., Sun, X., & Zhang, J. (2022). Prussian blue and its analogues as cathode materials for Na-, K-, Mg-, Ca-, Zn- and Al-ion batteries. *Nano Energy, 99,* 107424.

Yuan, Y. F., Xia, X. H., Wu, J. B., Chen, Y. B., Yang, J. L., & Guo, S. Y. (2011). Enhanced electrochromic properties of ordered porous nickel oxide thin film prepared by self-assembled colloidal crystal template-assisted electrodeposition. *Electrochimica Acta, 56,* 1208.

Zakaria, M. B., Ebeid, E. M., Chikyow, T., Bando, Y., Alshehri, A. A., Alghamdi, Y. G., Cai, Z., Kumar, N. A., Lin, J., Kim, H., & Yamauchi, Y. (2018). Synthesis of hollow Co—Fe Prussian blue analogue cubes by using silica spheres as a sacrificial template. *ChemistryOpen, 7,* 599.

Zhan, Y., Rui, M., Tan, J., Cheng, X., Ming, W., Tan, A., Cai, G. F., Chen, J. W., Kumar, V., Magdassi, S., & Lee, P. S. (2017). Ti-doped WO_3 synthesized by a facile wet bath method for improved electrochromism. *Journal of Materials Chemistry C, 5,* 9995—10000.

Zhao, C., Gai, P., Song, R., Chen, Y., Zhang, J., & Zhu, J. (2017). Nanostructured material-based biofuel cells: Recent advances and future prospects. *Chemical Society Reviews, 46,* 1545—1564.

CHAPTER 2

Electrochemical biosensing

Lamees Abbas[1], Maria Hany[1], Mariam Alnaqbi[1], Amani Al-Othman[1] and Muhammad Tawalbeh[2,3]

[1]Department of Chemical and Biological Engineering, American University of Sharjah, Sharjah, United Arab Emirates
[2]Sustainable and Renewable Energy Engineering Department, University of Sharjah, Sharjah, United Arab Emirates
[3]Sustainable Energy & Power Systems Research Centre, RISE, University of Sharjah, Sharjah, United Arab Emirates

2.1 Introduction

Biosensors are small portable devices that have risen significantly over the last few decades due to their ability to analyze elements using certain recognition factors. The mechanism of biosensors begins with the detection of an analyte, which is the substance that needs continuous monitoring in a solution. The level of the analyte is detected using a bioreceptor and transmitted as signals to the processor. Then, those signals are converted to digital ones, displayed, and analyzed. Hence, the electrochemical biosensor is a standard sensing device with the main operation of converting signals captured from biochemical mechanisms into electrical signals.

Although the history includes the demonstration of biosensors dates back to 1906, Leland C. Clark, Jr created the first "true" biosensor for oxygen detection in 1956 (Bhalla et al., 2016). Clark was known as the father of biosensors, and the first biosensor invented was named after him as the Clark electrode. Another discovery of an amperometric enzyme electrode by Clark was in 1962. It was the gateway to many more inventions, where the first potentiometric biosensor for tracking urea was uncovered by Guilbault and Montalvo, Jr in 1969. Since then, Yellow Spring Instruments (YSI) produced the first commercially available biosensor in 1975. This breakthrough enabled substantial progress to be achieved in the realm of biosensors (Bhalla et al., 2016).

Electrochemical biosensors are four main groups: amperometric, potentiometric, impedimetric, and conductimetric biosensors (Bahadir & Sezgintürk, 2015). The first one is amperometric biosensors, which

Electrochemistry and Photo-Electrochemistry of Nanomaterials
DOI: https://doi.org/10.1016/B978-0-443-18600-4.00002-8
© 2025 Elsevier Inc. All rights reserved, including those for text and data mining, AI training, and similar technologies.

35

work with a three-electrode arrangement. It detects analytes with a concentration lower than 10^{-12} M, comprising of a working electrode (WE), counter electrode (CE), and reference electrode (RE) (Alarcon-Angeles et al., 2018). It measures the current, which results from the reduction and oxidation reactions of electroactive substances and captures these signals with a constant potential maintained on the transducer (Malhotra & Ali, 2018). The second type is potentiometric biosensors used in conditions that establish zero-current, with an arrangement of two electrodes at an analyte concentration of 10^{-8} M (Alarcon-Angeles et al., 2018; Naresh & Lee, 2021). It detects the potential difference between the working electrode and the reverse electrode in the biosensor. Impedimetric biosensors are the ones in which the impedance, the ratio of voltage over current, is associated with the analyte concentration. An alternate voltage supply (AC) is applied across the two-electrode arrangement, in which the biosensors capture the signals from the interactions occurring at the electrode. The conductometric ones measure the change in the electrical conductivity resulting from chemical reactions, with the ability to operate at low voltages. Moreover, it resembles impedimetric biosensors in many aspects. However, each biosensor is used for a specific application (Kim et al., 2019).

Electrochemical biosensors are valuable and reliable tools for detecting and quantifying numerous substances. These sensors offer remarkable advancements in medical diagnoses, such as cancer and diabetic cell detection at earlier stages, environmental monitoring, food quality management, and biofuel cell manufacturing. In recent decades, biosensors have become more widely used in food quality control, with new development approaches focusing on enhancing real-time detection, thereby ensuring food reliability. Moreover, they play a crucial role in the medical field, especially in the transition from complex to simple and easy testing, such as using home testing devices. In addition, glucose strips based on enzymes and compact amperometric sensors have revolutionized blood glucose level testing, yielding five billion dollars per year as net profit. These strips can give real-time results in less than 10 seconds (Wang, 2006).

Its sensitivity, flexibility, compatibility, and economic feasibility make electrochemical biosensors more desirable and appealing compared to other traditional detection methods in treatment approaches (Mousa, 2010). Moreover, they can diagnose a larger spectrum of conditions more efficiently, showing results in less than minutes (Wang, 2006).

Its ability to sense and capture data immediately helps in obtaining accurate results. Furthermore, there are reduced costs associated with the biosensors ease its implementation in their numerous applications (Zhang et al., 2020).

On the other hand, there are limitations associated with the use of biosensors. Some of the challenges are related to the design and development of third-generation biosensors, involving the direct transfer of signals without the initiation of any redox reactions (Zhang & Li, 2004). Thus, to achieve this, it requires an effective molecular microenvironment on the electrodes' surface to transmit accurate signals. To overcome this, it requires physical, biological, and chemical disciplines to be simultaneously applied. Furthermore, to ensure maximized sensitivity, it is critical to choose an appropriate enzyme immobilization technique that ensures effective interferences in the sequence of activities occurring in the biosensor (Nguyen et al., 2019). Moreover, surface geometry affects the sensor response to detecting changes in the adsorbed molecules. This chapter aims to emphasize the fundamentals of electrochemical biosensors, classification, and applications in the industry, as well as examine advantages, benefits, and potential limitations. Fig. 2.1 shows the sequence through which the signals from the human body are converted to a signal that is converted and displayed (Bhalla et al., 2016). This helps to regulate a certain level of substance present by analyzing the signals displayed on the screen, it can be in the form of current.

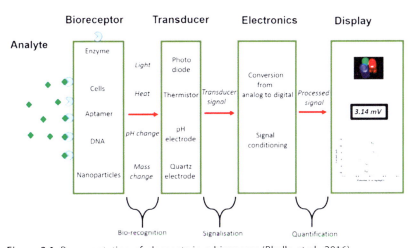

Figure 2.1 Representation of elements in a biosensor (Bhalla et al., 2016).

2.2 Types of biosensors
2.2.1 Amperometric biosensors

Amperometric biosensors are electrochemical biosensors that help capture signals with the help of amperometric transducers. It operates based on enzyme immobilization and attachment of the enzyme to the surface of an electrode. It is one of the vital steps in setting the appropriate microenvironment for biosensors. Therefore, the effectiveness of the biosensor and ability to sense change immediately depends on the immobilization technique used. Amperometric biosensors measure the electric current generated from the reduction and oxidation reactions initiated at the interface between the enzyme and the electrode, where signals are transmitted to generate a plot of current (I) versus time (t) on a screen. Accordingly, the current generated is associated with the analyte's concentration (Lojou & Bianco, 2006). The current produced due to the analyte's concentration versus time relationship is related to the amperometric biosensor mechanism through the Cottrell equation as follows Eq. (2.1) (Lojou and Bianco, 2006):

$$i = \frac{nFAc_0\sqrt{D}}{\sqrt{\pi t}} \quad (2.1)$$

where i is the current measure in Ampere; n being the number of electrons transferred; F is the Faraday constant which corresponds to 96,487 C/mol; c_0 is the initial concentration of the analyte (mol/mL); A is the area of planar electrode (cm^2); D is the diffusion coefficient (cm^2/s), and t is the time elapsed since the potential was applied (s).

The configuration of an amperometric electrode biosensor is shown in Fig. 2.2, which consists of three electrodes, working electrode (WE),

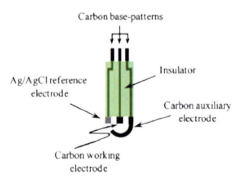

Figure 2.2 An amperometric electrode biosensor with carbon CE and WE electrodes and silver/silver chloride (Malhotra & Ali, 2018).

reference electrode (RE), and counter or auxiliary electrode (CE) (Malhotra & Ali, 2018). The working electrode is coated with carbon, the reference electrode is made up of silver or silver chloride with an auxiliary electrode made up of carbon so that the electrodes do not interfere with the reaction (Sardini et al., 2020). There is a constant voltage supply flowing through the arrangement, which results in a current generated due to the spontaneous redox reactions that occur in the electrochemical cell. The analyte is kept in the cell, in which the current generated is determined when it undergoes a reaction.

The use of amperometric biosensors contributed widely to numerous environmental and medical advancements. However, there are some challenges faced when these biosensors are in use. An inaccurate amperometric response is initiated if the contact between the enzyme and the electrode is not done properly. Hence, to enhance the accuracy of the signals captured and transferred, the most appropriate immobilization technique must be used to ensure that the electrode is fully covered with the enzyme for the particular application. Moreover, many studies have shown that the use of nanomaterials contributed vastly to improved electron transduction in a biochemical reaction. Therefore, nanomaterials, such as graphene improves the electrons transferred (Nauman Javed et al., 2022; Peña-Bahamonde et al., 2018). Graphene, in particular, has extraordinary properties such as the exceptional surface area, high mechanical strength, outstanding thermal conductivity, and very high electron mobility (Tawalbeh et al., 2022). It enhances the interactions between the analyte and the substrate when in contact with each other during redox reactions. High mobility in electron transfer enables the change in concentration of the analyte to be captured immediately.

2.2.2 Potentiometric biosensors

This category of biosensors mainly consists of a two-electrode arrangement with a working electrode (WE) and a reference electrode (RE). This biosensor detects changes in a system by determining the potential difference between both electrodes. The current value at its peak, over a linear potential, is directly proportional to the concentration of the analyte in the solution (Grieshaber et al., 2008). Potentiometric sensors use a zero-current technique for the potential across an interface, most commonly a membrane, which means a device to measure the current is not needed, making the arrangement less complex. In this biosensor, the working electrode detects the potential based on the ions accumulated due to a redox reaction in the solution at that specific electrode (Karimi-Maleh et al., 2021). The reference electrode is set to give a reference potential based on which a difference can be calculated (Pohanka & Skládal, 2008).

There is a proportional relationship between the potential difference measured and the ionic activity, which is a result of the ions generated from the redox reaction. This proportionality is true under the following conditions (Sardini et al., 2020):

- The concentration of ions interfering should be reduced or kept constant since these ions can give faulty readings.
- The selectivity of the membrane should be maximized for the respective analyte being tested.
- Only the potential difference at the interface between the membrane and the solution tested should be changed, according to the ion accumulation.

This relation is determined using a general equation known as the Nernst–Donnan Eq. (2.2) (Alarcon-Angeles et al., 2018):

$$E_{cell} = E^{0}{}_{cell} + \frac{RT}{nF}\ln Q \qquad (2.2)$$

where E_{cell} is the cell potential at equilibrium in Volts; $E^{0}{}_{cell}$ is the standard potential in volts; R is the universal gas constant; T is absolute temperature in Kelvin; n is the number of electrons; F is the Faraday constant, and Q is the mass action law.

Two types of transducers can be used to detect the accumulation of ions, which are ion-sensitive field effect transistors (ISFET) and ion-selective electrodes (ISE). However, the common one is the ISFETs, which works by placing a pH electrode. It is sensitive to the change in pH whenever there is an increase in the acidity of the analyte solution. An example of this is found in the detection of urea in the blood. The blood is used as the analyte with urease immobilized on the electrode, which initiates the breakdown of urea. Thus, a series of reactions take place involving redox reactions. The sensor captures a decrease in the pH due to an increase in the production of hydrogen ions. Therefore, as the concentration of urea in the analyte increases, more hydrogen ions are produced. As a result, the signals are transmitted to some electronic devices, where the signals are converted into useful displays. Accordingly, an action is taken to regulate the level of urea in the blood

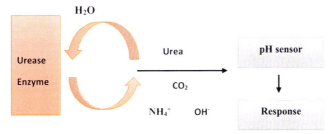

Figure 2.3 Schematic representation of urea detection in the blood (Pundir et al., 2019). Adapted from Pundir, C.S., Jakhar, S., & Narwal, V. (2019). Determination of urea with special emphasis on biosensors: A review. Biosensors and Bioelectronics, 123, 36–50.

(Singh et al., 2021). Fig. 2.3 shows a schematic representation of urea detection in the blood.
H_2O
Urea
pH sensor
CO_2
Urease
Enzyme
NH_4^+
OH^-
Response

Similarly, the same mechanism could be applied on various other applications, using other enzymes such as glucose oxidase, penicillinase, and acetylcholinesterase to regulate the level of penicillin and detect pesticides.

2.2.3 Conductometric biosensors

Conductometric biosensors are another type of biosensors that functions with potential difference applied, with no need for a reference electrode, having a two-electrode arrangement. Moreover, it is a low-cost technology. Conductometric biosensors have transducers that detect the ions moving toward the electrodes after the dissociation of the electrolyte. There is an electric field established in the system that makes the ions move in an ordered manner, where positively charged ions move toward the cathode, and the negatively charged ones move to the anode. These sensors use a sequence of electrical frequencies to measure the conductivity of a substance. The ions that are generated due to an enzyme-linked reaction, altering the conductivity of the solution. Thus, this change is sensed by the transducer (Jaffrezic-Renault & Dzyadevych, 2008). Enzyme processes that make or consume ionic species are influenced by the medium's overall ionic strength and could result in changes in the conductance/capacitance. In a conductometric transducer, various planar interdigitated electrode topologies are designed. Furthermore, within a narrow sensing region, the interdigitated architecture allows for lengthy electrode tracks. These biosensors can be used in detecting harmful substances that are present in the environment, food, and medicines (Chouteau et al., 2004). In this case, a microorganism in algae, known as Chlorella Vulgaris, consisting of numerous enzymes will be immobilized on the electrodes. The level of pollutants present is detected by their effect on living organisms, such as algae and the aquatic life that feeds on algae. The main metal used to construct the electrodes is platinum, an inert metal that will not interfere with the reaction. The main enzyme present in the microorganism is alkaline phosphatase, which breaks down phosphorus, which helps in their growth. Thus, the biosensor will be used to monitor the alkaline phosphate activity (APA) that is usually suppressed when pollutants are present. Certain substances, such as bovine serum albumin (BSA) and glutaraldehyde (GA), are used to immobilize the algal cell on the electrode. The reaction is initiated in the presence of light, with the biosensor immersed in the solution. The reaction that took place is given by the Eq. (2.3):

$$\text{Substrate} \rightarrow \text{Product} + PO_3^- \qquad (2.3)$$

The production of phosphate ions in this reaction changes the conductivity of the solution which is detected by the conductometric biosensor. If cadmium ions (Cd^{2+}), are added to the sample, it affects the enzymatic reaction. Therefore, the difference in the phosphate concentration will be detected and a signal will be transmitted (Chouteau et al., 2004). Fig. 2.4 displays a diagram of the conductometric setup for the pollutant's detection using APA (Chouteau et al., 2004).

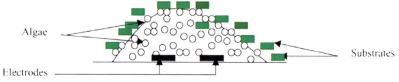

Figure 2.4 A diagram of the conductometric setup for the pollutant's detection using APA (Chouteau et al., 2004).

2.2.4 Impedimetric biosensors

This type of biosensor is the least common as compared to the other ones. It is used in the observation of various aspects from enzymatic reactions to the study of cell kinetics. It captures data based on the concept of impedance and capacitance involving real and imaginary elements, which differentiates it from other techniques. The calculation done to determine the impedance; includes the resistance when a voltage is applied in the circuit arrangement (Pohanka & Skládal, 2008). Then, the data captured is analyzed by electrochemical impedance spectroscopy (EIS), and the signals are displayed in the form of a sinusoidal curve, which is presented on Nyquist plots (Pohanka & Skládal, 2008). The disturbance in the electric field established due to the alternating current circuit (AC) between the electrodes is initiated by a biorecognition event, which will be sensed by the biosensor. For example, a specific antibody binding activity changes overall permittivity, resulting in a change in capacitance, which will be quantified. This change is detected and calculated using Eq. (2.4) (Kirchhain et al., 2020):

$$C = \frac{\varepsilon_o \varepsilon A}{d} \qquad (2.4)$$

where A is the area, d is plate separation, ε_o is dielectric constant, and ε is relative static permittivity. The main drawback is that it takes more time to get accurate plots when wider current frequencies are applied. Fig. 2.5 shows an impedimetric biosensor coated with platinum, an inert metal, that detects the binding of antibodies to the immobilized antigens on the electrode (Katz & Willner, 2003). The antigen−antibody interaction involves charge generation and impedance detection that are displayed on plots.

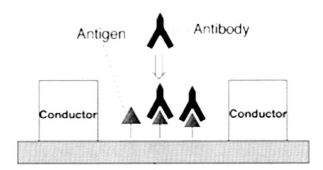

Figure 2.5 A Representation of an impedimetric biosensor between conductive electrodes (Katz & Willner, 2003).

2.3 Applications
2.3.1 Cancer detection

Biosensors are used for several medicinal purposes, one of which is in the detection and screening of cancer (Mousa, 2010). Cancer is a terminal disease, and most often, it is diagnosed at its late stages; this is because the existing methods of detection are not sensitive enough to detect cancer effectively at earlier stages, also known as the premalignant stage. The current methods can only detect the concentration of protein associated with its late stages, in which the symptoms begin to appear so patients do not test earlier (Zhang et al., 2020). By the time cancer is detected, the patient faces more difficulties in terms of getting cured or related to the cost of medications for the treatment. Therefore, more effective and real-time techniques are required to identify cancer and contribute to the reduction of death rates in a population. This is where one of the recent advancements in biological engineering comes into picture; biosensors can be one of the techniques that can initiate higher chances of fighting cancer since it involves the combination of biological and technological practices. This will aid in transmitting accurate results at a lower cost.

According to the International Agency for Research on Cancer (IARC), there are over 18.1 million suffer from cancer all across the globe (Choudhary & Arora, 2021). There are specific biological molecules in the human body, known as biomarkers, which consist of proteins or nucleic acids (Mousa, 2010). One of the methods for cancer detection is based on monitoring the presence or absence of biomarkers present in cancerous tissue or fluids such as serum, plasma, urine, and others. Therefore, the presence of specific biomarkers gives rise to higher chances of the patient having cancer. The different types of biosensors as discussed earlier such as a potentiometer, amperometer, impedimetric and conductometric biosensors, all work with the principle of detecting biomarkers. The development of cancer involves the formation of tumors that produce proteins. These proteins are identified by biosensors in various ways, then signals are transmitted and analyzed by doctors. Analytes associated with cancer are crucial to the early detection, treatment coordination, and the effectiveness of treatment of cancer. There is a specific set of biomarkers for each type of cancer; for instance, CEA (carcinoembryonic antigen) is associated with prostate cancer and AFP (alpha-fetoprotein) is associated with liver cancer. These biomarkers have specific biosensors that detect them, along with their associated electrochemical techniques

Electrochemical biosensing 45

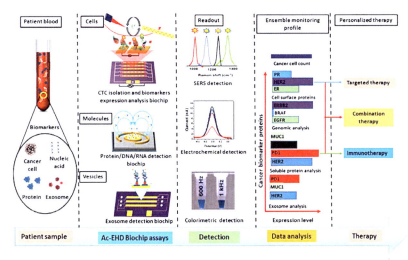

Figure 2.6 Sequence of steps in cancer detection (Choudhary & Arora, 2021).

(Choudhary & Arora, 2021). Fig. 2.6 shows the biomarkers in the human blood with the main focus of electrochemical detection using various biosensors as discussed in the chapter.

Biosensors can determine various aspects, whether a course of treatment has been introduced and if so, whether or not it was effective in eradicating cancer. Nevertheless, there are several methods in which electrochemical biosensors detect and treat cancer, one of which is using an impedimetric biosensor, the involvement of a cell impedance sensing is one of the more ideal methods. This works based on the phenomena of the growth of particles on top of microelectrodes that are capable of altering the resistance at the point of interaction between the microelectrode and other adhering electrodes. Therefore, vital information about cell activity can be obtained. The cell impedance sensing method can assess and determine the developments in cells moving or coming into contact with each other; hence this cell behavior can be observed numerically without posing any complications (Lojou & Bianco, 2006). The electrochemical biosensor comes into contact with the analyte of the tumor cell, detecting a chemical signal and transmitting the results to the transduction system. The signal is then converted into an electrochemical signal and displayed in the electronic system.

The amperometric biosensors can be one of the methods that help to detect biomarkers responsible for prostate cancer, known as the

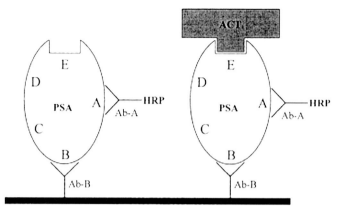

Figure 2.7 Amperometric biosensor for antigen (PSA) binding with antibody (HRP) (Sarkar et al., 2002).

prostate-specific antigen (PSA). Identifying PSA in the serum using biosensors can contribute to the detection of malignancies. The PSA level detected in cancerous tissue is 10 times more compared to a normal one. The antibody used to bind with PSA is horseradish peroxidase (HRP). The arrangement of the biosensor involves; WE made up of rhodinised carbon, RE made of Ag/AgCl and CE made up of carbon, as shown in Fig. 2.7 (Sarkar et al., 2002).

2.3.2 Glucose detection

Millions of people suffer from diabetes everyday all over the world. It is the main reason for comorbid diseases and high death rates. One of the main causes of diabetes is a disorder in the metabolic reaction that breaks down carbohydrates in the body, along with poor dietary habits and obesity. Diabetes is a disease that can be managed with the right medication and equipment. There is a universal increase in the need to manage diabetes, making the glucose analyte the most prevalent among other analytes tested clinically. Conventional methods such as the finger stick self-test have certain restrictions, including the number of times one can test in a day, performing it inaccurately can lead to a low-quality estimation of the blood glucose levels. Therefore, blood glucose levels need to be tracked constantly; it has been found that the continuous monitoring and rigorous management of diabetes reduce the number of health issues that may occur such as neuropathy. In addition, hypo- and hyperglycemia can be detected, and the right protocols can be activated such as regulating medicines and diets straight away depending on the levels (Wang, 2008; Yoo & Lee, 2010). Moreover, self-monitoring of blood glucose (SMBG) is proven to be the most effective method to manage diabetes and insulin injections as it can be implemented for all types of diabetes (Yoo & Lee, 2010).

Subcutaneous CGMS and blood CGMS are two of the most commonly known glucose monitoring systems. Nevertheless, subcutaneous CGMS is preferred because it reduces the threats related to electrode contamination and blood clotting. Hence, subcutaneous CGMS includes a needle electrode that can be implanted into the skin to measure blood glucose levels. Such devices have sensors that are disposable and offer real-time results, they display the glucose levels an average of every 2.5 minutes. Overall, CGMS can better and properly control diabetes promptly giving accurate real-time results of blood glucose concentration and amounts of insulin released into the body (Wang, 2008).

In the contemporary world, more inventions emphasized the use of biosensors in detecting blood glucose levels as well as acting as being considered as a continuous glucose monitoring system (CGMS). The commonly used biosensor for glucose level monitoring is amperometric. In addition to that, closed-loop glucose monitoring techniques detect blood glucose levels and inject insulin into the patient's body in a feedback loop for an ideal insulin dosage. This works by a "sense and act" system where the sensor detects the glucose levels and sends the data to a pump, which will release the correct dosage of insulin; this process can be seen in Fig. 2.7. The data is collected when there is a potential difference established between electrodes, which in turn generates current; it usually happens during oxidation-reduction reactions. Consequently, amperometric transducers measure the current. There are two enzymes used in the amperometric detection of blood glucose levels: glucose oxidase (GOx) and glucose-1-dehydrogenase (GDH) (Yoo & Lee, 2010).

The usage of GOx in biosensors is more common than GDH because it has a greater range of selectiveness when it comes to glucose. It is likewise easier and more cost-effective to acquire, and it has a higher endurance for a wider range of pH and temperatures, making it more suitable for harsher environments in the manufacturing and storing of biosensors. GOx acts as a catalyst with the cofactor, flavin adenine dinucleotide (FAD) in the oxidation reaction of glucose, in which hydrogen peroxide (H_2O_2) is released. The oxidation of H_2O_2 occurs at the anode coated. The movement of electrons is detected, hence, the number of

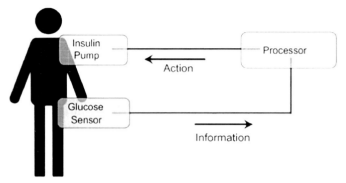

Figure 2.8 Amperometric biosensor used for the feedback control loop for detecting glucose level (Wang, 2008).

electrons captured reflects the blood glucose level (Kim et al., 2019; Naresh & Lee, 2021; Wang, 2008).

Glucose OxidaseAnode

Two steps take place for the oxygen released to be detected, which in turn helps in tracking the level of glucose through signals sent to a processor, as demonstrated in Fig. 2.8. Finally, the signals received by the digital system are transferred to the insulin pump, which acts based on the optimum level of glucose that should be in the blood (Wang, 2008).

2.3.3 Environmental contamination detection

Emerging contaminants and the toxicity exhibited when released into the atmosphere have become major issues that need to be tackled in the 21st century. Increased demand due to growth in the world's population, leads to an increase in manufacturing industries that deal with pharmaceuticals, pesticides, perfumes, and more. Thus, the release of toxic chemicals into oceans, lakes, and the atmosphere has been improved. This poses a great threat to the whole ecosystem, including people and the aquatic marine life that are directly exposed to the fumes and gases emitted from these chemicals, leading to the arousal of varied health conditions. These threats can be avoided with the use of control measures that enable real-time monitoring and using equipment that incorporates high sensitivity all at low cost. An example of contaminants emitted are titanium dioxide and diclofenac, these compounds affect reproduction and hormone activity in fish, as well as the process of photosynthesis in seaweed (Hernandez-Vargas et al., 2018). The conventional methods such as chromatography are no longer highly effective since they require pretreatment and preparation time before being used for detecting contaminant levels. They also have low sensitivity and a lower range in selectivity when low concentrations of the sample are taken for testing. This gives rise to electrochemical biosensors since they can be used for the lowest sample concentrations and volumes of the biological elements in the environment (Hernandez-Vargas et al., 2018).

The electrochemical biosensors can collect the information electronically, without the involvement of complexities. Hence, on-site inspections can take place by connecting the signals transmitted from the transducer used in the biosensor to a mobile system. This feature enables them to identify specific targets, making them more efficient in detecting analytes, monitoring progress, and prompting action when needed. Some of the examples of the level of certain analytes, heavy metals that can be detected are Pb^{2+} and Hg^{2+}. One of the biosensor arrangements can involve using DNA to aid in heavy metal detection. In this case, a single-stranded DNA (ss-DNA) will be used as a bioreceptors to help in binding with the heavy metals in the analyte. The WE will be coated with an inert metal along with the ss-DNA, a platinum wire used as the CE, and RE that consists of mercury that is in contact with mercury (Bhalla et al., 2016) chloride, known as saturated calomel, used due to its robust properties. Finally, any type of biosensor mechanism can be used as discussed in previous sections to regulate the level by detecting signals when these metals bind with the bases, such as Thymine {T} and Cytosine {C} in the ss-DNA used (Saidur et al., 2017).

2.3.4 Pathogen detection

Pathogens are microorganisms that cause diseases; these agents include bacteria, viruses, and fungi. They can infect people through various modes, as they can be transmitted through food, water, air, and blood. They vastly contribute to the death of millions of people across the globe, as they can attack the immune system, with the worst-case scenario of destroying organs. Therefore, specific techniques need to be implemented to improve the time at which they are detected, to be able to control them before their widespread. One of the greatest pandemics faced recently was COVID-19, which involved the coronavirus, that lead to a worldwide outbreak due to its rapid spread before being detected. The use of biosensors can be one of the prevention techniques.

The biosensors would function based on the principle of the body generating antibodies to bind to the region of the antigen called an

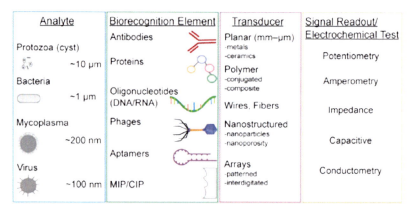

Figure 2.9 Diagram that summarizes the testing of a certain type of analyte with its corresponding recognition element, transducer, and electrochemical test (Cesewski & Johnson, 2020).

epitope, that is specific for each pathogen and destroy it before it spreads throughout the body and to other people. The setup is arranged based on having a two or three-electrode arrangement that depends on the transducer mechanism that will be integrated into the biosensor. The electrodes could be coated with conducting or semiconducting materials, most commonly would be platinum, which is an inert and thick metal that will aid in the detection of the pathogen present in the analyte. Usually, nanomaterials are used in biosensors to ensure high sensitivity and effective performance. Fig. 2.9 summarizes the types of pathogens that could be taken as analytes, the biorecognition elements that could be used to detect the target pathogen, along with using the most appropriate signal transmission technique (Cesewski & Johnson, 2020).

An example that can be illustrated is using *Escherichia coli* (*E. coli*), which can be detected using electrochemical impedance spectroscopy (EIS), with WE made up of indium tin oxide (ITO). The biorecognition element that will be used to bind with the antibody immobilized on the electrode will be anti-*E. coli*, thus capturing the presence of *E. coli* and sends signals to initiate the appropriate action needed (Cesewski & Johnson, 2020).

2.3.5 Herbicide contamination detection

Another important application for biosensors is in the detection of herbicides; it is a chemical substance used to monitor the growth of undesirable crops to obtain desirable ones. Sometimes, it is added to water bodies, to control the growth of weeds, since its excessive growth can harm the aquatic marine life. Therefore, the amount of herbicide used must be controlled to serve its purpose appropriately. One of the herbicides commonly used to control weed growth in farmlands is known as Chlorsulfuron; it is applied to inhibit the growth of weeds that compete with the desirable crops for nutrients (Badihi-Mossberg et al., 2007; Dzantiev et al., 2004). The arrangement of the biosensor is done using the principle of competing for vacant binding sites and accordingly determining the chlorsulfuron level. To ensure this is achieved, an analyte with chlorsulfuron will be chosen as the test sample. In addition to that, the electrode will be attached to the enzyme horseradish peroxidase with antibodies on a membrane linked to it. Another complex will be created between chlorsulfuron and the enzyme glucose oxidase.

Glucose is added to this setup; thus, it is catalyzed by the chlorsulfuron-glucose oxidase complex, releasing hydrogen peroxide, which further undergoes reduction by HRP. This process results in a shift captured in the current flowing through the electrodes due to electron transfer, this change will be signaled, determining the level of chlorsulfuron in the sample (Badihi-Mossberg et al., 2007; Dzantiev et al., 2004). This process takes up to 15 minutes.

2.3.6 Food contamination detection

Technological advancements have revolutionized how food is processed, controlled, and made safe for human consumption. Nowadays, electrochemical biosensors have been implemented to contribute to quality control, where pathogens can be detected as discussed in earlier sections. Moreover, the allergens and nutritional content of a food sample can easily be determined. Some of the pathogens that contribute to food contamination are *E. coli* and *Salmonella* (Cesewski & Johnson, 2020). They can be detected and treated before any food is packaged, which makes companies in the food industry achieve targets by continuously maintaining high standards.

2.4 Advantages

Electrochemical biosensors are much more desired than other types of sensors. This is because they have numerous advantages such as having higher sensitivity, fewer moving parts, and being more economically feasible (Bhalla et al., 2016). Electrochemical biosensors offer a wider range of diagnoses at a faster and more efficient rate, displaying results in minutes. Furthermore, they are smaller in size and cost-effective making them much more beneficial than conventional detection and treatment methods (Wang, 2006). Their small size enables them to be easily transported without any difficulties; it is also beneficial for home devices such as blood glucose detection, where they would be light and compact (Wang, 2008). The higher the sensitivity of a biosensor, the more likely the results obtained will be accurate, thus the better the selection and implementation of a treatment. In other words, the more sensitive they are, the clearer the margins of the cancerous tumors, hence enabling doctors to decide what treatments would work best with the type and size of the tumor. They are capable of detecting several types of cancers, offering a wider range of applications. These biosensors are installed with multiple transducers that are modified to detect a particular analyte produced by tumor cells (Lojou & Bianco, 2006).

Sensitivity also plays a role in nonmedicinal purposes such as in the environmental and food industries. When it comes to the detection of treatment of cancer and diabetes, they are highly sensitive enabling them to find and identify cancer and glucose analytes, making them more accurate, which means they can present the same result when the experiment is repeated (Bhalla et al., 2016; Mousa, 2010); they are simple to use, making them more efficient as time will be saved in implementing them. Moreover, the sensitivity of the biosensor can be increased by using a greater surface area, as more enzymes are immobilized. Titanium dioxide is found to be the most compatible nanomaterial with electrochemical biosensors, making it more fitting for their use (Choudhary & Arora, 2021; Mohamed et al., 2021). In addition to that, it exhibits high stability, as it can resist disturbances that can happen at any point in time.

2.5 Limitations

Biosensor design and development is challenging due to the multidisciplinary character of the industry, which is especially true for electrochemical biosensors; further research is made to identify ways in which the interaction between the surface of the biosensor and the analyte can be more effective in collecting real-time results. Thus, finding solutions to this issue enables further enhancement of the resilient, dependable, and sensitive characteristics of biosensors (Bizzotto et al., 2018). Some of the disadvantages related to amperometric biosensors, are highlighted in the following points:

- To ensure credible and accurate results, the geometry of the electrode used in the biosensor must be chosen appropriately, since it affects its sensitivity.
- The biosensor's response is affected by the mass transfer of the ions and other substances released during redox reactions.
- The incorrect choice of oxidizing agents used for the transducers can result in low stability of the biosensor.
- The data should be permanently corrected, as the sample can interfere with nonspecific particles formed during redox reactions.

To tackle the first challenge faced by biosensors is to chemically treat and polish the surface or apply heat. These can contribute to increasing the number of times the results are repeated. Usually, a biosensor loses its reproducibility ability when it is exposed to the sample solution being tested for an extended period (Bizzotto et al., 2018). Therefore, surface modification with high conductive organic molecules or redox polymers is a potential method for overcoming this barrier. Some organic salts such as N-methylphenazine (NMP) and tetracyanoquinodimethane (TCNQ) are conductive and were successfully used in dehydrogenase-based sensors which enabled effective electron transfer. Moreover, tetra-thiafulvalene (TTF), was used in oxidase-based sensors for the electrodes (Bizzotto et al., 2018).

There is a need for correcting the signal and dealing with ascorbic acid, uric acid, and more interfering particles. From this perspective, certain membranes can be used to improve stability and make sure that the ions do not interfere. Hence, examples of these membranes are acetate cellulose and polycarbonate film membranes. To solve another limitation related to poor choice of mediators, also known as an oxidizing agent, is working at a low potential difference maintained between both electrodes; this also minimizes various interferences. In some cases, carbon paste is used to prevent the issue of mediators' solubility influencing the results captured by the biosensor (Bizzotto et al., 2018).

2.6 Conclusions

In conclusion, electrochemical biosensors are small devices that exhibit various properties that make them one of the most effective biological

54 Electrochemistry and Photo-Electrochemistry of Nanomaterials

and technological advancements. It is used to detect and monitor analytes by capturing biological signals using biorecognition elements and bioreceptors and converting them into electrical signals displayed on a screen. There are several types of biosensors including amperometric, potentiometric, conductometric, and impedimetric biosensors. High sensitivity and reproducibility characteristics associated with biosensors make them applicable to be used in many medical, environmental, and food industries. Biosenors can be used in the detection of cancer through analytes such as DNA. Furthermore, measuring blood glucose levels in diabetic patients has become crucial with the use of biosensors. They also help in estimating the concentration of toxins in the environment and contributes to maintaining high food quality. Their ability to have wider selectivity ranges makes them more efficient than conventional monitoring methods. Their property of being small in size make them involved in the manufacturing of portable devices for home testing. However, further research is needed to eliminate certain limitations such as its effectiveness that can be affected by the environment and challenges faced related to design, development, and detection mechanisms.

References

Alarcon-Angeles, G., Álvarez-Romero, G. A., & Merkoçi, A. (2018). Electrochemical biosensors: Enzyme kinetics and role of nanomaterials. *Encyclopedia of Interfacial Chemistry: Surface Science and Electrochemistry*, 140−155. Available from https://doi.org/ 10.1016/B978-0-12-409547-2.13477-8, http://doi.org/10.1016/B978-0-12-409547-2.13477-8.

Badihi-Mossberg, M., Buchner, V., & Rishpon, J. (2007). Electrochemical biosensors for pollutants in the environment. *Electroanalysis*, *19*(19-20), 2015−2028. Available from https://doi.org/10.1002/elan.200703946, http://onlinelibrary.wiley.com/journal/10.1002/(ISSN)1521-4109.

Bahadir, E. B., & Sezgintürk, M. K. (2015). Electrochemical biosensors for hormone analyses. *Biosensors and Bioelectronics*, *68*, 62−71. Available from https://doi.org/10.1016/j.bios.2014.12.054, http://www.elsevier.com/locate/bios.

Bhalla, N., Jolly, P., Formisano, N., & Estrela, P. (2016). Introduction to biosensors. *Essays in Biochemistry*, *60*(1), 1−8. Available from https://doi.org/10.1042/ EBC20150001, http://essays.biochemistry.org/content/ppebio/60/1/1.full.pdf.

Bizzotto, D., Burgess, I. J., Doneux, T., Sagara, T., & Yu, H. Z. (2018). Beyond simple cartoons: Challenges in characterizing electrochemical biosensor interfaces. *ACS Sensors*, *3*(1), 5−12. Available from https://doi.org/10.1021/acssensors.7b00840, http://pubs.acs.org/journal/ascefj.

Cesewski, E., & Johnson, B. N. (2020). Electrochemical biosensors for pathogen detection. *Biosensors and Bioelectronics*, *159*. Available from https://doi.org/10.1016/j.bios.2020.112214, http://www.elsevier.com/locate/bios.

Choudhary, M., & Arora, K. (2021). Electrochemical biosensors for early detection of cancer. *Biosensor Based Advanced Cancer Diagnostics: From Lab to Clinics*, 123−151.

Available from https://doi.org/10.1016/B978-0-12-823424-2.00024-7, https://www.sciencedirect.com/book/9780128234242.

Chouteau, C., Dzyadevych, S., Chovelon, J. M., & Durrieu, C. (2004). Development of novel conductometric biosensors based on immobilised whole cell *Chlorella vulgaris* microalgae. *Biosensors and Bioelectronics*, *19*(9), 1089−1096. Available from https://doi.org/10.1016/j.bios.2003.10.012, http://www.elsevier.com/locate/bios.

Dzantiev, B. B., Yazynina, E. V., Zherdev, A. V., Plekhanova, Y. V., Rshetilov, A. N., Chang, S. C., & McNeil, C. J. (2004). Determination of the herbicide chlorsulfuron by amperometric sensor based on separation-free bienzyme immunoassay. *Sensors and Actuators, B: Chemical*, *98*(2-3), 254−261. Available from https://doi.org/10.1016/j.snb.2003.10.021.

Grieshaber, D., MacKenzie, R., Vörös, J., & Reimhult, E. (2008). Electrochemical biosensors—Sensor principles and architectures. *Sensors*, *8*(3), 1400−1458. Available from https://doi.org/10.3390/s8031400, http://www.mdpi.org/sensors/papers/s8031400.pdf.

Hernandez-Vargas, G., Sosa-Hernández, J. E., Saldarriaga-Hernandez, S., Villalba-Rodríguez, A. M., Parra-Saldivar, R., & Iqbal, H. M. N. (2018). Electrochemical biosensors: A solution to pollution detection with reference to environmental contaminants. *Biosensors*, *8*(2). Available from https://doi.org/10.3390/bios8020029, http://www.mdpi.com/2079-6374/8/2/29/pdf.

Jaffrezic-Renault, N., & Dzyadevych, S. V. (2008). Conductometric microbiosensors for environmental monitoring. *Sensors*, *8*(4), 2569−2588. Available from https://doi.org/10.3390/s8042569France, http://www.mdpi.org/sensors/papers/s8042569.pdf.

Karimi-Maleh, H., Orooji, Y., Karimi, F., Alizadeh, M., Baghayeri, M., Rouhi, J., Tajik, S., Beitollahi, H., Agarwal, S., Gupta, V. K., Rajendran, S., Ayati, A., Fu, L., Sanati, A. L., Tanhaei, B., Sen, F., Shabani-Nooshabadi, M., Naderi Asrami, P., & Al-Othman, A. (2021). A critical review on the use of potentiometric based biosensors for biomarkers detection. *Biosensors and Bioelectronics*, *184*. Available from https://doi.org/10.1016/j.bios.2021.113252.

Katz, E., & Willner, I. (2003). Probing biomolecular interactions at conductive and semi-conductive surfaces by impedance spectroscopy: Routes to impedimetric immunosensors, DNA-sensors, and enzyme biosensors. *Electroanalysis*, *15*(11), 913−947. Available from https://doi.org/10.1002/elan.200390114, http://onlinelibrary.wiley.com/journal/10.1002/(ISSN)1521-4109.

Kim, M., Iezzi, R., Shim, B. S., & Martin, D. C. (2019). Impedimetric biosensors for detecting vascular endothelial growth factor (VEGF) based on poly(3,4-ethylene dioxythiophene) (PEDOT)/gold nanoparticle (Au NP) composites. *Frontiers in Chemistry*, *7*. Available from https://doi.org/10.3389/fchem.2019.00234, https://www.frontiersin.org/articles/10.3389/fchem.2019.00234/full.

Kirchhain, A., Bonini, A., Vivaldi, F., Poma, N., & Di Francesco, F. (2020). Latest developments in non-faradic impedimetric biosensors: Towards clinical applications. *TrAC Trends in Analytical Chemistry*, *133*. Available from https://doi.org/10.1016/j.trac.2020.116073.

Lojou, E., & Bianco, P. (2006). Application of the electrochemical concepts and techniques to amperometric biosensor devices. *Journal of Electroceramics*, *16*(1), 79−91. Available from https://doi.org/10.1007/s10832-006-2365-9.

Malhotra, B. D., & Ali, M. A. (2018). *Nanomaterials in biosensors* (pp. 1−74). Elsevier BV. Available from https://doi.org/10.1016/b978-0-323-44923-6.00001-7.

Mohamed, O., Al-Othman, A., Al-Nashash, H., Tawalbeh, M., Almomani, F., & Rezakazemi, M. (2021). Fabrication of titanium dioxide nanomaterial for implantable highly flexible composite bioelectrode for biosensing applications. *Chemosphere*, *273*. Available from https://doi.org/10.1016/j.chemosphere.2021.129680.

Mousa, S. (2010). Biosensors: the new wave in cancer diagnosis. *Nanotechnology, Science and Applications, 4*(1). Available from https://doi.org/10.2147/NSA.S13465.

Naresh, V., & Lee, N. (2021). A review on biosensors and recent development of nanostructured materials-enabled biosensors. *Sensors, 21*(4). Available from https://doi.org/10.3390/s21041109.

Nauman Javed, R. M., Al-Othman, A., Tawalbeh, M., & Olabi, A. G. (2022). Recent developments in graphene and graphene oxide materials for polymer electrolyte membrane fuel cells applications. *Renewable and Sustainable Energy Reviews, 168*. Available from https://doi.org/10.1016/j.rser.2022.112836, https://www.journals.elsevier.com/renewable-and-sustainable-energy-reviews.

Nguyen, H. H., Lee, S. H., Lee, U. J., Fermin, C. D., & Kim, M. (2019). Immobilized enzymes in biosensor applications. *Materials, 12*(1). Available from https://doi.org/10.3390/ma12010121, https://www.mdpi.com/1996-1944/12/1/121/pdf.

Peña-Bahamonde, J., Nguyen, H. N., Fanourakis, S. K., & Rodrigues, D. F. (2018). Recent advances in graphene-based biosensor technology with applications in life sciences. *Journal of Nanobiotechnology, 16*(1). Available from https://doi.org/10.1186/s12951-018-0400-z, http://www.jnanobiotechnology.com/start.asp.

Pohanka, M., & Skládal, P. (2008). Electrochemical biosensors—Principles and applications. *Journal of Applied Biomedicine, 6*(2), 57−64. Available from https://doi.org/10.32725/jab.2008.008, http://www.zsf.jcu.cz/vyzkum/jab/6_2/pohanka.pdf.

Pundir, C. S., Jakhar, S., & Narwal, V. (2019). Determination of urea with special emphasis on biosensors: A review. *Biosensors and Bioelectronics, 123*, 36−50. Available from https://doi.org/10.1016/j.bios.2018.09.067, http://www.elsevier.com/locate/bios.

Saidur, M. R., Aziz, A. R. A., & Basirun, W. J. (2017). Recent advances in DNA-based electrochemical biosensors for heavy metal ion detection: A review. *Biosensors and Bioelectronics, 90*, 125−139. Available from https://doi.org/10.1016/j.bios.2016.11.039, http://www.elsevier.com/locate/bios.

Sardini, E., Serpelloni, M., & Tonello, S. (2020). Printed electrochemical biosensors: Opportunities and metrological challenges. *Biosensors, 10*(11). Available from https://doi.org/10.3390/bios10110166, https://www.mdpi.com/2079-6374/10/11/166.

Sarkar, P., Pal, P. S., Ghosh, D., Setford, S. J., & Tothill, I. E. (2002). Amperometric biosensors for detection of the prostate cancer marker (PSA). *International Journal of Pharmaceutics, 238*(1-2), 1−9. Available from https://doi.org/10.1016/S0378-5173(02)00015-7.

Singh, S., Sharma, M., & Singh, G. (2021). Recent advancementsin urea biosensors for biomedical applications. *IET Nanobiotechnology, 15*(4), 358−379. Available from https://doi.org/10.1049/nbt2.12050, https://ietresearch.onlinelibrary.wiley.com/journal/1751875X.

Tawalbeh, M., Nauman Javed, R. M., Al-Othman, A., & Almomani, F. (2022). The novel advancements of nanomaterials in biofuel cells with a focus on electrodes' applications. *Fuel, 322*. Available from https://doi.org/10.1016/j.fuel.2022.124237.

Wang, J. (2006). Electrochemical biosensors: Towards point-of-care cancer diagnostics. *Biosensors and Bioelectronics, 21*, 1887−1892. Available from https://doi.org/10.1016/j.bios.2005.10.027.

Wang, J. (2008). In vivo glucose monitoring: Towards 'Sense and Act' feedback-loop individualized medical systems. *Talanta, 75*(3), 636−641. Available from https://doi.org/10.1016/j.talanta.2007.10.023.

Yoo, E. H., & Lee, S. Y. (2010). Glucose biosensors: An overview of use in clinical practice. *Sensors*, *10*(5), 4558−4576. Available from https://doi.org/10.3390/s100504558, http://www.mdpi.com/1424-8220/10/5/4558/pdf.

Zhang, W., & Li, G. (2004). Third-generation biosensors based on the direct electron transfer of proteins. *Analytical Sciences*, *20*(4), 603−609. Available from https://doi.org/10.2116/analsci.20.603, http://www.jstage.jst.go.jp/browse/analsci.

Zhang, Z., Li, Q., Du, X., & Liu, M. (2020). Application of electrochemical biosensors in tumor cell detection. *Thoracic Cancer*, *11*(4), 840−850. Available from https://doi.org/10.1111/1759-7714.13353, http://onlinelibrary.wiley.com/journal/10.1111/(ISSN)1759-7714.

CHAPTER 3

Solar cells at the nanoscale

Leanne Shahin[1], Aya ElGazar[1], Taima Al Hazaimeh[1], Abdullah Ali[1], Amani Al-Othman[1] and Muhammad Tawalbeh[2,3]

[1]Department of Chemical and Biological Engineering, American University of Sharjah, Sharjah, United Arab Emirates
[2]Sustainable and Renewable Energy Engineering Department, University of Sharjah, Sharjah, United Arab Emirates
[3]Sustainable Energy & Power Systems Research Centre, RISE, University of Sharjah, Sharjah, United Arab Emirates

3.1 Introduction

There is an increasing demand for energy around the world. Experts speculate that by 2050, the world will require 30 TW of energy resources (Abdin et al., 2013). This huge amount of energy must be provided with minimal environmental harm and low CO_2 emissions. The growing use of fossil fuels has majorly affected the environment and caused global warming. This is mainly because of greenhouse gas emissions, in particular, CO_2 emissions, which are of great concern due to of their direct influence on climate change and global warming (Alami, Abu Hawili, et al., 2020; Tawalbeh et al., 2021). Therefore, increasing the world's energy dependence on renewable energy is essential (Kandeal et al., 2021). There are various types of renewable energy sources currently used, which include wind, solar, waves, hydro, biomass, and geothermal energy (Martis et al., 2020; Tawalbeh et al., 2020; Tawalbeh et al., 2023). Researchers consider solar energy to be one of the main catalysts to pilot global emission reductions given the already long attention it has garnered from researchers (Obaideen et al., 2021). The economic and industrial growth is predicted to drive solar energy generation up to 48% by 2050 (Maka & Alabid, 2022). Research conducted at the end of 2019 shows that more than 635 GW of energy is produced using photovoltaic systems. In the United States alone, at least two million solar panel systems are used, which can power over 16 million homes in the United States (Soto et al., 2022). This huge utilization of solar energy via photovoltaic cells is due to the numerous advantages it has compared to fossil fuels. Solar energy is abundant, renewable, more reliable, efficient, has a lower

Electrochemistry and Photo-Electrochemistry of Nanomaterials
DOI: https://doi.org/10.1016/B978-0-443-18600-4.00004-1
© 2025 Elsevier Inc. All rights are reserved, including those for text and data mining, AI training, and similar technologies.

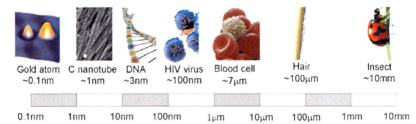

Figure 3.1 Visualizing nanoscale material (Serrano et al., 2009).

payback period, and reduces greenhouse emissions compared to producing electricity using fossil fuels (Salameh et al., 2021; Vatti et al., 2018). These advantages explain the increasing demand for solar cells nowadays, as well as the increasing research on improving their efficiency. Commercial solar panels have an efficiency range of 15%—20%, which is relatively low. To further increase this efficiency, researchers are working on new technologies that take advantage of the enhanced optical properties of nanosized materials. However, there are numerous environmental impacts related to the utilization of solar energy, such as the extensive use of land and the utilization of hazardous materials in the manufacturing process (Tawalbeh et al., 2021).

Nanotechnology is a wide area of research nowadays in physics, chemistry, biology, and engineering applications. This field of study encompasses utilizing the unique characteristic of nanoscale material from 1 to 100 nm, as shown in Fig. 3.1; nanoscale material could exhibit new properties that are absent in bulk material (Serrano et al., 2009). These properties are used in solar cells to increase light absorption, reduce reflection losses, increase efficiency, and reduce manufacturing costs. Therefore, in this book chapter, we will be explaining the history, mechanism, material used, advantages, and downsides of nanosized solar cells.

3.2 Solar cells
3.2.1 History of solar cells

The invention of solar cells set a change in motion in the field of energy. The history of this invention is an interesting story of innovation and development. Solar energy is not actually new; its history spans from the 7th century B.C. However, solar cell technology did not emerge until in 1839, when Edmond Becquerel discovered the photovoltaic effect while experimenting with an electrolytic cell exposed to light (Fraas & Partain,

2010). Later in 1873, Willoughby Smith observed that selenium can function as a photoconductor. Just a few years later, W.G. Adams and R.E. Day were able to detect the photovoltaic effect in a solidified selenium (Fraas, 2012). In 1883, and almost five decades after the discovery of the photovoltaic effect, the American inventor Charles Fritz created the first working selenium solar cell with an efficiency of less than 1% (Fraas, 2014). This period was just a discovery period without any scientific knowledge about the operation of these devices.

Further research on photovoltaics continued for the next several decades, especially after Albert Einstein published his paper on the photoelectric effect (Fraas et al., 2013). Einstein's paper brought more attention and acceptance for the idea of converting sunlight to electrical energy. Hence, leading to further development of different applications and theories, however, they were not commercially used. The next major advancement in solar cell technology came from the work of Bell Labs in 1954. Bell Labs found that semiconducting materials like silicon were more efficient than selenium. Daryl Chapin, Calvin Fuller, and Gerald Pearson managed to build a silicon solar cell with an efficiency of approximately 6%. Although it was considered the first device capable of converting solar energy to electricity, it was still very expensive to produce, and it was hard to commercialize it (Chapin et al., 1991). However, within a couple of years, scientists were able to increase solar cell efficiency and they were commonly used to power satellites, and in other space applications.

In 1973, University of Delaware built one of the world's first solar-powered buildings "Solar One." Instead of using solar panels, solar arrays were integrated into the rooftop. These arrays fed power to the building during the day, and during the night, and any excess power was purchased from the utility. "Solar One" was the first example of the current grid-connected systems (Cleveland & Morris, 2013). It was around the same time in the 1970s, when an energy crisis emerged in the United States because of the first Arab oil embargo. The federal government became more committed to making solar energy feasible and affordable to the public. Moreover, after the debut of "Solar One," people started to accept solar energy as an option for their houses (Fraas & Partain, 2010). Solar cells continued to experience rapid growth because of further cost reductions, and concerns about climate change. Today, solar cells are used in almost all sorts of devices, from calculators to rooftop solar panels, and research is continuing with the goal of making solar energy more competitive with fossil fuels.

3.2.2 How solar cells work

Solar cell is a device that directly converts light energy into electrical energy through photovoltaic effect. The photovoltaic effect relies on ideas from quantum theory. Light consists of packs of energy, called photons. Normally, when a matter absorbs light, the energy of these photons would excite electrons up to higher energy levels within the material, however, the excited electrons quickly relax back to their ground state. However, in a photovoltaic device, there is a built-in asymmetry, which pulls the excited electrons before they relax back and sends them to an external circuit. The excess energy of the electrons creates a potential difference that drives the electrons through a load, generating electricity (Leggett, 2012).

The asymmetry in solar cells is shaped through the process of doping in which impurities are injected into a silicon wafer leading to the formation of a p-n junction (Kim et al., 2012). The p-type silicon is made by atoms, such as boron or gallium, that contain one less electron in their outer energy level compared to silicon. These atoms contain one less electron and hence are not capable of forming bonds with the surrounding silicon, thereby, creating an electron vacancy or "hole." The n-type silicon, however, is produced by the addition of atoms, such as phosphorus, that have one more electron, in their outer level, than silicon. This electron will not be involved in bonding when these atoms bond with their neighboring silicon atoms; however, it will be freely mobile inside the silicon structure (Fernandez, 2014/2014).

A solar cell is made up of p-type silicon and n-type silicon layers placed beside each other. Beside the p-n junction, electrons on the n-type layer move into the holes on the other side of the junction, to the p-type layer, as seen in Fig. 3.2 (Simya et al., 2018). This migration of electrons, in pursuit of filling the holes, generates an area around the junction defined as the depletion zone (Soga, 2006). When all the holes in the depletion zone are saturated with electrons, the n-type side would contain positively charged ions, and the p-type side would contain negatively charged ions. An internal electric field is generated, by presence of these oppositely charged ions, that prevents electrons in the n-type layer from filling the p-type layer holes.

When sunlight strikes a solar cell, electrons are ejected, thereby, creating holes (vacancies). The occurrence of this process in an electric field will drive electrons to the n-type layer and the hole to the p-type layer. If

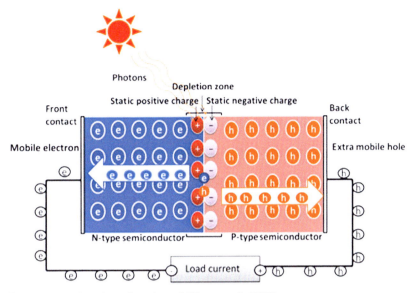

Figure 3.2 p-n junction of a solar cell (Simya et al., 2018).

a metallic wire connects the two layers, the depletion zone will be crossed as the electrons move from the n-type to the p-type layer. Next, the electrons will move back to the n-type layer via an external wire, generating current (Fernandez, 2014/2014).

3.2.3 Types of solar cells

Solar cells are typically named after the semiconducting materials that compose them. These materials should have such features that permit sunlight absorption. Some solar cells are designed to absorb sunlight reaching the Earth's surface, while other cells are designed for use in closed spaces. Solar cells could also be classified as silicon-based cells, compound semiconductors-based cells, dye-sensitized solar cells (DSSCS), and organic/polymer solar cells, along with other new concepts (Choubey et al., 2012).

Another method of solar cell classification stems from their generation. The conventional or the wafer-based cells are known to be the first-generation cells that are the oldest and the most popular technology because of their relatively high-power efficiency. They consist of crystalline silicon, and they are subdivided into single and polycrystalline silicon cells (Sharma et al., 2015). Amorphous silicon, CdTe, CdS, and CIGS cells compose the thin film cells that are classified as second-generation cells (Alami et al., 2018). Their production is cheaper than conventional silicon solar cells because they demand less construction materials. Third-generation solar cells provide relatively high efficiency with lower costs. They are developed using the thin layer deposition technique used for the second-generation cells. However, most third-generation cells have not yet been commercially used and are still in the research and development phase (Mohammad Bagher, 2015). Currently, new and different materials and approaches are being investigated in favor of improving solar cells' performance.

3.2.4 Nanotechnology in solar cells

Although solar cells are promising sustainable electricity generating systems, they have not been capable of competing economically with fossil fuels. Currently, research is focusing on integrating nanotechnology into solar cells to improve their efficiency and reduce their cost. Nanotechnology can improve the design and manufacture of second-generation solar cells. However, its major advantages are observed in third-generation solar cells, in which novel materials such as nanowires, nanofilms, and quantum dots (QDs) have shown encouraging results in improving solar cells efficiency (Bhatia, 2014).

3.2.5 Introduction to nanotechnology and nanomaterials

A nanomaterial is any material that has at least one dimension in the nanometer scale, which ranges from one to a few hundred nanometers. Nanotechnology refers to the creation of nanomaterials by humans. Nanomaterials can be made up of metals, metal oxides, or semiconductors (Alami, Faraj, et al., 2020). They could also be subdivided into two classes: natural nanomaterials and anthropogenic nanomaterials. The latter are the materials constructed by humans through power stations or manufacturing processes, while the natural nanomaterials are the ones found by themselves in nature like minerals, proteins, and viruses (Schaming & Remita, 2015).

Nanotechnology has many unique features, which distinguish it from other technologies. It offers the opportunity of creating materials with different novel properties. Materials at the nanoscale exhibit unique and different properties from their bulk molecules, such as new chemical, mechanical, electronic, and biological properties. For instance, nanoparticles have a much higher surface area to volume ratio compared to their larger scale particles. Moreover, devices at the nanoscale require less materials and energy to construct them, they use less energy, and their functions can be enhanced by decreasing their characteristic dimensions. Harnessing the novel properties of nanomaterials leads to considerable advancement in material and device performance, while utilizing less material and energy (Ramsden, 2016). As a result, these innovative nanotechnologies make nanomaterials very encouraging in the field of solar energy.

3.2.6 Applications of nanotechnology in solar cells

Nanotechnology can improve the efficiency, energy conversion, and storage of solar energy through various aspects. Nanotechnology can modify light and material interactions, which can lead to processing of semiconductors found in photovoltaics with low cost. It could also improve the efficiency of photocatalysts, which convert solar energy into fuels. The conversion of chemical fuels into electrical energy and vice versa can also be improved by nanotechnology, resulting in more compact batteries with higher power capacity. Nanotechnology also enhances the cell's surface efficiency through displays, friction, thermos electric, and solid-state lighting (Ahmadi et al., 2019).

Extensive research is currently being conducted worldwide on the applications of nanotechnology in solar cells. For instance, researchers at Los Alamos National Lab have tested copper-indium selenide sulfide QDs in solar cells. These quantum dots were found to be nontoxic with low cost. While researchers at MIT are investigating the efficiency of solar cells composed of graphene sheets. They are also studying the effect of coating the solar cells with zinc oxide nanowire to reduce the cost of production. Other researchers at Duke University are aiming to reduce the losses caused by reflection of light by dispersing silver nanotubes over a thin gold layer. Niels-Bohr Institute researchers are analyzing the resonance effect, in which sunlight can be concentrated in nanowires, to enhance the efficiency of solar cells (John et al., 2016).The field of nanotechnology is very wide, and its applications in solar cells are very promising. With further research, nanotechnology will have a significant contribution to the development of more efficient solar cells.

3.3 Nanotechnology materials in solar cells

Cost reduction and an enhancement in conversion efficiency of solar cells and nanotechnology are required to make photovoltaic energy competitive in the market to the extent that is capable of substituting fossil fuels. Wide-bandgap nanomaterials are now receiving more attention for their applications and their ability to advance solar cell performance. Nanomaterials in solar cells are of different geometrical shapes including nanowires, nanotubes, nanofilms, nanospheres, and porous 3D nanostructure networks. Additionally, different types of nanomaterials can serve solar cells. This section of the chapter discusses the different geometric shapes of nanostructured materials and the various material types utilized for nanotechnology in solar cells.

3.3.1 Forms of nanomaterials
3.3.1.1 Nanowires
Inner engineering of semiconductor materials and the nanomaterials used in solar cells was proved to enhance the power conversion efficiency (PCE) and, therefore, make solar cells more attractive. Semiconductor nanowires (NWs) have been of huge importance recently due to their role in enhancing the PCE. The NWs have a high aspect ratio (i.e., surface to volume ratio) and small dimensions, which magnetizes them for solar cell application (Sahoo & Kale, 2019). For instance, NW solar cells have increased the PCE from 5% to above 15% (Otnes & Borgström, 2017). They succeed at increasing the power conversion efficiency mainly due to their role with regard to light absorption and charge carrier separation and collection. To start with, the absorption of sunlight is the first step in transforming solar energy to electricity. An interesting characteristic of nanowires is their geometry-dependent absorption property that yields light interaction characteristics that are not observed in bulk materials and other nanostructures (Brongersma et al., 2014; Svensson et al., 2013). For instance, it was proved that an array of, subwavelength diameter, NWs absorb more light compared to a thin film that is composed of the identical material (Huang et al., 2012). Upon tuning their dimensions and given their geometry, NWs can enhance the sunlight absorption in a solar cell. As mentioned earlier, NWs can improve the performance of a solar cell by also improving the charge carrier separation and collection. The absorption of light leads to the creation of electrons and holes that are separated and collected through an external circuit. The electronic properties of the device determine the efficiency of charge collection (Kirchartz et al., 2015). There are several

central concepts that affect the electronic characteristics of nanowire array solar cells. These concepts include the crystal structure, junction geometry and design, contact information, surface passivation and cleaning, and doping control. These NWs-related concepts are of interest when understanding how the geometry of NWs affects the PCE by improving the charge carrier transport (Otnes & Borgström, 2017). These factors are currently under experimental studies. When these concepts are well understood, they can help in tuning the NW parameters in such a way that will improve the charge carrier separation and collection and therefore the PCE of the solar cell.

In conclusion, the structure of the nanowires solar cells can solve the carrier collection problems that face the conventional planar solar cells and can help improve the sunlight absorption. These advantages of nanowires can be utilized to increase the power conversion efficiency of a solar cell with relatively low costs (Ali & Rafat, 2017).

3.3.1.2 Nanotubes

Given their versatile features and special geometry, carbon nanotubes have been a topic of extensive research in recent years. In addition to carbon nanotubes, titanium dioxide nanotube arrays (TNAs) are also under study for potential applications in solar cells. These two types of nanotube arrays are the most common in photovoltaic energy; therefore, this section will address the role of such nanotubes in advancing the solar cell performance.

Carbon nanotubes (CNTs) benefit from a wide range of properties including excellent charge transport characteristics, chemical stability, and mechanical robustness. These features allow CNTs to both efficiently extract photogenerated charges and enhance perovskite solar cell stability (Habisreutinger et al., 2017). Additionally, CNTs are electroconductive materials; this property enables them to combine with other conducting polymers to form composites for use in organic solar cells (OSCs). The improved composite is expected to have a higher efficiency with reduced costs (Keru et al., 2014). Moreover, CNTs benefit from their 3D bendable structure in photovoltaic cells (Chen et al., 2012). Carbon materials were proved to increase the interaction between electrodes and electrolyte by enlarging the surface area and were also proved to improve the flexibility and durability of the cells (Yan et al., 2013).

In addition to carbon nanotubes, titanium dioxide nanotube arrays (TNAs) have been an area of research recently. For instance, Wang et al. (2012) fabricated high aspect ratio transparent TNAs and discovered that the solar cells made by the fabricated TNAs exhibit outstanding performance in solar cells, sensitized by the dye, resulting in a power conversion efficiency of 4.38%. The excellent performance of TNAs, especially in DSSCs is owed to their large surface specific areas, greater dye adsorption ability, and vectorial charge transport routine.

Overall, CNTs and TNAs are the most common nanotubes used in photovoltaic energy applications, and their geometry along with the different features that each possess allow them to be very promising in increasing the power conversion efficiency of the cell and improving its performance overall.

3.3.1.3 Nanofilms, nanosheets, or nanoplates

Thin film solar cells represent an encouraging approach for photovoltaic applications given their greater flexibility in device fabrication and design. Simple nanofilms, prepared with fewer materials at a lower temperature, are utilized instead of integrated electronics that results in decreased production cost (Rameshkumar et al., 2021). The efficiency of nanofilms mainly rely on the thickness of the nanofilm. An example of nanofilm solar cells is the transparent tin oxide (SnO_2) nanofilm solar cell. In SnO_2 nanofilm production, the nanofilm thickness and the heat treatment methods need to be given attention in optimizing the performance of the solar cell. A satisfactory performance of the cell is obtained by optimizing these factors (Rameshkumar et al., 2021). The SnO_2 films have other properties that make them stand out including their wide bandgap and excellent electrical and optical properties (Akkaya Arier, 2016). Overall, the device design and fabrication flexibility provided by the nanofilms structure make them one of the most promising materials in photovoltaic applications.

3.3.1.4 Nanospheres

Polymer-based OSCs represent promising approach of converting solar energy directly into electrical energy. They are photovoltaic cells with distinct characteristics such low production cost, mechanical flexibility, and room temperature processing (Bi & LaPierre, 2009). Although OSCs are very promising, they suffer from different problems that limit photon absorption energy, therefore, decreasing their power conversion efficiency (PCE). Incorporating noble metals in solar cells is a proposed solution for solving these problems in solar cells to increase the PCE. Recent studies are focused on incorporating these noble metals in spherical morphologies. A study was performed by Kozanoglu et al. (2013) where the effect on the efficiency was analyzed when gold particles, with distinct spherical morphologies (star, rod, and spherical), were instilled in the buffer layer of the OSC. Addition of the Au NPs with all three structures resulted in the device performance escalating. The increased efficiency is attributed to the surface plasmon resonance of gold nanoparticles and their capability of trapping the light for a longer time within the film. The addition of Au nanospheres in this study resulted in an 11% increase in the cell efficiency compared to the reference device. The Au nanostars open the room for future research to be conducted on them as they showed 29% increase in cell efficiency in this study; this high increase is attributed to the large size, sharp characteristics, and strong localized surface plasmon resonance effect associated with their morphology. Figs. 3.3 and 3.4 represent the Au nanospheres and nanostars embedded in the buffer of an OCS.

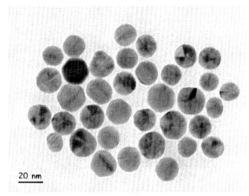

Figure 3.3 Au nanospheres embedded into the buffer layer of organic solar cell (OSC) (Kozanoglu et al., 2013).

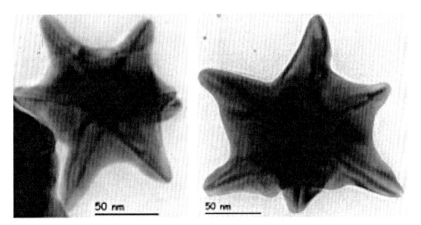

Figure 3.4 Au nanostars embedded into the buffer layer of organic solar cell (OSC) (Kozanoglu et al., 2013).

3.3.1.5 Porous 3D nanostructures networks

3D nanostructures networks provide enhancement to the cell efficiency due to the different properties that they have. Their interrelated pores and nanothickness suggest high molecular charge, enhanced mechanical characteristics with increased cyclical times, high-rate capacities, and little molecule scattering lengths (Anwar et al., 2020). Examples of these structures are graphene and its composites. Graphene is a flat thin sheet of carbon atoms; the special structure of graphene allows it to address the problem of power conversion efficiency in photovoltaic devices (Iqbal & Rehman, 2018). Additionally, the stability and flexibility of graphene are of a great importance that allow it to demonstrate great potential for photovoltaic applications (Koo et al., 2020).

Metal-organic frameworks (MOFs) are porous materials made up of organic linkers and metal nodes that are connected in coordination to yield a 3D porous structure (Baumann et al., 2019). The porous structure and considerably large surface area allow MOFs and their derivatives to be exploited in various applications (Al-Othman et al., 2021). They were recently utilized in thin-film solar cells yielding satisfactory results. As an example, they were applied in perovskite solar cells (PSCs) as an additive for electron and hole transport layers (HTL) resulting in PSCs with high efficiency and stability (Heo et al., 2020). In general, porous materials possess large surface areas providing smooth and rapid diffusion of electrons and ions through the pores. Therefore, the 3D porous structure of MOFs allows them to have promising applications in OSCs, inorganic solar cells, and DSSCs. Due to the promising results achieved, MOFs are currently an area of intensive research.

3.3.1.6 Nanomaterials in solar cells

There are several types of nanomaterials exploited in solar cells. Table 3.1 represents some nanomaterials utilized, the different cell types they are applied in, and the method in which they serve the solar cells.

3.4 Advantages of nanoscale solar cells

Currently, there are two main limitations for solar cells that need to be improved. First is the high cost of these solar cells and second is their low efficiency. To overcome these limitations, there are several advancements in nanotechnology that aid in producing more efficient and cheaper solar panels.

Table 3.1 Types of nanomaterials in solar cells.

Nanomaterial	The type of solar cells it serves	The properties it has	Its role in solar cells enhancement	Notes	Ref.
Silicon	Inorganic solar cells	— High aspect ratio — Excellent properties for light absorption, carrier transport, and charge separation	— Can be used as nanowires in solar cells — Fabricates high performance and cost-effective solar cells	—	Thomas et al. (2019)
Titanium dioxide (TiO_2)	DSSCs	— Large bandgap — Appropriate band edge levels for injection and charge extraction — Greater lifetime of excited electrons — Exhibit Photo corrosion resistance — Nontoxicity — Cost effective	— Has a hierarchical nanostructure that has the shape of nanoflower arrays and nanorods	—	Tala-Ighil, (2015); Thomas et al. (2019)
Zinc oxide (ZnO)	DSSCs	— Exhibit decomposition temperatures above 2200 K and superior stability in air up until 1300°C — Demonstrates good electrical conductivity over extensive temperature ranges	— Optimized in ZnO based DSSCs to achieve high efficiency	Poor chemical stability in acidic alkali media, and thus require attention in specific applications	Thomas et al. (2019)

(Continued)

Table 3.1 (Continued)

Nanomaterial	The type of solar cells it serves	The properties it has	Its role in solar cells enhancement	Notes	Ref.
Tin oxide (SnO_2)	DSSCs	— Transparent metal oxide — Superior electrical, thermal conducting and optical properties — Low in cost — Physical stability	— Various SnO_2 structures are employed as photoanodes in DSSCs — Operate for large surface area — Nanospheres function the best for light scattering and harvesting applications — 1D nanostructures (nanowires, nanotubes and nanorods), in the photoanode, advance the charge transport properties — 3D and 1D nanostructures yield superior charge transport properties and permit greater dye loadings	—	Rameshkumar et al. (2021); Thomas et al. (2019)
Niobium pentoxide (Nb_2O_5)	DSSCs		— Nb_2O_5 films improve the device performance by serving as an electron blocking layer in solar energy conversion devices — Nb_2O_5 films optimize oxygen evolution and decrease the recombination of electrons as compared to TiO_2	— Continuous research needs to be conducted to better understand the formation and the different morphologies and crystalline phases of Nb_2O_5	Thomas et al. (2019)

Metals

Cadmium telluride (CdTe)	Solar cells	— High absorption coefficient — Has an optimal band gap — It is relatively copious — Its production as a side product of important industrial metals makes it relatively copious in supply	— Used in CdTe- based thin film solar cells and records high efficiency with low cost — It is an absorbent PV material	— It is currently the second most used solar cell material in the world after silicon — They have lower manufacturing cost compared to silicon-based cells — Even though Cadmium lists as one of the top 6 hazardous toxic materials, CdTe, however, is less toxic at least in terms of acute exposure — Large-scale production of CdTe solar panels leads to the widely known disposal problems of CdTe	Thomas et al. (2019)
Copper indium gallium selenide (CuInxGa)	Photovoltaic applications	— High absorption: it can absorb a huge portion of the solar spectrum and therefore achieves the highest efficiency of any thin film technology	— Another type of absorbent that is widely used in photovoltaic applications.	—	Yan et al. (2013)

(Continued)

Table 3.1 (Continued)

Nanomaterial	The type of solar cells it serves	The properties it has	Its role in solar cells enhancement	Notes	Ref.
Gallium arsenide (GaAs)	Solar cells	— Very durable material because it is robust to moisture and ultraviolet radiation — Has a reduced temperature coefficient — Temperature coefficient evaluates efficiency loss versus temperature. GaAs-based cells therefore face minimal efficiency loss with temperature increase — Provide more efficient photon absorption and high-power output density because of wide and direct band gap — Has strong low light performance — Flexible and lightweight — For a given surface area more power is produced translating to high efficiencies and making it suitable when the surface area is limited	— Semiconductor used in solar cells	— Delivers greater amount of energy in low light or high heat, which are two of the most usual solar cell conditions — The major problem with GaAs-based cells is that the material is expensive	Yan et al. (2013)

Silver (Ag)	Solar cells	— Strong interaction with light — Their absorption and scattering properties can be tuned by controlling the particle size and shape	— Nanoparticles integrated in solar cells for efficiency enhancement	—	Yan et al. (2013)
Gold (Au)	OSCs	— High chemical and physical stability — Ease of surface functionalization with organic molecules	— Could be for example embedded in the buffer layer of OSCs in the form of nanospheres, nanorods or nanostars to enhance the efficiency of the cell	—	Kozanoglu et al. (2013); Yan et al. (2013)

Graphene and carbon nanotubes

Graphene	Solar cells	— Structurally, graphene is a single layer graphite with a thickness of one carbon atom — It has an outstanding chemical, physical, electrical, and optical properties — Reduced fabrication cost and increased efficiency	— Utilized as a transparent electrode for solar cells — Electron transporting layer for solar cells — HTL for solar cells — Electron acceptors in active layers of a solar cell	—	Thomas et al. (2019)
Carbon nanotubes	Solar cells	— Chemically and thermally stable — Mechanically robust — Conductive and porous material	— Used as transparent electrode for solar cells — HTL for solar cells	—	Yan et al. (2013)

To understand how implementing nanotechnology can improve the efficiency of a cell, it is important to understand the cause of inefficiency. To excite an electron, the photon absorbed by the matter must have the right energy, which is called the band gap energy. In the event of band gap energy not meeting, the extra energy is either lost in the form of heat, if there is excessive energy, or the photons pass through in case of low absorbed energy. Therefore, 70% of radiation energy from the sun is lost causing low efficiency in traditional solar cells (Sethi et al., 2011). Therefore, nanoscaled material is used due to its following advantages in different aspect such as:

3.4.1 Light absorption

The unique matter characteristics and properties of nanomaterials are used to solve the problem of wasted solar energy. Because nanoscale materials are very small, the surface interactions on the particle increase, which leads to multiple reflections. Having multiple reflections increases the amount of light absorption in the cell compared to regular material. Furthermore, an energy band gap must be determined when designing a solar cell. By using nanoscale material, we provide more flexibility for the designer to meet the desired energy gap by varying the size of the nanoparticle. Lastly, as discussed before in the mechanism of solar cells, the electrons must cross the depletion zone when moving from the n-type to the p-type layer. This movement of electrons happens on a much shorter path when nanomaterial is used, causing fewer recombination losses (Wang & He, 2017). Recombination losses occur when the excited electrons relax back to their normal state before reaching the solar cell contacts. According to Liu et al. (2018), this phenomenon is directly related to the electron's diffusion length, which is majorly reduced from a micrometer scale to a nanometer scale of thickness in nanoscale solar cells.

Moreover, in conventional solar cells, ultraviolet light is not utilized. This light is absorbed by the silicon layer and lost as a source of heat (Al-Nimr et al., 2020). Therefore, many solar cells are provided with cooling systems. However, if ultraviolet light interacts with nanoscale material such as a film of silicon nanoparticles, it can produce useful electricity. According to Sethi et al. (2011), utilizing a 1.0 nm film of silicon can improve the cell power performance by 60%. To increase the power performance to 67%, a 2.85 nm silicon film was used. It is essential to consider that such power enhancements occurred in the ultraviolet range.

However, the enhancements were 10% and 3% for the 1.0 nm and 2.85 nm films in the visible light range, respectively. This increase in efficiency is because nanoparticles can absorb shorter wavelengths (blue light) and longer wavelengths (red light) compared to regular-sized material.

3.4.2 Quantum dots

Nanosized semiconductors are called QDs. The QDs solar cells employ this type of material. The main advantage quantum dots (QD) provide is that unlike bulk semiconducting materials, which have a fixed band gap energy, the bandgap for QD is adjustable (Jasim, 2015). Therefore, the size can be designed to absorb longer wave light and increase the efficiency of the cell.

Using QD could also solve a major problem faced with traditional multijunction solar cells. Since each different material has a different band gap energy, the p-n junction for a specific type of material absorbs a specific wavelength of light. To utilize a large spectrum of different wavelengths, which increases the solar cell's efficiency, multiple semiconductor materials are used in multijunction solar cells. Using different types of materials can increase the cost of these solar cells. A solution to this problem is using the same material (QD), however, with different radius sizes to provide a wide range of band gap energy. Using the same material (QD) instead of multiple semiconductors reduces the manufacturing costs (Kerestes et al., 2014).

In addition to QD having a broader absorption range of wavelengths, QD excite more electrons. In bulk material, it is assumed that each incoming photon from sunlight excites only one electron. The excess energy is lost in lattice vibrations and heat energy. However, QD create a special phenomenon called the carrier multiplication effect. This effect refers to the occurrence of multiple electron excitation per one incoming photon. This special property of QD enhances the solar cell efficiency as it increases the current density produced per one photon of light. Furthermore, a 65% enhancement in the efficiency of the solar cell can be achieved using QD (Sethi et al., 2011).

3.4.3 Carbon nanotubes

Another novel advancement provided by nanotechnology is single-walled carbon nanotubes on titanium dioxide nanoparticles. The carbon nanotubes serve as an escape route for the excited electrons to reach the electrode. This route is vital since many electrons do not move to the current collector as expected. Consequently, the nanotubes create a better ballistic transport property. This technology aids in improving the solar cell's efficiency up to 50% in laboratory conditions using poly 3-octylthiophene (P3OT) polymer composite aligned with carbon nanotubes (Sethi et al., 2011). Other studies as well have shown an enhancement in the perovskite solar cell performance from 8.5% to 10.6% by using nanotubes (Zhang et al., 2016). The carbon nanotubes served as hole transport channels and increased the short-circuit current density from 14.54 to 16.83 mA/cm^2.

3.4.4 Self-cleaning

The performance of solar cells decreases in general from the accumulation of dirt and dust on the surface. The Eastern Mediterranean region experiences this problem the most given its rough climate. Regular surface cleaning is needed using manual, hydraulic, or robotic systems to maintain the solar cell performance. All methods add up to maintenance costs that must be always kept minimum. To solve this problem, self-cleaning nanostructured coatings on the surface of photovoltaic cells are used. The self-cleaning property of nanomaterial is related to a phenomenon in nature called the lotus leaf effect. This effect is found in lotus leaves that have a superhydrophobic and nanostructured surface that allows water droplets to pick up any dirt particles and slide down the leaf (Yamamoto et al., 2015). For instance, nano-coated solar cells of ZnO and CuO doped in polyvinyl alcohol solution increased the efficiency of the cells up to 50%–60% (Pawale et al., 2020). Moreover, it is essential to identify the accumulated dust's morphology and elemental properties to develop proper self-cleaning techniques for PV panels (Dhaouadi et al., 2021).

In another study, the self-cleaning method was studied for a nanomaterial called SurfaShield G coated on a monocrystalline solar panel. After 3 months in Jordan's climate condition, the power drop in the panel due to dust accumulation was measured. The coated panels had a power drop from 114.9 to 114 W whereas the uncoated panels dropped from 114.9 to 95 W. Therefore, the coated surface showed an increase of 20% for the maximum power produced and 2.3% for the overall panel efficiency (Al Bakri et al., 2021). The solar panels used in this study are shown in Fig. 3.5, where the right side illustrates the coated solar panel and the left side shows the uncoated panel.

Figure 3.5 Observing dust accumulation on a solar panel coated with nanomaterial called SurfaShield G (right side) versus an uncoated panel (left side) (Al Bakri et al., 2021).

3.4.5 Anti-reflective coating

Reflected light on the surface of solar panels is another issue that dents the solar cell efficiency. Solar panels use either a metal base or glass top to prevent the solar cells from environmental threats, damage, corrosion, and physical shock (Alam et al., 2018; Shelat et al., 2019). These glass coverings lose 8%–9% of sunlight by light reflection. These reflection losses are high since the refractive index of air (1.0) is different from the refractive index of glass (1.55) (Zou et al., 2018). Therefore, the application of antireflective nanosized coating is a promising solution for reflection losses.

There are many different types of nanomaterials used in antireflective coatings that showed promising results in improving solar cells' efficiency. For instance, metallic nanomaterial made of gold and silver showed a great ability to easily scatter incident light into the solar cell's absorption layer. Few papers work on other types of metallic materials such as indium. According to research conducted by Ghosh et al. (2017), they obtained an efficiency enhancement of 35.94% for double-layered nanosized indium coating, 34.94% for single-layered nanosized indium coating, and 26.67% for pure SiO_2 antireflective coating (with a thickness of 90 nm) compared to reference cells with no coating.

Other studies on polymer nanostructured antireflective coating obtained similar results to the metallic material. According to Thompson and Zou (2013), adding a polyvinylpyrrolidone layer on the surface creates a variation in surface roughness, which leads to graded index refraction, allowing the solar cell to trap more sunlight. The results show an ability to transmit 98.6% of incident light for polyvinylpyrrolidone enhanced silica nanoparticle film and 94.5% for silica nanoparticle films compared to only 90.7% for bare glass. Additionally, few studies focus on the effect of different structures of nanomaterial on increasing the cell's efficiency. For example, nanograting structures have drawn much attention recently in research due to their low reflection losses and high light trapping characteristics. In GaAs solar cells, it was found that nanostructured grating provided high conversion efficiencies (Das et al., 2020).

3.4.6 Cost reduction

In addition to an increase in efficiency, nanotechnology is more cost effective. The reduction in cost goes back to the fact that thin layers are used, which means less material for constructing solar cells is required. For example, nanoscale plastic photovoltaic cells can be coated on rooftops, which could provide enough energy to power the entire house with lower cost. Moreover, Nanosolar, a company famous for solar cells, managed to reduce the cost of production from 30 cents per watt to 3 cents per watt by using nanoscale solar coating (Sethi et al., 2011). Consequently, solar energy becomes a more attractive alternative for energy.

3.5 Challenges and limitations of nanotechnology in solar cells

As discussed throughout the different sections of this chapter, nanotechnology provides a variety of advantages in solar cell applications that include the improvement of light absorption and the power conversion efficiency of the cell along with other benefits discussed in the advantages section. These advantages are owed to the structure and geometry of the nanomaterials used along with the properties of the nanomaterials themselves. However, there are several limitations associated with each type of nanomaterial or nanostructure used in solar cells. For instance, the use of Au nanospheres, nanorods and nanostars in the buffer layer of OSCs to enhance the power conversion efficiency was previously discussed. However, the cost here is an issue when considering the commercialization of Au nanoparticles in solar cells. Since each type of material and structure used has limitations, the application of nanotechnology in solar cells is still under research.

Generally speaking, there are challenges that are linked to the use of nanotechnology in photovoltaic applications that are summarized below (Zhang, 2017):

1. Developing and optimizing materials that can aid the objective of enhancing the solar cell performance is probably the main challenge. Many of the materials proposed are still under research and development for them to be utilized in photovoltaic applications.

2. Finding a simple fabrication process with low manufacturing cost and complexities for large-scale production. This results in the commercialization of nanomaterials in solar cells being a major challenge. Most of the materials are still under study and are not yet commercialized.

3. Many of the present nanomaterials used in photovoltaic applications employ environmentally dangerous heavy metals that challenge the commercialization of such materials. Therefore, finding substitutes for these heavy metals that can provide the same performance without harming the environment is a challenge.

4. Cost of manufacturing and materials used is another issue that requires unraveling. Minimizing the cost of nanotechnology in solar cells is essential for commercialization of the technology.

5. Evaluation of the environmental risks and toxicity of the nanomaterials used is imperative for creating large-scale eco-friendly processes.

6. Overall, one may summarize that the challenges with nanotechnology in solar cells are mainly because this is a relatively new research topic. Most nanotechnology applications are still not commercialized yet and are still under development and study. However, the promising results that this technology is achieving in solar cell applications call for more research to be conducted in this topic to optimize the materials and achieve the best power conversion efficiency.

3.6 Conclusions

Making photovoltaic energy competitive in the market is hindered by several challenges. These challenges are mainly due to cost issues and conversion efficiency of the cell. Nanotechnology seems to be the way of improving the performance of solar cells. There are several nanostructures that are used for photovoltaic applications including nanowires, nanotubes, nanofilms, nanospheres, and 3D porous nanostructures. Each of these structures have their own properties that allow them to contribute to the solar cell efficiency improvement. However, one property that is common in all structures is the large surface area that nanoparticles provide, which allows for better performance of the cell. An additional property that was common is optimizing the structure of the nanoparticles in such a way to enhance the efficiency of the solar cells. In addition to nanostructures, there are several materials that can be employed in nanotechnology applications in solar cells including metals, carbon nanotubes and graphene, and many other materials. Each nanomaterial has its own properties that allow them to add to the efficiency of the cell, common properties include high light absorption and chemical and physical stability. However, nanotechnology in solar cells is hindered by several limitations. These limitations include finding and optimizing nanomaterials that can aid the purpose of improving solar cells efficiency. Another limitation is related to the commercialization of these nanotechnology-based solar cells; this is because most of the proposed materials are under research and some of them show environment-related issues. Additionally, many of the promising nanomaterials proposed are environmentally hazardous heavy metals that also have disposal issues. Finally, the cost of manufacturing is a main issue when considering the commercialization of nanotechnology-based solar cells. Many limitations are associated with nanotechnology in photovoltaic applications; nevertheless, the promising results that the technology is achieving in this area and the solar cell performance enhancement is providing a call for more research to be conducted to optimize nanomaterials and commercialize nanotechnology based solar cells.

References

Abdin, Z., Alim, M. A., Saidur, R., Islam, M. R., Rashmi, W., Mekhilef, S., & Wadi, A. (2013). Solar energy harvesting with the application of nanotechnology. *Renewable and Sustainable Energy Reviews, 26*, 837−852. Available from https://doi.org/10.1016/j.rser.2013.060.023.

Ahmadi, M. H., Ghazvini, M., Nazari, M. A., Ahmadi, M. A., Pourfayaz, F., Lorenzini, G., & Ming, T. (2019). Renewable energy harvesting with the application of nanotechnology: A review. *International Journal of Energy Research, 43*(4), 1387−1410. Available from https://doi.org/10.1002/er.4282, http://onlinelibrary.wiley.com/journal/10.1002/(ISSN)1099-114X.

Akkaya Aner, Ü. Ö. (2016). Optical and structural properties of sol-gel derived brookite TiO2-SiO2 nano-composite films with different SiO2:TiO2 ratios. *Optik, 127*(16), 6439−6445. Available from https://doi.org/10.1016/j.ijleo.2016.040.038, http://www.elsevier.com/journals/optik/0030-4026.

Al Bakri, H., Abu Elhaija, W., & Al Zyoud, A. (2021). Solar photovoltaic panels performance improvement using active self-cleaning nanotechnology of SurfaShield G. *Energy, 223*. Available from https://doi.org/10.1016/j.energy.2021.119908.

Al-Nimr, M. 'd, Milhem, A., Al-Bishawi, B., & Khasawneh, K. A. (2020). Integrating transparent and conventional solar cells TSC/SC. *Sustainability, 12*(18). Available from https://doi.org/10.3390/su12187483.

Al-Othman., Tawalbeh, M., Temsah, O., & Al-Murisi, M. (2021). *Industrial challenges of MOFs in energy applications. Encyclopedia of Smart Materials* (pp. 535–543). United Arab Emirates: Elsevier. Available from https://www.sciencedirect.com/book/9780128157336, https://doi.org/10.1016/B978-0-12-815732-9.00030-9.

Alam, K., Ali, S., Saher, S., Humayun, M. (2018). Silica nano-particulate coating having self-cleaning and antireflective properties for PV modules. In *Proceedings of the 21st International Multi Topic Conference*. Pakistan: Institute of Electrical and Electronics Engineers Inc. http://ieeexplore.ieee.org/xpl/mostRecentIssue.jsp?punumber = 8573782.

Alami, A. H., Aokal, K., Zhang, D., Tawalbeh, M., Alhammadi, A., & Taieb, A. (2018). Assessment of calotropis natural dye extracts on the efficiency of dye-sensitized solar cells. *Agronomy Research, 16*(4), 1569–1579. Available from https://doi.org/10.15159/AR.180.166, http://agronomy.emu.ee/wp-content/uploads/2018/05/Vol16No4_1.pdf#abstract-6273.

Alami, A. H., Abu Hawili, A., Tawalbeh, M., Hasan, R., Al Mahmoud, L., Chibib, S., Mahmood, A., Aokal, K., & Rattanapanya, P. (2020). Materials and logistics for carbon dioxide capture, storage and utilization. *Science of the Total Environment, 717*. Available from https://doi.org/10.1016/j.scitotenv.2020.137221, http://www.elsevier.com/locate/scitotenv.

Alami, A. H., Faraj, M., Aokal, K., Hawili, A. A., Tawalbeh, M., & Zhang, D. (2020). Investigating various permutations of copper iodide/FeCu tandem materials as electrodes for dye-sensitized solar cells with a natural dye. *Nanomaterials, 10*(4). Available from https://doi.org/10.3390/nano10040784.

Ali, N. M., & Rafat, N. H. (2017). Modeling and simulation of nanorods photovoltaic solar cells: A review. *Renewable and Sustainable Energy Reviews, 68*, 212–220. Available from https://doi.org/10.1016/j.rser.2016.090.114, https://www.journals.elsevier.com/renewable-and-sustainable-energy-reviews.

Anwar, H., Arif, I., Javeed, U., Mushtaq, H., Ali, K., & Sharma, S. K. (2020). Quantum dot solar cells. In S. K. Sharma, & K. Ali (Eds.), *Solar cells: From materials to device technology*. Springer.

Baumann, A. E., Burns, D. A., Liu, B., & Thoi, V. S. (2019). Metal-organic framework functionalization and design strategies for advanced electrochemical energy storage devices. *Communications Chemistry, 2*(1). Available from https://doi.org/10.1038/s42004-019-0184-6, nature.com/commschem/.

Bhatia, S. C. (2014). *Nanotechnology and solar power* (pp. 191–200). Informa UK Limited. Available from http://doi.org/10.1201/b18242-9.

Bi, H., & LaPierre, R. R. (2009). A GaAs nanowire/P3HT hybrid photovoltaic device. *Nanotechnology, 20*(46). Available from https://doi.org/10.1088/0957-4484/20/46/465205.

Brongersma, M. L., Cui, Y., & Fan, S. (2014). Light management for photovoltaics using high-index nanostructures. *Nature Materials, 13*(5), 451–460. Available from https://doi.org/10.1038/nmat3921.

Chapin, D. M., Fuller, C. S., & Pearson, G. L. (1991). *A new silicon p-n junction photocell for converting solar radiation into electrical power. Semiconductor Devices: Pioneering Papers* (pp. 969–970). Available from https://doi.org/10.1142/9789814503464_0138.

Chen, T., Qiu, L., Cai, Z., Gong, F., Yang, Z., Wang, Z., & Peng, H. (2012). Intertwined aligned carbon nanotube fiber based dye-sensitized solar cells. *Nano Letters*, *12*(5), 2568–2572. Available from https://doi.org/10.1021/nl300799d.

Choubey, P., Oudhia, A., & Dewangan, R. (2012). A review: Solar cell current scenario and future trends. *Environmental Engineering Science*, *4*.

Cleveland, C. J., & Morris, C. (2013). Handbook of energy: Volume II chronologies, top ten lists, and word clouds. United States: Elsevier. Retrieved from: https://www.sciencedirect.com/book/9780124170131, https://doi.org/10.1016/C2013-0-00172-7.

Das, N., Chandrasekar, D., Nur-E-Alam, M., & Khan, M. K. M. (2020). Light reflection loss reduction by nano-structured gratings for highly efficient next-generation GaAs solar cells. *Energies*, *13*(6). Available from https://doi.org/10.3390/en13164198, https://www.mdpi.com/1996-1073/13/16/4198.

Dhaouadi, R., Al-Othman, A., Aidan, A. A., Tawalbeh, M., & Zannerni, R. (2021). A characterization study for the properties of dust particles collected on photovoltaic (PV) panels in Sharjah, United Arab Emirates. *Renewable Energy*, *171*, 133–140. Available from https://doi.org/10.1016/j.renene.2021.020.083, http://www.journals.elsevier.com/renewable-and-sustainable-energy-reviews/.

Fernandez, A. (2014). *How a solar cell works*. American Chemical Society.

Solar cells: A brief history and introduction In L. Fraas, & L. Partain (Eds.), *Solar cells and their applications* (2nd edn., pp. 1–15). United States: John Wiley and Sons. Available from http://onlinelibrary.wiley.com/book/10.1002/9780470636886, https://doi.org/10.1002/9780470636886.ch1.

Fraas, L. M. (2012). Mirrors in space for low-cost terrestrial solar electric power at night. In *Proceedings of the Record of the IEEE Photovoltaic Specialists Conference* (pp. 2862–2867). United States: JX Crystals Inc.

Fraas, L. M. (2014). Low-cost solar electric power. United States: Springer International Publishing. Retrieved from https://doi.org/10.1007/978-3-319-07530-3, https://doi.org/10.1007/978-3-319-07530-3.

Fraas, L. M., Derbes, B., & Palisoc, A. (2013). Mirrors in dawn dusk orbit for low-cost terrestrial solar electric power in the evening. In *Conference: 51st AIAA Aerospace Sciences Meeting including the New Horizons Forum and Aerospace Exposition*. Grapevine.

Ghosh, S., Mallick, A., Kole, A., Chaudhury, P., Garner, S., & Basak, D. (2017). Study on AZO coated flexible glass as TCO substrate. In IEEE 44th Photovoltaic Specialist Conference (pp. 2664-2666). India: Institute of Electrical and Electronics Engineers Inc. http://ieeexplore.ieee.org/xpl/mostRecentIssue.jsp?punumber = 8360188.

Habisreutinger, S. N., Nicholas, R. J., & Snaith, H. J. (2017). Carbon Nanotubes in Perovskite Solar Cells. *Advanced Energy Materials*, *7*(10). Available from https://doi.org/10.1002/aenm.201601839, http://onlinelibrary.wiley.com/journal/10.1002/(ISSN)1614-6840.

Heo, D. Y., Do, H. H., Ahn, S. H., & Kim, S. Y. (2020). Metal-organic framework materials for perovskite solar cells. *Polymers*, *12*(9). Available from https://doi.org/10.3390/POLYM12092061, https://res.mdpi.com/d_attachment/polymers/polymers-12-02061/article_deploy/polymers-12-02061.pdf.

Huang, N., Lin, C., & Povinelli, M. L. (2012). Broadband absorption of semiconductor nanowire arrays for photovoltaic applications. *Journal of Optics*, *14*(2). Available from https://doi.org/10.1088/2040-8978/14/2/024004.

Iqbal, M. Z., & Rehman, A. U. (2018). Recent progress in graphene incorporated solar cell devices. *Solar Energy*, *169*, 634–647. Available from https://doi.org/10.1016/j.solener.2018.040.041, http://www.elsevier.com/inca/publications/store/3/2/9/index.htt.

Jasim, K. E. (2015). *Quantum dots solar cells*. InTech. Available from https://doi.org/10.5772/59159.

John, S. S., Malviya, P., Sharma, N., Manjhi, V. K., & Sudhakar, K. (2016). Nanotechnology for solar and wind energy applications recent trends and future development. In *Proceedings of the 2015 International Conference on Green Computing and Internet of Things*. India: Institute of Electrical and Electronics Engineers Inc.

Kandeal, A. W., Algazzar, A. M., Elkadeem, M. R., Thakur, A. K., Abdelaziz, G. B., El-Said, E. M. S., Elsaid, A. M., An, M., Kandel, R., Fawzy, H. E., & Sharshir, S. W. (2021). Nano-enhanced cooling techniques for photovoltaic panels: A systematic review and prospect recommendations. *Solar Energy*, 227, 259−272. Available from https://doi.org/10.1016/j.solener.2021.090.013, http://www.elsevier.com/inca/publications/store/3/2/9/index.htt.

Kerestes, S., Polly, D., Forbes, C., Bailey, A., Podell, J., Spann, P., Patel, B., Richards, P., Sharps, S., & Hubbard. (2014). Fabrication and analysis of multijunction solar cells with a quantum dot (In)GaAs junction. *Progress in Photovoltaics: Research and Applications*, 22(11), 1172−1179. Available from https://doi.org/10.1002/pip.2378, http://onlinelibrary.wiley.com/journal/10.1002/(ISSN)1099-159X.

Keru, G., Ndungu, P. G., & Nyamori, V. O. (2014). A review on carbon nanotube/polymer composites for organic solar cells. *International Journal of Energy Research*, 38(13), 1635−1653. Available from https://doi.org/10.1002/er.3194, http://onlinelibrary.wiley.com/journal/10.1002/(ISSN)1099-114X.

Kim, J., Kim, J., Kang, H., Yun, M., Jeon, B., Koo, J. E., Kwon, G.-G., & Cho, G. (2012). Doping of crystalline silicon solar cell by making use of atmospheric and subatmospheric plasma jet. *Abstracts IEEE International Conference on Plasma Science*. Available from https://doi.org/10.1109/PLASMA.2012.6384087, Edinburgh, UK.

Kirchartz, T., Bisquert, J., Mora-Sero, I., & Garcia-Belmonte, G. (2015). Classification of solar cells according to mechanisms of charge separation and charge collection. *Physical Chemistry Chemical Physics*, 17(6), 4007−4014. Available from https://doi.org/10.1039/c4cp05174b, http://pubs.rsc.org/en/journals/journal/cp.

Koo, S., Jung, J., Seo, G., Jeong, Y., Choi, J., Lee, S. M., Lee, Y., Cho, M., Jeong, J., Lee, J., Oh, C., & Yang, H. P. (2020). Flexible organic solar cells over 15% efficiency with polyimide-integrated graphene electrodes. *Joule*, 4(5), 1021−1034. Available from https://doi.org/10.1016/j.joule.2020.020.012, https://www.journals.elsevier.com/joule.

Kozanoglu, D. H., Apaydin, A., Cirpan, E. N., & Esenturk. (2013). Power conversion efficiency enhancement of organic solar cells by addition of gold nanostars, nanorods, and nanospheres. *Organic Electronics*, 14(7), 1720−1727. Available from https://doi.org/10.1016/j.orgel.2013.040.008, http://www.elsevier.com/locate/orgel.

Leggett, T. (2012). The physics of solar cells. *Contemporary Physics*, 53(5), 458−459. Available from https://doi.org/10.1080/00107514.2012.727031.

Liu, J., Yao, Y., Xiao, S., & Gu, X. (2018). Review of status developments of high-efficiency crystalline silicon solar cells. *Journal of Physics D: Applied Physics*, 51(12). Available from https://doi.org/10.1088/1361-6463/aaac6d, http://iopscience.iop.org/article/10.1088/1361-6463/aaac6d/pdf.

Maka, A. O. M., & Alabid, J. M. (2022). Solar energy technology and its roles in sustainable development. *Jamahiriya Clean Energy*, 6(3), 476−483. Available from https://doi.org/10.1093/ce/zkac023, https://academic.oup.com/ce/pages/About.

Martis, R., Al-Othman, A., Tawalbeh, M., & Alkasrawi, M. (2020). Energy and economic analysis of date palm biomass feedstock for biofuel production in UAE: Pyrolysis, gasification and fermentation. *Energies*, 13(22). Available from https://doi.org/10.3390/en13225877, https://www.mdpi.com/1996-1073/13/22/5877/pdf.

Mohammad Bagher, A. (2015). Types of solar cells and application. *American Journal of Optics and Photonics*, 3(5). Available from https://doi.org/10.11648/j.ajop.20150305.17.

Obaideen, K., AlMallahi, M. N., Alami, A. H., Ramadan, M., Abdelkareem, M. A., Shehata, N., & Olabi, A. G. (2021). On the contribution of solar energy to sustainable developments goals: Case study on Mohammed bin Rashid Al Maktoum Solar Park. *International Journal of Thermofluids*, *12*. Available from https://doi.org/10.1016/j.ijft.2021.100123.

Otnes, G., & Borgström, M. T. (2017). Towards high efficiency nanowire solar cells. *Nano Today*, *12*, 31−45. Available from https://doi.org/10.1016/j.nantod.2016.100.007, http://www.elsevier.com/wps/find/journaldescription.cws_home/706735/description#description.

Pawale, T. P., Motekar, R., Chakrasali, R.L., Hugar, P., & Halabhavi, S. B. (2020). Implementation of a novel self-cleaning roof top PV cells using protective film coating. In *Proceedings of the 3rd International Conference on Smart Systems and Inventive Technology*. India: Institute of Electrical and Electronics Engineers Inc. http://ieeexplore.ieee.org/xpl/mostRecentIssue.jsp?punumber = 9203793.

Rameshkumar, C., Ananth, D., Divyalakshmi, V., Balakrishnan, M., Senthilkumar, G., & Subalakshmi, R. (2021). An investigation of SnO2 nanofilm for solar cell application by spin coating technique. In *AIP Conference Proceedings*. India: American Institute of Physics Inc. http://scitation.aip.org/content/aip/proceeding/aipcp.

Ramsden, J. J. (2016). *Nanotechnology: An introduction* (2nd edn, pp. 1−325). United Kingdom: Elsevier Inc. Available from http://www.sciencedirect.com/science/book/9780323393119.

Sahoo, M. K., & Kale, P. (2019). Integration of silicon nanowires in solar cell structure for efficiency enhancement: A review. *Journal of Materiomics*, *5*(1), 34−48. Available from https://doi.org/10.1016/j.jmat.2018.110.007, https://www.journals.elsevier.com/journal-of-materiomics/.

Salameh, T., Tawalbeh, M., Juaidi, A., Abdallah, R., & Hamid, A. K. (2021). A novel three-dimensional numerical model for PV/T water system in hot climate region. *Renewable Energy*, *164*, 1320−1333. Available from https://doi.org/10.1016/j.renene.2020.100.137, http://www.journals.elsevier.com/renewable-and-sustainable-energy-reviews/.

Schaming, H., & Remita. (2015). Nanotechnology: from the ancient time to nowadays. *Foundations of Chemistry*, *17*(3), 187−205. Available from https://doi.org/10.1007/s10698-015-9235-y, http://link.springer.com/journal/10687.

Serrano, G., Rus, J., & García-Martínez. (2009). Nanotechnology for sustainable energy. *Renewable and Sustainable Energy Reviews*, *13*(9), 2373−2384. Available from https://doi.org/10.1016/j.rser.2009.060.003.

Sethi, V. K., Pandey, M., & Shukla, P. (2011). Use of nanotechnology in solar PV cell. *International Journal of Chemical Engineering and Applications*, 77−80. Available from https://doi.org/10.7763/IJCEA.2011.V2.79.

Sharma, S., Jain, K. K., & Sharma, A. (2015). Solar cells: In research and applications—A review. *Materials Sciences and Applications*, *06*(12), 1145−1155. Available from https://doi.org/10.4236/msa.2015.612113.

Shelat, N., Das, N., Khan, M.M.K., & Islam, S. (2019). Nano-structured photovoltaic cell design for high conversion efficiency by optimizing various parameters. In *29th Australasian Universities Power Engineering Conference*. Australia: Institute of Electrical and Electronics Engineers Inc. http://ieeexplore.ieee.org/xpl/mostRecentIssue.jsp?punumber = 9079162

Simya, O. K., Radhakrishnan, P., Ashok, A., Kavitha, K., & Althaf, R. (2018). Engineered nanomaterials for energy applications. In C. M. Hussain (Ed.), *Handbook of nanomaterials for industrial applications* (pp. 751−767). India: Elsevier. Available from: http://www.sciencedirect.com/science/book/9780128133514.

Fundamentals of solar cell. In T. Soga (Ed.), *Nanostructured materials for solar energy conversion*. Japan: Elsevier. Available from http://www.sciencedirect.com/science/book/9780444528445, https://doi.org/10.1016/B978-044452844-5/50002-0.

Soto, E. A., Arakawa, K., & Bosman, L. B. (2022). Identification of target market transformation efforts for solar energy adoption. *Energy Reports*, *8*, 3306−3322.

Svensson, J., Anttu, N., Vainorius, N., Borg, B. M., & Wernersson, L. E. (2013). Diameter-dependent photocurrent in InAsSb nanowire infrared photodetectors. *Nano Letters*, *13*(4), 1380−1385. Available from https://doi.org/10.1021/nl303751d.

Tala-Ighil, R. (2015). *Nanomaterials in solar cells* (pp. 1−18). Springer Science and Business Media LLC. Available from https://doi.org/10.1007/978-3-319-15207-3_26-1.

Tawalbeh, M., Salameh, T., Albawab, M., Al-Othman, A., Assad, M. E. H., & Alami, A. H. (2020). Parametric study of a single effect lithium bromide-water absorption chiller powered by a renewable heat source. *Journal of Sustainable Development of Energy, Water and Environment Systems*, *8*(3), 464−475. Available from https://doi.org/10.13044/j.sdewes.d7.0290, http://www.sdewes.org/jsdewes/dpebc7876dc96d6b068df90fea7f4b154f69db6f39.

Tawalbeh, M., Al-Ismaily, M., Kruczek, B., & Tezel, F. H. (2021). Modeling the transport of CO2, N2, and their binary mixtures through highly permeable silicalite-1 membranes using Maxwell − Stefan equations. *Chemosphere*, *263*. Available from https://doi.org/10.1016/j.chemosphere.2020.127935, http://www.elsevier.com/locate/chemosphere.

Tawalbeh, M., Mohammed, S., Alnaqbi, A., Alshehhi, S., & Al-Othman, A. (2023). *Analysis for hybrid photovoltaic/solar chimney seawater desalination plant: A CFD simulation in Sharjah, United Arab Emirates*. Renewable Energy (202, pp. 667−685). United Arab Emirates: Elsevier Ltd. Available from http://www.journals.elsevier.com/renewable-and-sustainable-energy-reviews/, https://doi.org/10.1016/j.renene.2022.11.106.

Tawalbeh, M., Al-Othman, A., Kafiah, F., Abdelsalam, E., Almomani, F., Alkasrawi, M. (2021). Environmental impacts of solar photovoltaic systems: A critical review of recent progress and future outlook. *Science of the Total Environmen*, *759*, 143528. Available from https://doi.org/10.1016/j.scitotenv.2020.143528.

Thomas, S., Kalarikkal, N., Wu, J., Sakho, E. H. M., & Oluwafemi, S. O. (2019). *Nanomaterials for solar cell applications* (pp. 1−743). India: Elsevier. Available from http://www.sciencedirect.com/science/book/9780128133378, https://doi.org/10.1016/C2016-0-03432-0.

Thompson, C. S. & Zou, M. (2013). Nanostructured PVP/SiO2 antireflective coating for solar panel applications. In *Proceedings of the IEEE Conference on Nanotechnology* (pp. 768−771). Beijing, China.

Vatti, N. R., Vatti, P. L., & Vatti, R. (2018). High efficiency solar energy harvesting using nanotechnology. In *4th International Conference for Convergence in Technology*. India: IEEE. http://ieeexplore.ieee.org/xpl/mostRecentIssue.jsp?punumber = 9046056.

Wang, H., Li, J., Wang, J., & Wu. (2012). High aspect-ratio transparent highly ordered titanium dioxide nanotube arrays and their performance in dye sensitized solar cells. *Materials Letters*, *80*, 99−102. Available from https://doi.org/10.1016/j.matlet.2012.040.053.

Wang, H. P., & He, J. H. (2017). Toward highly efficient nanostructured solar cells using concurrent electrical and optical design. *Advanced Energy Materials*, *7*(23). Available from https://doi.org/10.1002/aenm.201602385, http://onlinelibrary.wiley.com/journal/10.1002/(ISSN)1614-6840.

Yamamoto, M., Nishikawa, N., Mayama, H., Nonomura, Y., Yokojima, S., Nakamura, S., & Uchida, K. (2015). Theoretical explanation of the lotus effect: superhydrophobic property changes by removal of nanostructures from the surface of a lotus leaf. *Langmuir: The ACS Journal of Surfaces and Colloids*, *31*(26), 7355−7363. Available from https://doi.org/10.1021/acs.langmuir.5b00670, http://pubs.acs.org/journal/langd5.

Yan, J., Uddin, M. J., Dickens, T. J., & Okoli, O. I. (2013). Carbon nanotubes (CNTs) enrich the solar cells. *Solar Energy*, *96*, 239−252. Available from https://doi.org/10.1016/j.solener.2013.070.027.

Zhang, F. (2017). Grand challenges for nanoscience and nanotechnology in energy and health. *Frontiers in Chemistry*, *5*. Available from https://doi.org/10.3389/fchem.2017.00080, http://journal.frontiersin.org/journal/chemistry.

Zhang, H., Du, S., Chen, C., Luo, W., Liu, C., & Zhou, H. (2016). China Single-walled carbon nanotubes as efficient charge extractors in perovskite solar cell. In *16th International Conference on Nanotechnology*. IEEE.

Zou, X., Tao, C., Yan, L., Yang, F., Lv, H., Yan, H., Wang, Z., Li, Y., Wang, J., Yuan, X., & Zhang, L. (2018). One-step sol-gel preparation of ultralow-refractive-index porous coatings with mulberry-like hollow silica nanostructures. *Surface and Coatings Technology*, *341*, 57−63. Available from https://doi.org/10.1016/j.surfcoat.2018.010.013, http://www.journals.elsevier.com/surface-and-coatings-technology/.

CHAPTER 4

Nanomaterials for photo-electrochemical solar cells

Bahadur Ali

Department of Chemistry, College of Science, Mathematics, and Technology, Wenzhou-Kean University, Wenzhou, China

4.1 Introduction

The global energy landscape is at a pivotal juncture as the focus of the scientific and engineering communities shifts toward the development of clean, renewable, and sustainable energy sources. This shift has been propelled by concerns surrounding fossil fuel depletion, environmental pollution, and climate change. One of the most promising and extensively researched renewable energy sources is solar energy, harnessed using various types of solar cells. Among these, photo-electrochemical (PEC) solar cells have emerged as a viable contender due to their potential for high-energy conversion efficiency and cost-effectiveness (Bak et al., 2002).

4.1.1 Background and importance of photo-electrochemical solar cells

Since their invention in the 1970s, PEC solar cells have sparked tremendous interest within the scientific community due to their unique characteristics and potential to revolutionize the solar energy landscape. Also known as dye-sensitized solar cells (DSSCs) or Grätzel cells (named after Michael Grätzel, who significantly contributed to their development), PEC solar cells offer an innovative approach to solar energy conversion that is different from traditional photovoltaic (PV) technology.

Unlike conventional silicon-based solar cells, PEC solar cells do not rely on creating a p-n junction. Instead, they operate on photo-electrochemistry principles, where light absorption excites electrons in a photo-sensitizer, typically a dye molecule or semiconductor nanostructure adhered to a wide-bandgap metal oxide substrate. These excited electrons then transfer to the conduction band of the substrate, generating a photo-current. One of the most compelling reasons for the interest in PEC solar

Electrochemistry and Photo-Electrochemistry of Nanomaterials
DOI: https://doi.org/10.1016/B978-0-443-18600-4.00005-3
© 2025 Elsevier Inc. All rights are reserved, including those for text and data mining, AI training, and similar technologies.

89

cells is their potential for high power conversion efficiencies at lower costs than conventional PV technology. The materials used in PEC cells, particularly for the photoanode and the electrolyte, can be relatively inexpensive and abundant, making these devices potentially cheaper and more sustainable to produce at scale (Chandra, 2012).

Moreover, the assembly of PEC cells does not require high-temperature or high-vacuum processes, further contributing to their cost-effectiveness. PEC solar cells are also highly versatile and flexible. They can be designed to be semitransparent and can be integrated into windows, building facades, or loose surfaces, extending the possibilities for solar energy harvesting beyond traditional rooftop installations.

Importantly, PEC solar cells can operate efficiently under diffuse light conditions, making them suitable for indoor and low-light applications. This feature expands their usability in regions with less sunshine and indoor environments, increasing their overall potential for energy generation.

However, the development and commercialization of PEC solar cells are not without challenges. These include issues related to the dye's stability and durability, the liquid electrolyte's corrosiveness and leakage, and the cells' long-term performance. These challenges, among others, are driving ongoing research in the field, with nanomaterials emerging as a promising avenue for enhancing the performance and stability of PEC solar cells (Li et al., 2018).

In this chapter, we will explore the role of nanomaterials in addressing these challenges and improving the design, efficiency, and functionality of PEC solar cells (Figs. 4.1−4.6).

4.1.2 Nanomaterials: an overview

Nanomaterials are materials structured on the nanoscale, typically between 1 and 100 nanometers (nm) in at least one dimension. At this scale,

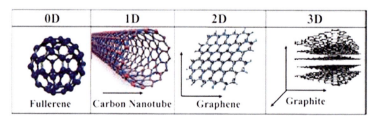

Figure 4.1 The schematic diagram of 0D, 1D, 2D, and 3D nanomaterials (Low et al., 2014).

Nanomaterials for photo-electrochemical solar cells 91

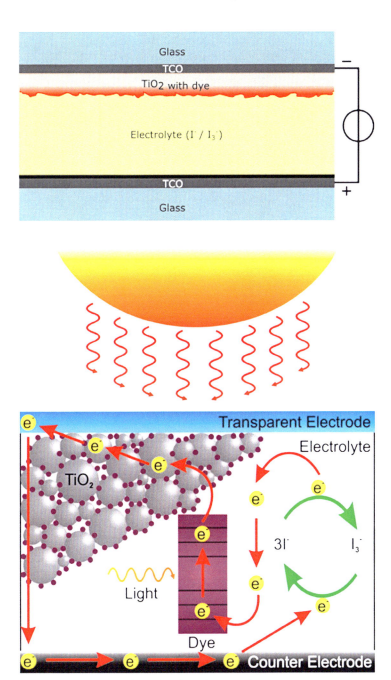

Figure 4.2 Dye-sensitized solar cell and working mechanism.

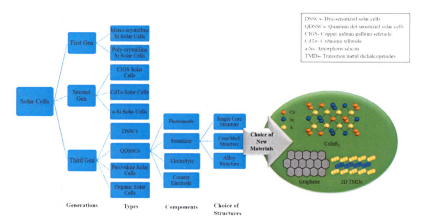

Figure 4.3 Different generations, types, components, and structures of solar cells.

materials often exhibit unique properties distinct from their macro-scale counterparts, including enhanced strength, lighter weight, increased control of the light spectrum, and greater reactivity. This is primarily due to their high surface area-to-volume ratio and quantum effects, which can significantly alter these materials' physical and chemical behavior.

Nanomaterials can be broadly categorized into the following types based on their dimensions:

- **Zero-dimensional nanomaterials (0D)** include quantum dots, nanoparticles, and nanoclusters. These materials are confined in all three dimensions, leading to discrete quantum states and size-dependent properties.
- **One-dimensional nanomaterials (1D)**: This group comprises nanostructures like nanowires, nanorods, and nanotubes, which are extended in one dimension and confined in the other two. These materials often demonstrate superior electrical and thermal conductivity properties.
- **Two-dimensional nanomaterials (2D)** include flat, sheet-like structures that extend in two dimensions and are confined in one, such as graphene and transition metal dichalcogenides. They often possess exceptional mechanical strength and conductivity.
- **Three-dimensional nanomaterials (3D)** include nanoporous materials, nanocomposites, and nanocoatings extending in all three dimensions. They are known for their high surface area and are often used in applications like catalysis and sensing.

Nanomaterials for photo-electrochemical solar cells 93

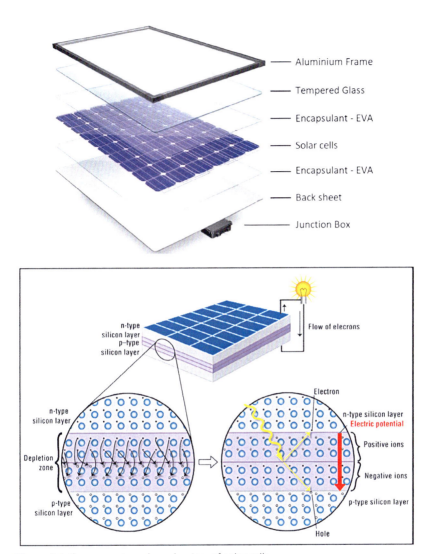

Figure 4.4 Components and mechanism of solar cells.

Nanomaterials have applications across various industries, including electronics, medicine, environmental science, and energy. Nanomaterials have been extensively researched in the energy field for their potential to enhance the performance of photovoltaics, batteries, supercapacitors, and fuel cells. Nanomaterials can contribute to improved light absorption, increased charge separation and transport, and enhanced stability when applied to solar cells,

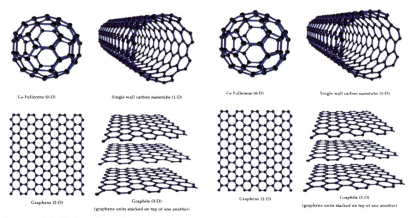

Figure 4.5 Carbon-based nanomaterials.

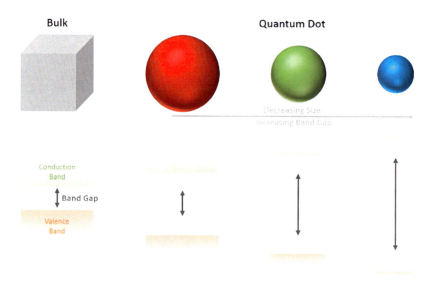

Figure 4.6 Effect of size of quantum dots on band gap.

specifically to electrochemical solar cells. In the following sections, we will explore in detail the role of different types of nanomaterials in enhancing the performance of PEC solar cells (Li et al., 2018).

4.1.3 The role of nanomaterials in solar cells

Nanomaterials have increasingly become a cornerstone of advancements in photovoltaic technology due to their unique physical, chemical, and optical properties, which can be manipulated at the nanoscale to optimize the efficiency of solar cells.

- **Enhanced light absorption**: Nanomaterials, due to their small size and tunable bandgaps, can absorb a broader spectrum of sunlight compared to bulk materials. For instance, quantum dots can be engineered to absorb different wavelengths of light by merely altering their size. This ability to capture more sunlight increases the chances of generating electron-hole pairs, leading to higher power outputs.
- **Improved charge separation and transport**: Nanomaterials also aid in efficiently separating and transporting charge carriers. In a solar cell, after light absorption, the excited electrons must be separated from the holes to prevent recombination, which would otherwise reduce the cell's efficiency. Nanomaterials, due to their high surface area-to-volume ratio, can enhance the speed and efficiency of this process. Nanowires and nanotubes, for example, can provide adequate pathways for charge transport, minimizing losses from recombination.
- **Increased cell flexibility and durability**: Nanomaterials make more flexible and durable solar cells. Nanomaterials like graphene have remarkable strength and flexibility, which can be harnessed to create lightweight, bendable solar cells. These cells can be integrated into various surfaces, broadening the application of solar technology.
- **Cost reduction**: Many nanomaterials can be synthesized from abundant and relatively inexpensive raw materials, potentially lowering the overall cost of solar cell production. Furthermore, applying nanomaterials can improve efficiency and reduce the price per unit of solar energy generated.
- **Novel device architectures**: Nanomaterials also enable the creation of new device architectures, such as quantum dot solar cells, plasmonic solar cells, and dye-sensitized solar cells. These innovative designs leverage nanomaterials' unique properties to overcome traditional photovoltaic materials' limitations and boost solar cell performance.

Nanomaterials fabricate photoanodes and counter electrodes and even design innovative electrolytes in PEC solar cells. By tailoring these nanomaterials, researchers can optimize the light absorption, electron transport, and overall stability of the cells, paving the way for efficient, affordable, and durable solar energy solutions (Xu et al., 2019).

4.1.4 Overview of photo-electrochemical solar cells

PEC solar cells, also known as Grätzel cells or dye-sensitized solar cells (DSSCs), are an essential class of photovoltaic devices that have generated significant interest due to their potential to achieve high power conversion efficiencies at low production costs. Unlike conventional solar cells, which are typically composed of a p-n junction, PEC solar cells are characterized by a photoactive electrode immersed in an electrolyte solution.

A standard PEC solar cell consists of three main components:

- **Photoanode**: This is the light-harvesting component of the cell and is typically composed of a semiconductor material doped with a photosensitive dye or quantum dots. These semiconductors are generally wide-bandgap metal oxides like titanium dioxide (TiO_2) or zinc oxide (ZnO).

- **Electrolyte**: This component serves as a medium for ion transport and replenishes the holes in the dye or quantum dots. The electrolyte usually contains a redox pair, most commonly iodide/triiodide (I^-/I_3^-), to ensure charge neutrality.

- **Counter electrode**: This is typically a conductive glass or metallic layer coated with a catalyst (often platinum- or carbon-based materials) to facilitate the reduction reaction of the redox pair.

The operation of a PEC solar cell begins when sunlight is absorbed by the photosensitive dye or quantum dots at the photoanode, exciting electrons and creating electron-hole pairs. The excited electrons then move from the photoanode, through the external circuit, to the counter electrode. Simultaneously, the electrolyte replenishes the holes left in the dye, maintaining charge balance.

The unique feature of PEC solar cells is their capacity to convert incident sunlight into electricity even under low-light conditions, such as on cloudy days or indoors. This characteristic, combined with their potential for low-cost fabrication and flexibility, makes PEC solar cells a promising technology for future solar energy harvesting. However, there are challenges to overcome, such as the dye's stability and the electrolyte's corrosiveness, which are active research areas.

Introducing nanomaterials into these systems presents a pathway to address these challenges and further enhance the performance of PEC solar cells (Ma et al., 2021).

4.1.5 The evolution of nanomaterials in solar energy harvesting

The utilization of nanomaterials in solar energy harvesting has witnessed a remarkable evolution over the past few decades. This has been facilitated by advancements in nanotechnology and the growing demand for efficient and sustainable solar energy solutions.

- **First-generation nanomaterials**: The journey began with simple nanoscale particles such as metal nanoparticles and semiconductor quantum dots. Metal nanoparticles enhanced solar cells' light absorption and scattering properties, while semiconductor quantum dots were exploited for their size-dependent optical and electronic properties. These first-generation nanomaterials allowed scientists to exceed the Shockley-Queisser limit, the theoretical maximum efficiency of a p-n junction solar cell, opening up new possibilities in photovoltaics.

- **Nanostructured films and layers**: The next phase of evolution saw the development of thin films and layers composed of nanoscale materials. For instance, nanoporous titanium dioxide (TiO_2) was widely used as a photoanode in dye-sensitized solar cells (DSSCs). These nanostructured films provided a larger surface area for dye adsorption, increasing harvesting and charge extraction capabilities.

- **One-dimensional nanostructures**: The focus shifted to one-dimensional nanowires and nanotubes. These structures offer excellent pathways for electron transport, improving charge separation and minimizing recombination losses. For example, zinc oxide (ZnO) nanowires were incorporated into DSSCs as photoanodes, enhancing their power conversion efficiency.

- **Advanced nanomaterials**: The current era of nanomaterials in solar energy harvesting is characterized by the use of advanced nanomaterials, including plasmonic nanoparticles, perovskite nanocrystals, and two-dimensional materials like graphene and transition metal dichalcogenides. Plasmonic nanoparticles can concentrate and absorb light in specific regions, known as hot spots, improving the light absorption capabilities of solar cells. Perovskite nanocrystals have emerged as highly efficient light absorbers, making them promising candidates for next-generation solar cells.

- **Hybrid and composite nanostructures**: Today, research is focused on designing hybrid and composite nanostructures that combine multiple materials to create systems with optimized properties. These include core-shell nanoparticles, mixed-dimensional nanostructures, and hybrid organic-inorganic systems, aiming to combine the advantages of different materials and overcome the limitations of individual components.

As the field progresses, the focus is now on addressing the stability, scalability, and environmental impact of nanomaterial-based solar cells and exploiting new concepts and materials to push the boundaries of solar energy harvesting further. Nanomaterials will continue to play a central role in the evolution of solar energy technologies. The following sections of this chapter will delve deeper into these advanced nanomaterials and their applications in PEC solar cells (Modestino & Haussener, 2015).

4.1.6 Objectives of chapter

This chapter aims to provide a comprehensive overview of the role and application of nanomaterials in PEC solar cells. The objectives are:

- **Exploring the fundamentals**: To familiarize readers with the fundamental concepts of nanomaterials and PEC solar cells. This includes understanding the basic principles of operation of PEC solar cells, the characteristics that distinguish them from conventional solar cells, and the essential features of nanomaterials.
- **Role of nanomaterials in solar cells**: To elucidate nanomaterials' unique roles in enhancing solar cell performance. This includes understanding how nanomaterials improve light absorption, charge separation, and transport and how they contribute to the durability and flexibility of solar cells.
- **In-depth study of specific nanomaterials**: To provide an in-depth study of different nanomaterials used in PEC solar cells. This will involve discussing the unique properties of these nanomaterials, the methods of their synthesis, their role in PEC solar cells, and the benefits they offer over conventional materials.
- **Challenges and future perspectives**: To discuss the current challenges facing the application of nanomaterials in PEC solar cells and to explore future perspectives. This involves addressing stability, cost, scalability, and environmental impact and envisioning future research directions.

Through achieving these objectives, this chapter seeks to provide a balanced, comprehensive, and up-to-date overview of nanomaterials in

PEC solar cells, serving as a valuable resource for researchers, students, and professionals interested in this exciting field of research (Tilley, 2019).

4.2 Fundamentals of photo-electrochemical solar cells

4.2.1 Basic concepts of solar cells

Solar cells, also known as photovoltaic cells, convert sunlight directly into electricity. This conversion process is based on the principles of the photovoltaic effect, which was first observed by Alexandre-Edmond Becquerel in 1839. Let us discuss the basic concepts that underlie the operation of solar cells.

4.2.1.1 Photovoltaic effect

The photovoltaic (PV) effect generates a voltage (or a direct current) in a material upon exposure to light. This phenomenon is primarily observed in certain types of semiconductors. The PV effect consists of three primary stages: light absorption, charge separation, and charge collection.

4.2.1.2 Light absorption and exciton generation

When light, composed of packets of energy known as photons, strikes the semiconductor material of the solar cell, it can transfer energy to the electrons in the material. If this energy is sufficient, it can "excite" the electrons, causing them to move from their regular positions (in the valence band of the semiconductor) to a higher energy level (the conduction band). This movement of electrons creates electron-hole pairs, where the "hole" represents the position the electron initially occupied. These electron-hole pairs are also referred to as excitons.

4.2.1.3 Charge separation and transport

Once the excitons are generated, separating the electrons and holes is crucial to prevent them from recombining. The built-in electric field of the p-n junction in the semiconductor facilitates this charge separation. Once separated, the electrons and holes move toward the n-type and p-type semiconductor, respectively.

4.2.1.4 Charge collection and current generation

The motion of these charge carriers (electrons and holes) toward their respective electrodes generates an electric current. This current, combined with the cell's voltage (defined by the built-in electric field), represents the solar cell's electrical power to deliver to an external circuit.

4.2.1.5 Energy conversion efficiency

The efficiency of a solar cell is the ratio of the electrical power it can output to the power of the incident sunlight. Several factors can affect the efficiency, including the material properties, the cell's construction quality, and the incident light's intensity and angle. Theoretical and practical limits also exist for the efficiency of solar cells, known as the Shockley-Queisser limit and the maximum achieved laboratory efficiency, respectively. The working principles of PEC solar cells are similar, with some modifications that will be discussed in the next section. However, these basic concepts provide a foundation for understanding how solar cells convert light energy into electrical energy (Turner et al., 2023).

4.2.2 Mechanism of photo-electrochemical solar cells

PEC solar cells, such as dye-sensitized solar cells (DSSCs) and quantum dot-sensitized solar cells (QDSSCs), operate based on principles similar to those of traditional photovoltaic cells, but with some crucial differences that allow them to overcome certain limitations of their conventional counterparts. Here, we outline the critical steps in operating a typical PEC solar cell.

4.2.2.1 Light absorption and exciton generation

Like conventional solar cells, the first step in operating a PEC solar cell is the absorption of light and the generation of excitons. However, in PEC cells, this process typically occurs in a separate light-absorbing material such as a dye or quantum dots coated onto the surface of a wide-bandgap semiconductor (the photoanode). When photons of light hit the dye or quantum dots, their energy is absorbed and used to excite electrons to a higher energy level.

4.2.2.2 Injection of excited electrons

In the next step, the excited electrons are injected from the dye or quantum dots into the conduction band of the semiconductor photoanode. This process effectively separates the electron-hole pairs and prevents their recombination.

4.2.2.3 Charge transport

The photoanode often comprises a nanostructured material, such as nanoporous titanium dioxide (TiO_2). The nanostructure provides a pathway for the transported electrons to reach the external circuit. Meanwhile, the holes remain in the dye or quantum dots.

4.2.2.4 Regeneration of the dye or quantum dots

In the next step, the electrolyte (a liquid, solid, or gel) provides ions to fill the holes in the dye or quantum dots, thereby recharging them and allowing them to generate more electron-hole pairs upon absorption of additional light. The electrons in the electrolyte then move toward the counter electrode.

4.2.2.5 Charge collection and current generation

Finally, the electrons recombine with the electrolyte ions at the counter electrode, and the process can start over. The flow of electrons from the photoanode to the counter electrode through the external circuit constitutes the electrical current generated by the solar cell.

This process, though similar to a traditional photovoltaic cell, offers the advantages of potentially lower cost and the ability to operate efficiently under varying lighting conditions. However, it also presents unique challenges, including efficient dye regeneration, preventing undesired reactions at the electrolyte/electrode interfaces, and the need for long-term stability and durability in real-world conditions (Turner et al., 2023).

4.2.2.6 Key components of photo-electrochemical solar cells and their functions

PEC solar cells have several key components, each with a specific function contributing to the cell's overall performance.

4.2.2.7 Photoanode

The photoanode typically comprises a comprehensive bandgap semiconductor material, such as titanium dioxide (TiO_2). This material is often structured at the nanoscale to maximize its surface area and increase the quantity of dye or quantum dots it can support. The primary function of the photoanode is to facilitate the transport of electrons from the dye or quantum dots to the external circuit, thus preventing their recombination with holes.

4.2.2.8 Dye or quantum dots

Dyes or quantum dots are light-absorbing materials in PEC solar cells. They are chosen for their ability to absorb a broad spectrum of sunlight and to inject excited electrons into the semiconductor photoanode. The choice of dye or quantum dots can significantly influence the cell's light-harvesting efficiency and, thus, its overall performance.

4.2.2.9 Electrolyte

The electrolyte, which can be in liquid, solid, or gel form, serves two primary functions. First, it provides ions to fill the holes in the dye or quantum dots, thereby regenerating them and allowing them to produce more excitons upon light absorption. Second, it transports the electrons that have recombined with the ions to the counter electrode.

4.2.2.10 Counter electrode

The counter electrode, often made of a catalyst such as platinum, facilitates the reduction of the electrolyte ions back to their original state. This process completes the cell's electrical circuit and allows the electrolyte to continue providing ions to the dye or quantum dots. Each of these components plays a vital role in the operation of a PEC solar cell. These components' design, selection, and integration are thus crucial considerations in developing and optimizing PEC solar cells. In the following sections, we will discuss how nanomaterials can enhance the performance and functionality of these components (Xu et al., 2019).

4.3 Types of nanomaterials in photo-electrochemical solar cells

Nanomaterials have exhibited remarkable capabilities in improving the efficiency, stability, and versatility of PEC solar cells. Due to their distinct properties at the nanoscale, they interact uniquely with light, electrons, and the surrounding environment. The following are the primary types of nanomaterials that have been investigated for their potential in PEC solar cells:

4.3.1 Metal oxide nanomaterials

Metal oxide nanoparticles have been at the forefront of materials research for PEC solar cells due to their unique physicochemical properties, stability, and potential for high performance. These nanoparticles offer benefits derived from their nanoscale dimensions, including a large surface-to-volume ratio, improved light scattering, and the ability to facilitate efficient charge transfer and separation. Here is a detailed overview of commonly used metal oxide nanoparticles:

4.3.1.1 Titanium dioxide (TiO$_2$)

Titanium dioxide (TiO$_2$), a white semiconducting metal oxide, has become a linchpin in PEC solar cells. Its inherent properties, such as a bandgap of around 3.2 eV in the anatase phase and an intrinsic stability, render it an attractive choice, particularly in dye-sensitized solar cells (DSSCs). The DSSCs exploit the high surface area of TiO$_2$ nanoparticles for efficient dye molecule adsorption. When these dye molecules absorb light, they transfer their electrons to TiO$_2$, shutting them to the electrode.

However, TiO$_2$ applications continue beyond DSSCs. The rise of perovskite solar cells has seen TiO$_2$ as an electron-transporting layer, bridging the gap between efficient charge extraction and reduced electron-hole recombination. Furthermore, there is an increasing interest in hybrid solar cells, where TiO$_2$ is married with organic semiconductors or other novel nanomaterials to harness the collective strengths of both entities.

Morphology, in the context of TiO$_2$, is key to its varied applications. While nanoparticles are prized for their surface area, especially in DSSCs, the structured geometry of nanowires and nanorods can channel electron transport more directionally, boosting efficiency by minimizing recombination. Mesoporous structures, on the other hand, bring forth advantages in dye-loading capacity and light scattering.

However, every material has its challenges. TiO$_2$ wide bandgap restricts its light absorption primarily to the UV spectrum, leaving a vast swathe of solar energy untapped. In DSSCs, this limitation is circumvented using dyes that absorb visible light. Additionally, the rapid recombination of its photo-excited charge carriers can dampen the efficiency, necessitating its pairing with efficient electron or hole transport materials or exploring doping strategies. Moreover, synthesizing TiO$_2$ with precise and consistent morphologies tailored for PEC applications demands sophisticated techniques like sol-gel and hydrothermal methods (Zheng et al., 2020).

4.3.1.2 Zinc oxide (ZnO)

Zinc oxide (ZnO), a versatile semiconducting material with a direct bandgap of about 3.3 eV, has garnered significant attention in PEC solar cells.

Its unique electronic and optical properties and inherent biocompatibility position ZnO as an attractive alternative and supplement to other semiconductor materials, notably titanium dioxide (TiO_2).

Central to ZnO's allure in PEC cells is its capacity for morphological versatility. From nanoparticles to nanorods, nanowires, and even more complex structures, the varied shapes that ZnO can adopt influence electron transport efficiency and light absorption in solar cells. In particular, ZnO nanorods and nanowires offer a direct pathway for electron transport, minimizing the chances of electron-hole recombination—a pivotal factor in enhancing solar cell efficiency.

Dye-sensitized solar cells (DSSCs) represent one of the most prominent applications of ZnO. ZnO nanoparticles provide a scaffold for dye molecules in these configurations, much like TiO_2. However, ZnO's ability to form diverse nanostructures gives it an edge in specific applications. When these dyes absorb sunlight, they relay electrons to ZnO, which then conducts them to the external circuit.

Yet, the application of ZnO is wider than DSSCs. Its potential as an electron-transporting layer in perovskite solar cells and in hybrid structures, where it is paired with organic materials or other nanomaterials, is being explored with vigor. These combinations aim to capitalize on the strengths of ZnO while addressing its limitations, such as its susceptibility to degrade organic materials upon UV exposure.

While ZnO boasts several advantages, its intrinsic electron mobility, though superior to TiO_2, is often compromised in the presence of native defects or impurities. Addressing this requires sophisticated doping strategies or surface treatments. Moreover, the stability of ZnO in certain acidic or basic conditions commonly used in PEC cells is another concern, necessitating protective strategy or the use of alternative electrolytes.

In closing, zinc oxide is a material of immense potential in the domain of PEC solar cells. Its strengths, from unique morphology to excellent electron transport properties, combined with the ongoing research to address its challenges, make it a cornerstone in the evolving tapestry of solar energy technologies.

4.3.1.3 Iron(III) oxide (Fe$_2$O$_3$ or hematite)

Iron(III) oxide, commonly known as hematite or Fe_2O_3, stands out as a prominent candidate in PEC solar cells. This naturally abundant, nontoxic metal oxide boasts a direct bandgap of approximately 2.1 eV, an attribute that allows it to absorb a significant portion of the visible light spectrum, making it a strong contender for efficient light harvesting. Moreover, hematite's intrinsic chemical stability under PEC conditions is a testament to its potential for prolonged and stable device operation, a quality essential for real-world solar cell applications. Factor in its widespread availability as one of Earth's most prevalent minerals, it is clear that hematite offers a cost-effective route to harnessing solar energy.

In practical applications, hematite's capacity to absorb visible light and its appropriately positioned band edges have propelled its use as a photoanode in PEC water splitting. Upon illumination, hematite facilitates the generation of photo-excited electrons and holes, which are instrumental in water splitting when incorporated into an optimized PEC system.

However, it is not without its challenges. Hematite grapples with inherent issues like low electron mobility and a pronounced propensity for electron-hole recombination. Such limitations have historically stymied efforts to realize high efficiencies. But the scientific community, ever resourceful, has explored avenues like doping and nanostructuring to bolster hematite's charge transport capabilities. Additionally, the short diffusion length of holes generated within hematite presents another hurdle. Strategies like crafting thin films, refining nanostructures, or integrating complementary materials have been proposed and studied to circumvent this issue. Furthermore, to tackle the problem of surface states acting as recombination hotspots, researchers are delving into surface modification and passivation techniques (Li et al., 2018).

Peering into the horizon, the future for hematite in PEC cells appears promising. Innovative strategies are continually emerging, whether it is integrating cocatalysts, designing novel nanostructures, or pairing hematite with other semiconductors to enrich its PEC repertoire. Tandem configurations, where hematite is synergized with another material to tap into the near-infrared or even UV regions of the solar spectrum, are also subjects of intensive research. While challenges persist, hematite's myriad advantages position it as a material of great promise. With sustained research and innovation, there is every reason to believe that hematite will carve a niche in the sustainable solar energy landscape of the future (Zhang et al., 2018).

4.3.1.4 Tungsten trioxide (WO₃)

Tungsten trioxide (WO_3) has emerged as a compelling contender in PEC solar cells, primarily driven by its unique semiconductor properties. With a moderately wide bandgap, around 2.5 to 2.8 eV, WO_3 can efficiently absorb a segment of the visible light spectrum, rendering it suitable for solar energy conversion processes. One of its most distinguishing features is its excellent photocatalytic ability, which is crucial for various applications, including PEC water splitting.

WO_3 natural n-type semiconducting property lends itself well to applications as a photoanode in PEC cells. When exposed to light, WO_3 can generate photo-induced electrons and holes. These charge carriers can then facilitate various PEC reactions, notably water oxidation, thereby contributing to the generation of clean, green hydrogen.

However, like many semiconductors in this domain, WO_3 is not without its set of challenges. A primary concern has been the relatively low charge carrier mobility, which can hinder the efficient transportation of generated electrons and potentially lead to increased recombination rates. This recombination can significantly undermine the overall efficiency of the PEC process. Nevertheless, the scientific community has been proactive in addressing these limitations. By designing nanostructured forms of WO_3, such as nanowires or nanorods, it is possible to provide direct pathways for electron movement, reducing transit times and mitigating recombination. Additionally, hybrid structures combining WO_3 with other materials or nanoparticles have been explored to enhance its performance further.

Another noteworthy aspect of WO_3 is its chromogenic property, which exhibits a color change in response to electrical or optical stimuli. This has spurred interest in its application in electrochromic devices, which can be integrated with solar cells for multifunctional energy-saving applications. With its suite of desirable properties and ongoing research dedicated to optimizing its performance, Tungsten trioxide holds substantial promise in PEC solar cells. As researchers unravel its potential and devise solutions to its challenges, WO_3 may well establish itself as a mainstay in the future of solar energy technologies.

4.3.2 Carbon-based nanomaterials

Carbon-based nanomaterials have been at the forefront of scientific research due to their unique structural, electronic, and mechanical properties. In the context of PEC solar cells, these nanomaterials offer a revolutionary approach as primary materials and components to enhance the performance of other semiconductor materials. This section sheds light on the prominent carbon-based nanomaterials employed in PEC solar cells.

4.3.2.1 Graphene

Graphene, often termed the "miracle material" of the 21st century, is a single layer of carbon atoms neatly arranged in a two-dimensional hexagonal lattice. Its discovery ushered in a new era of nanotechnology, redefining what was thought possible in various scientific domains, including PEC solar cells. The intrinsic properties of graphene, such as its unparalleled electrical conductivity, extraordinary mechanical strength, high thermal stability, and vast surface area, make it an invaluable asset in PEC systems.

In the context of PEC solar cells, graphene's conductive nature plays a pivotal role in facilitating efficient charge transfer. This efficient electron transport capability ensures minimal electron-hole recombination, a significant challenge in many solar cells, thereby enhancing the overall conversion efficiency. Moreover, its flexibility and strength can potentially lead to the development of bendable or foldable solar cells, paving the way for innovative applications beyond traditional rigid panels. Another notable property of graphene is its ability to act as a barrier against environmental factors. It can provide a protective shield when integrated into solar cells, mitigating the degradation of underlying photoactive materials due to external conditions. This protective quality is crucial for enhancing the operational longevity of PEC solar cells.

However, harnessing the full potential of graphene in PEC systems is a challenge. Ensuring uniform dispersion of graphene, especially in composite structures, and achieving optimal layer thickness for desired electron transport remain challenges. Additionally, tuning its electronic properties through controlled doping or functionalization to match the requirements of specific PEC applications is an area of active research. Graphene's inclusion in PEC solar cells represents a confluence of cutting-edge material science with sustainable energy harvesting. As researchers delve deeper into optimizing its integration and addressing its challenges, graphene stands poised to significantly contribute to the next generation of solar energy technologies.

4.3.2.2 Carbon nanotubes

Carbon nanotubes (CNTs) are nanoscopic cylindrical structures composed entirely of carbon atoms, resembling rolled-up sheets of graphene. Since their discovery, CNTs have garnered immense attention in the scientific community due to their exceptional electronic, mechanical, and thermal properties. In PEC solar cells, CNTs present a multifaceted platform with potential applications spanning various aspects of the cell's architecture and performance.

The intrinsic electronic properties of CNTs, characterized by their rapid charge transport capabilities, render them particularly valuable for PEC systems. These nanotubes can act as efficient electron highways, facilitating the swift movement of charge carriers. This is crucial for reducing electron-hole recombination, a significant efficiency bottleneck in many solar cells. Moreover, CNTs can be metallic or semiconducting based on their chirality, offering flexibility in designing tailored applications within solar cells.

From a structural standpoint, the high aspect ratio of CNTs, coupled with their impressive mechanical strength, provides added benefits. When incorporated into the active layers or electrodes of PEC solar cells, they can impart enhanced mechanical durability while offering pathways for charge transport, potentially boosting both the cell's resilience and efficiency.

Carbon nanotubes also possess an innate ability to form junctions with many materials, be it organic polymers or inorganic semiconductors. This makes them ideal candidates for hybrid structures in PEC cells. Such hybrids aim to synergistically combine the advantages of CNTs with other materials, optimizing light absorption, charge separation, and charge transport processes.

Yet, integrating CNTs into PEC systems is full of challenges. Ensuring their uniform dispersion and alignment, controlling their chirality for desired electronic properties, and functionalizing their surfaces for better compatibility with other materials are areas that demand meticulous research. Carbon nanotubes offer a promising avenue for advancements in PEC solar cell technology. Their unique attributes, combined with ongoing research endeavors to circumvent associated challenges, position CNTs as a pivotal component in the future evolution of solar energy solutions.

4.3.2.3 Fullerenes (buckyballs)

Fullerenes, often colloquially called "buckyballs" due to their distinctive spherical shape, are carbon allotropes composed entirely of carbon atoms, forming structures akin to hollow spheres, ellipsoids, or tubes. Discovered in the mid-1980s, fullerenes captivated the scientific community with their unique geometric configuration and intriguing electronic properties.

Their potential in the context of PEC solar cells has since become a focal point of research, with applications primarily in organic photovoltaics.

One of the standout attributes of fullerenes is their electron-accepting solid capability. This property is of particular significance in PEC solar cells, where separating electron-hole pairs, or excitons, is a crucial step in energy conversion. When paired with suitable donor materials, fullerenes can facilitate effective charge separation, generating free charge carriers ready for transport. This phenomenon enhances the overall efficiency of electron transfer within the solar cell, thereby contributing to improved energy conversion rates.

Furthermore, the versatility of fullerenes allows for chemical functionalization. This means their properties can be tailored to meet specific requirements of PEC cells, enhancing compatibility with other materials and boosting performance metrics. This adaptability gives researchers a valuable tool in optimizing organic solar cell architectures. In addition to their role in electron acceptance, fullerenes can also play a part in the stability and longevity of PEC solar cells. Their ability to form stable complexes with many organic semiconductors can lead to enhanced resistance against environmental factors, ensuring a prolonged operational lifespan of the cell.

However, integrating fullerenes into PEC solar cells is not free from challenges. Issues related to their dispersion in organic matrices, potential aggregation, and cost implications of large-scale synthesis are subjects of ongoing research. Fullerenes represent a fascinating intersection of advanced material science and renewable energy technology. As our understanding of their potential deepens and solutions to challenges emerge, fullerenes could play an even more central role in the next wave of PEC solar cell innovations.

4.3.2.4 Other carbon-based nanomaterials

Carbon-based nanomaterials, such as graphene oxide, reduced graphene oxide, and carbon dots, have also been explored for their potential in PEC solar cells. Their unique properties, including tunable bandgap, high surface area, and good compatibility with other materials, can be beneficial for enhancing the efficiency and stability of these cells (Yao et al., 2019).

4.3.3 Quantum dots

Quantum dots (QDs) are semiconductor nanoparticles, typically ranging from 2 to 10 nanometers, exhibiting quantum mechanical properties. Their minute size confers distinct electronic and optical characteristics that are highly tunable based on their size, shape, and material composition. This tunability, combined with other intrinsic properties, has propelled quantum dots to the forefront of research in PEC solar cells.

One of the standout features of quantum dots is their size-dependent bandgap. This means that by simply altering the size of the QD, one can tune its absorption and emission properties. In PEC systems, this allows for the absorption of specific wavelengths of light, or in some cases, a broad spectrum, to ensure optimal solar spectrum utilization.

Quantum dots also possess a phenomenon known as "multiple exciton generation" (MEG). In traditional solar cells, one photon typically generates one electron-hole pair. However, in QDs, a single high-energy photon, can generate multiple electron-hole pairs, promising a theoretical efficiency that surpasses traditional photovoltaic materials.

Integrating quantum dots into PEC cells often involves anchoring them onto wide-bandgap semiconductors. They can serve as light absorbers and inject photo-generated electrons into the semiconductor, driving the PEC processes. Given their nanoscale size, quantum dots also provide a vast surface area for reactions, promoting more efficient charge separation and reducing recombination losses. Despite their immense potential, applying quantum dots in PEC solar cells also comes with challenges. Ensuring long-term stability, mitigating potential toxicity (especially from cadmium-based QDs), and achieving scalable and cost-effective production methods are some areas where intensive research is ongoing.

4.3.3.1 Cadmium telluride (CdTe) quantum dots

Cadmium telluride (CdTe) quantum dots are semiconductor nanoparticles that have emerged as one of the most studied and promising materials in the quantum dot family, particularly for their application in the field of solar energy conversion. The popularity of CdTe quantum dots in PEC solar cells can be attributed to a combination of their intrinsic properties and the advanced techniques available for their synthesis and modification.

One of the primary advantages of CdTe quantum dots is their direct and nearly ideal bandgap of about 1.45 eV. This bandgap aligns well with the solar spectrum, allowing for efficient absorption of sunlight and hence maximizing the conversion of light into electricity. Furthermore, the size-dependent tunability of CdTe, like other quantum dots, provides flexibility in tailoring its optical properties for specific PEC applications.

Another strength of CdTe quantum dots is their relatively high quantum yield, which signifies the efficiency of photon-to-electron conversion. This property is crucial in PEC cells, where efficient charge generation directly impacts the overall energy conversion efficiency.

From a fabrication perspective, CdTe quantum dots can be synthesized using aqueous-based methods, which are relatively ecofriendly and cost-effective. This potential for green synthesis can be pivotal in scaling up production for commercial applications.

However, the use of CdTe quantum dots is not without concerns. The primary issue revolves around the toxic nature of cadmium, raising environmental and health concerns, especially during the synthesis and disposal stages. This has led to active research into developing cadmium-free alternatives, although CdTe remains prominent due to its superior photovoltaic performance.

There is also the challenge of ensuring the long-term stability of CdTe-based PEC cells, as quantum dots can degrade over time, especially under continuous illumination. Efforts are underway to develop protective coatings and strategies to mitigate this degradation.

Cadmium telluride quantum dots offer significant potential for enhancing the efficiency and versatility of PEC solar cells; it is imperative to balance their photovoltaic advantages with environmental considerations. As research progresses in this domain, CdTe quantum dots will likely see refinements that make them even more suitable for sustainable solar energy applications.

4.3.3.2 Lead selenide (PbSe) quantum dots

Lead selenide (PbSe) quantum dots have garnered significant attention in the photovoltaic community due to their remarkable optoelectronic properties and their potential in revolutionizing the design and performance of PEC solar cells.

One of the standout attributes of PbSe quantum dots is their infrared absorption capabilities. With a bandgap that can be tuned from the near-infrared to the mid-infrared range by varying the particle size, PbSe quantum dots allow for harnessing a broader portion of the solar spectrum than many conventional solar cell materials. This extended spectral sensitivity can be instrumental in achieving higher overall energy conversion efficiencies. Another advantage lies in the quantum dot's inherently high quantum efficiency, ensuring that a significant fraction of the absorbed photons contributes to generating electron-hole pairs. In the context of PEC cells, this can translate to enhanced photocurrents and, consequently, a boost in the cell's performance.

The synthesis of PbSe quantum dots, typically via colloidal routes, has matured over the years, allowing for consistent production of high-quality, monodisperse nanoparticles. The controllability of these synthetic methods ensures that the resulting quantum dots can be tailored for specific PEC configurations, optimizing their interaction with other materials in the system.

However, the path to integrating PbSe quantum dots into commercial PEC devices is full of challenges. One significant concern is the presence of lead, a toxic element, in the quantum dots. As with CdTe quantum dots, environmental and health implications are tied to the production, usage, and disposal of lead-based materials. This has motivated research into encapsulation methods and safe handling practices for PbSe quantum dots. Additionally, ensuring the stability of PbSe in PEC environments, particularly under prolonged illumination and in the presence of electrolytes, is a matter of ongoing research. Addressing potential degradation mechanisms will be crucial in ensuring the longevity of PbSe-based PEC devices.

4.3.3.3 Other quantum dot materials

Beyond the more well-known quantum dot materials such as CdTe and PbSe, a plethora of other semiconductor nanoparticles are being researched and developed for applications in PEC solar cells. These alternative materials aim to overcome some of the limitations or challenges of the mainstream quantum dots, especially concerning toxicity or narrow absorption spectra.

- **Cadmium sulfide (CdS) quantum dots**: CdS quantum dots are another member of the cadmium-based family with a bandgap that allows visible light absorption. Their compatibility with various substrates and other photovoltaic materials makes them a versatile choice for various PEC configurations. However, like CdTe, they present toxicity concerns due to the presence of cadmium.
- **Silicon (Si) quantum dots**: Silicon, abundant and relatively nontoxic, has been the mainstay of the solar industry in its bulk form. When reduced to the quantum dot size, silicon exhibits quantum confinement effects, offering tunable bandgaps. Si quantum dots represent an ecofriendlier alternative to some of their toxic counterparts, although challenges in efficiently harnessing their photovoltaic potential remain.

- **Copper indium selenide (CIS) and copper indium gallium selenide (CIGS) quantum dots**: These are less-toxic alternatives to cadmium or lead-based quantum dots. Bandgaps that can be tuned based on their composition provide a flexible platform for PEC solar cell design.
- **Perovskite quantum dots**: Emerging as a significant player in the solar cell domain, perovskite materials have been miniaturized to the quantum dot level, combining the advantages of perovskite solar cells with the quantum confinement effects of nanoparticles. These quantum dots offer high absorption coefficients and can be synthesized using relatively simple and low-cost methods. However, stability issues, especially in the presence of moisture, are areas of active research.
- **$AgBiS_2$ quantum dots**: As the photovoltaic community searches for nontoxic and environmentally friendly materials, $AgBiS_2$ has emerged as a potential candidate. These quantum dots are free from scarce and toxic elements, and preliminary studies indicate promising photovoltaic performance.
- **InP (indium phosphide) quantum dots**: As a direct bandgap semiconductor, InP quantum dots have been explored as a less toxic alternative to cadmium-based quantum dots. Their photoluminescent properties make them candidates for light absorption and emission applications.

In the broader context of PEC solar cells, introducing various quantum dot materials aims to address diverse challenges ranging from environmental concerns to the quest for higher efficiencies. The versatility and tunability of these nanoparticles provide a robust platform for innovation, ensuring that the realm of quantum dot-based PEC cells remains an active and dynamic field of research (Modestino & Haussener, 2015).

4.3.4 Conducting polymer nanomaterials

Conducting polymers, often referred to as conjugated or intrinsic conducting polymers, have seen an upsurge in interest for applications in a myriad of electronic devices, including PEC solar cells. Their unique blend of organic chemistry with conductive properties offers new avenues for tailoring the interface and performance of PEC cells. Conducting polymers, such as polyaniline and poly(3-hexylthiophene) (P3HT), are

4.3.4.1 Polyaniline

Polyaniline (PANI) is one of the most extensively researched conducting polymers due to its unique combination of properties that make it a potential candidate for PEC solar cells. PANI is renowned for its intrinsic conductivity, ability to switch between different oxidation states, and environmental stability. Its diverse range of conductivity, stemming from various oxidation states, allows for the tunability of its electronic properties. This adaptability can be leveraged to optimize its performance in PEC systems.

In the context of PEC solar cells, PANI can serve multifunctional roles. It can be employed as a hole transport layer, facilitating efficient charge transport and minimizing recombination events. This ensures that a higher proportion of generated charges contribute to the output, thus improving the overall efficiency of the solar cell. Additionally, PANI's chemical structure grants it a level of corrosion resistance, making it a potential candidate for protective layers in PEC systems. This protective role can significantly enhance device longevity by mitigating degradation from corrosive environments. Another notable feature of PANI is its ease of synthesis and processability. Through relatively simple chemical methods, PANI can be synthesized and subsequently processed into thin films suitable for device integration. Its compatibility with various solvents and substrates offers flexibility in device fabrication, paving the way for scalable and cost-effective manufacturing methods.

However, while PANI presents numerous advantages for PEC applications, challenges persist. These include ensuring consistent conductivity across the material, optimizing its interaction with adjacent layers in the PEC stack, and enhancing its long-term stability under operational conditions. Continued research is focused on addressing these challenges, potentially through chemical modifications or hybrid structures that combine PANI with other functional materials. As research progresses, PANI's role in the PEC landscape is anticipated to become even more pronounced, underlining its significance in the realm of sustainable energy solutions.

4.3.4.2 Poly(3-hexylthiophene) (P3HT)

Poly(3-hexylthiophene) (P3HT) is a conjugated polymer that has gained significant attention in the domain of organic electronics, especially in organic photovoltaics and, by extension, PEC solar cells. One of the key attributes that distinguish P3HT is its semicrystalline nature, allowing for enhanced charge transport, which is critical for efficient solar energy conversion.

The structural design of P3HT, characterized by a repeating thiophene ring connected by hexyl side chains, is fundamental to its optoelectronic properties. These hexyl chains aid in solubilizing the polymer, allowing for solution-based processing methods. As a result, thin films of P3HT can be deposited using scalable techniques such as spin-coating, printing, and even roll-to-roll processes. In PEC solar cells, P3HT often functions as a photoactive layer due to its suitable bandgap, which allows for the absorption of a significant portion of the solar spectrum. When paired with appropriate electron acceptor materials, such as fullerene derivatives, P3HT can facilitate efficient charge separation, leading to the generation of photocurrents. This pairing often forms the basis of the active layer in bulk heterojunction organic solar cells.

However, the journey of incorporating P3HT into high-performing PEC solar cells is not without challenges. The polymer's crystallinity, which is crucial for charge transport, can sometimes lead to phase separation issues in blended films. Moreover, ensuring efficient charge extraction and minimizing recombination remains an ongoing area of research. Despite these challenges, the potential of P3HT in the realm of PEC solar cells remains high. As researchers continue to explore innovative ways to optimize its properties and integrate it effectively within device architectures, P3HT stands as a promising material, pointing toward a brighter future for sustainable energy solutions (Ma et al., 2021)

4.3.4.3 Other conducting polymers

Several other conducting polymers, including polypyrrole, polythiophene, and poly(3,4-ethylenedioxythiophene) (PEDOT), are also used in PEC solar cells. They can improve the performance of these cells by enhancing charge transport, reducing charge recombination, and improving the stability of the cells.

- **Polypyrrole (PPy)**: Polypyrrole's high conductivity and environmental stability have garnered interest for PEC solar cells. Its ease of synthesis, usually through oxidative polymerization, allows for a scalable production process. In PEC cells, PPy can act as an electron transport layer, enhancing the charge collection at the electrode.
- **PEDOT: PSS (Poly(3,4-ethylenedioxythiophene) polystyrene sulfonate)**: This conducting polymer composite combines the

conductive properties of PEDOT with the solubility of PSS, allowing for easy processing. Its high transparency and good conductivity make it a popular choice as a hole transport layer in various solar cell architectures.

- **Polythiophenes and derivatives**: The polythiophene family has seen numerous derivatives designed to optimize its optoelectronic properties. These derivatives can offer broader absorption spectra or better charge transport characteristics, enhancing the overall efficiency of PEC cells.

4.3.5 Perovskite nanomaterials

Perovskite nanomaterials have burst onto the scene of solar energy research with an impact that is hard to overstate. Characterized by their unique crystal structure and versatile chemical compositions, perovskite materials have demonstrated rapid advancements in terms of efficiency and stability, making them one of the most promising candidates for next-generation PEC solar cells. At its core, the term "perovskite" refers to materials that adopt the crystal structure of the mineral calcium titanate ($CaTiO_3$). In the context of PEC solar cells, the most commonly studied perovskite materials are organic-inorganic lead halide perovskites, typically represented by the formula ABX_3, where A is a cation (often an organic cation like methylammonium or formamidinium), B is typically lead (Pb), and X is a halide anion (chlorine, bromine, or iodine).

Perovskite materials shine due to their exceptional optoelectronic properties. They possess high absorption coefficients, allowing them to absorb a significant portion of the solar spectrum despite being very thin. Their charge transport properties are also commendable, and they can be tuned to have direct bandgaps suitable for solar energy conversion. The chemistry of perovskite materials is highly adaptable. By changing the constituent cations or anions, researchers can fine-tune the bandgap and other material properties. This flexibility has been instrumental in achieving multi-junction solar cells with optimized bandgap combinations for enhanced efficiency. Many perovskite materials can be processed from solutions, allowing for the creation of thin films through methods like spin-coating, making the fabrication of perovskite-based PEC cells scalable and potentially cost-effective. Taking advantage of nanoscale dimensions can further optimize perovskite materials. Nanostructured perovskites, like nanowires or nanopillars, can facilitate more efficient charge transport and extraction, improving overall device performance.

However, the journey of integrating perovskite nanomaterials into PEC solar cells is accompanied by challenges. One of the most significant is the stability of these materials. Many perovskite compositions degrade in the presence of moisture, oxygen, or under prolonged UV exposure, which poses a challenge for long-term device operation. Additionally, concerns regarding the toxicity of lead and potential environmental impacts are subjects of ongoing research and debate. Yet, with these challenges come opportunities. Novel compositions replacing lead with less toxic alternatives, protective encapsulation strategies, and hybrid material systems are all active areas of research. As understanding and technology evolve, perovskite nanomaterials stand poised to redefine the landscape of PEC solar cells, offering a blend of high performance and manufacturability that could revolutionize solar energy harvesting.

4.3.5.1 Organic-inorganic hybrid perovskites

Organic-inorganic hybrid perovskites have garnered tremendous interest in the PEC solar cell community due to their distinctive combination of organic and inorganic components within their crystalline matrix. These hybrids have been recognized as revolutionary game changers in the world of solar cell research and have shown rapid advancements in both efficiency and versatility. Organic-inorganic hybrid perovskites typically have the formula ABX_3, where "A" is an organic cation such as methylammonium ($CH_3NH_3^+$) or formamidinium ($HC(NH_2)_2^+$), "B" is usually a metal ion, most commonly lead (Pb), and "X" is a halide anion like chlorine (Cl^-), bromine (Br^-), or iodine (I^-). The inorganic component contributes to the exceptional charge transport properties, while the organic moiety can be tailored to fine-tune the material's bandgap and other properties.

The dual nature of these perovskites offers a blend of properties. They exhibit high absorption coefficients, meaning they can effectively capture sunlight with relatively thin layers. Additionally, they have long carrier diffusion lengths, ensuring efficient charge separation and transportation, which is vital for high photovoltaic performance. One of the attractions of organic-inorganic hybrid perovskites is their ease of fabrication. These materials can often be synthesized and processed under ambient conditions using solution-based methods, such as spin coating or blade coating. This ease of processing presents opportunities for scalable manufacturing and reduces production costs.

The hybrid nature of these materials allows for a vast range of compositional variations. By substituting different organic cations or tweaking the halide content, researchers can modify the bandgap, stability, and other material characteristics, enabling the design of optimized materials for specific device architectures or performance criteria. Despite their impressive advantages, organic-inorganic hybrid perovskites are not devoid of challenges. Their sensitivity to moisture, oxygen, and UV radiation can lead to degradation over time, affecting device longevity. The presence of lead raises environmental and health concerns, prompting research into alternative, less toxic components. Encouragingly, the research community is proactive in addressing the challenges associated with these perovskites. Efforts include developing lead-free variants, creating protective barrier layers to shield the perovskite from environmental degradation, and exploring novel organic components to improve stability.

4.3.5.2 All-inorganic perovskites

All-inorganic perovskites have emerged as a fascinating subclass within the broader family of perovskite materials, primarily driven by the quest to address some of the stability issues associated with their organic-inorganic counterparts. By eschewing organic constituents, these perovskites offer the potential for enhanced robustness while maintaining many of the appealing optoelectronic properties characteristic of perovskite materials. All-inorganic perovskites follow the same general ABX_3 formula as their hybrid counterparts, but the "A" site is occupied by an inorganic cation, most commonly cesium (Cs_2). The "B" site typically features lead (Pb) or tin (Sn), while "X" remains a halide anion such as chloride (Cl^-), bromide (Br^-), or iodide (I^-).

Much like their organic-inorganic hybrid counterparts, all-inorganic perovskites boast impressive optoelectronic traits. They offer high absorption coefficients, direct band gaps, and efficient charge transport. Their absence of organic components does not significantly hinder their photo-conversion capabilities, making them strong contenders in the PEC solar cell arena.

One of the primary motivations behind the development of all-inorganic perovskites is the enhancement of material stability. Without organic constituents, these perovskites tend to exhibit better resistance to moisture, thermal degradation, and UV-induced decomposition. This robustness can translate to prolonged device lifetimes and reduced performance degradation over time. All-inorganic perovskites can be synthesized using methods similar to those employed for hybrid perovskites, including solution-based techniques. The nature of the inorganic components might necessitate specific synthesis or annealing conditions, but the overall fabrication remains relatively straightforward. While stability is enhanced, all-inorganic perovskites are not entirely free of challenges. Some compositions may still contain toxic elements like lead, and certain all-inorganic perovskites may exhibit suboptimal bandgaps or charge transport properties compared to their hybrid counterparts. Research into optimizing these materials for PEC applications is ongoing. Current research in the realm of all-inorganic perovskites is multifaceted. Efforts are underway to discover and design novel compositions with optimized bandgaps, improved charge transport, and further enhanced stability. Simultaneously, researchers are exploring innovative device architectures tailored to leverage the unique properties of these materials. All-inorganic perovskites present a promising pathway toward durable, efficient PEC solar cells. Their blend of desirable optoelectronic traits and potential for enhanced stability position them as noteworthy materials in the quest for sustainable, renewable energy solutions. As research progresses, it is conceivable that these materials will play a pivotal role in the future of solar energy harvesting (Ma et al., 2021).

4.3.5.3 Other perovskite materials

Beyond the commonly researched organic-inorganic hybrid and all-inorganic perovskites, a spectrum of alternative perovskite materials presents unique characteristics and hold promise for PEC solar cell applications. These diverse materials are at the forefront of exploration, seeking to combine the best attributes of perovskites while mitigating known challenges.

- **Lead-free perovskites**: Given the environmental and health concerns associated with lead, significant attention has been directed toward developing lead-free perovskites. Tin (Sn)-based perovskites, such as methylammonium tin iodide ($CH_3NH_3SnI_3$), have emerged as a notable contender. They offer similar electronic properties to lead-based perovskites but with less toxicity. However, their instability in ambient conditions remains a challenge.
- **Lower dimensional perovskites**: Traditional three-dimensional (3D) perovskites have been the focus of much research, but there is

growing interest in 2D and quasi-2D perovskites. These materials possess natural quantum well structures, leading to enhanced charge localization and potential resistance to defect-induced nonradiative recombination. They often exhibit enhanced stability compared to their 3D counterparts, though at times at the expense of reduced photovoltaic efficiency.

- **Double perovskites**: These materials incorporate two different cations in the "B" site of the ABX_3 structure, resulting in an $A2BB'X_6$ formula. By carefully selecting these cations, researchers aim to design materials that maintain the desirable properties of perovskites while eliminating or reducing toxic elements like lead.
- **Vacancy-ordered perovskites**: These are novel perovskite derivatives where specific atomic sites are intentionally left vacant, leading to unique crystalline structures. Preliminary research suggests potential benefits in charge transport and light absorption capabilities.

As with any emergent class of materials, these alternative perovskites come with their own set of challenges, from synthetic complexity to stability and performance optimization. However, their potential benefits, particularly concerning environmental sustainability and device longevity, make them worthy of continued exploration. As the perovskite community broadens its horizons, these alternative materials are expected to gain traction. Multidisciplinary efforts, combining the strengths of materials science, chemistry, and engineering, are likely to drive advancements in this space, pushing the boundaries of what perovskite solar cells can achieve.

The realm of perovskite materials extends well beyond the familiar territories of organic-inorganic hybrid and all-inorganic compounds. These "other" perovskite materials, though less conventional, may hold the keys to overcoming longstanding challenges in the field, unlocking new avenues for sustainable and efficient solar energy conversion. The coming years are set to witness exciting breakthroughs as researchers delve deeper into the untapped potential of these novel materials.

4.4 Fabrication methods for nanomaterial-based photo-electrochemical solar cells

4.4.1 Bottom-up methods

Bottom-up methods involve building larger structures from smaller components, such as atoms or molecules. These methods are advantageous for creating nanostructured materials with precise control over size, shape, and composition.

4.4.1.1 Hydrothermal/solvothermal synthesis
Hydrothermal and solvothermal methods are solution-based synthesis techniques in which the reaction occurs in a sealed vessel at elevated temperatures and pressures. These methods can be used to synthesize a variety of nanomaterials, including metal oxides, quantum dots, and carbon-based materials. The nanomaterials' size, shape, and crystallinity can be controlled by adjusting the reaction parameters.

4.4.1.2 Electrospinning
Electrospinning is a simple and versatile method for producing continuous polymer nanofibers. A high-voltage electric field is applied to a polymer solution or melt, creating a charged jet that solidifies into a nanofiber upon solvent evaporation. The resulting nanofibers can be photoactive or substrates in PEC solar cells.

4.4.1.3 Chemical vapor deposition (CVD)
CVD is a widely used method for synthesizing high-quality nanomaterials, such as carbon nanotubes and graphene. In this process, a volatile precursor decomposes or reacts on a heated substrate to form a thin film. The nanomaterials' size, morphology, and composition can be controlled by adjusting the process parameters.

4.4.2 Top-down methods
Top-down methods involve the fabrication of nanostructures from more extensive bulk materials. These methods are typically used for creating patterns or structures on surfaces.

4.4.2.1 Lithography
Lithography is a commonly used top-down method for creating patterns on a substrate. It involves using a mask to shield the substrate from radiation or chemical etching. The unshielded areas are then modified, allowing for the creation of patterns with nanometer precision.

4.4.2.2 Physical vapor deposition (PVD)
PVD is a vacuum deposition method used to produce thin films. In this process, a material is vaporized from a solid or liquid source and then

condensed onto a substrate. PVD techniques, such as sputtering and thermal evaporation, can deposit metal electrodes or other layers in PEC solar cells.

4.4.2.3 Sputtering
Sputtering is a PVD technique in which atoms are ejected from a target material due to bombardment with high-energy particles. This technique can deposit various materials onto various substrates, including metals, oxides, and nitrides.

4.4.3 Assembly and integration of nanomaterials
4.4.3.1 Layer-by-layer assembly
Layer-by-layer assembly involves the sequential deposition of oppositely charged materials to create a multilayer structure. This method is beneficial for creating thin films of nanomaterials or incorporating multiple materials into a single structure.

4.4.3.2 Self-assembly
Self-assembly is a process by which molecules or particles spontaneously organize into ordered structures due to specific interactions. This method can be used to create complex nanostructures or patterns of nanomaterials on a substrate.

4.4.3.3 Inkjet printing
Inkjet printing is a versatile method for depositing materials onto a substrate. In this process, an ink containing the desired material is ejected from a nozzle onto the substrate. This method is beneficial for creating different materials patterns or depositing nanomaterials onto flexible or large-area substrates.

4.4.4 Doping and surface functionalization of nanomaterials
4.4.4.1 Ion implantation
Ion implantation is a process by which ions of a dopant material are accelerated into a target material. This method can modify nanomaterials' electronic properties or create junctions in semiconductor materials.

4.4.4.2 Atomic layer deposition (ALD)
ALD is a thin film deposition method that involves the sequential exposure of a substrate to different precursor gases. This method can be used

to deposit thin and conformal films of materials onto nanomaterials, thereby modifying their surface properties or protecting them from degradation.

4.4.4.3 Plasma treatment

Plasma treatment involves exposing a material to a plasma to modify its surface properties. This method can be used to clean surfaces, introduce functional groups, or create a thin protective layer on the surface of nanomaterials. Each fabrication method offers unique advantages and potential challenges in constructing PEC solar cells. Understanding these methods' benefits and limitations is crucial for effectively designing and optimizing high-performance PEC solar cells.

4.5 Characterization of nanomaterials for photo-electrochemical solar cells

This part discusses the various techniques used to characterize the properties of nanomaterials for PEC solar cells. These techniques provide valuable information about the structure, composition, optical properties, electrical performance, and surface properties of the nanomaterials, which can be used to understand their behavior in solar cells and to optimize their performance. Each characterization technique, its principles, and its applications in the context of PEC solar cells will be discussed in detail.

4.5.1 Structural characterization

4.5.1.1 Scanning electron microscopy (SEM)

SEM is a versatile technique used to investigate the morphology and topography of a sample surface. SEM provides high-resolution images of nanomaterials and can reveal information about the nanomaterials' size, shape, and distribution in the solar cell.

4.5.1.2 Transmission electron microscopy (TEM)

TEM is a powerful tool used to characterize the internal structure of nanomaterials. It provides atomic-resolution images and can reveal detailed information about crystal structure, lattice spacing, and defects in nanomaterials.

4.5.1.3 X-ray diffraction (XRD)

XRD is widely used for phase identification and crystal structure analysis of nanomaterials. XRD can provide information about the sample's crystalline phases, crystallite size, and degree of crystallinity.

4.5.1.4 Atomic force microscopy (AFM)

AFM is a high-resolution scanning probe technique used to investigate the surface topography of nanomaterials. AFM can provide 3D images with nanometer resolution and reveal information about nanomaterials' surface roughness, particle size, and distribution.

4.5.2 Compositional characterization

4.5.2.1 Energy-dispersive X-ray spectroscopy (EDS)

EDS is an analytical technique used for elemental analysis of a sample. When combined with SEM or TEM, EDS can provide spatially resolved elemental information, revealing nanomaterials' composition and the sample's distribution of elements.

4.5.2.2 X-ray photoelectron spectroscopy (XPS)

XPS is a surface-sensitive analytical technique used to investigate the composition and chemical state of the elements in a sample. XPS can provide valuable information about the oxidation states of elements and the presence of any surface contaminants in the nanomaterials.

4.5.2.3 Inductively coupled plasma mass spectrometry (ICP-Ms)

ICP-Ms is a sensitive technique used for trace element analysis. It can measure the concentration of elements in the sample and provide information about nanomaterials' purity.

4.5.3 Optical characterization

4.5.3.1 UV-Vis spectroscopy

UV-Vis spectroscopy is a commonly used technique to measure a sample's absorption and transmission of light. It can provide information about the band gap energy and the light absorption capability of nanomaterials.

4.5.3.2 Photoluminescence spectroscopy (PL)

PL is a powerful tool for investigating the radiative recombination processes in a sample. It can provide information about the optical quality and electronic structure of nanomaterials.

4.5.3.3 Fourier transform infrared spectroscopy (FTIR)

FTIR is a technique to identify a sample's functional groups and chemical bonds. FTIR can reveal information about the surface chemistry and bonding states of nanomaterials.

4.5.4 Electrical characterization

4.5.4.1 Current-voltage (I-V) measurements

I-V measurements are fundamental to the characterization of solar cells. They can provide information about the solar cells' photocurrent, open-circuit voltage, fill factor, and power conversion efficiency.

4.5.4.2 Electrochemical impedance spectroscopy (EIS)

EIS is a technique used to investigate the impedance characteristics of a system over a range of frequencies. EIS can provide information about the charge transfer processes, recombination rates, and resistance components in solar cells.

4.5.4.3 Quantum efficiency measurements

Quantum efficiency measurements can provide information about the efficiency of light absorption and charge collection in solar cells. These measurements can reveal the wavelength-dependent response of the solar cells and can help identify any losses in the photovoltaic process.

4.5.5 Surface characterization

4.5.5.1 Contact angle measurements

Contact angle measurements can provide information about nanomaterials' surface-wetting properties and surface energy.

4.5.5.2 X-ray photoelectron spectroscopy (XPS)

As discussed above, XPS is used for compositional and chemical state analysis. It is an effective tool for surface analysis as the escape depth of the photoelectrons is typically a few nanometers.

4.5.5.3 Surface profilometry

Surface profilometry measures the surface roughness and thickness of thin films. It is an essential tool for assessing nanomaterial coatings' uniformity and quality.

These characterization methods, in combination, provide a comprehensive understanding of the nanomaterial properties that affect the performance of PEC solar cells. By characterizing these materials in detail, researchers can identify key performance factors and develop strategies to optimize solar cell performance.

4.6 Role of nanomaterials in enhancing the performance of photo-electrochemical solar cells

4.6.1 Nanomaterials for enhanced light absorption

Due to their unique properties and configurations, nanomaterials can manipulate light in novel ways, enabling enhanced absorption, increased scattering, and efficient light trapping. For example, semiconductor quantum dots exhibit size-dependent optical properties that allow for tunable light absorption across a broad spectrum, and metal nanoparticles can generate localized surface plasmon resonances that can further enhance the light absorption within the active layer.

4.6.2 Nanomaterials for efficient charge separation and reduced recombination

Nanomaterials can improve PEC solar cells' charge transport and separation properties due to their high surface area and unique electronic properties. One-dimensional nanomaterials such as nanotubes, nanowires, or nanorods provide adequate pathways for charge transport, facilitating efficient charge separation and reducing electron-hole recombination. Additionally, heterojunctions formed between different nanomaterials can further enhance charge separation and transport.

4.6.3 Bandgap engineering with nanomaterials

Through the quantum confinement effect, the electronic properties of nanomaterials can be tuned by altering their size and shape, which results in the modification of their bandgap energy. This allows the nanomaterials' absorption spectrum to be fine-tuned to match the solar spectrum better, leading to improved solar energy conversion efficiency.

4.6.4 Nanomaterials as catalysts

Due to their large surface area and unique electronic properties, nanomaterials can be effective catalysts for PEC reactions. For instance, noble metal nanoparticles, such as gold and silver, can lower the overpotential of reactions, improving the efficiency of the PEC conversion process.

4.6.5 Nanomaterials for increased stability and durability

Nanomaterials often exhibit improved chemical and thermal stability compared to their bulk counterparts. For instance, graphene and carbon nanotubes are chemically inert and thermally stable, which can enhance the longevity and durability of PEC solar cells. In addition, the high surface area of nanomaterials can lead to robust mechanical adhesion between layers, improving the structural stability of solar cells.

4.6.6 Nanomaterials for flexible photo-electrochemical solar cells

Nanomaterials can be incorporated into flexible substrates, leading to the development of flexible PEC solar cells. This flexibility enables the integration of solar cells into various platforms, including wearable electronics, building materials, and vehicles. In particular, graphene's excellent mechanical flexibility and high electrical conductivity make it a promising material for flexible PEC solar cells.

4.6.7 Cost-effective nanomaterials for solar cells

While some nanomaterials are expensive, others, such as metal oxide nanoparticles or carbon-based nanomaterials, are relatively low-cost and abundant. Using these materials can significantly reduce the manufacturing cost of PEC solar cells, making solar energy more economically viable on a large scale.

4.6.8 Nanomaterials for improved interfacial contacts

The interfaces between different layers in PEC solar cells play a crucial role in their overall performance. Due to their high surface area and excellent surface adhesion properties, nanomaterials can improve the quality of these interfacial contacts. This can facilitate efficient charge transfer across the interfaces, reduce interface-associated recombination losses, and thereby enhance the overall performance of the PEC solar cells.

4.7 Challenges and opportunities in the use of nanomaterials for photo-electrochemical solar cells

4.7.1 Introduction

4.7.1.1 Synthesis and fabrication

The synthesis of nanomaterials is often a complex process involving intricate control of various parameters such as temperature, reaction time, and concentration of reactants. The process needs to be precise and controlled, as minor deviations can significantly alter the properties of the resultant nanomaterials. Furthermore, synthesizing nanomaterials in large quantities while maintaining their unique properties can be challenging.

In addition, fabricating solar cells with these nanomaterials is another area of difficulty. Incorporating nanomaterials uniformly within the device structure and creating stable interfaces with other cell components are challenges. Moreover, existing industrial fabrication techniques may not be suitable or efficient for integrating these nanomaterials, necessitating the development of new methods that are both effective and economically viable.

4.7.1.2 Stability and durability

Due to their high surface area-to-volume ratio, nanomaterials often suffer from issues related to stability and durability. For instance, they can be susceptible to agglomeration or degradation over time, impairing their performance in solar cells. Environmental factors such as heat, light, and moisture can accelerate these processes, further undermining the long-term reliability of the devices.

4.7.1.3 Efficiency and performance

While nanomaterials offer the potential to enhance the efficiency of PEC solar cells, achieving this in practice is a challenge. Optimizing the properties of nanomaterials for efficient light absorption, charge separation, and minimization of recombination losses often involves a delicate balance. Even minor imperfections or inconsistencies can significantly impact device performance.

4.7.1.4 Economic viability

The cost associated with synthesizing and integrating nanomaterials is a significant challenge. Some nanomaterials require expensive raw materials or complex fabrication methods, which can make the resulting solar cells economically uncompetitive compared to traditional photovoltaic technologies. Therefore, cost-effective strategies for synthesizing and integrating nanomaterials are a crucial area of research.

4.7.2 Opportunities in the use of nanomaterials for photo-electrochemical solar cells

4.7.2.1 Efficiency enhancements

Nanomaterials offer significant opportunities to enhance the efficiency of solar cells. Their size and surface properties can be tuned for optimal light absorption. Additionally, they can aid in effective charge separation and minimize recombination losses, significantly enhancing power conversion efficiency.

4.7.2.2 Flexibility and integration

Nanomaterials can contribute to the development of flexible and lightweight solar cells. Traditional solar cells are rigid and heavy, limiting their use in specific applications. On the other hand, nanomaterial-based solar cells can be integrated into flexible substrates, opening up new opportunities for solar energy harvesting, such as in wearable technology or building-integrated photovoltaics.

4.7.2.3 Advanced functionalities

Beyond their direct impact on solar cell efficiency, nanomaterials can also introduce advanced functionalities to solar cells. For example, self-cleaning surfaces can be achieved using specific nanomaterials, reducing maintenance costs and improving long-term device performance. Similarly, the optical properties of solar cells could be tuned using nanomaterials, creating aesthetically pleasing or less visually obtrusive devices.

4.7.2.4 Sustainability

Finally, specific nanomaterials can contribute to more sustainable solar cell production processes. For instance, some nanomaterials can be synthesized from abundant and nontoxic raw materials, reducing the environmental impact of solar cell manufacturing. Furthermore, nanomaterial-based solar cells offer better opportunities for recycling at the end of their life cycle, contributing to a circular economy in the energy sector.

4.8 Summary and concluding remarks

4.8.1 Recapitulation of key points

PEC solar cells: Throughout the previous sections, we have dissected the concept, principles, and applications of PEC solar cells. PEC cells can directly convert sunlight into electricity as a promising alternative to traditional photovoltaics. They also present an opportunity to develop solar technology that is more efficient, sustainable, and versatile, with potential applications ranging from large-scale power generation to portable electronic devices.

The role of nanomaterials: Nanomaterials have been discussed in detail as one of the critical enabling technologies for next-generation PEC solar cells. Due to their unique physical and chemical properties, nanomaterials have the potential to enhance light absorption significantly, enable effective charge separation, and reduce electron-hole recombination in solar cells, thereby improving their overall performance.

Types of nanomaterials and their applications: The chapter detailed various nanomaterials, such as quantum dots, metal oxide nanoparticles, carbon nanotubes, and graphene. Each nanomaterial possesses distinct properties that can be exploited for specific applications within PEC solar cells. For instance, quantum dots have tunable band gaps, enabling more efficient light absorption, while graphene's exceptional electron mobility can facilitate charge transport.

Challenges and opportunities: Throughout the chapter, we have explored several challenges that need to be overcome to fully realize the potential of nanomaterials in PEC solar cells, including issues with synthesis and fabrication, stability, performance, and economic viability. However, we have also highlighted the substantial opportunities that nanomaterials provide, such as efficiency enhancements, increased device flexibility, the introduction of advanced functionalities, and increased sustainability in solar cell production.

4.8.2 Future prospects and directions

The potential of nanomaterials in PEC solar cells has yet to be exhausted. Advanced fabrication techniques could lead to more precise control over nanomaterial synthesis and integration into solar cells, improving device performance. Furthermore, developing novel nanomaterials, such as perovskite nanostructures or new types of quantum dots, could open up new pathways for solar cell enhancement. Finally, innovative solar cell architectures, such as plasmonic or tandem solar cells, offer exciting opportunities for further efficiency improvements.

4.8.3 The potential impact on society and the environment

The advances in nanomaterial-enhanced PEC solar cells carry profound implications for society and the environment. More efficient and cost-effective solar cells could accelerate the global transition to renewable energy, reducing our dependence on fossil fuels and helping to mitigate climate change. Additionally, these technologies could provide energy security and stimulate economic development, particularly in sun-rich, underdeveloped regions. Furthermore, developing more sustainable solar cell production processes could reduce the environmental footprint of the solar industry.

4.8.4 Final thoughts

The field of nanomaterial-enhanced PEC solar cells is inherently multidisciplinary, requiring the combined expertise of material scientists, chemists, physicists, and engineers. Moreover, the contribution of policy makers and economists is crucial to address regulatory and economic aspects, making the technology accessible and marketable. As such, collaboration across disciplines and sectors is essential. The chapter ends on an optimistic note, emphasizing the exciting potential of this field and the importance of continued research, development, and innovation in paving the way for a sustainable energy future.

References

Bak, T., Nowotny, J., Rekas, M., & Sorrell, C. (2002). Photo-electrochemical hydrogen generation from water using solar energy. Materials-related aspects. *International Journal of Hydrogen Energy, 27*(10), 991–1022.

Chandra, S. (2012). Recent trends in high efficiency photo-electrochemical solar cell using dye-sensitised photo-electrodes and ionic liquid based redox electrolytes. *Proceedings of the National Academy of Sciences, India Section A: Physical Sciences, 82*, 5–19.

Li, C., Luo, Z., Wang, T., & Gong, J. (2018). Surface, bulk, and interface: Rational design of hematite architecture toward efficient photo-electrochemical water splitting. *Advanced Materials, 30*(30), 1707502.

Low, J., Cao, S., Yu, J., & Wageh, S. (2014). Two-dimensional layered composite photo-catalysts. *Chemical Communications, 50*(74), 10768–10777.

Ma, D., Schneider, J., Lee, W. I., & Pan, J. H. (2021). Controllable synthesis and self-template phase transition of hydrous TiO2 colloidal spheres for photo/electrochemical applications. *Advances in Colloid and Interface Science, 295*, 102493.

Modestino, M. A., & Haussener, S. (2015). An integrated device view on photo-electrochemical solar-hydrogen generation. *Annual Review of Chemical and Biomolecular Engineering, 6*, 13–34.

Tilley, S. D. (2019). Recent advances and emerging trends in photo-electrochemical solar energy conversion. *Advanced Energy Materials, 9*(2), 1802877.

Turner, D., Li, M., Grant, D., & Ola, O. (2023). Recent enterprises in high-rate mono-lithic photo-electrochemical energy harvest and storage devices. *Current Opinion in Electrochemistry*, 101243.

Xu, X. T., Pan, L., Zhang, X., Wang, L., & Zou, J. J. (2019). Rational design and construction of cocatalysts for semiconductor-based photo-electrochemical oxygen evolution: a comprehensive review. *Advanced Science*, *6*(2), 1801505.

Yao, B., Zhang, J., Fan, X., He, J., & Li, Y. (2019). Surface engineering of nanomaterials for photo-electrochemical water splitting. *Small (Weinheim an der Bergstrasse, Germany)*, *15*(1), 1803746.

Zhang, X., Liu, M., Kong, W., & Fan, H. (2018). Recent advances in solar cells and photo-electrochemical water splitting by scanning electrochemical microscopy. *Frontiers of Optoelectronics*, *11*, 333−347.

Zheng, L., Teng, F., Ye, X., Zheng, H., & Fang, X. (2020). Photo/electrochemical applications of metal sulfide/TiO2 heterostructures. *Advanced Energy Materials*, *10*(1), 1902355.

CHAPTER 5

Nanomaterials for fuel cells' electrodes

Hafsah Azfar Khan[1], Bana Al Kurdi[1], Hind Alqassem[1], Abdullah Ali[1], Amani Al-Othman[1] and Muhammad Tawalbeh[2,3]

[1]Department of Chemical and Biological Engineering, American University of Sharjah, Sharjah, United Arab Emirates
[2]Sustainable and Renewable Energy Engineering Department, University of Sharjah, Sharjah, United Arab Emirates
[3]Sustainable Energy & Power Systems Research Centre, RISE, University of Sharjah, Sharjah, United Arab Emirates

5.1 Introduction

The first fuel cell was developed in 1838 by Sir William Grove who ideated that electricity can be produced by a reverse electrolysis process (Sharaf & Orhan, 2014). Upon testing this conjecture, he was able to develop a device that uses fuels, such as hydrogen, and an oxidizer, such as oxygen, to produce electricity. This device was called a gas battery; however, it is now known as a fuel cell. Subsequently, the first ever fully developed fuel cell was created by Francis Thomas Bacon in 1959. This device was then adopted in the 1960s by NASA in which proton exchange membrane fuel cells and alkaline fuel cells were used as part of the space program. The main working principle of a fuel cell is electrochemical energy conversion in which chemical energy is transformed into electrical energy. This process is efficient, clean, and flexible, which makes it a strong competitor in the field of renewable energy. Fig. 5.1 provides an overview a basic fuel cell structure that comprises of the anode, which is negatively charged, and the cathode, which is positively charged, and these electrodes consist of a porous material onto which a catalyst is embedded. The material and catalyst used depend on fuel and oxidant used along with the temperature of the fuel cell. For example, high-temperature fuel cells typically use cheap catalysts, such as nickel, whereas low-temperature fuel cells use more expensive catalysts, such as platinum. Furthermore, a liquid or solid electrolyte is sandwiched between the two electrodes that is conductive with respect to a specific ion, such as the hydrogen proton or hydroxide ions (Bocarsly & Niangar, 2009; Lim et al., 2022). Additionally, the incorporation of a gas diffusion layer

Electrochemistry and Photo-Electrochemistry of Nanomaterials
DOI: https://doi.org/10.1016/B978-0-443-18600-4.00006-5
© 2025 Elsevier Inc. All rights reserved, including those for text and data mining, AI training, and similar technologies.

133

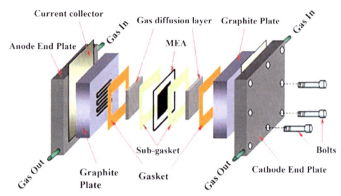

Figure 5.1 Schematic of a single fuel cell (Sharaf & Orhan, 2014).

ensures even dispersion of the fuel and oxidant on the electrode leading to a more efficient operation. This gas diffusion layer along with the electrodes and the membrane make up what is termed as the Membrane Electrode Assembly (MEA) (Mekhilef et al., 2012). Click or tap here to enter text. In turn, the MEA is sandwiched between bipolar plates, which conduct electrical current from one cell to another and prevents leakage of gases and electrolytes. Finally, an external wire is used that serves as a route for the electrons to travel to the anode from the cathode.

5.2 Types and applications of fuel cells

While all varieties of fuel cell have the same working principle of converting chemical energy to electrical energy, they vary in their operating temperature, efficiency, cost, and application (Guangul & Chala, 2020). Fuel cells can be broadly classified into six different types: alkaline fuel cells (AFCs), phosphoric acid fuel cells (PAFCs), solid oxide fuel cells (SOFCs), molten carbonate fuel cells (MCFCs), proton exchange membrane fuel cells (PEMFCs), and direct methanol fuel cells (DMFCs) (Tawalbeh, Alarab, et al., 2022). Wherein the membrane/electrolyte and the type of fuel dictate the fuel cell classification

5.2.1 Alkaline fuel cells

Two types of electrolytes can be used in an alkaline fuel cell: a liquid electrolyte that is 35—50 wt.% potassium hydroxide (KOH) and a more contemporary polymer electrolyte called an anion exchange membrane (Kalogirou, 2014; Wilberforce et al., 2016). The reactions taking place within the fuel cell, however, remain unchanged regardless of the electrolyte used. Oxygen is sent to the cathode where it is reduced to form hydroxide ions while hydrogen is sent to the anode reacts with the hydroxide ions react to produce water. Therefore, the operation of the alkaline fuel cell depends on the movement of the hydroxide ions from cathode to anode.

Although the achievable theoretical capability of this fuel cell is around 1.229 V, the produced voltage is rather lower due to activation losses and ohmic losses; therefore, the cell efficiency is 60%–70% (Burchardt et al., 2002). Furthermore, the classic and contemporary AFCs operate at atmospheric pressure and are low-temperature fuel cells as their operating temperature lies between 40°C–75°C and 50°C–90°C, respectively (Sharaf & Orhan, 2014; Sommer et al., 2016).

The catalyst mostly used for the cathode of AFCs is usually nickel at the anode and silver supported with carbon at the cathode (Fashedemi et al., 2022; Qussay et al., 2021/2021). The current cost of a typical alkaline fuel cell is estimated to be $175 per kW. Moreover, its lifetime is estimated to be larger than 5000 hours for a typical AFC and between 300 and 5000 hours for the contemporary AFC (Tomantschger et al., 1986).

5.2.2 Phosphoric acid fuel cells

Phosphoric acid is used as the electrolyte in this fuel cell. Since phosphoric acid is a liquid, it is stored in a ceramic called silicon carbide to prevent it from spilling (Kargupta et al., 2012). Ceramics are used for this purpose because phosphoric acid is highly corrosive; therefore, the material used to store the acid should be corrosion resistant. The operation of the PAFCs rely on the mobility of hydrogen protons; however, the ionic conductivity of phosphoric acid is quite low at low temperatures, which is why the fuel cell typically operates at high temperatures of 150°C–200°C. The cell operates as follows: hydrogen received at the anode is split into hydrogen protons and electrons, then the hydrogen protons travel to the cathode via the electrolytic medium and produce water as they react with oxygen (Mekhilef et al., 2012).

Considering that the PAFCs are high-temperature operating fuel cells, the catalysts typically used are platinum supported on carbon (Mohammed et al., 2019). Although PAFCs are the most commercially developed fuel cell to date, they still have many disadvantages such as its high cost; for instance, for stationary applications, the capital cost is around $1,500 per kW.

The PAFC also has a low efficiency of around 36%–45% (Abdi et al., 2017). In addition, since it operates at a high temperature it has a slow start-up time. It also has a low power density and a relatively large size. On a positive note, however, it is the most reliable fuel cell, produces high-grade heat, has a moderate lifetime of 5–20 years in stationary applications, and has a good tolerance to contaminants.

5.2.3 Solid oxide fuel cells

SOFCs are high-temperature fuel cells ($600°C-1200°C$) that can employ both hydrogen and carbon monoxide as fuel (Hussain & Yangping, 2020). First, oxygen is sent to the cathode where it is reduced to O^{2-} and this reaction remains the same regardless of the fuel used (Tawalbeh, Murtaza, et al., 2022). Next O^{2-} travels through the electrolyte and reacts with either hydrogen to produce water or reacts with carbon monoxide to produce carbon dioxide. Consequently, the operation of the SOFC depends on the movement of O^{2-} ions across the electrolyte. The electrolyte used in this fuel cell is solid yttria stabilized zirconia (YSZ) due to its high thermal and chemical stability, in addition to its enhanced oxygen conductivity (Son et al., 2020). Moreover, the catalyst used at the anode is nickel YSZ composite, whereas the catalyst used at the cathode is strontium-doped lanthanum manganite.

The disadvantages of this fuel cell are mainly due to its high operating temperature because it implies slow start up time and creates difficulty when selecting an appropriate material of construction. On the flip side, the advantage of an elevated operating temperature is that internal steam reforming is possible. Other disadvantages include low power density, high manufacturing cost, high operating cost of $1000 per kW for stationary applications and the underdevelopment of the SOFCs compared to other fuel cells (Acres, 2001; Mekhilef et al., 2012; Sharaf & Orhan, 2014). Whereas the advantages of employing SOFCs include high efficiency of around 60%–80%, fuel flexibility, noise-free operation, longer durability, expansive applications, inexpensive catalysts, since nobles metals are not required, and long life span of 40,000–80,000 hours (Gupta & Yadav, 2016; Malik et al., 2021; Mohammed et al., 2019; Qussay et al., 2021).

5.2.4 Molten carbonate fuel cells

Similar to SOFCs, MCFCs are high-temperature cells ($600°C-700°C$) that can use both hydrogen and carbon monoxide as fuel (Dincer & Bicer, 2018).

However, unlike SOFCs and the other fuel cells, MCFCs produce carbonate ions at the cathode triggered by the reaction between the injected O_2 and CO_2. The carbonate ions then move across the electrolyte and react with either hydrogen to produce water and carbon dioxide or with carbon monoxide to produce carbon dioxide. The most frequently used electrolyte in this cell is a liquid alkali carbonate suspended in a lithium aluminate ceramic to prevent any spillage. Furthermore, the most customarily used material at the anode is porous nickel hybridized with chromium or aluminum, whereas the material commonly used at the cathode is lithiated nickel oxide (Frangini et al., 2021).

The advantages of this cell include a high efficiency of around 55%—65% as it produces high-grade heat, consumes carbon dioxide, uses inexpensive catalysts, and has a moderate lifetime of about 5—20 years in stationary applications. Moreover, since it is operated at high temperatures, internal reforming is feasible and MCFCs also have fuel flexibility. Alternatively, some of the disadvantages include slow start-up (5—10 hours), low power density, higher likelihood of corrosion in the cell, high capital cost of around $1000 per kW in stationary applications, low operating life for highly intensive production systems and catalyst dissolution (Sharaf & Orhan, 2014; Uzunoglu & Alam, 2017).

5.2.5 Proton exchange membrane fuel cell

The operating temperature characterizes the PEMFCs into two types. The popularly used low-temperature PEMFCs are operated at a temperature range of 60°C—80°C and the electrolyte used is a sulfonated Teflon polymer called Nafion (Fan et al., 2021). Conversely, high-temperature PEMFCs, which operate at a temperature range of 150°C—200°C, are under development; therefore, the most appropriate electrolyte for them is still under research (Al-Othman, Nancarrow, et al., 2021; Ka'ki et al., 2021; Mohammed et al., 2021). However, polybenzimidazole (PBI) doped in phosphoric acid seems to be the most promising and is the only membrane that has passed the Department of Energy criteria for high-temperature fuel cells (Rosli et al., 2017). The electrolyte is switched from being water-based to being mineral acid-based (Lucia, 2014). Despite the operating temperature of the PEMFC, the fuel utilized remains the same.

In the low-temperature cell, hydrogen is injected at the anode where it splits to produce hydrogen protons and electrons. Next, the protons travel through the membrane and react with oxygen at the cathode to produce water. The commercial catalyst used in this cell is platinum supported on carbon black (Pt/C) for both the cathode and the anode (Show & Ueno, 2017). Furthermore, the operation of the low-temperature PEMFCs relies on the membrane water content as protons move from the anode to the cathode by the Grothuss mechanism from one water molecule to the next.

Alternatively, in the high-temperature fuel cell, hydrogen splits to form a hydrogen proton that reacts with $H_2PO_4^-$ at the anode to produce phosphoric acid, while $PBI^\cdot H^+$ separate to produce PBI and hydrogen proton at the cathode (Rosli et al., 2017). The catalysts frequently used in this cell are platinum-ruthenium supported on carbon for both the cathode and the anode (Sharaf & Orhan, 2014). The overall cell structure is not dissimilar to the low-temperature PEMFCs.

Some of the advantages of the low-temperature PEMFCs are fast start-up, high power density, moderate efficiency of around 40%−60%, and compact size. Moreover, the capital cost of the cell, in vehicle application, is approximately $100 per kW, which is usually considered cheap. The disadvantages of low-temperature PEMFCs include high manufacturing cost and significant reduction in proton conductivity (Tellez-Cruz et al., 2021). The advantages of the high-temperature PEMFC include high-grade heat production, faster kinetics, better heat and water management, better catalyst tolerance toward impurities, and higher efficiency (Nauman Javed, Al-Othman, Nancarrow, et al., 2022; Tawalbeh, Al-Othman, et al., 2022). However, the disadvantages include expensive catalysts and accelerated stack degradation. Fig. 5.2 shows a schematic diagram showing the components for PEMFCs (Nimir et al., 2023).

5.2.6 Direct methanol fuel cells

The DMFCs operate at temperatures ranging from 25°C to 110°C and can be considered as a type of PEMFCs in which methanol is used as the fuel instead of hydrogen. Methanol is injected at the anode where it reacts with water to produce carbon dioxide and hydrogen protons. The hydrogen proton then travels across the electrolyte and reacts with oxygen at the cathode to produce water. Additionally, this cell can be classified as active or passive in terms of operation. Active DMFCs consists of a methanol pump and an air-blowing device ensuring availability of methanol and air in the cell, whereas passive DMFCs do not consist of a methanol pump or an air-blowing device, which makes them less efficient (Mekhilef et al., 2012).

Nanomaterials for fuel cells' electrodes 139

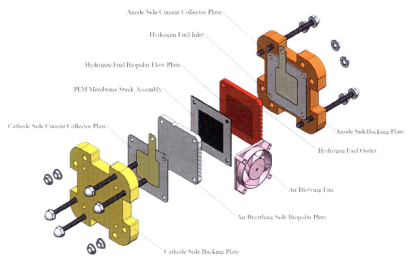

Figure 5.2 A schematic diagram for PEMFCs (Nimir et al., 2023). *PEMFCs*, proton exchange membrane fuel cells.

Other than the difference in the type of fuel used, DMFCs typically use platinum-ruthenium catalysts supported on carbon at both electrodes, because carbon dioxide produced in the cell exhibits a high affinity to a platinum only catalyst leaving a small area exposed to methanol, which reduces the efficiency of the cell (Sharaf & Orhan, 2014). However, carbon dioxide has higher affinity to ruthenium, as compared to platinum, leaving platinum exposed to the fuel and oxidant.

Merits of the DMFCs consist of easy fuel storage and transportation, simple system, high fuel energy density, and compact size. On the other hand, the disadvantages include reduced efficiency due to slow kinetics and fuel crossover and elevated capital costs due to the requirement of special fuel managing systems for the toxic and corrosive methanol (Mohammed et al., 2019; Tawalbeh et al., 2018). Fig. 5.2 summarizes features, reactants, products, electrolytes, and operating conditions of common types of fuel cells (Al-Othman et al., 2022).

5.3 Applications of fuel cells

Fuel cells can be used in various applications that can be divided into three main types: stationary, portable, and transportation. Portable applications can be categorized under two major markets. One under portable electric generators such as the outdoor lights used for camping and light mass-market applications that include surveillance and emergency relief efforts (Wilberforce et al., 2016) and the other under electronic devices like tablets, laptops, and phones. Merit such as large energy density, reusability, and compact size makes them excellent candidates for such applications. For example, PEMFCs and DMFCs are used in portable military applications courtesy their silent operation, increased energy density, and light weight. However, issues of durability and cost impede the expansive application of fuel cells in such applications.

Furthermore, there is a wide variety of stationary applications in which fuel cells can be used. For instance, the application of fuel cells can be exercised in emergency back-up power supply (EPS) courtesy their reusability, long operation time, high energy density, and compact size. PEMFCs and DMFCs are the most dominant and appropriate classes of fuel cell in the EPS domain. Additionally, energy from the fuel cells can be harnessed to power off grid areas; this application is popularly termed as remote area power supply (RAPS). Benefits such as high efficiency, low emissions, and eminent load-following allow the fuel cells to be used for residential electric power and Combined Heat and Power (CHP) distributed generation. For household CHP generations, PEMFCs and PAFCs are the most suitable varieties, whereas for a more voluminous residential block, high-temperature fuel cells are superior candidates (Sharaf & Orhan, 2014).

In the transportation sector, fuel cells have exhibited greater efficiency compared to that of conventional internal combustion engines (Chubbock & Clague, 2016). These cells provide advantages such as fuel flexibility, lower emissions, minimal maintenance, and extensibility. However, limitations such as durability, cost, and hydrogen storage impede switching to fuel cell—powered vehicles. Furthermore, PEMFCs and AFCs are excellent contenders for battery-powered vehicles (Acres, 2001).

5.4 State-of-the-art nanomaterials for fuel cell electrodes

Materials with at least one dimension measuring at a nanoscale, between 1 nm (diameter of a glucose molecule) and 100 nm (diameter of a virus), are defined as nanomaterials. They have been extensively analyzed for applications in the electrodes and membranes of the fuel cells given their exceptional catalytical, electrical, and mechanical features. This subchapter focuses on their applications as electrodes for varieties of fuel cells, specifically as catalysts and catalyst supports. To better understand the effects of nanomaterials on fuel cells, one must first understand the required specifications of fuel cell catalysts and catalyst supports.

Efficient fuel cell catalysts should have high electric conductivities, large specific surface areas, and high chemical and thermal stability (Abdalla et al., 2018). Among them, the property that considerably alters the catalyst performance is the electric conductivity (Logan, 2009). For instance, although platinum can provide a decent performance, its high cost is still an obstacle for the commercialization of fuel cells (Ghosh, 2017). Other disadvantages include the slow oxygen reduction reaction (ORR), at the cathode, and the degradation of the catalyst support. Hence, different strategies of enhancing fuel cell efficiencies are exercised to overcome these issues; one way is through the integration of nanomaterials. Saha et al. (2013) categorize the four ways in which nanomaterials may be integrated in fuel cells to advance the catalyst performance of the fuel cells. First is by simply optimizing the size and shape of catalyst to increase the total surface area per unit volume. Second is to generate composite materials, with doping, to allow higher catalytic performance. Third is to increase the interactions between the catalyst and catalyst support using suitable nanomaterials, therefore, benefitting from the synergistic effects such as increased conductivity. Finally, nanomaterials can be used to increase catalyst-support durability and stability. The relevance of nanomaterials will be unfolded for the following fuel cell types.

5.4.1 State-of-the-art nanomaterials in SOFCs

Nanomaterials have found immense applications in SOFCs, as electrolytes and as electrodes. This enables the fuel cell to operate at lower temperatures. Table 5.1 summarizes the selected nanomaterials that may be used in SOFCs (Abdalla et al., 2018). Fan et al. (2020) discovered that by using A-site-deficient SSFTR7020 perovskite oxide as electrode materials, for both the cathode and the anode, the polarization resistance of the cell decreases, while its stability is maintained. This is engendered by increased oxygen vacancies in the electrodes, allowing more active sites for catalyst operation. However, most applications of nanomaterials in SOFCs are in electrolytes, where researchers achieved reduced performance degradation, increased energy conversion efficiency, higher ionic conductivities, and higher power densities at lower operating temperatures (Wachsman et al., 2014).

Table 5.1 Summary of selected nanomaterials and their operating temperatures for SOFCs (Abdalla et al., 2018).

Anode	Cathode	Electrolyte	Substrate	Temperature (°C)
Ni	–	GDC	–	450–550
Ni	LSM-YSZ	ScSZ	–	700
	LSCF-GDC	GDC	–	650–850
Pt	LSCF	YSZ	Silicon wafer, Si_3N_4	450–500
Ni-SDC	SSC	ScSZ	–	600–700
Ru	Pt	CGO-YSZ	–	470–520
Pt	Pt	YSZ		350–500
Ni	Pt	YSZ	–	600
Pt	LSCF	YSZ	–	400–500
Pt-ZrO$_2$	LSM	YSZ	–	650–800
LSCF	LSCF	CGO	–	700
PSM	PSM	YSZ	–	500–800
SSC-NiO-YSZ	SSC-LSF-GDC	YSZ	–	700–800
LSM-YSZ	LSM-YSZ	YSZ	–	600–800
LSM-YSZ	LSM-YSZ	YSZ	–	650–850
NiO-YSZ	LSF-YSZ	YSZ	–	700–800
NiO-YSZ	LSM-GD	YSZ	–	750
LSM-YSZ	LSM-YSZ	YSZ	–	650–800

SOFCs, solid oxide fuel cells.

5.4.2 State-of-the-art nanomaterials in PEMFCs

The utilization of nanomaterials in PEMFCs resulted in higher power densities, larger electrochemical active surface areas, and lower H_2 crossovers. These effects were achieved when a single layer of graphene (SLG) was inserted between the electrodes and the electrolyte (Chen et al., 2022). One study investigated the effect of three types of N-doped carbon supports on the performance of Pt alloy nanoparticle catalyst of PEMFCs, Pt/V, Pt/N-V-400, and Pt/N-V-800. V represents Vulcan XC 72 R carbon, a trade name for black carbon powder, while 400 and 800 are the ammonolysis temperatures for synthesis of the support. The study found that Pt/V catalyst in liquid acidic media had the highest durability compared to Pt/C. However, the highest all-embracing performance was exhibited by Pt/N-V-800. The study concluded that carbon doping for catalyst support is only beneficial under operating voltages of around 0.6 to 0.9 V (Hornberger et al., 2022).

Another study by Sinniah et al. (2022) provides a table summarizing the durability and ORR activity of dissimilar classes of nanoparticulate Pt shored on carbon alternatives in an acidic medium, specifically carbides, nitrides, oxides, and Mxenes. The authors noted that highly pervious Pt/TiNiN and Pt/BN offered four times the mass activity of Pt/C, reaching 0.83 and 1.06 A/mg, respectively. Furthermore, both catalysts demonstrated a lower loss in electrochemical area over the lifetime of the cell. These properties are seen in most titanium-nitride-based materials given their high corrosion resistance and high conductivity. Furthermore, the authors noted that by depositing ultrafine Pt nanoparticles, 1.6−4 nm, on the analyzed materials, such as TiO_2, TiN, BN, and In, further enhanced the durability, because the loss in electrochemical surface area was lower. However, they also noted a power density loss similar to that of Pt/C, 79% as compared to 80%. Additionally, the poor hydration and heat management contributed to a greater decrease in cell performance. Finally, the authors suggests utilizing 1D-ordered nanostructures, such as orexfoliated 2D layer structures, holey structures, or TiN nanotube forests to increase the porosity and surface area of the non-carbon supports.

Another study aiming to reduce the amount of Pt needed in a PEMFC utilizes carbon-layer-protected fct-PtFe nanocatalysts shored on carbon black (PtFe@CS/C) (Kim et al., 2022). The catalyst indicated promising results, with a current density of 1.1 A/cm^2 at 0.6 V with considerably low Pt loading of 0.100 mg Pt/cm^2. At 0.8 A/cm^2, the catalyst showed a low activity loss of 1%, electrochemical surface area loss of 11%, and negligible loss in voltage after 30,000 accelerated stress testing (AST) cycles.

Oezaslan et al. (2013) investigated and discussed the synthesis and durability of different Pt-based core − shell nanoparticles for PEMFCs: Co-, Cu-, Ni-, and Au-based catalyst such as PtCu3@Pt, PtCo3@Pt, PtNi3@Pt, and Au@Pt3Fe. The loss in electrochemical surface area and activity for all these materials were summarized in a table, and it was observed that Au@FePt3 had the lowest values. Au@FePt3 of 10 nm particle size and operating at a voltage between 0.6 and 1.1 V exhibited electrochemical surface area loss of 6% and an activity loss of 7% after 60,000 testing cycles. The reduction in surface area is attributed to the increase of the Pt shell and the depletion of internal metals. However, compared to Pt/C fuel cells, the durability and catalytic activity of these fuel cells with different catalyst cores are higher.

5.4.3 State-of-the-art nanomaterials in DAFCs

A major problem with DAFCs is the accumulation of by-products on anode surface resulting in the poisoning of the anode catalyst. The use of nanomaterials is one way to unravel this issue. Fu et al. (2019) developed ultrathin Rh wavy nanowires as catalysts for DMFC anodes. The author obtained a high mass activity of 722 mA/mg and a high electrochemical surface area of 144.2 m²/g as compared to other researched Rh-based nanomaterials. In addition, Fu et al. developed ultrathin alloy Pt_3Ag wavy nanowire, which were used in direct ethanol fuel cell anodes as catalysts. This catalyst resulted in higher faradic efficiencies, higher activity, and lower poisoning from carbonaceous species. The surface area measured was 28.0 mA/cm² and an excellent mass activity of 6.1 A/mg was obtained.

5.5 Novel materials for nanoelectrodes in fuel cells

Nanomaterials are utilized in fuel cell electrodes to enhance the overall cell performance. A substantial improvement in performance is observed when nanosized catalysts are used within the electrode because when the surface area of the catalyst increases, the contact area between the catalyst and supplied fuel significantly boosts. Even though Pt-based electrodes have been the most commonly used in low-temperature fuel cells for decades, they are expensive, subjected to CO poisoning and fuel crossover, and are limited in supply due to their rare occurrence in nature. Thus, other materials, that can provide better or equivalent performance at a lower cost, need to be explored as a substitute for Pt-based catalysts. In particular, carbon nanotubes, nanostructured graphene and graphene oxide, metal−organic frameworks (MOFs), and metal-oxide nanosheets have garnered special interest for the development of nanoelectrodes with high efficiency, conductivity, and selectivity.

5.5.1 Carbon nanotubes

Carbon nanotubes (CNTs) are materials whose properties are dependent on their composition and geometry. These materials can be synthesized as single-walled CNTs (SWCNTs) or multiwalled CNTs (MWCNTs), and they exhibit a stark difference in properties. The SWCNTs are those which consist of layers of graphene that have been rolled up to create hollow tubes with walls that have a thickness of a singular atom, creating a tube that is considered to be one-dimensional and have a theoretical specific surface area of 1315 m^2/g (Birch et al., 2013; Raval et al., 2018). Whereas, MWCNTs comprise of two or more concentric cylindrical shells of graphene sheets that have been arranged along the same axis with a hollow and they have a theoretical specific area of 50−500 m^2/g, based on the number of tubes in the structure (Birch et al., 2013; Borghei et al., 2018). Fig. 5.3 structurally depicts the difference between SWCNT and MWCNT. Generally, both types of CNTs are doped with electron acceptors or electron donors to enhance conductivity, cyclability, and reversibility.

The MWCNTs and SWCNTs have been analyzed for their utilization as the carbon support in fuel cell electrodes due to their exceptional electrochemical durability and mechanical strength along with high surface area and electrochemical conductivity. However, the usage of MWCNTs is not preferred because achieving uniform deposition of the catalyst on the surface of the CNT support is considerably challenging. Nevertheless, deposition methods are focused on solving this issue as MWCNTs have a high surface area, they are chemically inert, and a strong bond is created

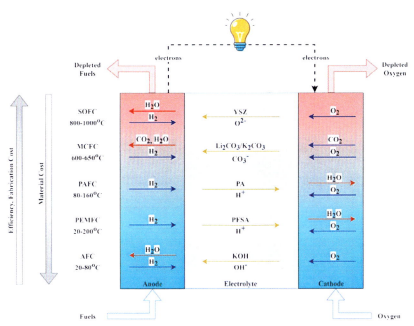

Figure 5.3 Features and operating conditions of common types of fuel cells (Al-Othman et al., 2022).

between the metal catalyst and the CNT during deposition (Haque et al., 2021). Since they have reduced defect density, high crystallinity, and an elevated aspect ratio, CNTs have also been proven to solve the issue of carbon oxidation in electrodes which reduces cell performance (Cha et al., 2018). Hence, the usage of CNT-supported electrodes also yields consistent cell potential, electrical resistance, and electrochemical surface area in the long term. In recent works, PBI/MWCNTs composites were developed and analyzed for their performance as catalyst support in high-temperature fuel cells (Haque et al., 2021). The composite material had a 23% higher specific surface area than pristine MWCNT and exhibited improved electrochemical properties in terms of high current density and current density retention over an extended period of operation. The composite also achieved a maximal power density of 112.1 mW/cm^2 ($\sim 10\%$ greater than that of pristine MWCNT) at an operating temperature of 150°C. Therefore, the performance of MWCNTs can be further improved by functionalizing them with polymers and they may be employed effectually in high-temperature fuel cells.

5.5.2 Graphene and graphene oxide

Graphene sheets have a crystalline structure, which comprise of honeycomb-shaped carbon sheets, resulting in a material that has desirable mechanical properties such as high flexibility and a lightweight structure (Dwivedi, 2022; Tawalbeh, Nauman Javed, et al., 2022). Moreover, they have excellent electrochemical properties including high specific surface area (2630 m^2/g), high electrical conductivity, high thermal conductivity, and chemical stability (Jamil et al., 2021). An added advantage of using a graphene-based electrode is that when graphene is utilized as the electrode support, the enhanced interfacial properties in the electrode-catalyst interface result in higher catalyst utilization and higher catalytic activity (Nauman Javed, Al-Othman, Tawalbeh, et al., 2022). Therefore, using a graphene-based electrode support would elevate the lifetime of the electrode in the fuel cell and reduce the overall cost. Additionally, the synthesis of metal composites comprising of graphene resulted in a better electrocatalytic activity and an enhancement in fuel oxidation and ORRs.

Despite the several benefits of using graphene nanosheets as the catalyst support in fuel cell electrodes, its surface does not allow water to permeate through since it is hydrophobic, which diminishes the adsorption of reactants and the electrolyte onto the electrode (Qiu et al., 2018). Hence, graphene oxide, which is a single-layered hydrophilic material that has been functionalized with oxygen-containing groups to alter different properties, has been explored as the catalyst support. Specifically, the electrical, mechanical, and electrochemical properties can be manipulated by functionalizing the graphene oxide with different oxygen-containing groups. However, graphene oxide is not thermally stable in the presence of reducing agents or oxygen, so it is unsuitable for high-temperature applications (Su & Hu, 2021).

In an attempt to control undue leaching of phosphoric acid and minimize fuel (hydrogen) crossover in high-temperature PEM fuel cells with a phosphoric acid-doped PBI membrane, the Pt catalyst was loaded onto single layer graphene that was of exceptional quality (Chen et al., 2022). The single layer graphene proved to significantly improve the endurance of the PEM fuel cell by effectually controlling phosphoric acid leaching in the cell. Moreover, fuel crossover was reduced by around 78% when the proposed Membrane Electrode Assembly (MEA) was employed. The unique structure of graphene has also proven to improve ion and electron transportation along with proton conduction within the operating mechanism of the fuel cell (Iqbal et al., 2020). Additionally, the high corrosion resistance along with the long-term durability also makes graphene promising candidate for future fuel cell design. Fig. 5.3 demonstrates the difference the difference in the chemical structures of graphene and graphene oxide.

Graphene oxide, on the other hand, is highly sought-after as the catalyst support for metal catalysts given their high surface area, hydrophilicity, and excellent electrical conductivity. The greatest challenge in relation to using graphene oxide as the catalyst support is the synthesis technique. In one approach, bimetallic Pt-Pd nanoflowers were embedded onto graphene oxide using an in situ overgrowth method by initiating a redox reaction between graphene oxide and the catalysts (Xu et al., 2020). It was noted that the proposed electrode maintained 70% electrocatalytic activity at temperatures as low as 5°C, which makes it highly suitable for application in countries with colder climate. Moreover, the material had higher durability and active surface area than commercial Pt/C electrodes. Nickel-based catalysts have also been an attractive alternative to Pt-based catalysts. Hence, pure nickel nanoparticles were embedded on reduced graphene oxide to study the electrochemical performance for methanol oxidation reaction (MOR) (Sun et al., 2018). To efficiently control the nickel nanoparticles load onto the reduced graphene oxide, laser ablation of nickel in graphene oxide solution was conducted. As a result, pure ultrafine nickel nanoparticles were successfully adhered onto reduced graphene oxide. The obtained electrode had an impressive current density of 1600 mA/mg and prevented catalyst poisoning, which is common in MORs. Additionally, fast reaction kinetics and high stability over 1000 cycles were observed, which were ascribed to the greater dispersion of nickel nanoparticles on the reduced graphene oxide surface. Another proposed method was to synthesize reduced graphene oxide hollow nanospheres that were in between layers of Pt nanoparticles on either side (Qiu et al., 2018). The proposed structure improved the electrocatalytic activity of Pt in ORR and MOR as it had higher electrochemically active specific area, improved long-term stability, and higher current density than commercial Pt/C electrodes. Hence, graphene oxide provides a highly suitable alternative to commercial Pt/C electrodes and provides better prospects for the use of other metallic catalysts.

5.5.3 Metal−organic frameworks

Metal−organic frameworks (MOFs) are highly pervious crystalline solids of coordination polymers that is composed of metal-containing nodes and organic binders. MOFs have created an avenue for improvements in fuel cell design on the account of their remarkably high surface area, tunable porosity, and nanostructure, greater thermal stability, uniformly structured cavities that range from the micro- to nanoscale, and ease of modification with a variety of functional groups to achieve desired properties (Al-Othman, Tawalbeh, et al., 2021; Kang Yoo & Kim, 2022). Fig. 5.2 demonstrates the formation of the MOF structure (Heo et al., 2020).

Recently, MOFs nanocomposites have been synthesized and tested to determine their effectiveness as catalyst supports in fuel cells. A novel nanocomposite was synthesized from titanium metal−organic frameworks and ultrathin lamellar nitrogen-doped graphene to study the impact on ORR kinetics and the catalytic reaction (Qin et al., 2019). The novel nanocomposite demonstrated significant improvements in ORR activity, higher durability, and low electrochemical impedance in both acidic and alkaline mediums, as compared to Pt/C electrodes and other MOF-derived materials such as doped zeolitic imidazolate frameworks (ZIL) and UiO-MOFs. In another attempt, nanocomposites (Pt-rGO/Fe-MOF) comprising reduced graphene oxide (rGO) and an N-doped Fe-based MOF were fabricated to analyze ORR activity and overall electrochemical performance

(Kang Yoo & Kim, 2022). The Fe-MOF facilitated the transportation of fuel and ORR products, while the rGO provided excellent electrical conductivity. The Fe-MOF prevented the stacking that occurs in rGO due to II-II interactions and the Pt-rGO/Fe-MOF nanocomposite had an electrochemical surface area of 261.83 m^2/g, which was significantly greater than Pt/C, Pt-rGO, and Pt-Fe-MOF. The novel material also demonstrated high current density, durability, and chemical stability, which were attributed to the synergistic impact of using a combination of rGO and Fe-MOF. Overall, MOFs provide prospects to reduce the issue of slugging ORR in Pt/C electrodes and increase the durability and stability of fuel cells.

5.5.4 Metal oxide nanosheets

Though platinum catalysts have proven to highly effective in fuel cells, metal oxide nanosheets have been explored to provide a cost-effective alternative with enhanced chemical stability and catalytic efficiency for the Hydrogen Evolution Reaction (HER), Oxygen Evolution Reaction (OER), MOR, and ORR. Metal oxide nanosheets exhibit a variety of exceptional properties such as flexibility, semiconductivity, photosensitivity, redox properties, atomic-level thickness, abundant active sites, noteworthy mechanical strength, along with high specific surface area and chemical stability (ten Elshof et al., 2016). As a result, metal oxide nanosheets have demonstrated promising results for improving the performance of fuel cells. Fig. 5.3 depicts the structures of metal oxide nanosheets (Figs. 5.4—5.7).

In recent works, ultrathin nickel oxide nanosheet containing abundant oxygen vacancies were embedded in the cathode to improve electrocatalytic efficiency of the MOR in DMFCs (Yang et al., 2019). The oxygen vacancies enabled a higher diffusion capacity, which resulted in the deposition of NiOOH on the catalyst surface, providing a higher number of active sites for MOR. Furthermore, the NiO nanosheets exhibited excellent electrocatalytic performance in terms of current density and MOR kinetics, as compared to bulk NiO. In another study, disordered cerium-modified cobalt oxide nanosheets ($Co_{3-x}Ce_xO_4$) were synthesized to analyze their performance as an electrocatalyst for OERs (Li et al., 2020). OER activity was increased by the rapid generation of intermediates, which was catalyzed by the boosted electronic transport between Co and Ce by utilizing the proposed $Co_{3-x}Ce_xO_4$ electrocatalyst. Additional active sites were available on the surface of the catalyst due to the defects

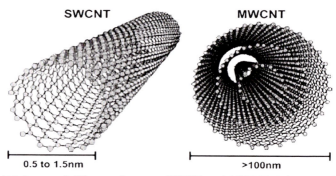

Figure 5.4 Structural difference between SWCNTs and MWCNTs (Ribeiro et al., 2017). *MWCNTs*, multiwalled carbon nanotubes; *SWCNTs*, single-walled carbon nanotubes.

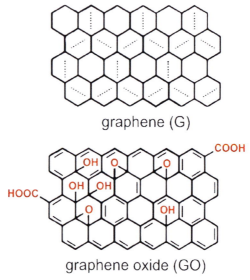

Figure 5.5 Chemical structures of graphene (G) and graphene oxide (GO) (Tadyszak et al., 2018).

in the structure, and it also had a higher faradaic efficiency and active surface area than that of Co_3O_4 nanosheets. Naik et al. (2020) synthesized hybrid anatase $Pt-TiO_2$ nanosheets using TiO_2 nanosheets as the catalyst support, along with $Pt-TiO_{2-x}$ (oxygen-deficient sample) nanosheets to compare their performance in ORRs. It was observed that introducingdefects such as oxygen vacancies helped increase the transfer of electrons between the titanium dioxide nanosheets and the platinum nanoparticles,

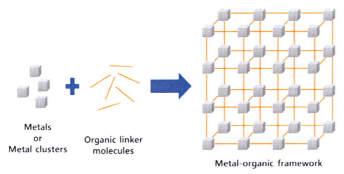

Figure 5.6 Formation of the MOF structure (Heo et al., 2020). *MOF*, metal−organic framework.

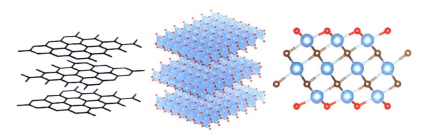

Figure 5.7 Structure of metal oxide nanosheets (Timmerman et al., 2020).

which enhanced the ORR activity. The increase in charge transfer was accredited to the Strong-Metal-Support Interaction (SMSI) with platinum that resulted in an alteration in the titanium dioxide's electronic structure, which helped increase the catalyst activity and stability. Furthermore, the Pt-TiO$_{2-x}$ nanosheets had a high power density of 958 mW/cm^2, as compared to the commercial Pt/C. Both Pt-TiO$_2$ nanosheets and Pt-TiO$_{2-x}$ nanosheets exhibited a similar surface area of ∼110 m^2/g, which is approximately 40% greater than the commercial Pt/C. Therefore, metal oxide nanosheets can effectively be applied as electrocatalysts in fuel cells to improve catalytic efficiency and chemical stability along with providing a cost-effective substitute to Pt-based catalysts.

5.6 Conclusions

The chapter outlines the applications of nanomaterials for electrodes in fuel cells. The six main types of fuel cells along with their differences in electrolytes, electrodes, fuels, and operating conditions, were explained. The advantages and disadvantages of each fuel cell were also discussed. PEMFCs and DMFCs seem to be on the rise commercially; however, challenges associated with their electrode materials still prevail. Nanomaterials are one way in which electrodes for various types of fuel cells may be enhanced. Incorporating nanomaterials may be done in four ways: decreasing the size of the electrodes, making composites of the electrode materials, simply adding nanomaterials, or using them as catalyst supports. These methods of using nanomaterials enhance the efficiency and activity of the fuel cell by increasing catalytic activity, electrochemical surface area, power density, and improving stability and durability. Additionally, nanomaterials may reduce the effects of fuel crossover, reduce catalyst poisoning, and decrease cell polarization. Furthermore, a variety of state-of-the-art nanomaterials and novel nanomaterials that have been experimented with as fuel cell electrodes were also discussed. State-of-the-art materials discussed include A-site-deficient SSFTR7020 perovskite oxide for SOFCs, Pt/V for PEMFCs, and ultrathin Rh wavy nanowires for DMFCs. The application of novel materials such as CNTs, graphene and graphene oxide, metal–organic frameworks, and metal oxide nanosheets was discussed along with their unique properties. Overall, understanding the mechanisms by which the nanomaterials enhance the MEA is key to choosing the optimal nanomaterial for a specific fuel cell design. Ultimately, nanomaterials provide a promising avenue for improving fuel cell efficiency and bring them closer to extensive commercialization.

References

Abdalla, A. M., Hossain, S., Azad, A. T., Petra, P. M. I., Begum, F., Eriksson, S. G., & Azad, A. K. (2018). Nanomaterials for solid oxide fuel cells: A review. *Renewable and Sustainable Energy Reviews, 82*, 353–368. Available from https://doi.org/10.1016/j.rser.2017.090.046, https://www.journals.elsevier.com/renewable-and-sustainable-energy-reviews.

Abdi, H., Rasouli Nezhad, R., & Salehimaleh, M. (2017). Fuel cells distributed generation systems. In G. B. Gharehpetian, & S. M. M. Agah (Eds.), *Design, operation and grid integration* (pp. 221–300). Iran: Elsevier. Available from https://www.sciencedirect.com/book/9780128042083.

Acres, G. J. K. (2001). Recent advances in fuel cell technology and its applications. *Journal of Power Sources, 100*(1-2), 60–66. Available from https://doi.org/10.1016/S0378-7753(01)00883-7.

Al-Othman, A., Tawalbeh, M., Martis, R., Dhou, S., Orhan, M., Qasim, M., & Olabi, A. G. (2022). Artificial intelligence and numerical models in hybrid renewable energy systems with fuel cells: Advances and prospects. *Energy Conversion and Management, 253*. Available from https://doi.org/10.1016/j.enconman.2021.115154.

Al-Othman, M., Tawalbeh, O., Temsah, M., & Al-Murisi. (2021). Industrial challenges of MOFs in energy applications. In A.-G. Olabi (Ed.), *Encyclopedia of smart materials* (pp. 535−543). United Arab Emirates: Elsevier. Available from https://www.sciencedirect.com/book/9780128157336, https://doi.org/10.1016/B978-0-12-815732-9.00030-9.

Al-Othman, P., Nancarrow, M., Tawalbeh, A., Ka'ki, K., El-Ahwal, B., El Taher, M., & Alkasrawi. (2021). Novel composite membrane based on zirconium phosphate-ionic liquids for high temperature PEM fuel cells. *International Journal of Hydrogen Energy, 46* (8), 6100−6109. Available from https://doi.org/10.1016/j.ijhydene.2020.020.112, http://www.journals.elsevier.com/international-journal-of-hydrogen-energy/.

Birch, M. E., Ruda-Eberenz, T. A., Chai, M., Andrews, R., & Hatfield, R. L. (2013). Properties that influence the specific surface areas of carbon nanotubes and nanofibers. *Annals of Occupational Hygiene, 57*(9), 1148−1166. Available from https://doi.org/10.1093/annhyg/met042.

Bocarsly, A. B., & Niangar, E. V. (2009). Fuel cells— Proton-exchange membrane fuel cells membranes: Elevated temperature. In J. Garche (Ed.), *Encyclopedia of electrochemical power sources* (pp. 724−733). United States: Elsevier. Available from http://www.sciencedirect.com/science/book/9780444527455, https://doi.org/10.1016/B978-044452745-5.00232-X.

Borghei, M., Lehtonen, J., Liu, L., & Rojas, O. J. (2018). Advanced biomass-derived electrocatalysts for the oxygen reduction reaction. *Advanced Materials, 30*(24). Available from https://doi.org/10.1002/adma.201703691, https://onlinelibrary.wiley.com/toc/15214095/20/9.

Burchardt, T., Gouérec, P., Sanchez-Cortezon, E., Karichev, Z., & Miners, J. H. (2002). Alkaline fuel cells: Contemporary advancement and limitations. *Fuel, 81*, 2151−2155.

Cha, B. C., Jun, S., Jeong, B., Ezazi, M., Kwon, G., Kim, D., & Lee, D. H. (2018). Carbon nanotubes as durable catalyst supports for oxygen reduction electrode of proton exchange membrane fuel cells. *Journal of Power Sources, 401*, 296−302. Available from https://doi.org/10.1016/j.jpowsour.2018.090.001, https://www.journals.elsevier.com/journal-of-power-sources.

Chen, J., Bailey, J. J., Britnell, L., Perez-Page, M., Sahoo, M., Zhang, Z., Strudwick, A., Hack, J., Guo, Z., Ji, Z., Martin, P., Brett, D. J. L., Shearing, P. R., & Holmes, S. M. (2022). The performance and durability of high-temperature proton exchange membrane fuel cells enhanced by single-layer graphene. *Nano Energy, 93*. Available from https://doi.org/10.1016/j.nanoen.2021.106829.

Chubbock, S., & Clague, R. (2016). Comparative analysis of internal combustion engine and fuel cell range extender. *Journal of Alternative Powertrains, 5*(1), 175−182. Available from https://doi.org/10.4271/2016-01-1188, http://saealtpow.saejournals.org/.

Dincer, I., & Bicer, Y. (2018). Electrochemical energy conversion. In I. Dincer (Ed.), *Comprehensive energy systems* (4-5). Canada: Elsevier Inc. Available from http://www.sciencedirect.com/science/book/9780128149256, https://doi.org/10.1016/B978-0-12-809597-3.00439-9.

Dwivedi, S. (2022). Graphene based electrodes for hydrogen fuel cells: A comprehensive review. *International Journal of Hydrogen Energy, 47*(99), 41848−41877. Available from https://doi.org/10.1016/j.ijhydene.2022.020.051, http://www.journals.elsevier.com/international-journal-of-hydrogen-energy/.

Fan, L., Tu, Z., & Chan, S. H. (2021). Recent development of hydrogen and fuel cell technologies: A review. *Energy Reports, 7*, 8421−8446. Available from https://doi.org/10.1016/j.egyr.2021.080.003, http://www.journals.elsevier.com/energy-reports/.

Fan, W., Sun, Z., Bai, Y., Wu, K., Zhou, J., & Cheng, Y. (2020). In situ growth of nanoparticles in A-site deficient ferrite perovskite as an advanced electrode for symmetrical solid oxide fuel cells. *Journal of Power Sources, 456*. Available from https://doi.org/10.1016/j.jpowsour.2020.228000.

Fashedemi, O. O., Bello, A., Adebusuyi, T., & Bindir, S. (2022). Recent trends in carbon support for improved performance of alkaline fuel cells. *Current Opinion in Electrochemistry*, *36*. Available from https://doi.org/10.1016/j.coelec.2022.101132, http://www.journals.elsevier.com/current-opinion-in-electrochemistry.

Frangini, S., Pietra, M. D., Seta, L. D., Paoletti, C., & Pérez-Trujillo, J. P. (2021). Degradation of MCFC materials in a 81 cm2 single cell operated under alternated fuel cell/electrolysis mode. *Frontiers in Energy Research*, *9*. Available from https://doi.org/10.3389/fenrg.2021.653531.

Fu, X., Zhao, Z., Wan, C., Wang, Y., Fan, Z., Song, F., Cao, B., Li, M., Xue, W., Huang, Y., & Duan, X. (2019). Ultrathin wavy Rh nanowires as highly effective electrocatalysts for methanol oxidation reaction with ultrahigh ECSA. *Nano Research*, *12* (1), 211−215. Available from https://doi.org/10.1007/s12274-018-2204-8, http://www.springer.com/materials/nanotechnology/journal/12274.

Ghosh, P. C. (2017). High platinum cost: Obstacle or blessing for commercialization of low-temperature fuel cell technologies. *Clean Technologies and Environmental Policy*, *9* (2), 595−601. Available from https://doi.org/10.1007/s10098-016-1254-4, https://link.springer.com/journal/10098.

Guangul, F. M., & Chala, G. T. (2020). A comparative study between the seven types of fuel cells. *Oman Applied Science and Engineering Progress*, *13*(3), 185−194. Available from https://doi.org/10.14416/J.ASEP.2020.040.007, http://ojs.kmutnb.ac.th/index.php/ijst/article/view/2680.

Gupta, N., & Yadav, G. D. (2016). Solid oxide fuel cell: a review. *International Research Journal of Engineering and Technology*, *3*, 1006−1011.

Haque, M. A., Sulong, A. B., Shyuan, L. K., Majlan, E. H., Husaini, T., & Rosli, R. E. (2021). Synthesis of polymer/MWCNT nanocomposite catalyst supporting materials for high-temperature PEM fuel cells. *International Journal of Hydrogen Energy*, *46*(5), 4339−4353. Available from https://doi.org/10.1016/j.ijhydene.2020.100.200, http://www.journals.elsevier.com/international-journal-of-hydrogen-energy/.

Heo, D. Y., Do, H. H., Ahn, S. H., & Kim, S. Y. (2020). Metal-organic framework materials for perovskite solar cells. *Polymers*, *12*(9). Available from https://doi.org/10.3390/POLYM12092061, https://res.mdpi.com/d_attachment/polymers/polymers-12-02061/article_deploy/polymers-12-02061.pdf.

Hornberger, E., Merzdorf, T., Schmies, H., Hübner, J., Klingenhof, M., Gernert, U., Kroschel, M., Anke, B., Lerch, M., Schmidt, J., Thomas, A., Chattot, R., Martens, I., Drnec, J., Strasser, P., & Hornberger, E. (2022). Impact of carbon N-doping and pyridinic-N content on the fuel cell performance and durability of carbon-supported Pt nanoparticle catalysts. *ACS Applied Materials and Interfaces*, *14*(16), 18420−18430. Available from https://doi.org/10.1021/acsami.2c00762, http://pubs.acs.org/journal/aamick.

Hussain, S., & Yangping, L. (2020). Review of solid oxide fuel cell materials: Cathode, anode, and electrolyte. *Energy Transitions*, *4*(2), 113−126. Available from https://doi.org/10.1007/s41825-020-00029-8.

Iqbal, M. Z., Siddique, S., Khan, A., Haider, S. S., & Khalid, M. (2020). Recent developments in graphene based novel structures for efficient and durable fuel cells. *Materials Research Bulletin*, *122*. Available from https://doi.org/10.1016/j.materresbull.2019.110674, http://www.sciencedirect.com/science/journal/00255408.

Jamil, M. F., Biçer, E., Kaplan, B. Y., & Gürsel, S. A. (2021). One-step fabrication of new generation graphene-based electrodes for polymer electrolyte membrane fuel cells by a novel electrophoretic deposition. *International Journal of Hydrogen Energy*, *46*(7), 5653−5663. Available from https://doi.org/10.1016/j.ijhydene.2020.110.039, http://www.journals.elsevier.com/international-journal-of-hydrogen-energy/.

Industrial process heat, chemistry applications, and solar dryers. In S. A. Kalogirou (Ed.), *Solar energy engineering: Processes and systems*. Elsevier BV. Available from 10.1016/b978-0-12-397270-5.00007-8.

Kang Yoo, P., & Kim, S. (2022). Preparation and electrochemical activity of platinum catalyst-supported graphene and Fe-based metal-organic framework composite electrodes for fuel cells. *Journal of Industrial and Engineering Chemistry*, 105, 259−267. Available from https://doi.org/10.1016/j.jiec.2021.090.027, http://www.sciencedirect.com/science/journal/1226086X.

Kargupta, K., Saha, S., Banerjee, D., Seal, M., & Ganguly, S. (2012). Performance enhancement of phosphoric acid fuel cell by using phosphosilicate gel based electrolyte. *Journal of Fuel Chemistry and Technology*, 40(6), 707−713. Available from https://doi.org/10.1016/s1872-5813(12)60026-7.

Ka'ki, A., Alraeesi, A., Al-Othman., & Tawalbeh, M. (2021). Proton conduction of novel calcium phosphate nanocomposite membranes for high temperature PEM fuel cells applications. *International Journal of Hydrogen Energy*, 46(59), 30641−30657. Available from https://doi.org/10.1016/j.ijhydene.2021.010.013, http://www.journals.elsevier.com/international-journal-of-hydrogen-energy/.

Kim, Y., Bae, H. E., Lee, D., Kim, J., Lee, E., Oh, S., Jang, J.-H., Cho, Y.-H., Karuppannan, M., Sung, Y.-E., Lim, T., & Kwon, O. J. (2022). High-performance long-term driving proton exchange membrane fuel cell implemented with chemically ordered Pt-based alloy catalyst at ultra-low Pt loading. *Journal of Power Sources*, 533. Available from https://doi.org/10.1016/j.jpowsour.2022.231378.

Li, L., Hu, Z., Tao, L., Xu, J., & Yu, J. C. (2020). Efficient electronic transport in partially disordered Co3O4 nanosheets for electrocatalytic oxygen evolution reaction. *ACS Applied Energy Materials*, 3(3), 3071−3081. Available from https://doi.org/10.1021/acsaem.0c00190, pubs.acs.org/journal/aaemcq.

Lim, J. H., Hou, J., Chun, J., Lee, R. D., Yun, J., Jung, J., & Lee, C. H. (2022). Importance of hydroxide ion conductivity measurement for alkaline water electrolysis membranes. *Membranes*, 12(6). Available from https://doi.org/10.3390/membranes12060556, https://www.mdpi.com/2077-0375/12/6/556/pdf?version = 1653568687.

Logan, B. E. (2009). Exoelectrogenic bacteria that power microbial fuel cells. *Nature Reviews. Microbiology*, 7(5), 375−381. Available from https://doi.org/10.1038/nrmicro2113.

Lucia, U. (2014). Overview on fuel cells. *Renewable and Sustainable Energy Reviews*, 30, 164−169. Available from https://doi.org/10.1016/j.rser.2013.090.025.

Malik, V., Srivastava, S., Bhatnagar, M.K., & Vishnoi, M. (2021). Comparative study and analysis between solid oxide fuel cells (SOFC) and proton exchange membrane (PEM) fuel cell—A review. In *Materials Today: Proceedings* (pp. 2270−2275). India: Elsevier Ltd. https://www.sciencedirect.com/journal/materials-today-proceedings.

Mekhilef, S., Saidur, R., & Safari, A. (2012). Comparative study of different fuel cell technologies. *Renewable and Sustainable Energy Reviews*, 16(1), 981−989. Available from https://doi.org/10.1016/j.rser.2011.090.020.

Mohammed, H., Al-Othman, A., Nancarrow, P., Tawalbeh, M., & El Haj Assad, M. (2019). Direct hydrocarbon fuel cells: A promising technology for improving energy efficiency. *Energy*, 172, 207−219. Available from https://doi.org/10.1016/j.energy.2019.010.105, http://www.elsevier.com/inca/publications/store/4/8/3/.

Mohammed, H., Al-Othman, A., Nancarrow, P., Elsayed, Y., & Tawalbeh, M. (2021). Enhanced proton conduction in zirconium phosphate/ionic liquids materials for high-temperature fuel cells. *International Journal of Hydrogen Energy*, 46(6), 4857−4869. Available from https://doi.org/10.1016/j.ijhydene.2019.090.118, http://www.journals.elsevier.com/international-journal-of-hydrogen-energy/.

Naik, K. M., Higuchi, E., & Inoue, H. (2020). Two-dimensional oxygen-deficient TiO2 nanosheets-supported Pt nanoparticles as durable catalyst for oxygen reduction reaction in proton exchange membrane fuel cells. *Journal of Power Sources*, *455*. Available from https://doi.org/10.1016/j.jpowsour.2020.227972, https://www.journals.elsevier.com/journal-of-power-sources.

Nauman Javed, R. M., Al-Othman, A., Tawalbeh, M., & Olabi, A. G. (2022). Recent developments in graphene and graphene oxide materials for polymer electrolyte membrane fuel cells applications. *Renewable and Sustainable Energy Reviews*, *168*. Available from https://doi.org/10.1016/j.rser.2022.112836, https://www.journals.elsevier.com/renewable-and-sustainable-energy-reviews.

Nauman Javed, R. M., Al-Othman, A., Nancarrow, P., & Tawalbeh, M. (2022). Zirconium silicate-ionic liquid membranes for high-temperature hydrogen PEM fuel cells. *International Journal of Hydrogen Energy*. Available from https://doi.org/10.1016/j.ijhydene.2022.050.009, http://www.journals.elsevier.com/international-journal-of-hydrogen-energy/.

Nimir, W., Al-Othman, A., Tawalbeh, M., Al Makky, A., Ali, A., Karimi-Maleh, H., Karimi, F., & Karaman, C. (2023). Approaches towards the development of heteropolyacid-based high temperature membranes for PEM fuel cells. *International Journal of Hydrogen Energy*, *48* (17), 6638−6656. Available from https://doi.org/10.1016/j.ijhydene.2021.110.174, http://www.journals.elsevier.com/international-journal-of-hydrogen-energy/.

Oezaslan, M., Hasché, F., & Strasser, P. (2013). Pt-based core-shell catalyst architectures for oxygen fuel cell electrodes. *Journal of Physical Chemistry Letters*, *4*(19), 3273−3291. Available from https://doi.org/10.1021/jz4014135.

Qin, X., Huang, Y., Wang, K., Xu, T., Wang, Y., Liu, P., Kang, Y., & Zhang, Y. (2019). Novel hierarchically porous Ti-MOFs/nitrogen-doped graphene nanocomposite served as high efficient oxygen reduction reaction catalyst for fuel cells application. *Electrochimica Acta*, *297*, 805−813. Available from https://doi.org/10.1016/j.electacta.2018.12.045, http://www.journals.elsevier.com/electrochimica-acta/.

Qiu, X., Yan, X., Cen, K., Sun, D., Xu, L., & Tang, Y. (2018). Achieving highly electrocatalytic performance by constructing holey reduced graphene oxide hollow nanospheres sandwiched by interior and exterior platinum nanoparticles. *ACS Applied Energy Materials*, *1*(5), 2341−2349. Available from https://doi.org/10.1021/acsaem.8b00452, pubs.acs.org/journal/aaemcq.

Qussay, R., Mahmood, M., Al-Zaidi, M. K., Al-Khafaji, R. Q., Al-Zubaidy, D. K., & Salman, M. M. (2021). A review: Fuel cells types and their applications. *International Journal of Applied Science and Engineering*, 2395−3470.

Raval, J. P., Joshi, P., & Chejara, D. R. (2018). Carbon nanotube for targeted drug delivery. In Inamuddin, A. M. Asiri, & A. Mohammadd (Eds.), *Applications of nanocomposite materials in drug delivery* (pp. 203−216). India: Elsevier. Available from https://www.sciencedirect.com/book/9780128137413, https://doi.org/10.1016/B978-0-12-813741-3.00009-1.

Ribeiro, E. C., Botelho, M. L., Costa, C. F., & Bandeira. (2017). Carbon nanotube buckypaper reinforced polymer composites: A review. *Polimeros*, *27*(3), 247−255. Available from https://doi.org/10.1590/0104-1428.03916, http://www.scielo.br/pdf/po/v27n3/0104-1428-po-0104-142803916.pdf.

Rosli, R. E., Sulong, A. B., Daud, W. R. W., Zulkifley, M. A., Husaini, T., Rosli, M. I., Majlan, E. H., & Haque, M. A. (2017). A review of high-temperature proton exchange membrane fuel cell (HT-PEMFC) system. *International Journal of Hydrogen Energy*, *42*(14), 9293−9314. Available from https://doi.org/10.1016/j.ijhydene.2016.060.211.

Saha, M. S., Neburchilov, V., Ghosh, D., & Zhang, J. (2013). Nanomaterials-supported Pt catalysts for proton exchange membrane fuel cells. *Wiley Interdisciplinary Reviews: Energy and Environment*, *2*(1), 31−51. Available from https://doi.org/10.1002/wene.47.

Sharaf, O. Z., & Orhan, M. F. (2014). An overview of fuel cell technology: Fundamentals and applications. *Renewable and Sustainable Energy Reviews*, *32*, 810−853. Available from https://doi.org/10.1016/j.rser.2014.010.012.

Show, Y., & Ueno, Y. (2017). Formation of platinum catalyst on carbon black using an in-liquid plasma method for fuel cells. *Nanomaterials*, 7(2). Available from https://doi.org/10.3390/nano7020031, http://www.mdpi.com/2079-4991/7/2/31/pdf.

Sinniah, J. D., Wong, W. Y., Loh, K. S., Yunus, R. M., & Timmiati, S. N. (2022). Perspectives on carbon-alternative materials as Pt catalyst supports for a durable oxygen reduction reaction in proton exchange membrane fuel cells. *Journal of Power Sources*, 534. Available from https://doi.org/10.1016/j.jpowsour.2022.231422, https://www.journals.elsevier.com/journal-of-power-sources.

Sommer, E. M., Vargas, J. V. C., Martins, L. S., & Ordonez, J. C. (2016). The maximization of an alkaline membrane fuel cell (AMFC) net power output. *International Journal of Energy Research*, 40(7), 924−939. Available from https://doi.org/10.1002/er.3483.

Son, M. J., Kim, M. W., Virkar, A. V., & Lim, H. T. (2020). Locally developed electronic conduction in a yttria stabilized zirconia (YSZ) electrolyte for durable solid oxide fuel cells. *Electrochimica Acta*, 353. Available from https://doi.org/10.1016/j.electacta.2020.136450, http://www.journals.elsevier.com/electrochimica-acta/.

Su, H., & Hu, Y. H. (2021). Recent advances in graphene-based materials for fuel cell applications. *Energy Science and Engineering*, 9(7), 958−983. Available from https://doi.org/10.1002/ese30.833, http://onlinelibrary.wiley.com/journal/10.1002/(ISSN)2050-0505.

Sun, H., Ye, Y., Liu, J., Tian, Z., Cai, Y., Li, P., & Liang, C. (2018). Pure Ni nanocrystallines anchored on rGO present ultrahigh electrocatalytic activity and stability in methanol oxidation. *Chemical Communications*, 54(13), 1563−1566. Available from https://doi.org/10.1039/c7cc09361f, http://pubs.rsc.org/en/journals/journal/cc.

Tadyszak, K., Wychowaniec, J. K., & Litowczenko, J. (2018). Biomedical applications of graphene-based structures. *Nanomaterials*, 8(11). Available from https://doi.org/10.3390/nano8110944, https://www.mdpi.com/2079-4991/8/11/944/pdf.

Tawalbeh, M., Al-Othman, A., & El Haj Assad, M. (2018). Graphene oxide—Nafion composite membrane for effective methanol crossover reduction in passive direct methanol fuel cells. In *5th International Conference on Renewable Energy: Generation and Application, ICREGA 2018* (pp. 192−196). United Arab Emirates: Institute of Electrical and Electronics Engineers Inc.

Tawalbeh, M., Murtaza, S. Z. M., Al-Othman, A., Alami, A. H., Singh, K., & Olabi, A. G. (2022). Ammonia: A versatile candidate for the use in energy storage systems. *Renewable Energy*, 194, 955−977. Available from https://doi.org/10.1016/j.renene.2022.060.015, http://www.journals.elsevier.com/renewable-and-sustainable-energy-reviews/.

Tawalbeh, M., Al-Othman, A., Ka'ki, A., Farooq, A., & Alkasrawi, M. (2022). Lignin/zirconium phosphate/ionic liquids-based proton conducting membranes for high-temperature PEM fuel cells applications. *Energy*, 260. Available from https://doi.org/10.1016/j.energy.2022.125237.

Tawalbeh, M., Nauman Javed, R. M., Al-Othman, A., & Almomani, F. (2022). The novel advancements of nanomaterials in biofuel cells with a focus on electrodes' applications. *Fuel*, 322. Available from https://doi.org/10.1016/j.fuel.2022.124237.

Tawalbeh, M., Alarab, S., Al-Othman, A., & Nauman Javed, R. M. (2022). The operating parameters, structural composition, and fuel sustainability aspects of PEM fuel cells: A mini review. *Fuels*, 3(3), 449−474. Available from https://doi.org/10.3390/fuels3030028.

Tellez-Cruz, M. M., Escorihuela, J., Solorza-Feria, O., & Compañ, V. (2021). Proton exchange membrane fuel cells (PEMFCS): Advances and challenges. *Polymers*, 13(18). Available from https://doi.org/10.3390/polym13183064, https://www.mdpi.com/2073-4360/13/18/3064/pdf.

ten Elshof, J. E., Yuan, H., & Gonzalez Rodriguez, P. (2016). Two-dimensional metal oxide and metal hydroxide nanosheets: synthesis, controlled assembly and applications in energy conversion and storage. *Advanced Energy Materials*, 6(23). Available from https://doi.org/10.1002/aenm.201600355, http://onlinelibrary.wiley.com/journal/10.1002/(ISSN)1614-6840.

Timmerman, M. A., Xia, R., Le, P. T. P., Wang, Y., & ten Elshof, J. E. (2020). Metal oxide nanosheets as 2D building blocks for the design of novel materials. *Chemistry—A European Journal, 26*(42), 9084−9098. Available from https://doi.org/10.1002/chem.201905735, http://onlinelibrary.wiley.com/journal/10.1002/(ISSN)1521-3765.

Tomantschger, K., McClusky, F., Oporto, L., Reid, A., & Kordesch, K. (1986). Development of low cost alkaline fuel cells. *Journal of Power Sources, 18*(4), 317−335. Available from https://doi.org/10.1016/0378-7753(86)80089-1.

Uzunoglu, M., & Alam, M. S. (2017). Fuel-cell systems for transportations. In M. H. Rashid (Ed.), *Power electronics handbook* (Fourth Edition, pp. 1091−1112). Turkey: Elsevier. Available from https://www.sciencedirect.com/book/9780128114070, https://doi.org/10.1016/B978-0-12-811407-0.00037-4.

Wachsman, E., Ishihara, T., & Kilner, J. (2014). Low-temperature solid-oxide fuel cells. *MRS Bulletin, 39*(9), 773−779. Available from https://doi.org/10.1557/mrs.20140.192, https://www.springer.com/journal/43577.

Wilberforce, T., Alaswad, A., Palumbo, A., Dassisti, M., & Olabi, A. G. (2016). Advances in stationary and portable fuel cell applications. *International Journal of Hydrogen Energy, 41*(37), 16509−16522. Available from https://doi.org/10.1016/j.ijhydene.2016.020.057.

Xu, L., Cui, Q., Zhang, H., Jiao, A., Tian, Y., Li, S., Li, H., Chen, M., & Chen, F. (2020). Ultra-clean PtPd nanoflowers loaded on GO supports with enhanced low-temperature electrocatalytic activity for fuel cells in harsh environment. *Applied Surface Science, 511.* Available from https://doi.org/10.1016/j.apsusc.2020.145603.

Yang, W., Yang, X., Jia, J., Hou, C., Gao, H., Mao, Y., Wang, C., Lin, J., & Luo, X. (2019). Oxygen vacancies confined in ultrathin nickel oxide nanosheets for enhanced electrocatalytic methanol oxidation. *Applied Catalysis B: Environmental, 244,* 1096−1102. Available from https://doi.org/10.1016/j.apcatb.2018.120.038, http://www.elsevier.com/inca/publications/store/5/2/3/0/6/6/index.htt.

CHAPTER 6

Nanoelectrochemistry in microbial fuel cells

Yunfeng Qiu, Yanxia Wang, Xusen Cheng, Yanping Wang,
Qingwen Zheng, Zheng Zhang, Zhuo Ma and Shaoqin Liu
Faculty of Life Science and Medicine, School of Medicine and Health, Harbin Institute of Technology,
Harbin, Longjiang Hei, P.R. China

6.1 Introduction

In the past few decades, industrialization and urbanization resulted in rapid socioeconomic growth, but the accompanying problems of energy shortage and environmental pollution have also become more and more serious. Energy, as the lifeline of socioeconomic development, is closely related to human production and life (Chen et al., 2021). According to the calculation of the Ministry of Ecology and Environment of China, in 2022, the total energy consumption in China was nearly 5.41 billion tons of standard coal, an increase of 2.9% compared to last year. Coal consumption accounted for 56.2% of the total energy consumption, an increase of 0.3% points compared to last year. From the above, it can be seen that the energy consumption trend in China is still dominated by nonrenewable energy. Although the proportion of clean energy is increasing year by year, the development and utilization are still slow, and the energy problem urgently needs to be solved.

From the perspective of sustainable development, wastewater can be seen as an intermediate between energy and the environment. The global average consumption of wastewater treatment is 0.75 KW/m, which is energy intensive and expensive. However, wastewater contained tremendous thermal and chemical energy (Gude, 2015). Therefore, seeking a reasonable technique to effectively reduce energy consumption in wastewater treatment while efficiently utilizing the energy in wastewater plays a positive role in alleviating energy shortages and environmental protection. Microbial fuel cells (MFCs) are an environmentally friendly biotechnology (Lovley, 2012; Schröder et al., 2015) that utilize electrically active microorganisms to consume organic matter and transfer metabolic electrons to the extracellular anode (Deng et al., 2023).

Electrochemistry and Photo-Electrochemistry of Nanomaterials
DOI: https://doi.org/10.1016/B978-0-443-18600-4.00007-7
© 2025 Elsevier Inc. All rights reserved, including those for text and data mining,
AI training, and similar technologies.

Oxygen, potassium ferrocyanide, and high oxidation potential substances can all serve as electron acceptors (Li et al., 2017). MFCs operate without additional aeration and have strong environmental adaptability. The organic matter degradation rate in the anode chamber is fast, the mineralization is thorough, and there is no need for subsequent sludge treatment processes. If the cathode is matched with pollutants with high oxidation potential as electron acceptors, it can also simultaneously treat cathode pollutants, which is an important application prospect for sewage treatment and synchronous production technology (Logan et al., 2019). In recent years, the application of MFCs in black and odorous river sediment has provided a good idea for treating urban water pollution while recovering electricity (Blatter et al., 2021). Marine MFCs provide an important energy system for the long-term operation of unmanned servo sensors (Feregrino-Rivas et al., 2023). The 1000 L MFCs system can operate year-round to treat municipal wastewater, with a COD content of less than 50 mg/L in the final discharged water and an output power density of up to 7−60 W/m (Lovley, 2012), demonstrating strong engineering application prospects (Liang et al., 2018). Using MFCs as the core unit, it can also be coupled with technologies such as hydrogen production (Kuleshova et al., 2022; Sogani et al., 2021), desalination (Hemalatha et al., 2020; Sibi et al., 2023), and sensing to construct new devices such as microbial hydrogen production electrolysis cells, microbial fuel cell synchronous desalination, high toxicity metal ion removal (Shi et al., 2023; Zhang et al., 2020; Zhang et al., 2021), and self-powered sensors. Therefore, as seen in Fig. 6.1, it has received widespread global attention in the past 20 years (Bird et al., 2022; Ma et al., 2023).

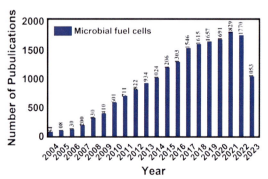

Figure 6.1 Publications on MFCs during 2004−2023. *Data were acquired from Web of Science.*

6.2 Overview of microbial fuel cells

6.2.1 The history of microbial fuel cells

The concept of MFCs can be traced back to 1911, when British botanists Potter (1911) placed two platinum electrode plates in a culture medium containing *Escherichia coli* and *yeast*, and found that the external ammeter pointer was deflected, meaning that microorganisms oxidize and decompose organic compounds to produce current. However, this discovery did not attract attention at the time. Twenty years later, Barnett Cohen et al. created microbial semifuel cells, where a current of 2 mA could generate a voltage exceeding 35 V by adding electronic mediators, allowing MFCs to develop to some extent (Cohen, 1931). By the 1990s, researchers had isolated dissimilatory metal reducing bacteria from surface sediments of hydrocarbon contaminated ditches and found that they could use their own secreted media as electron carriers, creating the possibility of improving the performance of MFCs. Oh and Logan (2005) proposed for the first time the combination of MFCs for electricity generation and wastewater treatment, which could not only generate electricity but also treat wastewater. This attracted the attention of researchers and became a research hotspot in the fields of environment and energy. Today, MFCs also exhibit significant potential advantages in biological hydrogen production (Wang et al., 2017), biosensing (Kharkwal et al., 2017), and bioremediation (Chen et al., 2016; Sivasankar et al., 2019).

6.2.2 Basic principles of microbial fuel cells

The most common structures of MFCs are single and dual chambers (Guo et al., 2020). Fig. 6.2 is typical schematic diagrams of dual chamber MFCs devices, mainly composed of three parts: anode, cathode, and proton exchange membrane (Yaqoob et al., 2021). Anodes are places where electrically active microorganisms grow and colonize, producing electrons through the oxidation of substrates, which can serve as electron output bodies (Dey et al., 2022). Cathodes generally use highly oxidizing substances such as potassium permanganate (Wang et al., 2019), potassium ferrocyanide (Lawson et al., 2020), and oxygen as electron acceptors (Estrada-Arriaga et al., 2021). Among them, oxygen is used in the experimental stage of engineering due to its low price, environmental friendliness, and pollution-free properties (Sonawane et al., 2017; Yaqoob et al., 2021). Potassium ferrocyanide is easy to reduce and has stable chemical properties and is so often used for theoretical research in the laboratory.

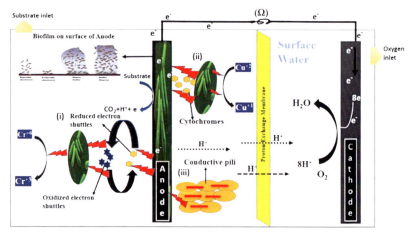

Figure 6.2 Schematic diagram of a typical dual chamber microbial fuel cell. *From Yaqoob, A. A., Ibrahim, M. N. M. & Guerrero-Barajas, C. (2021). Modern trend of anodes in microbial fuel cells (MFCs): An overview. Environmental Technology & Innovation, 23.*

Taking sodium acetate as the anode electron donor and potassium ferrocyanide as the cathode electron acceptor as an example, the chemical reaction equations of the anode and cathode are shown in Eqs. (6.1) and (6.2):

$$\text{Anode: } 2HCO_3^- + 9H^+ + 8e^- \rightarrow CH_3COO^- + 4H_2O, \quad (6.1)$$

$$\text{Cathode: } Fe(CN)_6^{3-} + e^- \rightarrow Fe(CN)_6^{4-}. \quad (6.2)$$

Nernst equation for anode potential:

$$E_{an} = E_{an}^\ominus - \frac{RT}{F} \ln \frac{[CH_3COO^-]}{[HCO_3^-]^2 [H^+]^9}. \quad (6.3)$$

Nernst equation for cathode potential:

$$E_{ca} = E_{ca}^\ominus - \frac{RT}{F} \ln \frac{[Fe(CN)_6^{4-}]}{[Fe(CN)_6^{3-}]}. \quad (6.4)$$

In above equations:
E_{an} is theoretical potential of anode (V).
$E_{an}^\ominus$ is the redox potential of sodium acetate under standard conditions (0.187 V, vs. SHE).
E_{ca} is theoretical potential of cathode (V).

E_{ca}^{Θ} is the redox potential of potassium ferrocyanide under standard state (0.361 V, vs. SHE).

R is molar gas constant (8.314 J/mol/K).

F is Faraday constant (96485 C/mol).

T is absolute temperature (K).

The concentration of feed substrate sodium acetate is 1 g/L, the pH of the anode solution is 7, and the concentration of bicarbonate is set to 5 mM. After calculation, the theoretical potential "E" of the anode can be obtained $E_{an} = -0.30$ V. On the cathode $\left[\mathrm{Fe(CN)}_6^{4-}\right]/\left[\mathrm{Fe(CN)}_6^{4-}\right] = 1$, then the theoretical potential E of the cathode $E_{ca} = E_{ca}^{\Theta} = 0.361$ V, the maximum theoretical voltage output by MFCs is $E_{cell} = E_{ca} - E_{an} = 0.661$ V. In fact, due to various polarization phenomena, the theoretical voltage is often lower than the actual voltage (Wang et al., 2020).

6.2.3 Extracellular electron transfer between microorganisms and anodes

Electroactive microorganisms in MFCs are currently the most complex and poorly understood field, naturally attracting great attention. Electroactive microorganisms, also known as exoelectrogens, can colonize the surface of electrodes and generate electrons by oxidizing organic matter. Exoelectrogens mainly come from surface sediments such as sludge and wastewater (Ueki, 2021). At present, the most extensively and systematically studied microorganisms are *Geobacter* and *Shewanella*, which often appear as models for constructing extracellular electron transfer (EET) (Hu et al., 2021). In fact, the efficiency of MFCs' production capacity is constrained by the EET rate, especially the electron transfer at the interface between electrically active microorganisms and anodes (Zhou et al., 2022). Therefore, a clear understanding of the electron transfer process between the two interfaces is a necessary condition for improving the electrical energy output of MFCs in the later stage.

The electron transfer from the producing bacteria to the anode surface requires two processes: intracellular and EET, and the former one can be divided into aerobic and anaerobic types. As shown in Fig. 6.3, *Shewanella* is used as an example to introduce the intracellular electron transfer process under anaerobic conditions. There are over 300,000 cytochrome C in the narrow periplasmic space, forming a dense electron transport network. The electrons generated by the oxidation of organic matter are

164 Electrochemistry and Photo-Electrochemistry of Nanomaterials

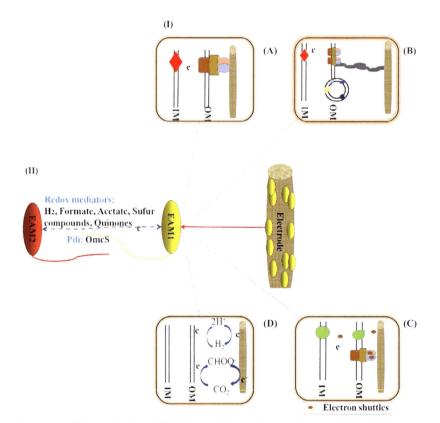

Figure 6.3 *EET mechanisms.* (A) DET via OMCs and bound flavin, as well as (B) nanowires. (C, D) MET via redox mediators between anodes and microorganisms. *Reproduced with permission from Zhao, J., Li, F., Cao, Y., Zhang, X., Chen, T., Song, H. & Wang, Z. (2021). Microbial extracellular electron transfer and strategies for engineering electroactive microorganisms. Biotechnology Advances, 53. https://doi.org/10.1016/j.biotechadv.2020.107682.*

transferred to CymA through the conversion of quinone (MQ) and hydroquinone (MQH$_2$) in the menaquinone pool through nicotinamide adenine dinucleotide (NADH). Then, under the coordinated action of various proteins such as FccA, MCC, STC, etc., it passes through the periplasm to reach MtrA on the outer membrane of the cell, and finally passes through the transmembrane channel composed of MtrA MtrB MtrC/OmcA to transfer to the extracellular space. Among them, CymA and MtrA are located in the inner and outer membranes of cells, respectively, and are important channel proteins that enter and exit the periplasmic membrane (Light et al., 2018). In addition, researchers discovered through model prediction and gene knockout methods that multiple redox proteins were also involved in transmembrane transmission.

The EET mechanism of electrically active bacteria is a key component of the electron transfer process and an important determinant of the power generation performance of microbial fuel cells. There are two main mechanisms for electron transfer between microorganisms and MFCs anodes: (1) direct electron transfer (DET), in which microorganisms use Cytochrome C or cell surface attachments (nanowires) on the outer membrane to transfer electrons to the anode surface after close contact with the anode surface; (2) indirect electron transfer (MET) refers to the transfer of electrons between microorganisms and anodes through electronic mediators such as riboflavin, melanin, or their own metabolites such as hydrogen and formic acid (Zhao et al., 2021).

The schematic diagram of the DET process is shown in Fig. 6.3A. After the oxidation of organic matter within microbial cells, electrons are transferred across the periplasmic membrane to the outer membrane, and then carried by cytochrome MtrA through the cell membrane. Finally, they are transferred to the electron acceptor through MtrC and OmcA on the outer membrane, G. *Sulfurreducens* utilizes the inner membrane proteins ImcH and CbcL to receive electrons dissociated from coenzyme Q and transfer them to the periplasmic proteins PpcA or PpcD. After crossing the periplasmic space, the electrons are then transferred to the outer membrane proteins OmaB, OmaC, OmcB, and OmcC, and finally to the extracellular solid electron acceptor to complete the outward transfer of electrons. From this, it can be seen that the biocompatibility and specific surface area of the anode material will directly affect the electron transfer efficiency during the DET process. Therefore, improving the biocompatibility of the anode material and increasing the specific surface area of the anode material can improve the EET efficiency.

The schematic diagram of the MET process is shown in Fig. 6.3B. Some flavin adenine dinucleotides (FAD) produced by microorganisms in the intracellular decomposition of organic matter cross the membrane and enter the periplasm. On the one hand, they bind to periplasm proteins (such as FccA) as cofactors to participate in some metabolic processes. On the other hand, it can also be hydrolyzed into flavin mononucleotide (FMN) and adenosine monophosphate (AMP) under the action of UshA. FMN directly dephosphorizes to riboflavin through the outer membrane, or obtains electrons from MtrC and is reduced to FMN_{red}. Then, through diffusion, the electrons are transferred to the extracellular electron acceptor, which in turn becomes FMN_{ox}, mediating the electron transfer process between active microorganisms and extracellular receptors.

There are tens of nanometers of extracellular polymers on the surface of electrically active microorganisms, including electron mediator molecules. It can enter the cell interior as an electron carrier, transport electrons to the electrode surface and reenter the cell interior, working alternately in a reduced and oxidized state. Common electronic mediators include riboflavin (Okamoto et al., 2013), flavin mononucleotides (von Canstein et al., 2008), and phenazine (Zhang et al., 2011), among others. Marsili et al. (2008) found a 70% decrease in electron transfer rate to the electrode by removing riboflavin from the membranes of *S. oneidensis Mr-1* and *S. oneidensis Mr-4* cells, indirectly confirming that MET is the main electron transfer mode. Microorganisms can also accelerate the kinetic reaction between microorganisms and anodes by adding artificial electronic mediators. It is usually able to pass through the cell membrane, receive electrons from intracellular electron carriers, leave the cell in a reduced form, and then transfer electrons to the electrode surface to enhance anodic reaction kinetics and increase current density (Hou et al., 2013; Jayapriya & Ramamurthy, 2012). However, compared to the electronic mediators secreted by microorganisms themselves, their application is greatly limited. Mainly because most systems run in open systems and require continuous addition of artificial electronics, which will also increase costs. More importantly, many artificial electronic intermediaries are toxic to humans and cannot be released into the environment.

Cytochrome C is widely present in most bacteria and archaea and is considered one of the most important electron transfer pathways currently. It can specifically recognize redox proteins on the cell membrane and receive electron transfer to terminal receptors (Strycharz-Glaven et al., 2013). Cytochrome C contains heme, which can form conjugated porphyrins around metal ions and serve as the active center for redox reactions. By knocking out the corresponding gene fragments in cytochrome c, it was found that there were significant changes in the surface and electrical properties of microorganisms, confirming that cytochrome c is an important pathway for EET (Strycharz et al., 2011). In recent years, some studies have shown that the pili of *Geobater* and *Shewanella* strains have high conductivity and can serve as biological nanowires to participate in electron transfer between anode surfaces and bacteria, making them a new type of DET process. Richter et al. (2009) discovered the electrochemical activity of *G. sulfurreducens* biofilm using cyclic voltamme-try, and the results showed the presence of an electron transfer medium bound conductive network at the biofilm electrode interface. In addition, they found that type IV pili are important in this electron transfer process. More interestingly, the pili often wrap around the cytochrome binding to the cell membrane, which increases the possibility of electron transfer between cells. Currently, it is generally believed that nanowires are formed by the extrusion and extension of the outer membrane of con-ductive bacteria, with a conductivity of up to 277 S/cm, three orders of magnitude higher than a single cell (Tan et al., 2017). Therefore, nano-wires are expected to become the optimal dynamic electron path in EET. However, the conductive mechanism of nanowires is still controversial, and currently there are relatively accepted hypotheses. The first one is that they favor electronic transitions. And the second one is relying on over-lapping aromatic amino acids π-π Orbits to transmit electrons.

6.2.4 The dissimilation reduction mechanism of exoelectrogens and metal compounds

The research on the interaction between bacteria and interfaces origi-nated from the interaction between exoelectrogens and metal com-pounds (Yu et al., 2021). This is mainly due to the presence of a large number of transition metal compounds such as iron, cobalt, and manga-nese in the environment where exoelectrogens exist in nature. Later research showed that transition metals act as electron acceptors around extracellular bacteria, storing and mediating electron transfer. Minerals also provide an energy source for cell metabolism (Bose et al., 2014; Emerson & Craig, 1997; Jiao et al., 2005; Shelobolina et al., 2012). Therefore, the interaction between extracellular bacteria and metal compounds is also an important component of the study of the bio-tic—abiotic interface. The dissimilatory reduction mechanism of metal compounds is similar to the EET mode of electrogenic bacteria, with the following three types: (1) The DET mechanism can spontaneously express flagella or pili attachment to the metal surface, constructing a direct pathway for electron transfer. (2) The electronic mediator mech-anism shuttles back and forth between the electrogenic bacteria and the metal surface. The oxidized electron mediator receives electrons from the producing bacteria and becomes a reduced state, which then diffuses to the metal compound and is reduced to an oxidized state (Lower et al., 2001). (3) Chelating solubilization mechanism. Some small molecule substances coordinate with metal ions to form chelates, and solid metal compounds exist in the form of liquid ions, making them easy to receive electrons (Weber et al., 2006).

Similar to the EET between bacteria and electrode surfaces, nanowires are also an important mechanism for the differential reduction between bacteria and metal compounds. Researchers have always believed that the attachment of electrogenic bacteria to the surface of metal compounds promotes the expression of fimbriae and flagella, while later studies have shown that mycelium does not need to attach to the surface of solid metal. On the contrary, fimbriae play a role in transferring electrons to insoluble Fe (III) oxides and other solid terminal electron acceptors (Childers & Susan, 2018). The formation of these conductive "nanowires" expands the available space area outside the cell membrane, allowing for the infiltration of nanopore spaces in soil and sediment that were previously thought to be physically unavailable for cells. These "nanowires" have the potential to establish bridges between individual cells, making communication between cells possible, and enabling cells attached to Fe (III) oxides or other electron receptors to perform electron shuttle functions. Iron reducing microorganisms can use alternative active mechanisms to reduce Fe (III) oxides without direct contact with the solid surface. Electronic mediators reduce contact between the two and play a role in the combination of microbial catalysis and abiotic processes. Common redox active organic compounds are widely present in soil and sediment, such as humic acids, plant exudates, and antibiotics. Although these compounds serve as electron mediators for microbial mediated Fe (III) reduction in eutrophic environments, their molecular level mechanism of action is still unclear. The chelation solubilization mechanism is a unique dissimilatory reduction mechanism between metals and bacteria, and the production of complex ligands helps to dissolve solid Fe (III) oxides, making it easier for microorganisms to obtain soluble Fe (III) forms. It is worth noting that electricity producing bacteria typically transmit electrons through multiple means, not limited to one.

The microbial electrode interface microenvironment is complex. As shown in Fig. 6.4, on the one hand, the electrode needs to provide an attachment site and substrate mass transfer channel for the electroactive microbial membrane, and at the same time, it needs to transfer the electrons generated by metabolism to the external circuit. The EET process relies on microbial transmembrane electron transfer, interspecific electron

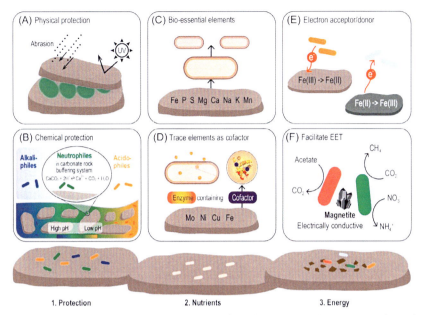

Figure 6.4 *The relationship between minerals and microorganisms.* (A, B) Physical and chemical protection by minerals. (C, D) Bioessential and trace elements from minerals to provide microorganisms growth. (E) Electron acceptors provided by minerals. (F) Electrical conductors for facilitating EET. *From Dong, H. L., Huang, L. Q., Zhao, L. D., Zeng, Q., Liu, X. L., Sheng, Y. Z., Shi, L., Wu, G., Jiang, H. C., Li, F. R., Zhang, L., Guo, D. Y., Li, G. Y., Hou, W. G. & Chen, H. Y. (2022). Critical review of mineral-microbe interaction and co-evolution: Mechanisms and applications.* National Science Review, 9(10), nwac128.

transfer, DET between conductive cytochrome c and electrodes, and MET between cytochrome c/electron mediator complexes and anodes (Frauke et al., 2015). Although researchers analyzed the crystal structure of cytochrome c and conductive pili using cryoelectron microscopy (Gu et al., 2021; Wang et al., 2019) and simulated the electron transfer energy barrier between nanomaterials and microorganisms using theoretical calculations (You et al., 2017), existing nanomaterial-modified anodes still cannot efficiently transfer extracellular electrons, resulting in MFCs having output power far lower than their theoretical output power. Therefore, exploring new nanomaterials to modify MFCs anodes, enhancing microbial adhesion, and improving electron transfer between nanomaterials and cytochrome c, conductive pili, or mediators is of great research significance for improving the output power of MFCs.

6.3 Factors influencing anodic nanoelectrochemistry

The above domestic and international studies indicate that MFCs have shown great potential in energy development and environmental protection, but the low power density output seriously limits their application. At the beginning of the experiment, researchers often borrowed electrode materials from other batteries for use in MFCs anodes, such as commercial carbon cloth (CC), carbon felt (CF), carbon brushes (CB), and so on. However, these carbon materials have relatively poor conductivity and small surface area, which is not conducive to bacterial adhesion and electron transfer, thereby affecting the performance of MFCs. In recent years, many scientists have attempted to modify the anode surface using physical and chemical methods to improve the overall performance of MFCs, and have achieved good results. If natural three-dimensional (3D) biomass materials are used, through simple carbonization and modification, anode materials with diverse structures, good mechanical properties, and rich nitrogen content can be obtained.

The research on biofilms mainly focuses on the formation kinetics, electron transfer, growth, and metabolic activity. It is generally believed that the formation of biofilms can be divided into the following steps (Korth et al., 2015; Speers & Reguera, 2012): (1) bacteria attach to the electrode surface; (2) form single-layer and multilayer bacteria; (3) form polymer scaffolds; (4) the biofilm matures to form a 3D structure. Among them, the attachment and reproduction of monolayer bacteria are key steps that determine the subsequent formation of biofilms. It is worth noting that the adhesion and overall electrical output of monolayer bacteria are influenced by the synergistic effects of electrode surface properties (surface roughness, wettability) and structure (porosity, micro-/nanostructure). Therefore, understanding the physical and chemical properties and structure of electrode materials is of great guiding significance for the subsequent development of anodes.

At present, the overall performance of MFCs is mainly improved from two perspectives: one is to adjust the composition of the material, including modifying the surface of carbon-based materials with nanocarbon, metal nanoparticles, conductive polymers, and doping certain heteroatoms (such as N, P, S, etc.), which not only enhances the conductivity of the anode material but also constructs a direct or indirect pathway for EET between the producing bacteria and the anode interface. By controlling the composition of the material, the electron transfer rate between the bacterial anode interface is accelerated, creating conditions for improving the performance of MFCs. The second is to adjust the surface properties and structure of the electrode, such as roughness, hydrophilicity/hydrophobicity, pore size of the electrode, micro-/nanostructure, etc. The use of electrochemical redox treatment to activate electrode materials can increase roughness, improve hydrophilicity, and greatly enhance the adhesion of the anode surface to bacteria, accelerate the colonization of electricity producing bacteria and the formation of biofilms. The adjustment of micro-/nanostructures effectively expands the electrochemical active area of the electrode and increases the active sites for bacterial attachment. A suitable pore size not only increases the surface area of the electrode but also facilitates mass transfer of nutrients and metabolites, preventing the occurrence of biological blockages. Numerous experiments have shown that these factors significantly affect the power generation and decontamination capabilities of MFCs.

6.3.1 Surface roughness

The surface roughness of electrode materials is one of the key factors determining bacterial adhesion. The microbial electrode interface can be modified through chemical and physical means on the electrode surface, changing the surface roughness of the electrode, thereby affecting the growth of biofilm and the generation of current (Wei et al., 2020). Generally speaking, rough surfaces can accelerate bacterial adhesion and colonization, promoting biofilm formation. Because it can not only form surface grooves, shorten the distance of ion diffusion to the internal structure, promote rapid transport of charge carriers but also provide a larger electrochemical active area and increase bacterial adhesion (Chinnaraj et al., 2021; Crawford et al., 2012; Truong et al., 2010). However, excessive gaps in the biofilm can also lead to incomplete contact between bacteria and the anode surface, reducing the conductivity of the biofilm and resulting in a decrease in anode performance. Therefore, in the process of designing electrodes, it is necessary to comprehensively consider the effects of roughness and voids.

The research on the impact of roughness on microbial systems is currently mainly focused on stainless steel—based electrodes (Pons et al., 2011). Pons et al. investigated the effect of adjusting the surface roughness of stainless steel electrodes on the formation of sulfur reducing *Geobacterium* biofilms. When the surface roughness (R_a) increases from 2.0 to 4.0 μm, the output current increases by 1.6 times. It can be seen that changes in roughness have a significant impact on output power. Generally speaking, the roughness close to the surface of microbial cells is conducive to bacterial adhesion and growth. Lu et al. (2020) used different processing techniques to prepare ceramic surfaces with varying roughness. Their research found that when the surface roughness decreased from submicron to nanoscale, the surface state gradually shifted from hydrophobicity to hydrophilicity, which was not conducive to the adhesion of hydrophobic *Staphylococcus aureus*. In addition, it was further found that bacterial adhesion fixed points gradually decreased or even disappeared, leading to a significant decrease in the binding strength between bacteria and the surface. On the surface with $R_a = 1$ nm, the number of bacterial adhesion is only 2% of the conventional surface ($R_a = 205$ nm). From this, it can be seen that a smooth nanosurface can effectively inhibit the initial adhesion and proliferation of *Staphylococcus aureus* on the ceramic surface.

6.3.2 Surface wettability

Adjusting the surface wettability of electrode materials through appropriate surface treatment is a basic requirement for improving the production capacity of MFCs. The ability of a liquid to keep the electrode surface moist or maintain interactions through weak intermolecular forces is called wettability. The contact angle is the main parameter that measures the hydrophilicity and hydrophobicity of the anode surface (Santoro et al., 2014). Contact angle ($90° > \theta > 0°$) is relatively low, indicating that the material surface has hydrophilicity. If the contact angle ($> 90°$) is too high, it indicates that the material surface has hydrophobicity (Chen et al., 2020). Being too hydrophilic or too hydrophobic can affect bacterial adhesion in Fig. 6.5. The surface of materials that are too hydrophilic can easily form a hydration layer, which hinders the interaction between bacteria and the anode interface, thereby hindering bacterial adhesion. The interaction between the surface of excessively hydrophobic materials and bacterial extracellular proteins is weak, and the adhesion is greatly reduced. The surface roughness of the electrode has a significant impact on its hydrophilicity. Generally speaking, the hydrophilicity of the surface increases with a decrease in electrode roughness. In addition, the presence of impurities or the porosity of the electrode surface also significantly affects wettability. Finally, due to different extracellular proteins and strains, the contact angles also vary.

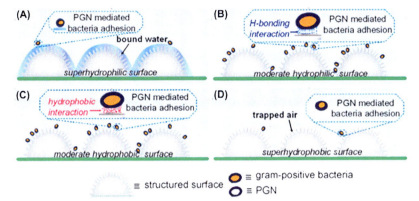

Figure 6.5 *Surface wettability effect on bacterial adhesion.* (A−D) The bacterial adhesion on different wettability surface (Dou et al., 2015). *From Dou, X.Q., Zhang, D., Feng, C. & Jiang, L. (2015). Bioinspired Hierarchical surface structures with tunable wettability for regulating bacteria adhesion. ACS Nano, 9(11), 10664−10672. https://doi.org/10.1021/acsnano.5b04231.*

Conductive polymer−modified electrodes can improve the hydrophilicity and hydrophobicity of the electrodes, synergize with the electrostatic capture effect of external power sources, and effectively reduce the charge transfer resistance of MFCs. Overall, appropriately increasing the hydrophilicity or wettability of the electrode surface can improve the performance of MFCs. Du, An, et al. (2017) coated a layer of polydopamine (PDA) on the surface of stainless steel mesh to make an anode electrode. The study found that after the electrode was subjected to PDA, the startup time of MFCs decreased from 88 to 76 hours, the maximum power density increased from 613 ± 9 mW/m^2 to 803 ± 6 mW/m^2, an increase of 31%, and the coulombic efficiency also increased from 19% to 48%. Mainly due to the introduction of abundant amino and catechol functional groups by PDA, the surface hydrophilicity and EET of the material are improved. In addition, PDA also increased the proportion of *Proteobacteria* and *Firmicutes*, indicating that PDA has a selective enrichment effect on the anode microbial community to a certain extent.

6.3.3 Hierarchical porous structures

In microbial systems, porous surfaces not only effectively increase the surface area of bacterial attachment but also facilitate the expression of pili by electric producing bacteria, construct pathways for DET, and reduce the internal resistance of microbial systems (Angelaalincy et al., 2018).

Therefore, increasing the porosity can increase the surface area, but the decrease in conductivity of high porosity electrodes can have a negative impact on the overall performance of MFCs (Tao et al., 2016; Wu et al., 2023). In addition, the diameter of bacteria is generally between 0.4 and 4 μM, mostly concentrated in 1−2 μm. Compared to nanostructured materials with smaller pore sizes, macroporous surfaces are more suitable for microbial systems. The surface of the macropores can directly promote the colonization of microorganisms and the transport of nutrients and metabolites within the anode, thereby improving the overall performance of the system (Flexer & Jourdin, 2020). Moreover, due to differences in bacterial strains, reactor structures, and operating modes, the appropriate pore size also varies, so it is necessary to design and select according to the actual situation. Therefore, pore size is also an important factor to consider when preparing anode materials.

Karthikeyan et al. prepared conductive electrodes using mushrooms, wild mushrooms, and corn stalks as raw materials using a simple carbonization process. The relationship between three electrode structures and adhesion was systematically studied using electrochemical redox probes and electroactive biofilms. The corn grain anode with pores of 2−7 μm has bacteria attached both internally and externally, indicating that this pore size is suitable for the early survival of exoelectrogens. But as the internal biofilm grows, pore blockage may occur. Once the micropores are blocked by the biofilm, the substrate supply and metabolite removal inside the electrode will be hindered, inhibiting the growth of the biofilm and causing internal bacterial death. The thickness of mature biofilms can reach tens or hundreds of micrometers, and porous electrodes with visible pore sizes within a few micrometers are still insufficient to provide space for biofilm growth. But wild mushroom anodes with pores ranging from 75 to 200 μm can provide sufficient space for microorganisms to adhere without clogging. It can be seen that pore sizes of around hundreds of micrometers can effectively eliminate pore blockage caused by biofilm growth. Recently, we used high-temperature carbonized gluten (NCF) as a carrier, loaded with MOF materials, and carbonized to obtain 3D porous MFC anodes (HPCF) (Li et al., 2023). Carbonized MOF is distributed both inside and outside the NCF pores, and its dispersion is very good. Moreover, the presence of MOF improves the graphitization degree of the composite material and significantly reduces the material's resistance, which is only 2.22 Ω, much smaller than the blank CC (1823.00 Ω). After loading the MFC, Fig. 6.6 shows that the startup cycle of the battery is very short, and the final power density reaches 11.21 W/m^3.

Nanoelectrochemistry in microbial fuel cells 175

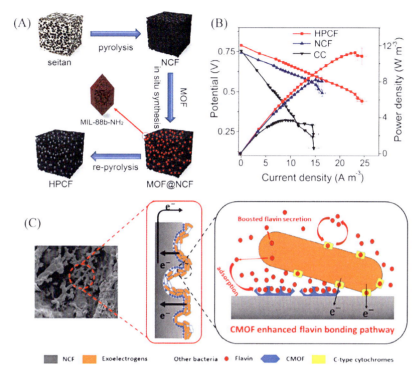

Figure 6.6 *Hierarchical porous anode.* (A) Schematic procedure of hierarchical porous anode of HPCFs. (B) The power densities of MFCs equipped by HPCFs and control groups. (C) Scheme of microorganism adhesion and flavin excretion. *From Li, H. D., Zhang, L., Wang, R. W., Sun, J. Z., Qiu, Y. F. & Liu, S. Q. (2023). 3D hierarchical porous carbon foams as high-performance free-standing anodes for microbial fuel cells. Ecomat, 1, e12273.*

Zou et al. (2017) prepared a novel polyaniline-doped mesoporous carbon nanocomposite material with good biocompatibility by using nanostructured $CaCO_3$ templates to assist in carbonization of natural plant extracts. The prepared PANI-LMC hybrid can achieve rapid transfer of extracellular electrons through both endogenous flavin electron mediators and bacterial outer membrane redox centers, significantly enhancing DET at the interface between bacteria and electrodes. The maximum output power density of the PANI-LMC hybrid anode is 1280 mW/m², which is one order of magnitude higher than the output power density of traditional CC. This work not only provides a new, efficient, and low-cost anode material for MFCs but also provides a foundation for in-depth understanding of the physical structure and chemical properties of electrode materials, which play a crucial role in EET from microorganisms to electrodes.

6.3.4 Surface micro-/nanostructures

The micro- and nanoscale surface morphology affects bacterial adhesion, biofilm formation, and electron transfer reaction mechanisms, thereby affecting the performance of microbial systems. Therefore, adjusting the surface micro-/nanostructure is an effective way to enhance bacterial interactions (Encinas et al., 2020; Ye et al., 2017). Generally, surface structure is changed through physical and chemical methods. However, not all micro-/nanostructures have a positive effect in Fig. 6.7 (Cheng et al., 2019), and only micro-/nanostructures with specific sizes can effectively improve the performance of MFCs.

Alatraktchi et al. (2012) reported that a product with a nanoroughness of 100 nm was obtained by deep reactive ion etching of gold electrodes and sputtering gold nanoparticles on carbon paper as MFCs electrodes. The output power density of nanoparticle sputtering is much higher than

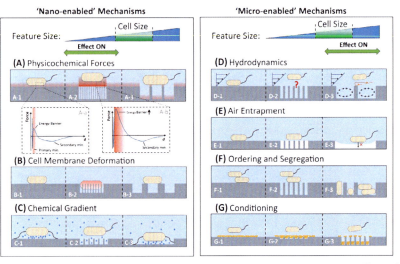

Figure 6.7 *Micro-/nanostructured surface effect on bacterial adhesion.* The effect of (A–C) nanostructured and (D–G) microstructured surface morphology on bacterial adhesion *From Cheng, Y., Feng, G., & Moraru, C. I. (2019). Micro- and nanotopography sensitive bacterial attachment mechanisms: A review. Frontiers in Microbiology, 10, 191. https://doi.org/10.3389/fmicb.2019.00191.*

that of ion etching (346 $\pm$ 2.5 mW/m^2). Researchers altered the microstructure of the anode to investigate the effect of surface morphology on the attachment of electrogenic bacteria and the electrical energy output of MFCs. Nanostructures with heights of 115 nm and 300 nm, and aspect ratios of 0.3 were successfully prepared using template method on glassy carbon electrodes, with smooth glassy carbon electrodes as the control. In a three-electrode system, comparing the current density and bacterial attachment density of patterned and nonpatterned electrodes, it was found that the 115 nm mode current density and bacterial attachment density significantly increased by 40% and 78%, respectively. At the same time, the fluorescence confocal microscope image shows that the height of the patterned prism is 115 nm, which is conducive to bacterial adhesion and the formation of anodic biofilm. In addition, unlike the method of directly constructing nanostructures on the substrate, various techniques can be used to modify nanoparticles onto the substrate to enhance bacterial adhesion and colonization, which is also an effective method for improving surface micro-/nanostructures. For example, adding or depositing nanoparticles on the electrode surface typically exhibits better performance than bare electrodes.

6.4 Anodic modification materials

The performance of MFCs is related to various factors, such as the degradation rate of organic substrates, the EET rate between electrically active microorganisms and the electrode surface, the circuit resistance of the device, and external operating conditions. The rate of EET is largely influenced by the properties of the electrode material, which in turn affects the overall performance of MFCs. Therefore, developing efficient anodes is the key to improving the performance of MFCs. At present, anode materials can be divided into commercial carbon−based anodes, biomass carbon materials, graphene and carbon nanotube−modified carbon material anodes, biomimetic composite anodes, and so on.

6.4.1 Commercial anodes

An ideal anode material should have advantages such as strong conductivity, high biocompatibility, and strong chemical stability, and carbon materials are widely used in anode materials to meet these requirements to a certain extent. CC, CB, CF, and graphite sheet are currently relatively mature anode materials (Cai et al., 2020). CC is the most commonly used carbon material in MFCs anodes, which not only has a large surface area but also a relatively high porosity, providing sufficient living space for bacterial growth (Hindatu et al., 2017). However, unstable chemical properties affect the long-term operation of the battery. Carbon brushes are a very interesting carbon material with a large surface area and optimal area to volume ratio, which can be conductive by compounding with titanium materials. But the surface of carbon fiber is too smooth, which is not conducive to bacteria adhering to the surface (Sonawane et al., 2017). CF is widely used as an electrode due to its excellent conductivity and mechanical stability, providing abundant redox reaction sites and relatively low cost (Zhao et al., 2017). However, the thick CF hinders the diffusion of substrate from the outside to the inside after the formation of the biofilm, which to some extent affects the microbial colonization inside the anode. Graphite has a crystalline form of sp^2 hybrid carbon atoms, resulting in good conductivity. The potential properties such as biocompatibility and mechanical strength of graphite materials also make them very suitable as anode materials for MFCs. However, due to its smooth surface, the accumulation of biofilm is relatively low, which severely limits the power density output.

6.4.2 Biomass-derived anodes

On this basis, researchers have shifted their focus towards low-cost natural or available resources in daily life, which provide a potential green method for producing valuable bioenergy from natural waste. Plant waste is the most commonly used natural precursor for carbon-based anodes, commonly referred to as anode materials derived from natural waste. Porous carbon materials are usually obtained by directly carbonizing plant precursors at very high temperatures. As seen in Fig. 6.8A, during the preparation process, plant heteroatoms act as natural dopants, and water evaporation leads to the formation of porous structures in the carbon material throughout the entire carbonization process.

Chen, Tang, et al. (2016) used chestnut shell waste as a precursor to produce multilevel hierarchical anodes. The brush like structure of the spherical shell provides a large surface area for the growth of biofilm on the anode, with an output power density of 759 mW/m^2 and a coulombic efficiency of 75%. Hung et al. (2019) studied porous carbon anodes derived from natural waste and obtained a power density of 3927 mW/m^2, which is higher than that of activated carbon anodes. The increase in

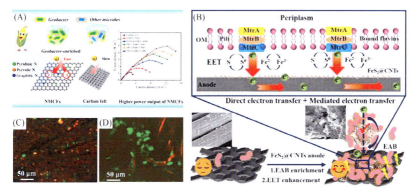

Figure 6.8 *Biomass anode and CNT-modified anode.* (A) Schematic illustration of NMCF promoting MFCs performance. (B) Mechanism of FeS$_2$@CNT anode improving the performance of MFCs (C, D) CLSM images of CC and FeS$_2$@CNT anodes after the attachment of microorganisms (Wang et al., 2020). *From Wang, Y., He, C., Li, W., Zong, W., Li, Z., Yuan, L., Wang, G., & Mu, Y. (2020). High power generation in mixed-culture microbial fuel cells with corncob-derived three-dimensional N-doped bioanodes and the impact of N dopant states.* Chemical Engineering Journal, 399. https://doi.org/10.1016/j.cej.2020.125848. *Liu, Y., Sun, Y., Zhang, M., Guo, S., Su, Z., Ren, T. & Li, C. (2023). Carbon nanotubes encapsulating FeS$_2$ micropolyhedrons as an anode electrocatalyst for improving the power generation of microbial fuel cells.* Journal of Colloid and Interface Science, 629, 970–979. https://doi.org/10.1016/j.jcis.2022.09.130.

anode power density derived from coffee waste is due to its high conductivity and suitable pore size, which promotes electron transfer and bacterial colonization on the anode. In addition, the long-term performance of MFCs using coffee-derived anodes was tested, operating continuously for 100 hours without adding any nutrients, achieving a power density of 2000 mW/m^2. These results indicate that coffee waste–derived anodes are an excellent power generation method that can significantly reduce the overall operating cost of MFCs.

6.4.3 Carbon nanotubes-based anodes

Since its discovery in 1991, carbon nanotubes (CNTs) have been widely favored by researchers due to their unique mechanical, electrical, and chemical properties. According to the characteristics of the tubular structure, it can be divided into single-walled CNTs and multiwalled CNTs. Due to its high elastic modulus, tensile strength, and thermal/electrical conductivity, CNTs are widely used in various fields, including MFCs. As seen in Fig. 6.8B–D, the presence of carbon nanotubes not only promotes biofilm adhesion but also enhances the EET process

(Liu et al., 2023). Liang et al. (2011) added CNTs powder and *G. sulfurreducens* to the anode cavity of MFCs, forming a composite biofilm on the anode, and studied the startup time and steady-state power generation performance of MFCs under different CNTs dosage conditions. The results showed that the optimal dosage was when the concentration of CNTs powder was 4 mg/mL, and the battery startup time and anode resistance were significantly reduced. Based on the results of phospholipid analysis, it was found that the addition of CNTs may promote an increase in mass transfer rate. Ren et al. (2015) compared the effects of three different arrangements of CNTs on the performance of MFCs. The results showed that the anode prepared by layer-by-layer rotary spraying method could attract more electricity producing bacteria *G. sulfurreducens* and form a thicker biofilm. This study lays the foundation for future research on optimizing MFCs using two-dimensional (2D) and 3D nanostructured electrodes.

CNT-modified anodes not only improve the conductivity and electrocatalytic activity of the overall material but also greatly increase the specific surface area, providing more attachment sites for bacterial adhesion. Xie et al. (2011) designed a porous CNTs textile composite material, where the macroscopic porous structure of the wrapped CNTs fibers creates an open 3D space, resulting in an interface area of the anode liquid biofilm anode that is 10 times larger than the projected surface area of the CNTs fibers, which facilitates efficient substrate transportation and internal colonization of various microbial communities. Compared with traditional CC anodes, the performance of CNTs textile anodes is significantly improved: the maximum current density is increased by 157%, the maximum power density is increased by 68%, and the energy recovery rate is increased by 141%. Our team has recently designed composite anode materials such as CNT-modified 3D substrate-loaded metal oxides, sulfides, etc. to address the issues of poor adhesion of existing anode microorganisms and low EET rates (Wang et al., 2022).

From this, it can be seen that CNT-modified MFCs anode materials can accelerate direct and MET compared to traditional MFCs anode materials, thereby optimizing MFCs output, with high conductivity, large surface area, and fiber structure connecting anode and cytochrome c on the bacterial outer membrane. In addition to one-dimensional carbon nanomaterials such as CNTs, carbon nanowires and carbon nanofibers (CNFs) also belong to the category of one-dimensional carbon nanomaterials, which will not be elaborated here.

6.4.4 Graphene and reduced graphene oxides-based anodes

On the basis of one-dimensional carbon nanomaterials, 2D carbon nano-materials are gradually being used in the field of MFCs anode research. Among them, the most prominent 2D nanomaterial is graphene, which has good conductivity, stability, and biocompatibility. Compared with 1D carbon nanomaterials, the planar structure of 2D carbon nanomaterials can provide a larger contact area for bacterial adhesion and EET. Zhou et al. developed a biosynthetic graphene using eucalyptus leaf extract as a reducing agent. When using biosynthetic graphene as the anode material for MFCs, they found that the maximum power density was 70% higher than that of electrochemically reduced grapheme-modified MFCs. When using graphene as the anode material, the potential biological toxicity of graphene needs to be considered, but the increase in maximum power density indicates that biosynthetic graphene induces its surface hydrophi-licity and biocompatibility. On the basis of graphene, reduced graphene particles and reduced graphene nanosheets are gradually applied to MFCs anode materials. Among them, due to the antiaggregation and open structure of reduced graphene particles, the specific surface area, electrochemical kinetics, and mass transfer ability of MFCs anode mate-rials are effectively enhanced. The reduced graphene nanosheets also exhibit excellent electrochemical performance, which may be related to the π-π electron transition between graphene nanosheets.

Recently, Wei Feng's team designed and synthesized a series of gra-phene-/nitrogen-doped CNTs aerogels with 3D network structure by using the self-assembly characteristics of 1D CNTs and 2D graphene nanomaterials (Jin et al., 2023). It has the 3D conductive structure net-work characteristics required for the anode of microbial fuel cells. The 3D carbon—based anode structure can promote the loading of high-density bacterial biofilms and facilitate mass transfer, diffusion, and electron trans-port. This 3D anode structure is directly related to the output voltage and power of the battery. Future researchers can further optimize the structure and composition of 3D graphene-/nitrogen-doped CNTs anodes to regu-late the output performance of MFCs.

6.4.5 Heteroatom-doped carbonaceous anodes

The modification of pure carbon materials has been proven to effectively improve the chemical properties of MFCs. However, the pure carbon material obtained by high-temperature calcination has hydrophobicity, which will affect the bacterial adhesion on the anode material and the formation of biofilm, thereby affecting the EET mechanism between the biofilm and the anode material. To address these issues, on the one hand, chemical treatment of pure carbon materials, such as acid treatment, can effectively increase the hydrophilic functional groups, on the other hand, doping heteroatoms into carbon materials. Heteroatoms can generate hydrophilic defects and break the chemical inertness of original carbon materials, which is beneficial for improving the hydrophilicity of bacterial adhesion and biocompatibility. The hydrophilic defects generated by heteroatoms can promote the EET between the bacterial outer membrane anode material and cytochrome c.

Among various types of heteroatom doping, the most common is the carbon material doped with nitrogen (N) in the anode material. Ci et al. developed a special type of N-doped CNTs (bamboo like NCNTs). Due to N-doping and bamboo like structure, bamboo like NCNTs provide better biocompatibility and larger active area for EETs, ultimately resulting in a 1.6-fold increase in peak current density of MFCs compared to CNTs. You et al. developed a 3D macroporous N-rich graphite carbon (NGC) scaffold for MFCs anodes. The N content in the NGC scaffold is as high as 10.89%. The enhanced interaction between pyrrole nitrogen structure and bacterial extracellular membrane active centers can accelerate the occurrence of EET. Natural biomass is another precursor that integrates the functions of carbon and heteroatom sources. For example, N-doped biomass carbon prepared from mango wood as an anode material for MFCs has a maximum power density of 589.8 mW/m^2. Cheng and Logan treated the anode CC with ammonium, which increased the surface charge of the CC electrode and improved the performance of MFCs. Ammonium-treated CC could increase the power density to 1970 mW/m^2, and improve bacterial adhesion and system performance at higher current densities $(0.5-0.9 \text{ mW/m}^2)$. The research results indicate that this is a new method to improve the power generation of MFCs. Yu et al. (2015) developed a nitrogen-doped carbon nanoparticle (NDCN) anode. The test found three pairs of redox peaks on the NDCN anode, indicating the existence of multiple pathways mediating electron transfer between the electrogenic bacteria and the anode interface, and the highest catalytic current caused by MET. This is because the use of NDCN increases the absorption of riboflavin by the anode, thereby improving the performance of MFCs. Yuan et al. synthesized a porous nitrogen-doped anode with a high N/C ratio of 3.9% and a large electrochemical active surface area of 145.4 cm^2, enriching bacterial attachment sites and providing more active sites for EET. Interestingly, the riboflavin content on the anode surface has also increased, indicating that N-doped-modified anode electrodes are a feasible strategy to improve the performance of MFCs (Yuan et al., 2019).

Bi et al. (2018) designed a 3D N-doped porous carbon through one-step pyrolysis of sodium citrate and melamine, and coated it on the surface of CC. The results show that its maximum power density can reach 2.777 mA/cm^2, which is twice that of commercial CC electrodes. The main reason is that the nitrogen-doped porous carbon structure increases the biocompatibility of the anode, promotes the growth and reproduction of microorganisms, and reduces the internal resistance of the modified anode. In addition, as shown in Fig. 6.9, Li et al. used density functional theory (DFT) to confirm from a theoretical calculation perspective that the matching of graphite nitrogen and nitrogen oxide with the reaction site atoms of riboflavin mononucleotides (FMN) is more conducive to the adsorption of flavin and facilitates rapid direct electrochemical processes (Wu et al., 2020). However, some people have proposed different views, believing that graphitized nitrogen and pyrrole nitrogen are the key to promoting the uptake of electrons from cytochrome c by carbon electrodes.

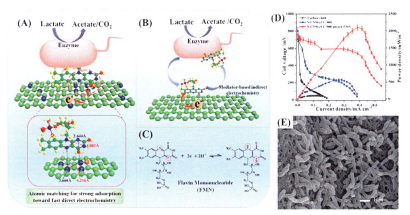

Figure 6.9 *N doping carbon anode.* (A−C) Schematic illustration of N doping promoting the adhesion of flavin. (D) Power density of N-CNWs anode. (E) SEM images of anode covered by exoelectrogens biofilm. *From Wu, X., Qiao, Y., Guo, C., Shi, Z. & Li, C. M. (2020). Nitrogen doping to atomically match reaction sites in microbial fuel cells.* Communications Chemistry, 1, 1−9.

Our research team recently used SnS_2 as a sacrificial template to directly grow N-, S-codoped carbon microflowers on CC substrates (Cheng et al., 2023). The SnS_2 template can be removed, resulting in the in situ doping of S in carbon and regulating electrochemical activity by introducing a synergistic effect of N and S heteroatoms. The power density and COD removal efficiency of MFC with N, S-CMF@CC anode are 2.50 W/m^2 and 91.68%, respectively, which are 1.71 and 1.12 times higher than those of CC. In addition, it is worth noting that its power density and COD removal efficiency are also higher than previously reported anodes. Compared with traditional CC dense nanomaterial arrays, our anode has a hierarchical porous micro-/nanostructure, which can effectively promote microbial colonization, *Geobacter* enrichment, electron/ion transport, and flavin excretion. The results indicate that dual-atom doping can effectively form suitable wetting and positively charged microregions on carbon sheets, which facilitates cell adhesion. At the same time, carbon nanosheets provide an ideal environment and material transport pathway for cell colonization and grow seamlessly on CC filaments, ensuring rapid electron transfer. These results indicate that sacrificial polymer coating templates are a promising method for preparing hierarchical micro-/nanostructures with the required characteristics as MFC anodes.

6.4.6 Metal- and metal compound-derived anodes

Wang et al. (2018) has proposed a novel FeS_2 nanoparticle-coated graphene (rGO) electrode as the anode for MFCs, as shown in Fig. 6.10. The introduction of FeS_2 nanoparticles not only facilitates bacterial adhesion but also enriches electrically active *Geobacterium* species on the electrode surface. Nanoparticle FeS_2 also promotes the interaction between microbial biofilm and anode surface, increasing the transfer rate of extracellular electrons from electroactive bacteria to the electrode surface, effectively improving the performance of MFCs, with a maximum power density of 3200 mW/m^2, and removing 1319 $\pm$ 28 mg/L of chemical oxygen demand (COD) from brewery wastewater.

In order to improve the power generation performance and operational stability of MFCs, our research team recently constructed multilevel nanostructured anodes of iron cobalt metal and N-doped carbon nanotubes (Wang et al., 2022). An integrated self-supporting anode FeCo with hierarchical nanostructures was prepared by modifying iron/cobalt nanosheets on the surface of N-doped carbon nanotubes using chemical

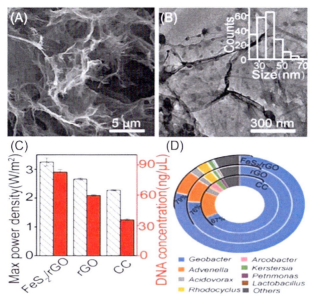

Figure 6.10 *FeS₂/rGO MFCs anode.* Structure and performance of FeS₂/rGO MFCs anode. (A) SEM and (B) TEM images. (C) Correlation diagram between maximum power density and biomass of MFCs. (D) Microbial community structure of different anodes. *From Wang, R., Yan, M., Li, H., Zhang, L., Peng, B., Sun, J., Liu, D. & Liu, S. (2018). FeS₂ nanoparticles decorated graphene as microbial-fuel-cell anode achieving high power density.* Advanced Materials, 30(22). https://doi.org/10.1002/adma.201800618.

vapor deposition and electrodeposition techniques. The maximum output power density for FeCo/NCNT@CF abide is 3.04 W/m² at a current density of 5.59 A/m², and the daily COD removal rate is 220.97 mg/L/day, which is 1.5 and 3.9 times that of bare CF anode, respectively. The proportion of electricity producing bacteria *Geobacter* is as high as 75.3%. CV and EIS after inoculation indicate FeCo/NCNT@CF has a high specific capacitance and low internal resistance. SEM and CLSM also show that bacteria have a strong survival state on the electrode, forming a dense biofilm. It is assumed that multilevel nanostructures can provide a huge specific surface area and rich active sites, and an integrated self-supporting connection method can reduce the internal resistance of the electrode, thereby accelerating electron transfer. FeCo/NCNT@CF has higher specific capacitance and larger electrochemical active area, which can store and release more electrons. Moreover, there are multiple pathways of EET, accelerating the entire electrochemical cycle process. N-doped carbon nanotubes construct a direct pathway for electron transfer between bacteria and interfaces, while iron and cobalt metals promote the secretion of nanowires by *Geobacter*, building bridges for electron transfer among different bacterial species.

6.4.7 Polymer-based anodes

The main reason why conductive polymers have received widespread attention in the anode modification of MFCs is their excellent conductivity, which is in the range of $10-10^{-6}$ S/cm. They can also be adjusted according to the manufacturing process and doping concentration to achieve conductivity of up to 10^5 S/cm, which can accelerate the EET rate between microorganisms and the anode interface (Kaur et al., 2015; Lakard, 2020; Le et al., 2017). There are three common conductive polymers in MFCs: polyaniline (PANI), polypyrrole (PPy), and poly3,4-ethylenedioxythiophene (PEDOT) (Boeva & Sergeyev, 2014; Reza, 2006). These polymers also have good chemical and environmental stability, biocompatibility, and are relatively easy to synthesize. In order to achieve high bioelectricity rates, these conductive polymers are usually codoped with other nanomaterials to modify the anode.

PANI is a positively charged conductive polymer that can generate electrostatic attraction with negative charges, while bacteria are negatively charged and can interact with each other. Using this characteristic, Wu et al. (2018) prepared CNTs and PANI nanocomposite anodes by grafting polymerization of PANI onto ITO substrates. Research has found that compared to bare ITO electrodes, PANI-modified anodes exhibit low charge transfer resistance and excellent electrochemical behavior. In the *Shewanella loihica PV-4* MFC system, the maximum current density of the CNT-/PANI-modified ITO electrode reached $6.98\,\mu A/cm^2$, 26 times that of the bare ITO electrode, and its power density was $34.51\,mW/m^2$, 7.5 times that of the bare ITO electrode. Sonawane, Al-Saadi, et al. (2018) carefully screened the concentrations of acid and aniline monomers, and applied electro polymerization technology to coat polyaniline film on the surface of stainless steel plate (SS-P) to prepare conductive SS/PANI-P anode electrodes. After running for a period of time, it was found that its power density can reach up to $780\,mA/m^2$. On this basis, due to the excellent electrocatalytic performance, good environmental stability, convenient synthesis, and low economic cost of PPy, Sonawane, Patil, et al. (2018) proposed using a conductive polymer—coated stainless steel wool (SS-W) as a low-cost anode for MFCs. The highest power densities of SS/PANI-W and SS/PPy-W anodes are 0.288 $\pm$ 0.036 mW/cm^2 and 0.187 $\pm$ 0.017 mW/cm^2, respectively, which are far superior to the 0.127 $\pm$ 0.011 mW/cm^2 of SS W anodes. This indicates that PANI coating is more suitable for anode modification of MFCs than PPy coating. Moreover, compared with SS-P-based anodes, under the same experimental conditions, the power density of all SS-W-based anodes increased by at least 70%.

PEDOT has been widely used in fields such as solar cells, biosensors, and organic electrochemical semiconductor materials due to its advantages of simple molecular structure, good conductivity, biocompatibility, and stability (Ghasemi et al., 2011). PEDOT has special oxidation and reduction characteristics, with a certain linear relationship between its charge state and potential, and can serve as an electron donor and acceptor (Winther-Jensen & MacFarlane, 2011). Kang et al. (2015) modified a certain amount of PEDOT (2.5 mg/m^2) on three different substrates: graphite plate, CC, and graphite felt. The PEDOT coating greatly improved the electrochemical performance and electron transfer ability of the anode. In 2016, he also found that after PEDOT modification, the EAB abundance reached 83%. At the same time, the microbial diversity of the anode biofilm indicates that compared to the previous inoculum, the microbial community structure of the anode biofilm has undergone significant changes, proving that PEDOT coating the anode is beneficial for microbial attachment and reproduction (Kang et al., 2017). Later, someone prepared PEDOT/MnO$_2$/CF composite anodes, where MnO$_2$ acts as an electron mediator, and PEDOT coatings can reduce the capacitance loss of MnO$_2$ due to biological metabolism and adhesion. Both jointly improve the conductivity and electron transfer rate of the anode material (Liu et al., 2019).

Kim et al. (2023) used polydopamine (PDA) and PPy to modify graphite felt electrodes. The preparation of polymer coatings can be easily achieved using electrochemical deposition methods. The study found that the power density of MFC with modified PDA/PPY-GF reached 920 mW/m^2, which was 1.5, 1.17, and 1.18 times higher than the control group GF, PDA-GF, and PPY-GF, respectively. Inspired by mussels, PDA has excellent hydrophilicity and adhesion, which is conducive to the formation of thicker biofilms. Meanwhile, PPy has excellent conductivity and can provide rich electrochemical active sites for EET in microorganisms. Researchers characterized the physical and chemical properties of polymer-modified anodes using Raman spectroscopy, Fourier transform infrared spectroscopy (FTIR), Brunauer Emmett Teller (BET) surface area measurement, and contact angle analysis. The research results indicate that the use of PDA/PPy comodification strategy can obtain point active biofilms, while improving microbial adhesion and bioelectrochemical performance in actual MFC reactors.

6.4.8 MXene-based anodes

After the discovery of Graphene, 2D materials have attracted great attention from researchers with theoretical and experimental backgrounds due to their fascinating properties. In recent years, a new 2D metal carbide- or nitride-layered material, called MXene, has become a promising material. The most commonly used MXene ($Ti_3C_2T_x$) for research was obtained by selectively etching certain atomic layers from MAX precursor (Ti_3AlC_2). In addition to MXene based on $Ti_3C_2T_x$, many other MXene families have also been explored to form a wide range of 2D materials. This 2D material is also widely used in MFCs. MXene has better hydrophilicity and biocompatibility than graphene, which promotes MXene as an excellent anode modification material.

Multilayer $Ti_3C_2T_x$ (m-MXene) is an alternative anode-modified material due to its high specific surface area and conductivity. However, the negative charge on the surface of the $Ti_3C_2T_x$ with a multilayer structure can limit its modification effect. The solution-phase flocculation method (ammonium ion method) was used to restack and polymerize MXene nanosheets as anode-modified materials (n-MXene). The n-MXene-modified anode has a higher specific surface area, and its surface hydrophilicity and surface positivity are better than those of the m-MXene anode. The maximum current density of the n-MXene-modified anode is 2.1 A/m^2, which is higher than that of the bare carbon fiber cloth anode (1.3 A/m^2). The performance improvement is attributed to the reduction of charge transfer resistance and diffusion resistance, as well as a significant increase in anode surface biomass, while promoting the secretion of a large number of nanowires (Dong et al., 2022). Researchers have found that MXene was modified using a superhydrophilic cationic polymer (polydiallyldimethylammonium chloride) (PDDA) to prepare MXene@PDDA/CC serves as the anode for MFCs. This unique anode combines the advantages of high conductivity, superhydrophilicity, and biocompatibility. This anode promotes the enrichment of electrochemical active bacteria and improves the efficiency of EET. The MFCs based on MXene@PDDA/CC anode can generate significant power output and satisfactory water removal efficiency through an external series resistor of 1 kΩ. The maximum output voltage is 585 mV, and the maximum power density is 811 mW/m^2. After 12 h, the decolorization efficiency and COD efficiency of methyl orange reached 79% and 84%, respectively. MXene anode modification can also be carried out using simple impregnation and hydrothermal methods. The results of electrochemical impedance spectroscopy and cyclic voltammetry confirm that NiFe$_2$O$_4$-MXene@CF Composite anodes have superior bioelectrochemical activity, attributed to low charge transfer resistance, high conductivity, and abundant catalytic active sites. Microbial community analysis shows that the relative abundance of electrically active bacteria on the in the NiFe$_2$O$_4$-MXene@CF anode biofilm is relatively high.

6.4.9 Biomimetic anodes

Nanomaterial-modified anodes can improve microbial adhesion and EET, but as mentioned earlier, the power density reported in current research is still far lower than its theoretical power density. Inspired by the crystal structure analyzed by cryoelectron microscopy (Gu et al., 2021; Wang et al., 2019), cytochrome C distributed in the outer membrane of microorganisms and conductive pili is an important protein responsible for DET between electrodes, with its redox active center being heme (iron porphyrin). Relying on proteins as scaffolds, hemes are stacked and arranged with their assistance to form electron transfer pathways, completing the transmembrane and EET processes. As seen in Fig. 6.11A and B, Yujie Feng's research group modified carbon substrates with iron phthalocyanine and iron porphyrin polycationic ammonium salt to improve the spatial structure of microbial membranes (Li, Liu, Chen, Wu, et al., 2021; Li, Liu, Chen, Yuan, et al., 2021). By utilizing the strong interaction between cytochrome C and FeN$_4$ active sites, the EET was accelerated, and the power density was increased by 4.32 and 1.92 times, respectively. Liu Xing et al. constructed a graphene/heme coating with a maximum current density 10.2 times higher than the control group.

However, molecules such as heme, porphyrin, and phthalocyanine self-aggregate severely in solvents, and their disordered deposition or film conductivity is usually lower than 10^{-8} S/cm, far lower than the conductivity of microbial conductive pili (El-Naggar et al., 2010; Yalcin et al., 2020),

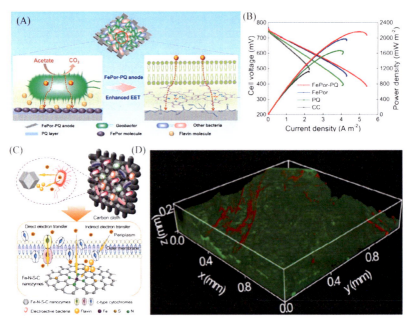

Figure 6.11 *FePor-PQ and Fe-N-S-C Biomimetic anodes.* (A) Schematic illustration of FePor-PQ anode improving EET. (B) The power density of various anodes. (C) EET mechanism of Fe-N-S-C biomimetic anode. (D) CLSM image of Fe-N-S-C anode after the attachment of microorganisms. *From Li, Y., Liu, J., Chen, X., Wu, J., Li, N., He, W., & Feng, Y. (2021). Tailoring surface properties of electrodes for synchronous enhanced extracellular electron transfer and enriched exoelectrogens in microbial fuel cells. ACS Applied Materials & Interfaces, 13(49), 58508−58521. https://doi.org/10.1021/acsami.1c16583. From Xiang, Y., Liu, T., Jia, B., Zhang, L., & Su, X. (2023). Boosting bioelectricity generation in microbial fuel cells via biomimetic Fe-N-S-C nanozymes. Biosensors and Bioelectronics, 220, 114895. https://doi.org/10.1016/j.bios.2022.114895.*

making it unsuitable for ideal artificial nanowires. Further analysis of C-type cytochrome revealed that it is a cytochrome oxidase located on the cytoplasmic membrane. Okamoto et al. (2014) reported on the mechanism of its action in electron transfer from the inner membrane to the outer membrane. As shown earlier, the interface nanoelectrochemical process of MFCs anodes is closely related to the c-Cyts containing heme in the outer membrane, which can mediate electron transfer in the EET process of electroactive bacteria. Therefore, the reported literature confirms that simulating the electronic shuttle function of c-Cyts can effectively improve the energy conversion efficiency of MFC (Li, Liu, Chen, Wu, et al., 2021). Heme contains an Fe-N-C active center structure. Inspired by this biological enzyme system, Xiang et al. (2023) developed a method for preparing biomimetic nanoenzyme MFCs anodes. MOFs are a type of porous framework materials formed by metal and organic ligands. After high-temperature calcination, the M-N-C structure of enzymes can be obtained. By introducing metal Zn, metal single atom catalysts can be obtained, which are very similar to the active centers of heme. ZIF-8 adsorbs Fe ions and Bis (4-hydroxyphenyl) sulfone, which can form an Fe-N-S-C enzyme structure after high-temperature calcination, effectively enhancing the EET process and promoting the performance of MFCs in Fig. 6.11C and D. The starting time of traditional carbon cloth is 5.6 days, and the power density is $1009 \pm 26 \, mW/m^2$. Compared with CC, the start-up time of Fe-N-S-C nanoenzyme-modified MFC anode was reduced to 3.5 days, and the power density increased to $2366 \pm 34 \, mW/m^2$.

6.5 Reconstructing EET pathways in microorganisms

6.5.1 Polymer coated microorganisms

Researchers have utilized the excellent conductivity and hydrophilicity of conjugated polymers to enhance the EET and metabolic activity of electrogenic bacteria. The reported conjugated polymers include PPy, PANI, and PDA. After coating a layer of polymer on the surface of bacteria, their conductivity and hydrophilicity were improved, and their adhesion was stronger, significantly improving their EET efficiency and operational stability. Yu et al. (2011) reported that a protective and conductive PPy shell is formed in situ on the surface of *oneidensis Mr-1*. The introduced PPy membrane provides great flexibility in controlling the number and vitality of cells, as well as the morphology and thickness of biofilms. The experimental group coated with PPy showed an 11 times higher power output in MFC compared to natural biofilms. It can be seen that PPy not only enhances the DET process but also enhances the metabolic activity of bacteria. It is worth noting that PPy can also be applied to the surface of other electrically active bacteria, which can also improve their EET process and maintain long-term survival ability. Considering the hydrophobicity of PPy, later research can improve bacterial adhesion by mixing with other polymers.

Inspired by natural mussels, PDA has excellent biocompatibility and good conductivity. Through metal ion oxidation and other means, uniform thin films can be formed in situ on the surface of bacteria, and PDA can serve as a collector for anchoring electrons. By optimizing the polymerization time and conditions, researchers successfully prepared PDA nanoshells on the cell surface, reconnecting the transmembrane electron channels of bacteria, which is beneficial for improving the electron transfer efficiency at the interface between electroactive bacteria and anodes. In response to the issue of poor tolerance of electroactive biofilms in acidic environments, Du, Li, et al. (2017) utilized biocompatible PDA to encapsulate electroactive bacteria and resist extreme acid shocks. Electroactive bacteria are completely encapsulated in a ~ 50 nm nanofilm formed by PDA, which can protect their metabolic activity and electricity production ability even after pH 0.5 shock. After 30 minutes of strong acid shock, the current density outputted by the anode containing bacteria using PDA was 0.20 ± 0.05 A/m^2, while the exposed control group only had 0.01 ± 0.01 A/m^2. Under strong acid conditions, without PDA encapsulation, the biofilm is easily partially damaged, with a thickness ranging from 72 µM drops to 23 µM, causing 92% of cell death. From this, it can be seen that PDA encapsulated bacteria can protect their electrical activity under extreme conditions, providing feasible research ideas for expanding the application of microbial electrochemical systems, especially as sensors for monitoring water quality in acidic environments in the future.

If multiple polymers are used to modify bacteria together, the advantages of different conjugated polymers can be fully utilized, while improving the adhesion and conductivity of bacteria. For example, Wang et al. (2021) found that comodification of bacteria with PDA and PPy significantly increased cell adhesion performance. Research has found that the maximum power density of MFC coated with PDA/PPy can reach 2520.2 mW/m^2, which is 1.6 times and 11 times higher than the control group coated with a single polymer, respectively. As mentioned earlier, the enhanced power generation performance of MFC coated with double copolymers can be attributed to the synergistic effect of PPy and PDA coatings. On the one hand, high conductivity PPy promotes EET between bacteria and anodes, inducing bacterial secretion of riboflavin. On the other hand, hydrophilic PDA can regulate the infiltration of PPy and promote bacterial adhesion on its surface.

6.5.2 Microorganisms internalize or biomineralize nanomaterials

Carbon dots (CDs) have unique electronic and physicochemical properties, such as high electron transfer efficiency, excellent chemical stability, good biocompatibility, and unique optical properties (Guo et al., 2021). Research has found that coating CDs on the surface of bacteria can promote cell adhesion on the anode surface, facilitate the formation of thicker biofilms, and promote EET. As seen in Fig. 6.12A, Yang et al. (2020)

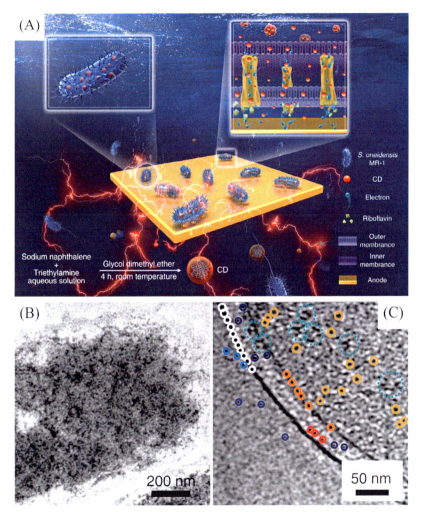

Figure 6.12 *CDs-fed* S. oneidensis *Mr-1*. (A) Schematic illustration of CDs-fed *S. oneidensis* Mr-1 to improve the generation of electricity. (B, C) TEM and HRTEM images of intracellular uptake of the CDs by *S. oneidensis* Mr-1. *From Yang, C., Aslan, H., Zhang, P., Zhu, S., Xiao, Y., Chen, L., Khan, N., Boesen, T., Wang, Y., Liu, Y., Wang, L., Sun, Y., Feng, Y., Besenbacher, F., Zhao, F. & Yu, M. (2020). Carbon dots-fed* Shewanella oneidensis *Mr-1 for bioelectricity enhancement.* Nature Communications, 11(1), 1379.

prepared hydrophilic carbon dots and coincubated them with *S. oneidensis Mr-1* and found that their bioelectricity performance was significantly improved. CDs have abundant surface functional groups that can adsorb on the surface of bacteria. Its particle size is less than 10 nm and can also be ingested by bacteria. Through TEM in Fig. 6.12B and C and various enzyme activity experiments, it has been proven that CDs can bind to the enzymes inside the bacteria, improving the electron transfer rate. CDs can serve as a "fast channel" for transmembrane/EET, and it has been found that CDs can also repair the electron transfer ability of mutant bacteria lacking conductive proteins. Based on the excellent biocompatibility of CDs, the surface of electroactive bacteria that absorb CDs has higher negative charges and adhesion, which can form a thicker biofilm in the short term, which is conducive to the enrichment and EET of electroactive microorganisms.

By utilizing the optical properties of CDs, researchers can also endow electrically active bacteria with photoresponsive functions. Guo et al. (2019) used positively charged N- and S-doped carbon dots (m-NSCDs) to modify negatively charged electrogenic bacteria, endowing them with enhanced energy conversion processes. Conductive m-NSCDs not only enhance the conductivity of bacterial outer membranes but also serve as photosensitizers to enhance bacterial photocatalytic function. Under continuous light irradiation conditions, the maximum power density of MFC modified with m-NSCDs reaches 1697.9 mW/m^2, which is approximately 2.6 times higher than that of unmodified bacteria.

Previous studies have shown that metals and their compounds have excellent biocompatibility and abundant electrocatalytic active sites, making them a potential material for constructing bacterial EET pathways (Jiang et al., 2023). If metal-based nanomaterials are combined with bacteria, it is expected to improve the power generation and wastewater treatment performance of MFC. Deng et al. (2020) found that conductive iron sulfide nanoparticles are biosynthesized within and on the surface of sulfate reducing bacteria cells. Research has confirmed that FeS NP can directly absorb electrons from extracellular electrodes, generating more metabolic energy, providing an important theoretical basis for understanding the micro interaction mechanism between biomineralized nanoparticles and bacteria. There are a large number of "dead" ducts at the end of the bacterial peripheral cytoplasm, which significantly inhibit the transmembrane transport of electrons. The iron sulfide NPs synthesized by bacterial biomineralization cannot only connect to c-type cytochrome, but also promote transmembrane electron transfer by connecting to the polysulfide reductase PsrABC. TEM confirmed that the embedded FeS in the membrane is located in the peripheral cytoplasm, forming a conductive network that can serve as an artificial transmembrane electron conduit to restore the conductive function of the "dead" conduit in Fig. 6.13. Yu et al. (2020) constructed FeS conductive conduits in situ on a single cell membrane, serving as electron collectors that form close contact with the extracellular membrane, providing a rich electron transfer interface. The interface charge transfer resistance of the *oneidensis Mr-1* biological FeS-modified anode was reduced by about 25 times, and the maximum power density of the assembled MFCs reached 3210 mW/m².

Recently, Bocheng Cao et al. embedded Ag nanoparticles (Ag NPs) into the outer cell membrane of bacteria, significantly promoting the transfer of electrons from internal electron carriers to the negative electrode (Cao et al., 2021). In the field of antibacterial materials, Ag NPs are widely used as fungicides. However, this work innovatively loads Ag NPs onto the surface of 3D reduced graphene oxide. This composite anode can promote the formation of biofilms and form a continuous conductive path of silver nanoparticles on the outer membrane, significantly improving EET and effectively increasing bacterial turnover frequency. It should be pointed out that Ag NPs distributed in the outer film can act as conductive sites and directly contact the collector of the anode, significantly reducing the energy barrier for extracting charges. Assembling the battery,

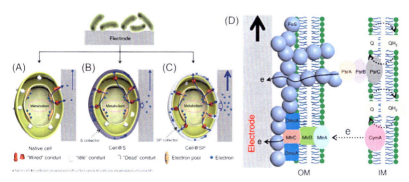

Figure 6.13 *Reconstruct EET paths.* (A—C) Schematic illustration of EET between anode and various cells. (D) Scheme of EET process for the SW@SP cell at electrode interface. *From Yu, Y.-Y., Wang, Y.-Z., Fang, Z., Shi, Y.-T., Cheng, Q.-W., Chen, Y.-X., Shi, W. & Yong, Y.-C. (2020). Single cell electron collectors for highly efficient wiring-up electronic abiotic/biotic interfaces.* Nature Communications, 11, 4087.

it was found that the rGO anode modified with Ag NPs can output a maximum power density of 6600 mW/m^2, which is much higher than the blank control group. After calculation, the single cell cycle frequency is 8.6×10^5 s^{-1}, which is the best reported turnover frequency for MFCs so far.

The above work indicates that constructing transmembrane and extracellular electron direct transfer channels can construct nonredox centers or multiple redox cycle electron hopping conductive pathways, improve charge transfer efficiency, and provide a feasible research approach for breaking through the electron transfer limitations of electrogenic bacteria and increasing output power.

In order to further expand the functions of bacteria, researchers attempted to modify them with multiple materials to explore their multifunctional properties and endow MFC with photoresponsive electricity production or photosynthesis performance (Cestellos-Blanco, Chan et al., 2022; Cestellos-Blanco, Friedline, et al., 2021; Cestellos-Blanco, Zhang, et al., 2020). Li, Jiang, et al. (2021) used an electrostatic assembly strategy to deposit conductive gold NPs and photoresponsive cadmium sulfide (CdS) NPs onto the surface of *E. coli*, which is the anode with photoresponsive function. Positive-charged Au can regulate the charge of the bacterial outer membrane and promote negative-charged CdS NP modification. By using two types of nanomaterials for modification, not only can the hindrance of electron transfer in photoelectric reactions be solved but also the light irradiation tolerance of bacteria can be improved. The results showed that compared with the MFC of natural *Escherichia coli*, the output power of Au- and CdS-modified bacteria MFC increased by 1.9 and 2.5 times, respectively. And it has been demonstrated through activity testing that CdS NPs modification can inhibit bacterial damage during light irradiation.

The previous article described single function enhancement and dual function expansion strategies. Recently, Jiang et al. (2020) developed a three-function modification strategy to further expand the functionality of microbial fuel cell anodes. They used the mineralization of bacteria to reduce $AuCl_4^-$ to prepare Au NPs, and then adsorbed iron tetroxide NPs through electrostatic interaction to prepare Au NPs on the surface of bacteria $Au@Fe_3O_4$ composite nanoparticles. Function one is to utilize highly conductive Au NPs to improve the electron transfer efficiency between biofilms and electrodes. The second function is to enhance the long-term metabolic activity of electrically active bacteria by utilizing biocompatible iron oxide NPs. Function three is to quickly enrich electrically active bacteria comodified with iron oxide NPs using magnetic electrodes, which can enrich a large number of microorganisms in a short period of time. The maximum power density output is 1792 mW/m^2, which is higher than control groups, respectively.

6.6 Conclusions and perspectives

In this chapter, a detailed analysis and a review of the latest progress in electron transfer between microorganisms and anode interfaces were conducted, with a focus on two aspects: anodic microbial attachment and EET. From the extensive literature review mentioned above, it can be seen that researchers have made a series of important progress in modifying MFCs anodes with nanomaterials, revealing EET mechanisms at the micro-/nanoscale and even molecular level, providing important guidance for the rational design of electrode materials in the future.

Although MFCs have achieved some remarkable achievements in recent years, they are still constrained by the following aspects (Ding et al., 2021; Ibrahim et al., 2020; Shanthi Sravan et al., 2021; Vishwanathan, 2021): (1) Electron transfer involves multiple processes, including the interface between electrically active bacteria/electrodes and the electrode/external circuit, resulting in slow electron transfer rates and low output power density. (2) The mechanisms of extracellular and interspecific electron transfer in electroactive microorganisms need to be further explored, and the coordination between electron transport proteins is still being explored. (3) At present, research mainly focuses on the modification of small-sized electrodes in the laboratory stage. It is difficult to prepare large area electrodes and cannot achieve large-scale commercial applications, especially in the design and application of battery stacks in practical environments, which still face many problems. (4) The cost of modifying electrodes with nanomaterials or hybrid microbial anodes is high, which is not conducive to large-scale promotion and use. Therefore, researchers suggest focusing on the following aspects in the study of MFCs anodes:

1. Microbial adhesion mainly involves the initial adhesion and colonization of microorganisms on the anode surface, the enrichment of electrically active microbial membranes, and the metabolism and growth during long-term operation. The isoelectric point of microbial outer membrane protein is below 7, and its surface is negatively charged in a neutral

solution. Positive potential applied by the anode can enrich suspended microorganisms. Modifying chitosan, cationic polymers, and quaternary ammonium salt molecules on the anode surface can increase their surface potential and attract microbial attachment through electrostatic interactions. Nanomaterial-modified electrodes can promote microbial attachment and competitive enrichment of electrically active microorganisms. A large amount of literature has shown that the modification of rough surfaces such as nanowires, nanotubes, and inverse opal on the anode surface endows the electrode with the ability to "capture" bacteria and can also promote the adhesion and colonization of microorganisms (Bian, Shi, et al., 2018; Fang et al., 2020). In practical water systems or sludge environments, it is necessary to focus on the long-term chemical stability of nanomaterials, the corrosion or biomineralization of nanomaterials by microorganisms, and the environmental toxicity of nanomaterial-modified anodes.

2. EET involves complex direct and MET processes between microorganisms and electrode interfaces. Nanomaterials can reduce the body resistance of electrodes, provide direct contact with microorganisms, and provide electron transfer pathways. Conductive polymers such as CNTs and polyaniline also exhibit excellent microbial anode performance due to their positive charge characteristics and intrinsic conductivity. They can serve as artificial nanowires to transfer microbial electrons over long distances to electrodes. High capacitive anode materials such as Mn and Ru also demonstrate the ability to store electrons released by microorganisms (Deeke et al., 2012), which can improve the energy recovery rate. Metals and metal oxides, sulfides, MXene, and other conductive hyphae have similar electrical conductivity, which can improve the microbial electrode interface and promote electron transport through cytochrome C media (Liu et al., 2018). The introduction of transmembrane and outer membrane silver nanoparticles improves the charge extraction efficiency of Shigella MFCs, significantly accelerating the transmembrane and EET of electrons (Cao et al., 2021; Gaffney & Minteer, 2021). Microorganisms can simultaneously adsorb and absorb carbon nanoparticles, accelerate electron transfer rate by binding with enzymes/peptides/proteins within cells, construct fast transmembrane/EET channels, and replace the electron transfer function of redox active proteins, significantly improving the power density of MFCs (Yang et al., 2020). In addition to simply improving the conductivity or capacitance of anode

nanostructures (Bian, Jiang, et al., 2018), enhancing the interface interaction between the modified nanostructures on the anode surface and cytochrome C or mediators can further accelerate DET or MET. Future anode material design can attempt the design and preparation of biomimetic anode materials, explore the special electron transfer mechanism between microorganisms and biomimetic interfaces, clarify the interface electrochemical mechanism, and provide important theoretical support and electrode materials for the design and promotion of practical applications of MFC.

3. Modifying multiple functional nanomaterials outside or inside bacteria can further improve the EET and power output of MFC, which is a research direction worth further exploration. Especially endowing microbial anode with light response function is conducive to the development of artificial photosynthetic systems. However, although existing research has preliminarily explored the impact of photosensitizers on bacterial metabolic activity, relevant research is still weak, especially ignoring the vitality of biofilms during long-term operation (Sun et al., 2023). Therefore, it is necessary to develop conductive nanomaterials with better biocompatibility or multilevel nanostructures such as conductivity and photosensitivity (Kornienko et al., 2018) and to improve the direct contact electron transfer between high redox activity proteins in cell membranes or nanowires and nanomaterial scaffolds. The long-term operation of MFCs anode generates a thicker biofilm, which contains bacteria and extracellular polysaccharide matrix. The dispersion of multifunctional nanomaterials in the extracellular polysaccharide matrix also significantly affects electron transfer between bacteria and electrodes, but there is still limited research on this topic.

In summary, although researchers have developed some successful strategies to improve the performance of MFC, the direct interface electrochemical microenvironment between microorganisms and anodes still needs to be optimized, and the mechanism of interface electron transfer process still requires the development of in situ electrochemical characterization, microbial metabolism analysis, and theoretical modeling calculations. As a result, the output power density and sewage treatment capacity of the existing MFCs are still far below their theoretical values. In short, there is still a long way to go for MFCs anode and related research, which relies on continuous exploration and original innovation by researchers. In the future, MFCs will definitely be able to move towards practical applications.

References

Alatraktchi, F. A., Zhang, Y., Noori, J. S., & Angelidaki, I. (2012). Surface area expansion of electrodes with grass-like nanostructures and gold nanoparticles to enhance electricity generation in microbial fuel cells. *Bioresource Technology*, *123*, 177−183.

Angelaalincy, M. J., Navanietha Krishnaraj, R., Shakambari, G., Ashokkumar, B., Kathiresan, S., & Varalakshmi, P. (2018). Biofilm engineering approaches for improving the performance of microbial fuel cells and bioelectrochemical systems. *Frontiers in Energy Research*, *6*, 1−12.

Bi, L., Ci, S., Cai, P., Li, H., & Wen, Z. (2018). One-step pyrolysis route to three dimensional nitrogen-doped porous carbon as anode materials for microbial fuel cells. *Applied Surface Science*, *427*, 10−16.

Bian, B., Shi, D., Cai, X., Hu, M., Guo, Q., Zhang, C., Wang, Q., Sun, A. X., & Yang, J. (2018). 3d Printed porous carbon anode for enhanced power generation in microbial fuel cell. *Nano Energy*, *44*, 174−180.

Bian, R., Jiang, Y., Wang, Y., Sun, J.-K., Hu, J., Jiang, L., & Liu, H. (2018). Highly boosted microbial extracellular electron transfer by semiconductor nanowire array with suitable energy level. *Advanced Functional Materials*, *28*(19), 1707408.

Bird, H., Heidrich, E. S., Leicester, D. D., & Theodosiou, P. (2022). Pilot-scale microbial fuel cells (MFCs): A meta-analysis study to inform full-scale design principles for optimum wastewater treatment. *Journal of Cleaner Production.*, *346*, 131227.

Blatter, M., Delabays, L., Furrer, C., Huguenin, G., Cachelin, C. P., & Fischer, F. (2021). Stretched 1000-L microbial fuel cell. *Journal of Power Sources*, *483*, 229130. Available from https://doi.org/10.1016/j.jpowsour.2020.229130.

Boeva, Z. A., & Sergeyev, V. G. (2014). Polyaniline: Synthesis, properties, and application. *Polymer Science Series C*, *56*(1), 144−153.

Bose, A., Gardel, E. J., Vidoudez, C., Parra, E., & Girguis, P. R. (2014). Electron uptake by iron-oxidizing phototrophic bacteria. *Nature Communications*, *5*, 3391.

Cai, T., Meng, L., Chen, G., Xi, Y., Jiang, N., Song, J., Zheng, S., Liu, Y., Zhen, G., & Huang, M. (2020). Application of advanced anodes in microbial fuel cells for power generation: A review. *Chemosphere*, *248*, 125985.

von Canstein, H., Ogawa, J., Shimizu, S., & Lloyd, J. R. (2008). Secretion of flavins by *Shewanella* species and their role in extracellular electron transfer. *Extracellular Electron Transfer*, *74*(3), 615−623.

Cao, B., Zhao, Z., Peng, L., Shiu, H.-Y., Ding, M., Song, F., Guan, X., Lee, C. K., Huang, J., Zhu, D., Fu, X., Wong, G. C. L., Liu, C., Nealson, K., Weiss, P. S., Duan, X., & Huang, Y. (2021). Silver nanoparticles boost charge-extraction efficiency in *Shewanella* microbial fuel cells. *Science*, *373*(6561), 1336−1340.

Cestellos-Blanco, S., Chan, R. R., Shen, Y. X., Kim, J. M., Tacken, T. A., Ledbetter, R., Yu, S., Seefeldt, L. C., & Yang, P. D. (2022). Photosynthetic biohybrid coculture for tandem and tunable CO_2 and N_2 fixation. *Proceedings of the National Academy of Sciences of the United States of America*, *119*(26). Available from https://doi.org/10.1073/pnas.2122364119.

Cestellos-Blanco, S., Friedline, S., Sander, K. B., Abel, A. J., Kim, J. M., Clark, D. S., Arkin, A. P., & Yang, P. D. (2021). Production of PHB from CO_2-derived acetate with minimal processing assessed for space biomanufacturing. *Frontiers in Microbiology*, *12*. Available from https://doi.org/10.3389/fmicb.2021.700010.

Cestellos-Blanco, S., Zhang, H., Kim, J. M., Shen, Y. X., & Yang, P. D. (2020). Photosynthetic semiconductor biohybrids for solar-driven biocatalysis. *Nature Catalysis*, *3*(3), 245−255. Available from https://doi.org/10.1038/s41929-020-0428-y.

Chen, B. Y., Ma, C. M., Han, K., Yueh, P. L., Qin, L. J., & Hsueh, C. C. (2016). Influence of textile dye and decolorized metabolites on microbial fuel cell-assisted bioremediation. *Bioresource Technology*, *200*, 1033–1038. Available from https://doi. org/10.1016/j.biortech.2015.10.011, http://www.elsevier.com/locate/biortech.

Chen, H., Li, Q., Benbouzid, M., Han, J., & Aït-Ahmed, N. (2021). Development and research status of tidal current power generation systems in China. *Journal of Marine Science and Engineering*, *9*(11), 1286.

Chen, S., Tang, J., Jing, X., Liu, Y., Yuan, Y., & Zhou, S. (2016). A Hierarchically structured urchin-like anode derived from chestnut shells for microbial energy harvesting. *Electrochimica acta*, *212*, 883–889.

Chen, X. P., Li, Y. F., Yuan, X. L., Li, N., He, W. H., Feng, Y. J., & Liu, J. (2020). Surface modification by Ll -cyclodextrin/polyquaternium-11 composite for enhanced biofilm formation in microbial fuel cells. *Journal of Power Sources*, *480*, 228789.

Cheng, X., Liu, B., Qiu, Y., Liu, K., Fang, Z., Qi, J., Ma, Z., Sun, T., & Liu, S. (2023). Enhanced microorganism attachment and flavin excretion in microbial fuel cells via an N,S-codoped carbon microflower anode. *Journal of Colloid and Interface Science*, *648*, 327–337.

Cheng, Y. F., Feng, G. P., & Moraru, C. I. (2019). Micro- and nanotopography sensitive bacterial attachment mechanisms: A review. *Frontiers in Microbiology*, 10.

Childers., & Susan, E. (2018). Ciufo; Stacy; Lovley; Derek, R. *Geobacter Metallireducens* accesses insoluble Fe(Iii) oxide by chemotaxis. *Nature*, *416*(6882), 767–769.

Chinnaraj, S. B., Jayathilake, P. G., Dawson, J., Ammar, Y., Portoles, J., Jakubovics, N., & Chen, J. J. (2021). Modelling the combined effect of surface roughness and topography on bacterial attachment. *Journal of Materials Science & Technology*, *81*, 151–161.

Cohen, B. (1931). The bacterial culture as an electrical half-cell. *Journal of Bacteriology*, *21*(1), 18–19.

Crawford, R. J., Webb, H. K., Truong, V. K., Hasan, J., & Ivanova, E. P. (2012). Surface topographical factors influencing bacterial attachment. *Advances in Colloid and Interface Science*, *179*, 142–149.

Deeke, A., Sleutels, T. H. J. A., Hamelers, H. V. M., & Buisman, C. J. N. (2012). Capacitive bioanodes enable renewable energy storage in microbial fuel cells. *Environmental Science & Technology*, *46*(6), 3554–3560.

Deng, S., Wang, C., Ngo, H. H., Guo, W., You, N., Tang, H., Yu, H., Tang, L., & Han, J. (2023). Comparative review on microbial electrochemical technologies for resource recovery from wastewater towards circular economy and carbon neutrality. *Bioresource Technology*, *376*. Available from https://doi.org/10.1016/j.biortech.2023.128906.

Deng, X., Dohmae, N., Kaksonen, A. H., & Okamoto, A. (2020). Biogenic iron sulfide nanoparticles to enable extracellular electron uptake in sulfate-reducing bacteria. *Angewandte Chemie International Edition*, *59*(15), 5995–5999.

Dey, N., Samuel, G. V., Raj, D. S., & Gajalakshmi, B. (2022). Nanomaterials as potential high performing electrode materials for microbial fuel cells. *Applied Nanoscience*, 1–16.

Ding, Q., Cao, Y., Li, F., Lin, T., Chen, Y., Chen, Z., & Song, H. (2021). Construction of conjugated polymer-exoelectrogen hybrid bioelectrodes and applications in microbial fuel cells. *Sheng Wu Gong Cheng Xue Bao*, *37*(1), 1–14.

Dong, H. L., Huang, L. Q., Zhao, L. D., Zeng, Q., Liu, X. L., Sheng, Y. Z., Shi, L., Wu, G., Jiang, H. C., Li, F. R., Zhang, L., Guo, D. Y., Li, G. Y., Hou, W. G., & Chen, H. Y. (2022). A critical review of mineral-microbe interaction and co-evolution: Mechanisms and applications. *National Science Review*, *9*(10).

Dou, X. Q., Zhang, D., Feng, C. L., & Jiang, L. (2015). Bioinspired hierarchical surface structures with tunable wettability for regulating bacteria adhesion. *Acs Nano*, *9*(11), 10664–10672.

Du, Q., An, J. K., Li, J. H., Zhou, L. A., Li, N., & Wang, X. (2017). Polydopamine as a new modification material to accelerate startup and promote anode performance in microbial fuel cells. *Journal of Power Sources*, *343*, 477–482.

Du, Q., Li, T., Li, N., & Wang, X. (2017). Protection of electroactive biofilm from extreme acid shock by polydopamine encapsulation. *Environmental Science & Technology Letters, 4*(8), 345−349.

El-Naggar, M. Y., Wanger, G., Leung, K. M., Yuzvinsky, T. D., Southam, G., Yang, J., Lau, W. M., Nealson, K. H., & Gorby, Y. A. (2010). Electrical transport along bacterial nanowires from *Shewanella oneidensis MR-1*. *Proceedings of the National Academy of Sciences, 107*(42), 18127−18131.

Emerson, D., & Craig, M. (1997). Isolation and characterization of novel iron-oxidizing bacteria that grow at circumneutral Ph. *Applied and Environmental Microbiology, 63*(12), 4784−4792.

Encinas, N., Yang, C. Y., Geyer, F., Kaltbeitzel, A., Baumli, P., Reinholz, J., Mailander, V., Butt, H. J., & Vollmer, D. (2020). Submicrometer-sized roughness suppresses bacteria adhesion. *ACS Applied Materials & Interfaces, 12*(19), 21192−21200.

Estrada-Arriaga, E. B., Guadarrama-Pérez, O., Silva-Martínez, S., Cuevas-Arteaga, C., & Guadarrama-Pérez, V. H. (2021). Oxygen reduction reaction (Orr) electrocatalysts in constructed wetland-microbial fuel cells: Effect of different carbon-based catalyst biocathode during bioelectricity production. *Electrochimica acta, 370*, 137745. Available from https://doi.org/10.1016/j.electacta.2021.137745, http://www.journals.elsevier.com/electrochimica-acta/.

Fang, X., Kalathil, S., Divitini, G., Wang, Q., & Reisner, E. (2020). A three-dimensional hybrid electrode with electroactive microbes for efficient electrogenesis and chemical synthesis. *Proceedings of the National Academy of Sciences, 117*(9), 5074.

Feregrino-Rivas, M., Ramirez-Pereda, B., Estrada-Godoy, F., Cuesta-Zedeño, L. F., Rochín-Medina, J. J., Bustos-Terrones, Y. A., & Gonzalez-Huitron, V. A. (2023). Performance of a sediment microbial fuel cell for bioenergy production: Comparison of fluvial and marine sediments. *Biomass Bioenergy, 168*, 106657. Available from https://doi.org/10.1016/j.biombioe.2022.106657, http://www.journals.elsevier.com/biomass-and-bioenergy/.

Flexer, V., & Jourdin, L. (2020). Purposely designed hierarchical porous electrodes for high rate microbial electrosynthesis of acetate from carbon dioxide. *Accounts of Chemical Research, 53*(2), 311−321.

Frauke, K., Igor, V., & Krömer, J. O. (2015). Microbial electron transport and energy conservation—The foundation for optimizing bioelectrochemical systems. *Frontiers in Microbiology, 6*, 575.

Gaffney, E. M., & Minteer, S. D. (2021). A silver assist for microbial fuel cell power. *Science, 373*(6561), 1308−1309.

Ghasemi, M. L., Prabhakaran, M. P., Morshed, M., Nasr-Esfahani, M. H., Baharvand, H., Kiani, S., Al-Deyab, S. S., & Ramakrishna, S. (2011). Application of conductive polymers, scaffolds and electrical stimulation for nerve tissue engineering. *Journal of Tissue Engineering Regenerative Medicine, 5*(4), 17−35.

Gu, Y., Srikanth, V., Salazar-Morales, A. I., Jain, R., O'Brien, J. P., Yi, S. M., Soni, R. K., Samatey, F. A., Yalcin, S. E., & Malvankar, N. S. (2021). Structure of *Geobacter* pili reveals secretory rather than nanowire behaviour. *Nature, 597*(7876), 430−434.

Gude, V. G. (2015). Energy and water autarky of wastewater treatment and power generation systems. *Renewable and Sustainable Energy Reviews, 45*, 52−68. Available from https://doi.org/10.1016/j.rser.2015.01.055, https://www.journals.elsevier.com/renewable-and-sustainable-energy-reviews.

Guo, D., Wei, H.-F., Song, R.-B., Fu, J., Lu, X., Jelinek, R., Min, Q., Zhang, J.-R., Zhang, Q., & Zhu, J.-J. (2019). N,S-doped carbon dots as dual-functional modifiers to boost bio-electricity generation of individually-modified bacterial cells. *Nano Energy, 63*, 103875.

Guo, R. T., Li, L., Wang, B. W., Xiang, Y. G., Zou, G. Q., Zhu, Y. R., Hou, H. S., & Ji, X. B. (2021). Functionalized carbon dots for advanced batteries. *Energy Storage Materials*, *37*, 8−39.

Guo, Y., Wang, J., Shinde, S., Wang, X., Li, Y., Dai, Y., Ren, J., Zhang, P., & Liu, X. (2020). Simultaneous wastewater treatment and energy harvesting in microbial fuel cells: An update on the biocatalysts. *RSC Advances*, *10*(43), 25874−25887. Available from https://doi.org/10.1039/d0ra05234e.

Hemalatha, M., Shanthi Sravan, J., & Venkata Mohan, S. (2020). Self-induced bioelectro-potential influence on sulfate removal and desalination in microbial fuel cell. *Bioresource Technology*, *309*, 123326. Available from https://doi.org/10.1016/j.biortech.2020.123326.

Hindatu, Y., Annuar, M. S. M., & Gumel, A. M. (2017). Mini-review: Anode modification for improved performance of microbial fuel cell. *Renewable & Sustainable Energy Reviews*, *73*, 236−248.

Hou, H., Chen, X., Thomas, A. W., Catania, C., Kirchhofer, N. D., Garner, L. E., Han, A., & Bazan, G. C. (2013). Conjugated oligoelectrolytes increase power generation in *E. coli* microbial fuel cells. *Advanced Materials*, *25*(11), 1593−1597.

Hu, Y., Wang, Y., Han, X., Shan, Y., Li, F., & Shi, L. (2021). Biofilm Biology and engineering of *Geobacter* and *Shewanella Spp.* for energy applications. *Frontiers in Bioengineering and Biotechnology*, *9*, 786416. Available from https://doi.org/10.3389/fbioe.2021.786416.

Hung, Y. H., Liu, T. Y., & Chen, H. Y. (2019). Renewable coffee waste-derived porous carbons as anode materials for high-performance sustainable microbial fuel cells. *ACS Sustainable Chemistry & Engineering*, *7*(20), 16991−16999.

Ibrahim, R. S. B., Zainon Noor, Z., Baharuddin, N. H., Ahmad Mutamim, N. S., & Yuniarto, A. (2020). Microbial fuel cell membrane bioreactor in wastewater treatment, electricity generation and fouling mitigation. *Chemical Engineering & Technology*, *43* (10), 1908−1921.

Jayapriya, J., & Ramamurthy, V. (2012). Use of non-native phenazines to improve the performance of pseudomonas aeruginosa Mtcc 2474 catalysed fuel cells. *Bioresource Technology*, *124*, 23−28.

Jiang, Y., Li, P., Wang, Y., Jiang, L.-P., Song, R.-B., Zhang, J.-R., & Zhu, J.-J. (2020). Trifunctional modification of individual bacterial cells for magnet-assisted bioanodes with high performance in microbial fuel cells. *Journal of Materials Chemistry A*, *8*(46), 24515−24523.

Jiang, Y.-J., Hui, S., Tian, S., Chen, Z., Chai, Y., Jiang, L.-P., Zhang, J.-R., & Zhu, J.-J. (2023). Enhanced Transmembrane electron transfer in *Shewanella Oneidensis MR-1* using gold nanoparticles for high-performance microbial fuel cells. *Nanoscale Advances*, *5*(1), 124−132.

Jiao, Y., Kappler, A., Croal, L. R., & Newman, D. K. (2005). Isolation and characterization of a genetically tractable photoautotrophic Fe(Ii)-oxidizing bacterium, rhodopseudomonas palustris strain Tie-1. *Applied and Environmental Microbiology*, *71*(8), 4487−4496.

Jin, S., Feng, Y., Jia, J., Zhao, F., Wu, Z., Long, P., Li, F., Yu, H., Yang, C., Liu, Q., Zhang, B., Song, H., & Feng, W. (2023). Three-dimensional N-doped carbon nanotube/graphene composite aerogel anode to develop high-power microbial fuel cell. *Energy & Environmental. Materials*, *6*(3), e12373.

Kang, Y. L., Ibrahim, S., & Pichiah, S. (2015). Synergetic effect of conductive polymer poly(3,4-ethylenedioxythiophene) with different structural configuration of anode for microbial fuel cell application. *Bioresource Technology*, *189*, 364−369.

Kang, Y. L., Pichiah, S., & Ibrahim, S. (2017). Facile reconstruction of microbial fuel cell (MFC) anode with enhanced exoelectrogens selection for intensified electricity generation. *International Journal of Hydrogen Energy*, *42*(3), 1661−1671.

Kaur, G., Adhikari, R., Cass, P., Bown, M., & Gunatillake, P. (2015). Electrically conductive polymers and composites for biomedical applications. *RSC Advances*, *5*(47), 37553−37567.

Kharkwal, S., Tan, Y. C., Lu, M., & Ng, H. Y. (2017). Development and long-term stability of a novel microbial fuel cell bod sensor with Mno_2 catalyst. *International Journal of Molecular Sciences*, *18*(2), 1−10. Available from https://doi.org/10.3390/ijms18020276, http://www.mdpi.com/1422-0067/18/2/276/pdf.

Kim, M., Li, S., Kong, D. S., Song, Y. E., Park, S.-Y., Kim, H.-I., Jae, J., Chung, I., & Kim, J. R. (2023). Polydopamine/polypyrrole-modified graphite felt enhances biocompatibility for electroactive bacteria and power density of microbial fuel cell. *Chemosphere*, *313*, 137388.

Kornienko, N., Zhang, J. Z., Sakimoto, K. K., Yang, P. D., & Reisner, E. (2018). Interfacing nature's catalytic machinery with synthetic materials for semi-artificial photosynthesis. *Nature Nanotechnology*, *13*(10), 890−899. Available from https://doi.org/10.1038/s41565-018-0251-7.

Korth, B., Rosa, L. F. M., Harnisch, F., & Picioreanu, C. (2015). A framework for modeling electroactive microbial biofilms performing direct electron transfer. *Bioelectrochemistry*, *106*, 194−206.

Kuleshova, T. E., Ivanova, A. G., Galushko, A. S., Kruchinina, I. Y., Shilova, O. A., Udalova, O. R., Zhestkov, A. S., Panova, G. G., & Gall, N. R. (2022). Influence of the electrode systems parameters on the electricity generation and the possibility of hydrogen production in a plant-microbial fuel cell. *International Journal of Hydrogen Energy*, *47*(58), 24297−24309. Available from https://doi.org/10.1016/j.ijhydene.2022.06.001, http://www.journals.elsevier.com/international-journal-of-hydrogen-energy/.

Lakard, B. (2020). Electrochemical biosensors based on conducting polymers: A review. *Applied Sciences*, *10*(18), 1−24.

Lawson, K., Rossi, R., Regan, J. M., & Logan, B. E. (2020). Impact of cathodic electron acceptor on microbial fuel cell internal resistance. *Bioresource Technology*, *316*, 123919. Available from https://doi.org/10.1016/j.biortech.2020.123919, http://www.elsevier.com/locate/biortech.

Le, T. H., Kim, Y., & Yoon, H. (2017). Electrical and electrochemical properties of conducting polymers. *Polymers (Basel)*, *9*(4), 1−32.

Li, H. D., Zhang, L., Wang, R. W., Sun, J. Z., Qiu, Y. F., & Liu, S. Q. (2023). 3D hierarchical porous carbon foams as high-performance free-standing anodes for microbial. *Fuel Cells. Ecomat*, *5*(1), e12273.

Li, P., Jiang, Y., Song, R.-B., Zhang, J.-R., & Zhu, J.-J. (2021). Layer-by-layer assembly of au and cds nanoparticles on the surface of bacterial cells for photo-assisted bioanodes in microbial fuel cells. *Journal of Materials Chemistry B*, *9*(6), 1638−1646.

Li, S., Cheng, C., & Thomas, A. (2017). Carbon-based microbial-fuel-cell electrodes: From conductive supports to active catalysts. *Advanced Materials*, *29*(8), 1602547. Available from https://doi.org/10.1002/adma.201602547, http://onlinelibrary.wiley.com/journal/10.1002/(ISSN)1521-4095.

Li, Y., Liu, J., Chen, X., Wu, J., Li, N., He, W., & Feng, Y. (2021). Tailoring surface properties of electrodes for synchronous enhanced extracellular electron transfer and enriched exoelectrogens in microbial fuel cells. *ACS Applied Materials & Interfaces*, *13*(49), 58508−58521.

Li, Y., Liu, J., Chen, X., Yuan, X., Li, N., He, W., & Feng, Y. (2021). Tailoring spatial structure of electroactive biofilm for enhanced activity and direct electron transfer on iron phthalocyanine modified anode in microbial fuel cells. *Biosensors and Bioelectronics*, *191*, 113410.

Liang, P., Duan, R., Jiang, Y., Zhang, X., Qiu, Y., & Huang, X. (2018). One-year operation of 1000-L modularized microbial fuel cell for municipal wastewater treatment. *Water Research*, *141*, 1−8. Available from https://doi.org/10.1016/j.watres.2018.04.066.

Liang, P., Wang, H., Xia, X., Huang, X., Mo, Y., Cao, X., & Fan, M. (2011). Carbon nanotube powders as electrode modifier to enhance the activity of anodic biofilm in microbial fuel cells. *Biosensors and Bioelectronics, 26*(6), 3000–3004.

Light, S. H., Su, L., Rivera-Lugo, R., Cornejo, J. A., Louie, A., Iavarone, A. T., Ajo-Franklin, C. M., & Portnoy, D. A. (2018). A flavin-based extracellular electron transfer mechanism in diverse Gram-positive bacteria. *Nature, 562*(7725), 140–144.

Liu, D., Wang, R. W., Chang, W., Zhang, L., Peng, B. Q., Li, H. D., Liu, S. Q., Yan, M., & Guo, C. S. (2018). Ti$_3$C$_2$ Mxene as an excellent anode material for high-performance microbial fuel cells. *Journal of Materials Chemistry A, 6*(42), 20887–20895.

Liu, P., Zhang, C., Liang, P., Jiang, Y., Zhang, X., & Huang, X. (2019). Enhancing extra-cellular electron transfer efficiency and bioelectricity production by vapor polymerization poly (3,4-ethylenedioxythiophene)/Mno$_2$ hybrid anode. *Bioelectrochemistry, 126*, 72–78.

Liu, Y. F., Sun, Y. X., Zhang, M., Guo, S. Q., Su, Z. J., Ren, T. L., & Li, C. J. (2023). Carbon nanotubes encapsulating Fes$_2$ micropolyhedrons as an anode electrocatalyst for improving the power generation of microbial fuel cells. *Journal of Colloid and Interface Science, 629*, 970–979.

Logan, B. E., Rossi, R., Ragab, A., & Saikaly, P. E. (2019). electroactive microorganisms in bioelectrochemical systems. *Nature reviews. Microbiology, 17*(5), 307–319. Available from https://doi.org/10.1038/s41579-019-0173-x, http://www.nature.com/nrmicro/index.html.

Lovley, D. R. (2012). Electromicrobiology. *Annual Review of Microbiology, 66*, 391–409.

Lower, S. K., Hochella, M. F., Jr, & Beveridge, T. J. (2001). Bacterial recognition of mineral surfaces: Nanoscale interactions between *Shewanella* and Alpha-FeOOH. *Science (New York, N.Y.), 292*(5520), 1360–1363.

Lu, A., Gao, Y., Jin, T., Luo, X., Zeng, Q., & Shang, Z. (2020). Effects of surface roughness and texture on the bacterial adhesion on the bearing surface of bio-ceramic joint implants: An in vitro study. *Ceramics International., 46*(5), 6550–6559.

Ma, J., Zhang, J., Zhang, Y., Guo, Q., Hu, T., Xiao, H., Lu, W., & Jia, J. (2023). Progress on anodic modification materials and future development directions in microbial fuel cells. *Journal of Power Sources, 556*, 232486. Available from https://doi.org/10.1016/j.jpowsour.2022.232486.

Marsili, E., Baron, D. B., Shikhare, I. D., Coursolle, D., Gralnick, J. A., & Bond, D. R. (2008). *Shewanella* secretes flavins that mediate extracellular electron transfer. *Proceedings of the National Academy of Sciences of the United States of America., 105*(10), 3968–3973.

Oh, S. E., & Logan, B. E. (2005). Hydrogen and electricity production from a food processing wastewater using fermentation and microbial fuel cell technologies. *Water Research, 39*(19), 4673–4682. Available from https://doi.org/10.1016/j.watres.2005.09.019, http://www.elsevier.com/locate/watres.

Okamoto, A., Hashimoto, K., & Nealson, K. H. (2014). Flavin redox bifurcation as a mechanism for controlling the direction of electron flow during extracellular electron transfer. *Angewandte Chemie International Edition., 53*(41), 10988–10991.

Okamoto, A., Hashimoto, K., Nealson, K. H., & Nakamura, R. (2013). Rate enhancement of bacterial extracellular electron transport involves bound flavin semiquinones. *Proceedings of the National Academy of Sciences of the United States of America., 110*(19), 7856–7861.

Pons, L., Delia, M. L., & Bergel, A. (2011). Effect of surface roughness, biofilm coverage and biofilm structure on the electrochemical efficiency of microbial cathodes. *Bioresource Technology, 102*(3), 2678–2683.

Potter, C. M. (1911). Electrical effects accompanying the decomposition of organic compounds. *Proceedings of the Royal Society of London, 84*(571), 260–276.

Ren, H., Pyo, S., Lee, J. I., Park, T. J., Gittleson, F. S., Leung, F. C. C., Kim, J., Taylor, A. D., Lee, H.-S., & Chae, J. (2015). A high power density miniaturized microbial fuel cell having carbon nanotube anodes. *Journal of Power Sources*, *273*, 823−830.

Reza, A. (2006). Polypyrrole conducting electroactive polymers: Synthesis and stability studies. *E-Journal of Chemistry*, *3*(4), 186−201.

Richter, H., Nevin, K. P., Jia, H., Lowy, D. A., Lovley, D. R., & Tender, L. M. (2009). Cyclic voltammetry of biofilms of wild type and mutant *Geobacter Sulfurreducens* on fuel cell anodes indicates possible roles of Omcb, Omcz, Type Iv Pili, and protons in extracellular electron transfer. *Energy & Environmental Science*, *2*(5), 506−516.

Santoro, C., Guilizzoni, M., Baena, J. P. C., Pasaogullari, U., Casalegno, A., Li, B., Babanova, S., Artyushkova, K., & Atanassov, P. (2014). The effects of carbon electrode surface properties on bacteria attachment and start up time of microbial fuel cells. *Carbon*, *67*, 128−139.

Schröder, U., Harnisch, F., & Angenent, L. T. (2015). Microbial electrochemistry and technology: Terminology and classification. *Energy & Environmental Science*, *8*(2), 513−519. Available from https://doi.org/10.1039/c4ee03359k, http://pubs.rsc.org/en/journals/journal/ee.

Shanthi Sravan, J., Tharak, A., Annie Modestra, J., Seop Chang, I., & Venkata Mohan, S. (2021). Emerging trends in microbial fuel cell diversification-critical analysis. *Bioresource Technology*, *326*, 124676.

Shelobolina, E., Xu, H. F., Konishi, H., Kukkadapu, R., Wu, T., Blothe, M., & Roden, E. (2012). Microbial lithotrophic oxidation of structural Fe(Ii) in biotite. *Applied and Environmental Microbiology*, *78*(16), 5746−5752.

Shi, X., Duan, Z., Jing, W., Zhou, W., Jiang, M., Li, T., Ma, H., & Zhu, X. (2023). Simultaneous removal of multiple heavy metals using single chamber microbial electrolysis cells with biocathode in the micro-aerobic environment. *Chemosphere*, *318*, 137982. Available from https://doi.org/10.1016/j.chemosphere.2023.137982.

Sibi, R., Sheelam, A., Gunaseelan, K., Jadhav, D. A., & Gangadharan, P. (2023). osmotic microbial fuel cell for sustainable wastewater treatment along with desalination, bioenergy and resource recovery: A critical review. *Bioresource Technology Reports*, *23*, 101540. Available from https://doi.org/10.1016/j.biteb.2023.101540.

Sivasankar, P., Poongodi, S., Seedevi, P., Sivakumar, M., Murugan, T., & Loganathan, S. (2019). Bioremediation of wastewater through a Quorum sensing triggered MFC: A sustainable measure for waste to energy concept. *Journal of Environmental Management*, *237*, 84−93. Available from https://doi.org/10.1016/j.jenvman.2019.01.075.

Sogani, M., Pankan, A. O., Dongre, A., Yunus, K., & Fisher, A. C. (2021). Augmenting the biodegradation of recalcitrant ethinylestradiol using *Rhodopseudomonas palustris* in a hybrid photo-assisted microbial fuel cell with enhanced bio-hydrogen production. *Journal of Hazardous Materials*, *408*, 124421. Available from https://doi.org/10.1016/j.jhazmat.2020.124421, http://www.elsevier.com/locate/jhazmat.

Sonawane, J. M., Al-Saadi, S., Singh Raman, R. K., Ghosh, P. C., & Adeloju, S. B. (2018). Exploring the use of polyaniline-modified stainless steel plates as low-cost, high-performance anodes for microbial fuel cells. *Electrochimica acta*, *268*, 484−493.

Sonawane, J. M., Patil, S. A., Ghosh, P. C., & Adeloju, S. B. (2018). Low-cost stainless-steel wool anodes modified with polyaniline and polypyrrole for high-performance microbial fuel cells. *Journal of Power Sources*, *379*, 103−114.

Sonawane, J. M., Yadav, A., Ghosh, P. C., & Adeloju, S. B. (2017). Recent advances in the development and utilization of modern anode materials for high performance microbial fuel cells. *Biosensors and Bioelectronics*, *90*, 558−576. Available from https://doi.org/10.1016/j.bios.2016.10.014, http://www.elsevier.com/locate/bios.

Speers, A. M., & Reguera, G. (2012). Electron donors supporting growth and electroactivity of geobacter sulfurreducens anode biofilms. *Applied and Environmental Microbiology*, 78(2), 437−444.

Strycharz, S. M., Glaven, R. H., Coppi, M. V., Gannon, S. M., Perpetua, L. A., Liu, A., Nevin, K. P., & Lovley, D. R. (2011). Gene expression and deletion analysis of mechanisms for electron transfer from electrodes to *Geobacter sulfurreducens*. *Bioelectrochemistry*, 80(2), 142−150.

Strycharz-Glaven, S. M., Glaven, R. H., Wang, Z., Zhou, J., Vora, G. J., & Tender, L. M. (2013). Electrochemical investigation of a microbial solar cell reveals a nonphotosynthetic biocathode catalyst. *Applied and Environmental Microbiology*, 79(13), 3933−3942.

Sun, J., Wang, R., Li, H., Zhang, L., & Liu, S. (2023). Boosting bioelectricity generation using three-dimensional nitrogen-doped macroporous carbons as freestanding anode. *Materials Today Energy*, 33, 101273. Available from https://doi.org/10.1016/j.mtener.2023.101273, https://www.sciencedirect.com/science/article/pii/S2468606923000291.

Tan, Y., Adhikari, R. Y., Malvankar, N. S., Ward, J. E., Woodard, T. L., Nevin, K. P., & Lovley, D. R. (2017). Expressing the *Geobacter Metallireducens* pila in *Geobacter Sulfurreducens* yields pili with exceptional conductivity. *mBio*, 8(1).

Tao, Y. F., Liu, Q. Z., Chen, J. H., Wang, B., Wang, Y. D., Liu, K., Li, M. F., Jiang, H. Q., Lu, Z. T., & Wang, D. (2016). hierarchically three-dimensional nanofiber based textile with high conductivity and biocompatibility as a microbial fuel cell anode. *Environmental Science & Technology*, 50(14), 7889−7895.

Truong, V. K., Lapovok, R., Estrin, Y. S., Rundell, S., Wang, J. Y., Fluke, C. J., Crawford, R. J., & Ivanova, E. R. (2010). The influence of nano-scale surface roughness on bacterial adhesion to ultrafine-grained titanium. *Biomaterials*, 31(13), 3674−3683.

Ueki, T. (2021). Cytochromes in extracellular electron transfer in geobacter. *Applied and Environmental Microbiology*, 87(10), 03109−03120.

Vishwanathan, A. S. (2021). Microbial fuel cells: A comprehensive review for beginners. *3 Biotech*, 11(5), 1−14.

Wang, D., Pan, J., Xu, M., Liu, B., Hu, J., Hu, S., Hou, H., Elmaadawy, K., Yang, J., Xiao, K., & Liang, S. (2021). Surface modification of *Shewanella Oneidensis MR-1* with polypyrrole-dopamine coating for improvement of power generation in microbial fuel cells. *Journal of Power Sources*, 483, 229220.

Wang, F., Gu, Y., O'Brien, J. P., Yi, S. M., Yalcin, S. E., Srikanth, V., Shen, C., Vu, D., Ing, N. L., Hochbaum, A. I., Egelman, E. H., & Malvankar, N. S. (2019). Structure of microbial nanowires reveals stacked hemes that transport electrons over micrometers. *Cell*, 177(2), 361−369.

Wang, L., Liu, W., He, Z., Guo, Z., Zhou, A., & Wang, A. (2017). Cathodic hydrogen recovery and methane conversion using Pt coating 3d nickel foam instead of Pt-carbon cloth in microbial electrolysis cells. *International Journal of Hydrogen Energy*, 42 (31), 19604−19610. Available from https://doi.org/10.1016/j.ijhydene.2017.06.019.

Wang, R. W., Yan, M., Li, H. D., Zhang, L., Peng, B., Sun, J., Liu, D., & Liu, S. (2018). Fes$_2$ nanoparticles decorated graphene as microbial-fuel-cell anode achieving high power density. *Advanced Materials*, 30(22), 1800618.

Wang, S., Tian, S., Zhang, P., Ye, J., Tao, X., Li, F., Zhou, Z., & Nabi, M. (2019). Enhancement of biological oxygen demand detection with a microbial fuel cell using potassium permanganate as cathodic electron acceptor. *Journal of Environmental Management*, 252, 109682. Available from https://doi.org/10.1016/j.jenvman.2019.109682.

Wang, W., Zhang, Y., Li, M., Wei, X., Wang, Y., Liu, L., Wang, H., & Shen, S. (2020). Operation mechanism of constructed wetland-microbial fuel cells for wastewater treatment and electricity generation: A review. *Bioresource Technology*, 314(2020), 123808. Available from https://doi.org/10.1016/j.biortech.2020.123808.

Wang, Y. P., Cheng, X. S., Liu, K., Dai, X. F., Qi, J. T., Ma, Z., Qiu, Y. F., & Liu, S. Q. (2022). 3D hierarchical Co8fes8-Feco2o4/N-Cnts@Cf with an enhanced microorganisms-anode interface for improving microbial fuel cell performance. *ACS Applied Materials & Interfaces, 14*(31), 35809—35821.

Wang, Y. X., He, C. S., Li, W. Q., Zong, W. M., Li, Z. H., Yuan, L., Wang, G. M., & Mu, Y. (2020). High power generation in mixed-culture microbial fuel cells with corncob-derived three-dimensional n-doped bioanodes and the impact of N dopant states. *Chemical Engineering Journal, 399*.

Weber, K. A., Achenbach, L. A., & Coates, J. D. (2006). Microorganisms pumping iron: Anaerobic microbial iron oxidation and reduction. *Nature reviews. Microbiology, 4*(10), 752—764.

Wei, J., Li, Y., Dai, D., Zhang, F., Zou, H., Yang, X., Ji, Y., Li, B., & Wei, X. (2020). Surface roughness: A crucial factor to robust electric double layer capacitors. *ACS Applied Materials & Interfaces, 12*(5), 5786—5792.

Winther-Jensen, B., & MacFarlane, D. R. (2011). New generation, metal-free electrocatalysts for fuel cells, solar cells and water splitting. *Energy & Environmental Science, 4*(8), 2790—2798.

Wu, J. X., Liu, R. J., Dong, P. F., Li, N., He, W. H., Feng, Y. J., & Liu, J. (2023). Enhanced electricity generation and storage by nitrogen-doped hierarchically porous carbon modification of the capacitive bioanode in microbial fuel cells. *The Science of the Total Environment, 858*.

Wu, W. G., Niu, H., Yang, D. Y., Wang, S. B., Jiang, N. N., Wang, J. F., Lin, J., & Hu, C. Y. (2018). Polyaniline/carbon nanotubes composite modified anode via graft polymerization and self-assembling for microbial fuel cells. *Polymers, 10*(7).

Wu, X., Qiao, Y., Guo, C., Shi, Z., & Li, C. M. (2020). Nitrogen doping to atomically match reaction sites in microbial fuel cells. *Communications Chemistry, 3*(1), 1—9.

Xiang, Y., Liu, T., Jia, B., Zhang, L., & Su, X. (2023). Boosting bioelectricity generation in microbial fuel cells via biomimetic Fe-N-S-C nanozymes. *Biosensors and Bioelectronics, 220*, 114895.

Xie, X., Hu, L., Pasta, M., Wells, G. F., Kong, D., Criddle, C. S., & Cui, Y. (2011). Three-dimensional carbon nanotube-textile anode for high-performance microbial fuel cells. *Nano Letters, 11*(1), 291—296.

Yalcin, S. E., O'Brien, J. P., Gu, Y., Reiss, K., Yi, S. M., Jain, R., Srikanth, V., Dahl, P. J., Huynh, W., Vu, D., Acharya, A., Chaudhuri, S., Varga, T., Batista, V. S., & Malvankar, N. S. (2020). Electric field stimulates production of highly conductive microbial Omcz nanowires. *Nature chemical biology, 16*(10), 1136—1142.

Yang, C., Aslan, H., Zhang, P., Zhu, S., Xiao, Y., Chen, L., Khan, N., Boesen, T., Wang, Y., Liu, Y., Wang, L., Sun, Y., Feng, Y., Besenbacher, F., Zhao, F., & Yu, M. (2020). Carbon dots-fed *Shewanella Oneidensis MR-1* for bioelectricity enhancement. *Nature Communications, 11*(1), 1379.

Yaqoob, A. A., Ibrahim, M. N. M., & Guerrero-Barajas, C. (2021). Modern trend of anodes in microbial fuel cells (MFCs): An overview. *Environmental Technology & Innovation, 23*, 101579.

Ye, Z., Ellis, M. W., Nain, A. S., & Behkam, B. (2017). Effect of electrode sub-micron surface feature size on current generation of *Shewanella Oneidensis* in microbial fuel cells. *Journal of Power Sources, 347*, 270—276.

You, S., Ma, M., Wang, W., Qi, D., Chen, X., Qu, J., & Ren, N. (2017). 3D macroporous nitrogen-enriched graphitic carbon scaffold for efficient bioelectricity generation in microbial fuel cells. *Advanced Energy Materials, 7*(4), 1601364.

Yu, S. S., Chen, J. J., Cheng, R. F., Min, Y., & Yu, H. Q. (2021). Iron cycle tuned by outer-membrane cytochromes of dissimilatory metal-reducing bacteria: Interfacial dynamics and mechanisms in vitro. *Environmental Science & Technology, 55*(16), 11424—11433.

Yu, Y.-Y., Chen, H.-l, Yong, Y.-C., Kim, D.-H., & Song, H. (2011). Conductive artificial biofilm dramatically enhances bioelectricity production in shewanella-inoculated microbial fuel cells. *Chemical Communications*, *47*(48), 12825−12827.

Yu, Y. Y., Guo, C. X., Yong, Y. C., Li, C. M., & Song, H. (2015). Nitrogen doped carbon nanoparticles enhanced extracellular electron transfer for high-performance microbial fuel cells anode. *Chemosphere*, *140*, 26−33.

Yu, Y.-Y., Wang, Y.-Z., Fang, Z., Shi, Y.-T., Cheng, Q.-W., Chen, Y.-X., Shi, W., & Yong, Y.-C. (2020). Single cell electron collectors for highly efficient wiring-up electronic abiotic/biotic interfaces. *Nature Communications*, *11*(1), 4087.

Yuan, H. R., Deng, L. F., Qian, X., Wang, L. F., Li, D. N., Chen, Y., & Yuan, Y. (2019). significant enhancement of electron transfer from *Shewanella Oneidensis* using a porous N-doped carbon cloth in a bioelectrochemical system. *The Science of the Total Environment*, *665*, 882−889.

Zhang, J., Cao, X., Wang, H., Long, X., & Li, X. (2020). Simultaneous enhancement of heavy metal removal and electricity generation in soil microbial fuel cell. *Ecotoxicology and Environmental Safety*, *192*, 110314. Available from https://doi.org/10.1016/j.ecoenv.2020.110314.

Zhang, J., Sun, Y., Zhang, H., Cao, X., Wang, H., & Li, X. (2021). Effects of cathode/anode electron accumulation on soil microbial fuel cell power generation and heavy metal removal. *Environmental Research*, *198*, 111217. Available from https://doi.org/10.1016/j.envres.2021.111217.

Zhang, T., Zhang, L., Su, W., Gao, P., Li, D., He, X., & Zhang, Y. (2011). The direct electrocatalysis of phenazine-1-carboxylic acid excreted by pseudomonas alcaliphila under alkaline condition in microbial fuel cells. *Bioresource Technology*, *102*(14), 7099−7102.

Zhao, C. E., Gai, P. P., Song, R. B., Chen, Y., Zhang, J. R., & Zhu, J. J. (2017). Nanostructured material-based biofuel cells: Recent advances and future prospects. *Chemical Society Reviews*, *46*(5), 1545−1564.

Zhao, J. T., Li, F., Cao, Y. X., Zhang, X. B., Chen, T., Song, H., & Wang, Z. W. (2021). Microbial extracellular electron transfer and strategies for engineering electroactive microorganisms. *Biotechnology Advances*, 53.

Zhou, E., Lekbach, Y., Gu, T., & Xu, D. (2022). Bioenergetics and extracellular electron transfer in microbial fuel cells and microbial corrosion. *Current Opinion in. Electrochemistry*, *31*, 100830.

Zou, L., Qiao, Y., Zhong, C., & Li, C. M. (2017). Enabling fast electron transfer through both bacterial outer-membrane redox centers and endogenous electron mediators by polyaniline hybridized large-mesoporous carbon anode for high-performance microbial fuel cells. *Electrochimica acta*, *229*, 31−38.

CHAPTER 7

Nanoelectrochemistry in next generation lithium batteries

Moon San[1], Do Youb Kim[1], Myeong Hwan Lee[1], Jungdon Suk[1,2] and Yongku Kang[1,2]

[1]Advanced Energy Materials Research Center, Korea Research Institute of Chemical Technology (KRICT), Daejeon, Republic of Korea
[2]Department of Chemical Convergence Materials, University of Science & Technology (UST), Daejeon, Republic of Korea

7.1 Introduction

Rechargeable Li batteries have led to remarkable revolution in consumer electronics, electric vehicles (EV), and energy storage system (ESS). In particular, EVs are increasingly recognized as a viable option for autonomous vehicles playing a critical role in mitigating CO_2 emissions. While rechargeable Li batteries have proven successful in portable electronics, their widespread application in EVs faces challenges such as limited energy density leading to a shorter driving range, constrained cycle life, the necessity to decrease charging times, and safety concerns inherent to Li batteries. Overcoming these obstacles is essential to fulfill the evolving requirements of EVs.

Nanotechnology emerges as a promising approach for addressing the limitations of Li batteries in materials level. One illustration of this is the modifying the nanostructure of battery materials which facilitates the design of novel battery materials featuring high specific capacity and rapid electrochemical reactions, leading to Li batteries with increased capacity and enabling fast charging. Furthermore, adjusting the interfacial layer between the electrode and electrolyte interphase region not only has the potential to reduce degradation of electrode materials, thereby enhancing cycle life, but also to mitigate abnormal parasitic exothermal reaction.

In this chapter, we briefly review examples how nanotechnology is applied to mitigate the limitations associated with the four essential materials in Li batteries: anode materials, cathode materials, electrolytes, and separators.

Electrochemistry and Photo-Electrochemistry of Nanomaterials
DOI: https://doi.org/10.1016/B978-0-443-18600-4.00008-9
© 2025 Elsevier Inc. All rights reserved, including those for text and data mining, AI training, and similar technologies.

7.2 Nanoelectrochemistry in the cathode materials
7.2.1 Fundamental properties of cathode materials

Cathode materials are pivotal in dictating the performance of batteries. Their fundamental properties, both chemical and physical, determine the efficiency, capacity, and longevity of the battery. This exploration of these properties includes the role of nanoscale characteristics in enhancing electrochemical performance and a comparative analysis between traditional and nanoscale cathode materials. Cathode materials primarily consist of lithium metal oxides, where the choice of metal and its oxidation state plays a significant role in the battery's characteristics. Traditional cathodes like lithium cobalt oxide (LiCoO$_2$), lithium manganese oxide (LiMn$_2$O$_4$), and lithium iron phosphate (LiFePO$_4$) vary in voltage, capacity, and thermal stability based on their chemical makeup. LiCoO$_2$ offers high energy density but poses safety risks, LiMn$_2$O$_4$ provides high thermal stability, and LiFePO$_4$ is known for its safety and long-life cycle (Fig. 7.1) (Julien, 2016; Pan, 2018).

The crystal structure of these materials affects how lithium ions intercalate and deintercalate during charging and discharging. For instance, the layered structure of LiCoO$_2$ facilitates smooth lithium-ion movement, while the olivine structure of LiFePO$_4$ offers high structural stability. Particle size,

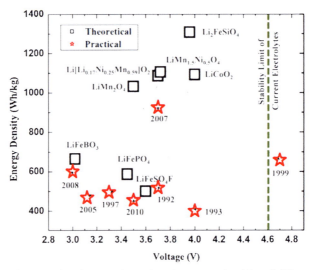

Figure 7.1 Theoretical and practical gravimetric energy densities of different cathode materials. *Permission from Xu, B., Qian, D., Wang, Z., & Meng, Y. S. (2012). Recent progress in cathode materials research for advanced lithium ion batteries. Materials Science and Engineering: R: Reports, 73(5–6), 51.*

morphology, and surface area of cathode materials also play crucial roles, influencing their performance. Smaller particles provide a higher surface area-to-volume ratio, enhancing reaction kinetics for faster charging and discharging, albeit potentially increasing side reactions with the electrolyte. The advent of nanotechnology in cathode material development has led to significant improvements in battery performance. Nanoscale cathode materials exhibit increased surface area, leading to better electrode—electrolyte contact and faster electrochemical reactions. The reduced size of nanomaterials shortens the diffusion pathway for lithium ions, improving rate capability. Nanostructured cathodes can better accommodate strain during ion insertion and extraction, enhancing cycling stability. Furthermore, nanoscale modifications to cathode materials can result in higher capacities, improved power densities, and enhanced safety profiles.

Comparing traditional and nanoscale cathode materials reveals significant differences in performance. While traditional cathodes like $LiCoO_2$ offer high energy densities, nanoscale cathodes push these boundaries further due to their enhanced electrochemical properties. Nanoscale cathodes can charge and discharge faster due to reduced lithium-ion diffusion lengths, which is crucial in high-power applications where rapid energy delivery is essential. Nanostructuring often improves the life cycle of cathode materials by mitigating the effects of repeated charge—discharge cycles. However, despite these advantages, nanoscale cathodes face challenges in uniformity, scalability, and potential cost implications. There is also the concern of increased reactivity due to larger surface areas, which could impact long-term stability.

The fundamental chemical and physical properties of cathode materials largely determine the performance of batteries. The shift toward nanoscale cathode materials has shown considerable promise in enhancing these properties, thereby improving the overall efficiency, capacity, and safety of batteries. A balanced approach, considering the advantages and limitations of both traditional and nanoscale materials, is crucial for the ongoing development of advanced battery technologies. As research continues to evolve in this field, the exploration of nanoscale cathode materials is expected to play a pivotal role in shaping the future of energy storage systems.

7.2.2 Synthesis of nanostructured cathode materials

The synthesis of nanostructured cathode materials is a cornerstone in advancing battery technology. This chapter explores the various methods and techniques for synthesizing nanomaterials, the principles of nanoengineering for optimizing cathode performance, and provides insights through case studies of successful nanomaterial syntheses.

The creation of nanostructured cathode materials involves several sophisticated techniques, each offering unique advantages in tailoring material properties. The sol–gel method is a versatile technique for producing oxide materials, involving hydrolysis and polycondensation reactions, leading to the formation of a colloidal suspension (sol) that transforms into a gel-like network. This method offers precise control over the composition and morphology of the resulting nanomaterial (Liu, 2004). Hydrothermal/solvothermal synthesis, conducted in high-pressure and temperature conditions in aqueous or organic solvents, is ideal for synthesizing crystalline materials with controlled size and shape and is useful for producing complex, multicomponent oxide materials (Kang, 2023). The co-precipitation technique involves the precipitation of a solid material from a solution containing the desired metal cations, allowing for homogeneous mixing of elements at the molecular level, essential for complex oxide cathode materials (Dong & Koenig, 2020). Template-based methods utilize templates to control the morphology and size of the nanostructures, effective for creating nanostructured cathodes with specific architectures like nanowires or nanotubes. Mechanical milling, a physical method used to reduce particle size and mix materials at the nanoscale, can be used to synthesize nanocomposite cathode materials (Chen, 2016).

Engineering cathode materials at the nanoscale is aimed at optimizing their electrochemical properties. Nanosized particles offer a larger surface area to volume ratio, improving the kinetics of electrochemical reactions and enhancing battery performance. Smaller particle sizes result in shorter lithium-ion diffusion pathways, leading to better rate capabilities and faster charging. Nanostructuring can improve the structural stability of cathode materials during cycling, enhancing the durability and lifespan of batteries. Through nanoengineering, it is possible to tailor the electronic and ionic conductivity of cathode materials, optimizing them for specific battery applications.

Several case studies highlight the successful synthesis and application of nanostructured cathode materials. Nanostructured lithium iron phosphate ($LiFePO_4$) has been synthesized using hydrothermal methods, resulting in materials with improved rate performance and cycling stability, demonstrating success in electric vehicle applications due to its safety, stability, and longevity. Layered lithium nickel cobalt manganese oxide (NCM) produced through coprecipitation offers high energy density and capacity and is utilized in consumer electronics and electric vehicles, showcasing improvements in energy density and battery life. Lithium-rich layered oxides synthesized using sol–gel methods yield high-capacity cathode materials, showing potential in extending the range of electric vehicles due to their high energy capacities. Lithium manganese oxide ($LiMn_2O_4$) Nanoparticles, created using mechanical milling, result in nanoscale particles with enhanced performance in terms of power output and thermal stability, applied in high-power applications like power tools and hybrid vehicles (Chen, 2016).

The synthesis of nanostructured cathode materials represents a significant advancement in battery technology. Through various synthesis techniques, researchers have successfully developed nanomaterials that exhibit superior electrochemical properties compared to their bulk counterparts. These advancements not only lead to batteries with higher energy densities, faster charging capabilities, and longer lifespans but also open new avenues for innovation in energy storage solutions. As the demand for more efficient and sustainable batteries grows, the continued exploration and optimization of nanostructured cathode materials will remain a key focus in the field of energy research.

7.2.3 Types of advanced cathode materials

In the dynamic and ever-progressing realm of battery technology, the creation and refinement of advanced cathode materials stand as a critical area of focus. A range of sophisticated cathode materials is covered, encompassing layered oxides and their nanoscale adaptations, phosphate-based cathodes, and novel materials such as sulfides and air cathodes. These materials are at the cutting edge of current research, holding the promise of substantially improving battery performance due to their unique and enhanced properties.

Layered oxide cathode materials, particularly those based on lithium, have been the backbone of modern lithium-ion batteries. They are known for their high energy density and good cycling stability. Traditional layered oxides, such as lithium cobalt oxide ($LiCoO_2$) and lithium nickel manganese cobalt oxide (NMC), are widely used and offer high capacity and energy density but face challenges like cost, safety, and limited lithium resource availability. Nanoscale variants of these materials have led to improved performance, with nanostructured NMC exhibiting enhanced rate capability and cycle ability. The reduced particle size in these nanoscale variants leads to faster lithium-ion diffusion and better accommodation of volume changes during cycling.

Phosphate-based cathodes, such as lithium iron phosphate ($LiFePO_4$), have gained attention due to their safety, thermal stability, and long cycle life. $LiFePO_4$, with its olivine structure, offers a stable and safe cathode choice, not releasing oxygen at high temperatures, a common issue with layered oxides. While it has a lower energy density compared to layered oxides, its safety profile makes it suitable for applications like electric vehicles and large-scale energy storage systems. Nanostructuring $LiFePO_4$ has significantly improved its electrochemical performance, particularly in terms of rate capability, due to enhanced surface area and shorter lithium-ion diffusion pathways in the nanoscale material.

Research in battery technology continually pushes boundaries, exploring new materials that offer unique advantages. Lithium sulfides in lithium-sulfur (Li-S) batteries promise high energy densities, potentially exceeding those of traditional lithium-ion batteries. However, challenges like the polysulfide shuttle phenomenon and the low electrical conductivity of sulfur need to be addressed for practical applications. Lithium-air (Li-air) batteries have the theoretical potential to achieve extremely high energy densities, using oxygen from the air as the active material in the cathode, which reacts with lithium to form lithium peroxide. Challenges include slow reaction kinetics, moisture and CO_2 sensitivity, and the development of suitable electrolytes. Research is also exploring other materials like lithium-rich layered oxides (LLO), which offer higher capacities and manganese-rich cathodes for their cost-effectiveness and abundance (Fig. 7.2) (Kurzweil, 2015).

While advanced cathode materials offer promising avenues for enhanced battery performance, they also present challenges. The complex synthesis of advanced materials can be a barrier to their commercialization, and developing cost-effective and scalable synthesis methods is crucial. Many high-capacity materials face challenges in terms of long-term stability and cycle life, which must be addressed for practical applications. Additionally, the environmental impact and safety of these advanced cathodes need careful consideration, especially when scaling for widespread use.

7.2.4 Nanoelectrochemical mechanisms in cathode materials

Investigating nanoelectrochemistry in cathode materials is essential in cutting-edge battery research, intertwining the fields of material science

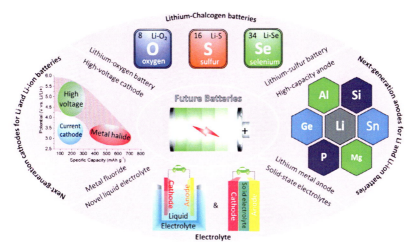

Figure 7.2 The selected most popular research trends in rechargeable Li and Li-ion battery field for achieving high energy densities in future batteries. Permission from Wu, F., Maier, J., & Yu, Y. (2020). Guidelines and trends for next-generation rechargeable lithium and lithium-ion batteries. Chemical Society Reviews, 49(5), 1569

and electrochemistry. The complex nanoelectrochemical activities within cathode materials are crucial, particularly when considering aspects such as charge transfer, ion transportation, and diffusion in nanostructured cathodes. Equally important are the comprehensive effects of nanostructuring on their electrochemical performance, which play a pivotal role in advancing battery technology.

The efficiency of charge transfer processes in battery cathodes is fundamentally altered at the nanoscale. Nanostructuring increases the surface area of cathode materials, leading to a higher number of active sites for redox reactions. This increase in reactive surface area enhances the overall charge transfer rate, crucial for high-performance batteries. At the nanoscale, quantum effects can influence the electronic properties of materials, potentially altering band gaps and electronic conductivities, impacting the material's ability to facilitate charge transfer. Nanostructuring can also modify the pathways available for electron transport within the cathode material, leading to more efficient electron movement, particularly in materials where bulk pathways are less conducive (Fig. 7.3).

Nanostructuring significantly influences how ions move within cathode materials, a critical factor for battery performance. By reducing the size of cathode particles to the nanoscale, the pathways for lithium-ion diffusion are shortened, leading to faster ion transport and enhancing the

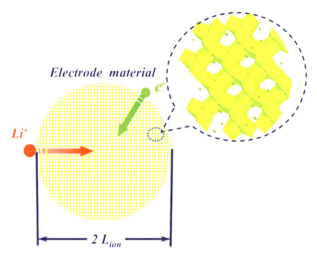

Figure 7.3 Representative sketch of a charge transfer process that occurs within an intercalation compound particle. *Permission from Wang, Y., Li, H., He, P., Hosono, E., & Zhou, H. (2010). Nano active materials for lithium-ion batteries. Nanoscale, 2(8), 1294.*

rate capability of the battery. Nanostructured cathodes can offer more uniform lithium-ion distribution within the material, beneficial for maintaining consistent performance across the entire cathode surface. At the nanoscale, the interaction between lithium ions and the cathode material can be more pronounced, affecting how ions are trapped and released during the charge and discharge cycles, impacting both the capacity and longevity of the battery.

The overall effect of nanostructuring on the electrochemical performance of cathode materials is profound, encompassing various aspects of battery functionality. The enhanced surface area and improved ion transport pathways in nanostructured cathodes can lead to higher specific capacities, as more active material is accessible for reaction. Batteries with nanostructured cathodes can charge and discharge at higher rates, owing to faster electron and ion transport, advantageous for applications requiring quick energy delivery, such as in electric vehicles. Nanostructured cathodes often exhibit improved structural stability during cycling, enhancing the lifespan of the battery, crucial for maintaining performance over numerous charge—discharge cycles. However, nanostructuring can also lead to challenges in thermal management, as the increased surface area can accelerate thermal runaway processes, necessitating careful attention to these thermal aspects for the safe application of nanostructured cathode materials.

The exploration of nanoelectrochemical mechanisms in cathode materials represents a significant advancement in battery technology. Understanding and optimizing these mechanisms at the nanoscale is key to developing next-generation batteries with higher energy densities, faster charging capabilities, and improved longevity. As research delves deeper into the nanoscale world of cathode materials, the potential for creating batteries that surpass current limitations grows, paving the way for a future with more efficient and reliable energy storage solutions.

7.2.5 Nanoengineering for high-performance cathodes

The advancement of battery technology hinges significantly on the nanoengineering of cathode materials. Delving into the design and optimization of nanostructures for enhanced capacity, this field also focuses on strategies to improve rate capability and cycle life, and critically, addresses challenges in stability and safety at the nanoscale.

The capacity of a battery cathode can be significantly enhanced through meticulous nanoengineering. Tailoring particle size and shape is key; reducing particle size to the nanoscale increases the surface-to-volume ratio, providing more active sites for lithium-ion intercalation. Optimizing shape, such as adopting nanowires or nanosheets, can further enhance the electrochemical reactivity and accessibility of the cathode material. Engineering porosity introduces pathways for easier ion transport and accommodates volume changes during cycling, which enhances capacity. Porous nanostructures can be designed through templating methods, sol−gel processes, or controlled etching. Additionally, composite and hybrid structures that combine different materials at the nanoscale can result in synergistic effects, such as embedding conductive nanomaterials like carbon nanotubes in cathode compounds to enhance electrical conductivity and capacity (Zhang, 2017).

Improving rate capability and cycle life is vital for high-performance batteries and can be optimized through nanoengineering. Designing nanostructures that shorten lithium-ion diffusion paths significantly improves charge and discharge rates. This involves structuring materials so that ions have a more direct route during intercalation and deintercalation processes. Enhancing electronic conductivity by incorporating highly conductive nanomaterials or creating conductive pathways within the cathode material can enhance rate capability. This might involve creating a nanoscale network of conductive materials within the cathode matrix.

Furthermore, nanostructuring can be used to improve the structural integrity of cathode materials during the repeated swelling and shrinking that occurs during battery cycling, using techniques like coating or doping with stable materials to maintain the structural integrity of nanoscale cathodes (Poizot, 2000).

While nanoengineering offers numerous benefits, it also brings forth challenges in terms of stability and safety, which are crucial to address. Nanostructured materials can be more prone to thermal runaway due to their high reactivity, making thermal stability through material selection and design essential. Advanced thermal management systems and careful engineering of the battery architecture can mitigate these risks. Chemical stability is also critical, as nanostructured cathodes may be more reactive with electrolytes or susceptible to degradation. Coatings or surface modifications can enhance chemical stability, and choosing materials that are inherently stable at the nanoscale is also a critical aspect of the design process. Mechanical stability under the stress of repeated charging and discharging cycles is vital, and engineering materials with a certain degree of flexibility or resilience at the nanoscale can help withstand these mechanical stresses.

Nanoengineering stands at the forefront of developing high-performance cathode materials for advanced batteries. By meticulously designing and optimizing nanostructures, substantial improvements in capacity, rate capability, and cycle life can be achieved. Concurrently, addressing the challenges of stability and safety at the nanoscale is paramount to the successful implementation of these advanced materials. The continued innovation and refinement in nanoengineering approaches are pivotal in realizing the next generation of efficient, durable, and safe battery technologies.

7.2.6 Challenges and future perspectives in nanoelectrochemistry of cathodes

While nanoelectrochemistry presents a promising avenue for advancing cathode materials in batteries, it is accompanied by significant challenges and prospects. The integration of nanomaterials in cathodes, despite its advantages, faces several hurdles such as scalability and uniformity, synthesis and processing costs, stability and durability, and compatibility with battery systems. Scaling up the production of nanomaterials while maintaining their quality and uniformity is a significant challenge. The methods for synthesizing nanomaterials are often complex and costly, which can impede their commercial viability. Nanostructured cathode materials often exhibit issues with long-term stability and durability under operational conditions, including challenges like capacity fading, structural degradation, and thermal instability. Ensuring compatibility of nanomaterials with other battery components is critical, as nanomaterials can sometimes exhibit unpredictable interactions with other components, affecting overall battery performance (Schipper, 2017; Xu, 2012).

The production of nanomaterials for cathodes has environmental and economic implications that need careful consideration. The production of nanomaterials often requires significant amounts of energy and resources, leading to a larger environmental footprint. Additionally, waste generated during synthesis can pose environmental hazards. The recyclability of nanomaterials is a growing concern, and developing effective recycling methods is crucial for sustainable battery technology. Lifecycle analysis of nanomaterials is necessary to fully understand their environmental impact, from production to disposal. The economic feasibility of implementing nanotechnology in batteries must be justified by a substantial improvement in battery performance and lifespan, balancing cost with performance benefits (Lu, 2016).

The field of nanoelectrochemistry in cathode materials is rapidly evolving, with several trends and potential advancements on the horizon. Innovations in synthesis methods that are both cost-effective and environmentally friendly are anticipated. The development of nanostructures that can perform multiple functions is an emerging trend, leading to more efficient and compact battery designs. There is a growing emphasis on using sustainable and earth-abundant materials in nanoelectrochemistry, including exploring alternatives to rare or toxic elements currently used in cathode materials. Nanomaterials are expected to play a crucial role in emerging battery technologies, such as solid-state batteries and lithium-sulfur batteries. AI and machine learning tools are being increasingly used to predict and optimize the properties of nanomaterials for cathodes (Qiu, 2023).

In conclusion, the nanoelectrochemistry of cathode materials stands at a crossroads, with immense potential for revolutionizing battery technology but also facing significant challenges. Overcoming these challenges requires a multifaceted approach, addressing scalability, stability, environmental impact, and economic viability. The future trends in nanoelectrochemistry are geared toward sustainability, innovation in material synthesis, and alignment with emerging battery technologies, paving the way for the next generation of energy storage solutions. As the field progresses, it holds the promise of unlocking new capabilities in battery performance, making energy storage more efficient, durable, and environmentally friendly.

7.3 Nanoelectrochemistry in the anode materials

Anode materials play a key role in determining the performance of rechargeable batteries, akin to the way of cathode materials. Upon charging, lithium ions migrate from the cathode to the anode, where they are inserted into the anode materials. This insertion reaction traditionally involves engaging in one of three novel insertion mechanisms: intercalation, alloying, conversion. Historically, carbon-based materials, particularly graphite, have been predominantly used as the conventional anode material in commercial lithium-ion batteries, relying on the intercalation mechanism (Nzereogu, 2022). However, traditional graphite-based anodes encounter limitation due to their low capacity and energy density. To meet future energy demands while significantly reducing reliance on fossil fuels, next-generation battery systems must surpass the theoretical energy density achieved by state-of-the-art conventional lithium-ion batteries. Driven by the need for higher energy density, a surge of research has focused on the development of novel anode materials exploiting alloying and conversion mechanisms (Liu & Cao, 2010; Majdi, 2021; Qi, 2017; Wang, 2010). Notably, pure lithium metal also has recently emerged as a promising contender, offering a distinct pathway toward next-generation battery systems. To engineer high-performance next-generation anode materials, nanotechnology has been employed to precisely control lithium-ion dynamics at the atomic level through tailored nanostructures (Jung, 2020; Maier, 2005; Majdi, 2021; Qi, 2017). Nanosized materials have accelerated performance improvements by enhancing the sluggish diffusion kinetics of lithium ions and facilitating charge transfer across the limited surface area of conventional anode materials. This leads to outstanding performance characterized by enhanced ionic and charge transfer kinetics (Jung, 2020; Maier, 2005). These advancements herald the advent of a new era of "nanoelectrochemistry."

Alloying and conversion mechanisms unlock new paradigms for high-energy density batteries, and nanotechnology plays a crucial role in harnessing their full potential. This section delves into the latest nanotechnological advancements applied to these mechanisms, paving the way for next-generation battery systems. While pure lithium metal also presents a promising avenue, it will not be explored in this chapter.

7.3.1 Alloy-type materials and nanoelectrochemistry

Alloy-type materials are defined by metallic or semimetallic elements that undergo an alloying reaction with lithium ions, as described by equation (Julien, 2016):

$$xLi^+ + xe^- + M \leftrightarrow Li_xM$$

Group IV and V metals or semimetals, denoted as M in reaction (Julien, 2016), are commonly employed as alloying anode materials due to their potential for high energy density. Among them, silicon (Si), germanium (Ge), tin (Sn), and antimony (Sb) have been extensively investigated for next-generation lithium-ion batteries, demonstrating promising theoretical capacities significantly exceeding that of the standard graphite anode (372 mAh/g) (Majdi, 2021; Nzereogu, 2022; Qi, 2017). Despite their promising theoretical capacity, alloy-type materials face a critical challenge related to large volume changes during lithiation and delithiation. These changes induce a substantial irreversible capacity loss during the initial cycle and hinder long-term cycling stability (Majdi, 2021). Furthermore, these substantial volume changes induce extensive internal fractures within the active particles, disrupting the electrical pathways between active materials and the current collector/conducting network. Additionally, electrolyte infiltration through these fractures triggers significant side reactions on the increased surface area, which diminishes capacity retention and overall battery performance. To address the critical challenges of volume expansion, particle cracking, and electrical disconnection, researchers have extensively explored various nano-based strategies. Herein, we primarily focus on Si anode materials due to their potential for achieving near-term commercialization as the next generation of high-performance batteries. However, we will also provide a brief overview of Ge- and Sn-based anode materials for a broader perspective.

7.3.1.1 Silicon

Silicon (Si) has garnered significant interest as a next-generation anode material due to its advantageous combination of a low average operating voltage (below 0.4 V vs. Li/Li$^+$) and a remarkable theoretical capacity (3579 mAh/g upon complete lithiation to $Li_{15}Si_4$;). Impressively, its maximized volumetric energy density can reach up to 1200 Wh/L, a 75% increase compared to the 750 Wh/L offered by the conventional graphite-based system. This enables a significantly higher mass loading in a pure Si ultrathin film electrode compared to a conventional graphite anode with the same areal capacity with facilitating higher energy density. Despite promising theoretical advantages of Si anode, pure Si anodes remain at the research and development stage. Currently, the commercial applications of Si anodes are primarily focused on Si-graphite hybrids, taking advantage of both materials' properties. Studies of pure Si have focused on addressing the significant challenges associated with its massive volume expansions exceeding 300% during battery cycling, a phenomenon also observed in other alloying materials (Gonzalez, 2017; Je, 2023; Sun, 2022, 2016; Zhang, 2022; Zhu, 2021). This repeated cycling induces substantial internal stress within the Si particles, ultimately leading to particle cracking, fracturing, and even pulverization, as illustrated in Fig. 7.4A. The progressive particle fragmentation triggered by fracturing and pulverization triggers delamination of the active materials from the current collector. Additionally, these newly exposed, Si surfaces readily react with the electrolyte, contributing to the formation of a thick, unstable, and high-resistance solid-electrolyte interface (SEI) layer (as shown in Fig. 7.4A)

Nanotechnologies ranging from particle size control to intricate nanostructuring have been employed to mitigate the challenges posed by volume expansion in Si anodes (Liu, 2012). Studies focusing on Si nanoparticle sizing demonstrate the influence of particle size on internal stress accumulation. When considering both the stress state and the associated lithiation-induced swelling and plastic deformation, this analysis reveals a critical threshold size below 150 nm (Fig. 7.4B). Si nanoparticles smaller than this threshold exhibit enhanced mechanical resilience, effectively accommodating the cycling-induced stresses without significant damage. While nanosizing effectively mitigates internal stress, it does not fully address other critical challenges hindering Si anode performance. These include limited electronic transport within the particle, rapid capacity fade due to an unstable SEI layer, and inadequate strain accommodation during cycling. To address these remaining issues, researchers have explored various multidimensional (1D to 3D) nanostructuring strategies for Si anodes, offering promising avenues for unlocking their full potential (Gonzalez, 2017; Jain, 2022; Je, 2023; Sun, 2022; Zhang, 2022). Cui groups investigated Si nanowires (NWs) as a 1D nanostrategy, designing and growing electrodes to alleviate volume expansion through radial and axial expansion, thereby mitigating internal stress build-up (Chan, 2008). The direct growth of NWs offers efficient lithium-ion diffusion channels,

Nanoelectrochemistry in next generation lithium batteries 225

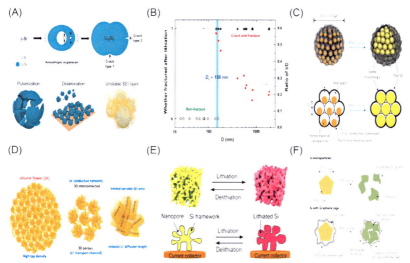

Figure 7.4 Silicon-based anode materials with nanoelectrochemistry. (A) Anisotropic volume expansion and cracking during Si lithiation reaction and the schematic depiction of main degradation mechanism of Si anodes induced by the large volume expansion of Si lithiation. (B) Critical size (D$_c$) of Si particles to resolve the internal stress and minimizing the crack and fracture during cycling. (C) Schematic illustration of the pomegranate-inspired Si anode. (D) Schematic of silicene flowers (SF): Combining high tap density, three-dimensional electron/lithium-ion transport channels, reduced lithium-ion diffusion length, and limited variable SEI formation for enhanced battery performance. (E) Schematic depiction of lithiation/delithiation in ant-nest-like microscale porous Si particles. This unique architecture facilitates inward volume expansion, minimizing stress and preserving the Si framework during repeated cycling. (F) Schematic depiction of the design and structure of graphene cage encapsulation for the bulk Si engineering with Si microparticles. *Permission from (A) Je, M., Han, D.-Y., Ryu, J., & Park, S. (2023). Constructing Pure Si Anodes for Advanced Lithium Batteries. Accounts of Chemical Research, 56(16), 2213. (B) Liu, X. H., Zhong, L., Huang, S., Mao, S. X., Zhu, T., & Huang, J. Y. (2012). Size-Dependent Fracture of Silicon Nanoparticles During Lithiation. ACS Nano, 6(2), 1522. (C) Liu, N., Lu, Z., Zhao, J., McDowell, M. T., Lee, H.-W., Zhao, W., & Cui, Y. (2014). A pomegranate-inspired nanoscale design for large-volume-change lithium battery anodes. Nature Nanotechnology, 9 (3), 187. (D) Zhang, X., Qiu, X., Kong, D., Zhou, L., Li, Z., Li, X., & Zhi, L. (2017). Silicene Flowers: A Dual Stabilized Silicon Building Block for High-Performance Lithium Battery Anodes. ACS Nano, 11(7), 7476. (E) An, W., Gao, B., Mei, S., Xiang, B., Fu, J., Wang, L., Zhang, Q., Chu, P. K., & Huo, K. (2019). Scalable synthesis of ant-nest-like bulk porous silicon for high-performance lithium-ion battery anodes. Nature Communications, 10(1), 1447. (F) Li, Y., Yan, K., Lee, H.-W., Lu, Z., Liu, N., & Cui, Y. (2016). Growth of conformal graphene cages on micrometre-sized silicon particles as stable battery anodes. Nature Energy, 1(2), 15029.*

further enhanced by the transition to an amorphous $Li_{4.4}Si$ alloy during lithiation. This work demonstrates an optimal NW design that facilitates efficient alloying reactions and enhances Si anode performance, offering a promising avenue for overcoming the challenges of Si anodes.

While Si nanowires show promise with their high resistance to pulverization and improved conductivity, their SEI remains unstable due to outward expansion during lithiation, hampering long-term cycling performance. To address this, a yolk-shell structure was designed, where Si nanoparticles (≈ 100 nm) serve as the core ("yolk") encapsulated by a $5-10$ nm thick amorphous carbon shell ("shell") (Liu, 2012). Each Si yolk is anchored to the carbon shell, leaving a strategically designed void space on the opposite side. This void space effectively accommodates the volume expansion during lithiation without compromising the structural integrity of the carbon shell, allowing for the formation of a robust and stable SEI on the shell surface and preventing repetitive SEI rupture and reformation. Taking inspiration from the segmented compartments of a pomegranate, researchers have further developed microstructured Si anodes, shown in Fig. 7.4C, offering promising avenues for practical applications (Liu, 2014). Similar with yolk-shell structure, the Si particles were assembled into hierarchical porous structures coated with a conductive and protective amorphous carbon layer for enhanced ion transport and electrolyte stability, as described in Fig. 7.4C. These optimized electrodes demonstrated impressive retention of gravimetric capacity, reaching 1160 mAh/g after 1000 cycles at a high current density of 635 mA/g.

While the advantages of nanostructured Si are undeniable, its practical application faces hurdles due to limitations such as low tap density, hindering efficient packing in battery modules, and poor Coulombic efficiency, leading to significant capacity loss and reduced energy output. Furthermore, the complex and costly fabrication processes of nanostructures pose additional challenges. The development of microstructures that mimic the beneficial properties of nanoscale Si while overcoming these limitations holds immense promise for unlocking the practical potential of Si anodes in high-performance batteries. An intriguing microscale structure called a silicene flower (SF) has been reported, composed of interlocked silicene nanoplates arranged in a 3D, flower-like configuration (Fig. 7.4D) (Zhang, 2017). These one-atom-thick silicon sheets, akin to graphene, provide numerous advantages. Their flexible, interconnected nature within the flower structure accommodates the significant volume changes Si undergoes during battery cycling, minimizing "frequent cracking and reformation of the SEI" for improved interfacial stability. Additionally, the short distance for Li-ion diffusion through the interconnected nanoplates, combined with their inherent conductivity, allows for robust electron and Li-ion transport. Interestingly, this flower-like arrangement also boasts a high tap density, crucial for efficient packing in practical battery modules. However, direct contact between the electrolyte and Si could still lead to two potential issues: the formation of an unstable SEI layer and morphological deformation of the structure. Addressing these challenges through surface modifications or electrolyte engineering remains an important area of ongoing research. Addressing the drawbacks of microscale Si, researchers have explored designing 3D porous structures to mitigate volume expansion and improve lithium-ion transport. Yet, low tap density often hinders their practical application. A significant breakthrough arrived with the development of antnest-like microscale porous Si (AMPSi) as shown in Fig. 7.4E (An, 2019). This ingenious design features interconnected Si nanoligaments intertwined with a bicontinuous nanoporous network, confirmed by synchrotron radiation tomography and electron microscopy. It demonstrably overcomes the tap density hurdle, achieving a remarkable 0.84 g/cm^3, significantly exceeding typical values for Si anodes. Furthermore, AMPSi boasts a high areal capacity of 5.1 mAh/cm^2 (2.9 mg/cm^2) and a low electrode swelling ratio of 17.8%. The economic and scalable top-down fabrication process underscores AMPSi's potential for industrial-scale battery production. It paves the way for further advancements in rational design of high-performance Si anodes, offering promising avenues for next-generation lithium-ion batteries.

While microstructured Si exhibits promising performance, the cost-effectiveness and high energy density of bulk Si microparticles (SiMPs) hold significant commercial appeal (Chen, 2021; Gonzalez, 2017; Jain, 2022; Je, 2023; Ryu, 2016, 2018; Sun, 2022, 2016; Sung, 2021; Zhang, 2020; Zhang, 2022; Zhu, 2021). However, their larger size than the critical 150 nm threshold exacerbates volume expansion-induced hoop stress, exceeding both yield and ultimate strengths and leading to inevitable fracture and pulverization during cycling. Even stress mitigation strategies cannot fully prevent SiMP fracture, and progressive pulverization eventually compromises performance. Material design for bulk SiMPs must address the electrical isolation of fractured Si fragments to prevent delamination and side reactions with the electrolyte. Cui group implemented a novel nano-/

microstructured encapsulation strategy by layering conformal graphene cages with etching-induced void space on the SiMPs (Li, 2016). This design utilizes the mechanical strength and flexibility of graphene to physically confine Si fragments within the multilayer structure while maintaining electrical connectivity and fostering a reliable SEI (Fig. 7.4F). The internal void further reduces stress on the SiMP, resulting in superior stability and comparable performance compared to conventional SiMP anodes, despite utilizing the same loading mass. Nevertheless, completely preventing destructive reactions linked to SiMP and SEI ruptures remains challenging. Therefore, multipronged strategies are crucial to effectively suppress the detrimental volume expansion at both the particle and electrode levels, promoting structural integrity even in the presence of Si particle fracture.

7.3.1.2 Other alloying type materials

Nonsilicon alloying anodes, particularly Ge and Sn, have emerged as promising candidates for high-energy-density next-generation batteries due to their high theoretical capacities (Fig. 7.5A) (Cheng, 2021; Liang, 2020; Park, 2010). Their potential stems from the abundance of these

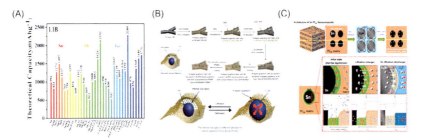

Figure 7.5 Other alloying materials with nanoelectrochemistry. (A) Theoretical capacity landscape of diverse alloying materials for Sn, Sb, and Ge anodes. (B) Synthesis and preparation process of Ge-QD@NG/NGF nanoarchitecture and schematic illustration of their charge/discharge process mechanism. (C) Schematic depiction of the Sn-PC60 nanocomposite matrix and their reaction mechanism during charge/discharge process. Permission from (A) Liang, S., Cheng, Y.-J., Zhu, J., Xia, Y., & Müller-Buschbaum, P. (2020). A Chronicle Review of Nonsilicon (Sn, Sb, Ge)-Based Lithium/Sodium-Ion Battery Alloying Anodes. Small. Methods, 4(8), 2000218. (B) Mo, R., Rooney, D., Sun, K., & Yang, H. Y. (2017). 3D nitrogen-doped graphene foam with encapsulated germanium/nitrogen-doped graphene yolk-shell nanoarchitecture for high-performance flexible Li-ion battery. Nature Communications, 8(1), 13949. (C) Ardhi, R. E. A., Liu, G., Tran, M. X., Hudaya, C., Kim, J. Y., Yu, H., & Lee, J. K. (2018). Self-Relaxant Superelastic Matrix Derived from C60 Incorporated Sn Nanoparticles for Ultra-High-Performance Li-Ion Batteries. ACS Nano, 12(6), 5588.

elements and their potential to exceed the theoretical capacity of Li-ion batteries based solely on graphite anodes. However, similar to Si, these materials suffer from significant volume changes during cycling and are prone to pulverization, hindering their practical application. As with Si anodes, various nanotechnology strategies have been developed to address these challenges in Ge and Sn anodes (Liang, 2020). Despite boasting a high theoretical specific capacity of 1624 mAh/g and superior Li-ion diffusivity and electronic conductivity compared to Si, Ge faces a critical challenge: its large volume expansion during lithiation. This issue has limited its practical application, similar to Si. To address this, carbon matrix composites have emerged as the most promising strategy. Yang et al. developed a 3D nitrogen-doped graphene foam encapsulating Ge quantum dots within an N-doped graphene yolk-shell (Ge-QD@NG/NGF) nanoarchitecture (Fig. 7.5B) (Mo, 2017). This innovative design offers several advantages: (Mo, 2017) internal void space within the yolk-shell accommodates the significant volume expansion of Ge during lithiation, (Pan, 2018) numerous open channels facilitate electrolyte access and fast Li-ion and electron diffusion, and (Liu, 2004) the N-doped graphene framework maintains high electrical conductivity throughout the electrode. As a result, this anode exhibits impressive performance, including a high specific reversible capacity of 1220 mAh/g, ultrahigh-rate capability exceeding 800 mAh/g at 40°C, and exceptional cycling stability with over 98% capacity retention after 1000 cycles in flexible LIBs.

Tin (Sn) also presents itself as a promising anode material due to its ability to form various alloys with lithium through electrochemical reactions. This behavior was documented as early as 1910 with the identification of compounds like Li_4Sn, Li_3Sn_2, and Li_2Sn_5 (Park, 2010). Subsequent years saw extensive research into the physical and electrochemical properties of Li_xSn alloys ($0.4 \leq x \leq 4.4$). Particularly noteworthy is the formation of the lithium-rich phase $Li_{22}Sn_5$ at $x = 4.4$, boasting a remarkable theoretical capacity of 994 mAh/g, the highest among all Li−Sn alloys (Liang, 2020; Park, 2010). This outstanding theoretical capacity has driven significant interest in Sn-based anode materials for lithium-ion batteries. However, a major hurdle remains: the pronounced volume expansion of Sn during cycling in LIBs, reaching up to 259%. Similar to the challenges faced with Si anodes, this substantial volume change has been a key focus of research efforts since the very beginning. To address the notorious volume expansion of Sn anodes, researchers adopted a nanostructuring approach with Sn nanoparticles embedded within a polymerized C60 (PC60) matrix (Fig. 7.5C) (Ardhi, 2018; Kravchyk, 2013; Zhang, 2017). This innovative Sn-PC60 nano-composite combines two key features. (1) Stress buffering: the PC60 matrix, with its self-relaxant superelasticity, absorbs the substantial stress generated by Sn's 259% volume change during cycling. (2) Sn nanoparticle stability: Homogeneously dispersed Sn nanoparticles (~ 10 nm) with optimal interparticle spacing and a protective SnO_2 layer resist aggregation, maintaining structural integrity over numerous cycles.

7.3.2 Conversion materials and nanoelectrochemistry

Alongside alloying strategies, conversion-type materials have emerged as attractive candidates for replacing current intercalation-based electrodes due to their significantly higher theoretical capacities ($500-1500$ mAh/g), often achieved through complete transformation during reaction, and potentially lower cost or higher cost-effectiveness per unit capacity (Armand & Tarascon, 2008; Goriparti, 2014; Yu, 2018). In general, the conversion reaction can be expressed as follows (Pan, 2018):

$$MX + xLi^+ + xe^- \leftrightarrow M + Li_xX$$

Where M denotes a transition metal and X an anionic species. Intriguingly, transition metal compounds like oxides, sulfides, and nitrides readily partake in conversion reactions (Eq. 2), producing nanoparticle phases during charge–discharge cycles. This phenomenon, unique to conversion materials compared to intercalation or alloying, occurs regardless of the initial particle size. Notably, discharge typically results in the formation of nanosized metallic phases (M) dispersed within the LiX matrix (Armand & Tarascon, 2008). While the equilibrium potential predicted from the Gibbs free energy of the bulk material may suggest specific operating voltages, the actual electrochemical behavior of conversion-type electrodes often deviates significantly (~ 1.0 V). In the Fig. 7.6A, the first three charge–discharge profiles of a Li/FeF_3 cell displayed with the redox reactions involving Fe^{3+}/Fe^{2+} and Fe^{2+}/Fe° according to Li insertion into the FeF_3 (Liu, 2012). While the initial lithiation stage aligns with the expected Fe^{3+}/Fe^{2+} redox potential, the subsequent reduction from Fe^{2+} to Fe° exhibits a significantly lower discharge voltage. This voltage discrepancy can be attributed to the formation of nanosized Fe metal particles within the LiF matrix, as shown in Fig. 7.6B. The large surface area of these nanoparticles introduces an

Nanoelectrochemistry in next generation lithium batteries 231

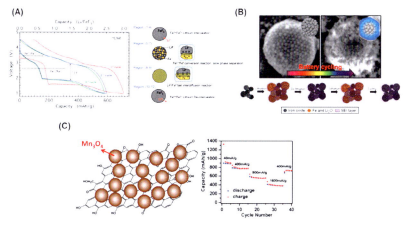

Figure 7.6 Conversion type materials with nanoelectrochemistry. (A) Charge–discharge profiles of the Li/FeF$_3$ systems during the initial three cycles and their proposed reaction mechanisms during the discharge process. (B) Schematic depiction of the reaction mechanism and SEI layer formation on mesoporous iron oxide nanoparticle clusters (MIONCs) with carbon coating materials. (C) Schematic illustration of the Mn$_3$O$_4$/RGO structure and their rate capability performance. *Permission from (A) Liu, P.; Vajo, J. J.; Wang, J. S.; Li, W.; Liu, J. Thermodynamics and Kinetics of the Li/FeF3 Reaction by Electrochemical Analysis. The Journal of Physical Chemistry C 2012, 116 (10), 6467. (B) Lee, S. H.; Yu, S.-H.; Lee, J. E.; Jin, A.; Lee, D. J.; Lee, N.; Jo, H.; Shin, K.; Ahn, T.-Y.; Kim, Y.-W. et al. Self-Assembled Fe3O4 Nanoparticle Clusters as High-Performance Anodes for Lithium Ion Batteries via Geometric Confinement. Nano Letters 2013, 13 (9), 4249. (C) Wang, H.; Cui, L.-F.; Yang, Y.; Sanchez Casalongue, H.; Robinson, J. T.; Liang, Y.; Cui, Y.; Dai, H. Mn3O4-Graphene Hybrid as a High-Capacity Anode Material for Lithium Ion Batteries. Journal of the American Chemical Society 2010, 132 (40), 13978.*

additional surface energy penalty to the overall reaction, causing the equilibrium voltage to deviate from that of the bulk FeF$_3$ conversion reaction. This phenomenon highlights the complex interplay between nanoscale morphology and electrochemical behavior in conversion-type electrode materials (Jung, 2020).

Such a conversion-reaction-based anode materials, especially transition metal oxides, have shown promising properties as replacements for the currently employed graphite anodes (Goriparti, 2014; Lee, 2013; Wang, 2015; Yu, 2018). In particular, iron oxides (Fe$_2$O$_3$, Fe$_3$O$_4$) and manganese oxide (Mn$_3$O$_4$) are attractive electrode materials due to their abundance, environmental friendliness, and high theoretical capacity (Lee, 2013; Wang, 2010). However, their practical application is hindered by limitations in electrical conductivity (e.g., $<10^{-6}$ S/cm), Li-ion diffusion, and cyclability. These issues often arise from high volume expansion and the aggregation of nanosized metallic phases during cycling, leading to rapid capacity fading. To address these challenges, recent research has focused on nanotechnologies such as nanocomposites, surface coating, and element doping, aiming to enhance conductivity, stabilize morphology, and optimize Li-ion transport in transition metal oxide-based electrodes (Goriparti, 2014; Yu, 2018). T. Hyeon group presented a novel bottom-up self-assembly approach to synthesize mesoporous iron oxide nanoparticle clusters (MIONCs) with carbon coatings (Lee, 2013). These MIONCs exhibited significantly improved cyclic stability compared to conventional iron oxide materials (random aggregates and bare nanoparticles) due to their 3D-ordered and mesoporous structure. This structure facilitates the formation of a stable solid electrolyte interphase (SEI) layer and enhances electrolyte accessibility, as visualized in Fig. 7.6C. Additionally, the versatile bottom-up strategy shows potential for application to other metal oxide nanoparticles with high specific capacities, suggesting that self-assembly of active materials could be a promising route for improved battery performance. In the case of another conversion type transition metal oxide Mn_3O_4, it has attracted attention for its high theoretical capacity (936 mAh/g) but suffers from extremely low electrical conductivity (10^{-7} to 10^{-8} S/cm), limiting its practical capacity to ~ 400 mAh/g (Wang, 2010; Yu, 2018). To address this bottleneck, H. Dai's group developed a two-step solution-phase method to grow Mn_3O_4 nanoparticles on reduced graphene oxide (RGO), creating a Mn_3O_4/RGO hybrid material (Fig. 7.6D) (Wang, 2010). This hybrid exhibited an impressive capacity of ~ 900 mAh/g based on Mn_3O_4 mass (~ 810 mAh/g based on total hybrid mass), alongside good rate capability and cycling stability. This growth-on-graphene approach demonstrates potential for significantly boosting the electrochemical performance of highly insulating electrode materials, offering a promising and convenient technique for improving specific capacities and rate capabilities in batteries.

7.4 Nanoelectrochemistry in the electrolytes

7.4.1 Localized high-concentration electrolytes

The conventional electrolyte for Li rechargeable batteries typically consists of ca. $1-1.2$ M Li salts in carbonate base organic solvents. However, this electrolyte is highly susceptible to degradation and reacts with electrode materials, such as Li metal, due to its low electrochemical and chemical stability (Xu, 2014). Recently, high concentration electrolytes (HCE) reported to reduce the electrolyte reactivity. In HCE, the Li salt concentration exceeded 3 M, and the presence of free organic solvent was minimized. Thus, the HCE facilitated the formation of stable SEI layer and the parasitic side reactions are minimized (Ding, 2017; Wang, 2016). However, the use of the high concentration of Li salts led to an increase of viscosity, resulting in a decrease of ionic conductivity.

To overcome the disadvantages of HCE, recently, Chen (2018) reported localized high-concentration electrolytes (LHCEs) by diluting an HCE with bis(2,2,2-trifluoroethyl) ether as an "inert" diluent solvent. The diluent solvent does not dissolve the Li salt but is miscible with the Li solvating solvent in the HCE. The prepared LHCE exhibits low concentration, low cost, low viscosity, improved conductivity, and good wettability with Li metal electrode. The LHCE demonstrated dendrite-free Li plating/stripping with a high Coulombic efficiency and significantly improved the cycling stability of Li||NMC batteries.

Recently, the structure of the LHCEs was proposed by means of Raman analysis and MD simulations that the salt−solvent clusters in LHCE exhibit micelle-like behavior in which the solvent acts as a surfactant between an insoluble salt in a diluent solvent (Efaw, 2023).

He (2024) reported the nonflammable LHCE for high-efficiency Li metal batteries. By using 1,2-difluorobenzene (DFB) as a diluent solvent and lithium bis(fluorosulfonyl)imide (LiFSI) to regulate nonflammable dimethylacetamide (DMAC)-based LHCE, the formation of robust LiF-rich solid electrolyte interphase (SEI) and cathode electrolyte interphase (CEI) was facilitated, thereby enhancing performances of Li metal batteries.

7.4.2 Polymer electrolytes

Low ionic conductivity and poor mechanical stability are key issues for poly(ethylene oxide) (PEO)-based solid polymer electrolytes (SPE) for solid state Li polymer batteries. Block copolymer with soft segment of PEO with rigid polymer has been known a promising strategy for realization to increase the ionic conductivity as well as the mechanical stability. Block copolymer of polystyrene (PS) and PEO (SEO) have been studied the polystyrene-block-poly(ethyleneoxide) copolymers containing a Li[N(SO$_2$CF$_3$)$_2$] (LiTFSI) salt (Panday, 2009). It was found that block copolymer electrolytes formed spontaneous lamellar structure and the ionic conductivity was strongly dependent on the grain size of a lamellar block copolymer electrolyte (Chintapalli, 2014). Kim (2022) studied nanostructure of the block copolymer to enhance the ionic conductivity and electrochemical stability by blending SEO block copolymer with nitrile-end-functionalized PEG.

Guo (2023) reported the in situ polymerization method to prepare block copolymer foe SPE. SPEs were prepared by combining RAFT polymerization of poly(ethylene glycol) methyl ether acrylate (PEGA) and carboxylic acid-catalyzed ROP of ε-CL. This "one-pot fabrication" method can efficiently improve electrode/electrolyte interface characteristic by the penetration of polymer electrolyte into the cathode during polymerization.

Melodia (2023) studied polymerization-induced microphase separation (PIMS) technique to fabricate bicontinuous nanostructured solid polymer electrolytes (Fig. 7.7). A rigid poly(isobornyl acrylate-stat-trimethylpropane triacrylate) and soft poly(oligoethylene glycol methyl ether acrylate) were spontaneous phase separated during crosslinking reaction, and ionic liquid 1-butyl-3-methylimidazolium bis-(trifluoromethyl sulfonyl)imide selectively

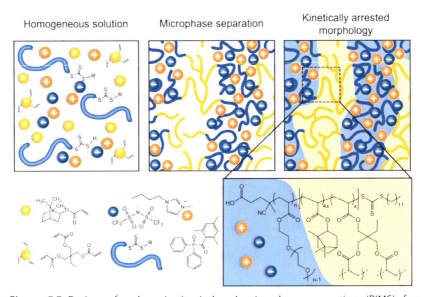

Figure 7.7 Design of polymerization-induced microphase separation (PIMS) for bicontinuous nanostructured solid polymer electrolytes. *Permission from Melodia, D., Bhadra, A., Lee, K., Kuchel, R., Kundu, D., Corrigan, N., & Boyer, C. (2023). 3D Printed Solid Polymer Electrolytes with Bicontinuous Nanoscopic Domains for Ionic Liquid Conduction and Energy Storage. Small, 19(50), 2206639.*

migrates to polar polymer domain. The resulting polymer electrolyte provided high shear modulus and ionic conductivity of 1.2 mS/cm at 30°C.

PEO polymers were modified with polyrotaxen by a self-assembled supramolecular structure to increase ionic conductivity. Seo (2021) reported cyclodextrine (CD) was mechanically interlocked in the partially crosslinked linear PEG polymer. The resulting SPEs has high ionic conductivity (5.93 × 10^{-3} S/cm at 25°C) and Li ion transference number. It was suggested that the hydrogen bonding in the hydroxyl group of CD promotes the dissociation of the Li salts effectively, thus, increasing the population of free Li ions, and also facilitate the transport of Li ions through the ether oxygen pathway in the movable CD shuttles.

Ding (2023) reported the molecular self-assembled ether-based polyrotaxne for a solid polymer electrolyte. The polyrotaxne incorporated solid polymer electrolyte was prepared by threading cyclic 18-crown ether-6 (18C6) into linear poly(ethylene glycol) (PEG) via intermolecular hydrogen bonds and terminating with hexamethylene diisocyanate trimer. The prepared electrolyte significant increased ionic conductivity about 30 times compared to that without assembling polyrotaxane functional units (Fig. 7.8). The assembly of rotaxane in SPE contributed to enhancing the cycle life of the polymer battery.

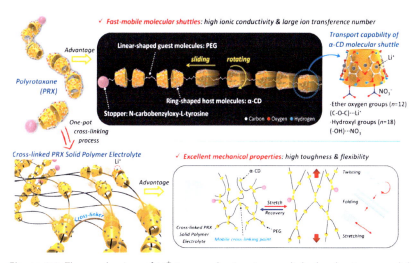

Figure 7.8 The mechanism of Li$^+$ ion conduction in crosslinked polyrotaxane solid polymer electrolyte. *Permission from Seo, J., Lee, G.-H., Hur, J., Sung, M.-C., Seo, J.-H., & Kim, D.-W. (2021). Mechanically Interlocked Polymer Electrolyte with Built-In Fast Molecular Shuttles for All-Solid-State Lithium Batteries. Advanced Energy Materials, 11(44), 2102583.*

7.4.3 Inorganic nanoparticle incorporated composite polymer electrolytes

As mentioned before, solid polymer electrolyte is promising candidate for an all solid-state Li battery. However, low ionic conductivity, poor electrochemical stability, and dimensional integrity are key issues to develop practical batteries. Composite polymer electrolytes (CPE), incorporating inorganic nanoparticles into polymer electrolyte, have been studied to increase electrochemical stability as well as mechanical properties.

Different Li ion nonconducting inorganic nanoparticles such as SiO_2, TiO_2 and $Al2O_3$ were added to decrease the crystallinity of PEO-based polymer electrolyte, thus increase ionic conductivity and electrochemical stability (Liu, 2023; Nunes-Pereira, 2015). However, the nonconducting inorganic fillers open hinder the Li ion passage thus, the large amount of inorganic filler results on decrease of ionic conductivity.

Recently, different Li ion conducting inorganic nanoparticles such as lithium aluminum titanium phosphate (LATP), lithium lanthanum zirconium oxide (LLZO) and lithium lanthanum tantalum zirconium oxide (LLZTO) were studied for CPE (Liu, 2023; Yang, 2023). Among them, garnet type Li ion conducting oxides were widely studied owing to their high ionic conductivity and compatibility with polymer electrolyte (Feng, 2024; Yang, 2023). The main mechanism of enhanced ionic conductivity is hypothesized that the Li ion conducting inorganic nanoparticles facilitate the Li ion transport by creation of ion channel on the interface region of polymer and inorganic particles (Zheng & Hu, 2018; Zheng, 2016). Therefore, the importance of surface interaction between polymer and inorganic nanoparticles has been critical to improve ionic conductivity as well as electrochemical properties of CPE.

Li (2024) used silane coupling agent with different functional group modifying surface of LLZO/PEO CPE. The amino-terminated silane coupling agent was found to improve interfacial compatibility by forming hydrogen bond between amino groups and PEO and resulted on enhancement of mechanical strength and conductivity. Polydopamine with the amino groups has also been employed to modify the surface of garnet-type inorganic particles, aiming to enhance compatibility with PEO. CPE incorporating polydopamine-modified garnet-type inorganic particles demonstrated high ionic conductivity, high Li ion transference number, and expanded electrochemical stability window (Huang, 2019; Mengesha, 2023). Polydopamine-modified garnet type inorganic particles

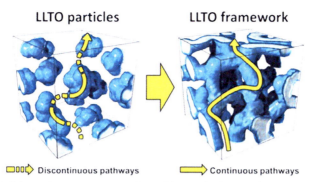

Figure 7.9 Schematic representation of possible conduction mechanism in composite electrolytes with agglomerated nanoparticles and 3D continuous framework. *Permission from Bae, J., Li, Y., Zhang, J., Zhou, X., Zhao, F., Shi, Y., Goodenough, J. B., & Yu, G. (2018). A 3D Nanostructured Hydrogel-Framework-Derived High-Performance Composite Polymer Lithium-Ion Electrolyte. Angewandte Chemie International Edition, 57(8), 2096.*

were found to be strong adsorption of Li ions at the garnet-polydopamine interface. This phenomenon contributes to the preferential migration of Li ion through the interface (Xu, 2023).

A CPE with three-dimensional (3D) nanostructured inorganic framework has been known as a promising approach for enhancing ionic conductivity and battery performances. The 3D nanostructured hydrogel-derived $Li_{0.35}La_{0.55}TiO_3$ (LLTO) framework was used for preparing CPE. The improved ionic conductivity and electrochemical stability was attributed to formation of continuous Li-ion path through interconnected 3D framework (Bae, 2018) (Fig. 7.9). Recently, electrospinning technique has been applied to prepare the nanofiberous 3D framework for CPE. The structure of 3D framework can be easily controlled by changing the precursor solution characteristics, electrospinning parameters and posttreatment conditions (Wang, 2023). Utilizing an electrospun 3D inorganic framework, the CPE not only facilitate lithium-ion transport but also effectively minimize dendrite growth on the lithium metal surface. These CPEs have been successfully applied in the development of all-solid-state Li metal polymer batteries, leading to improved battery performance (Mengesha, 2023; Zhao, 2023).

7.5 Nanoelectrochemistry in the separators

The separator in modern lithium-ion batteries (LIBs) is a vital component, working in concert with the cathode, anode, and electrolyte. Its primary function is to prevent electron short circuits by preventing direct contact between the cathode and anode, all the while serving as a channel for ion transport between these electrodes (Babiker, 2023; Zou, 2023). Although historically downplayed in comparison to other key components, the separator in LIBs is currently gaining heightened attention. LIB separators commonly utilize thin membranes based on polyolefin materials such as polyethylene (PE) and/or polypropylene (PP). However, the intrinsic limitations of polyolefins have highlighted with updates in electrode and electrolyte materials, as well as alterations in cell manufacturing processes, aimed at increasing the energy density of LIB cells. Polyolefin separators demonstrate low interfacial compatibility and insufficient wettability to organic electrolytes due to the absence of polar functional groups (Babiker, 2023; Liu & Chuan, 2021). This ultimately leads to a reduction in cell performance. On the other hand, coupled with the low thermal stability of these substrates with melting points approximately 130°C for PE and 160°C for PP, the decreasing thickness of the separator, in response to the high energy density of LIB cells, poses a risk of severe internal short circuits under elevated temperature or external impacts (Kim, 2021; Liu & Chuan, 2021; Xing, 2022). To overcome these drawbacks of polyolefin separators, various effective approaches have been proposed. Among these approaches, separators for next-generation batteries utilizing nanostructures are reviewed in this chapter.

7.5.1 General characteristics of LIB separator

Prior to the detailed review, a brief overview of the general characteristics of the LIB separators will be provided. The porosity, pore size, and thickness of the separator are crucial feature determining its properties and the overall cell performance. LIB separators should have a porosity of 30%–50%, with pore sizes ranging from 30 to 100 nm to store sufficient electrolyte and ensure permeability (Orendorff, 2012; Prasanna & Lee, 2013). Furthermore, maintaining a uniform and thin thickness in the separator is imperative to uphold the power and energy density of the battery. It is crucial to note, however, that an excessively thin separator may compromise safety and mechanical strength. Adequate mechanical strength is prerequisite in separators to withstand tension and pressure during battery assembly, coupled with electrochemical/chemical stability for prolonged battery operation. Additionally, ensuring good wettability to electrolytes is essential in the separator, facilitating the uniform transport of Li ions across it. Thermal stability is also a critical characteristic, preventing short circuits resulting from the shrinkage of the separator at elevated temperatures during battery operation (Lee, 2014). Moving beyond the previously discussed general characteristics, recent research has introduced additional functionalities into separator design. Noteworthy features include the implementation of the shutdown effect to prevent thermal runaway, achieve a high transference number, inhibit lithium dendritic growth, impart flame retardancy, and exhibit self-healing properties (Babiker, 2023).

7.5.2 Nanocomposite separators

In the realm of nanocomposite separators, the coating of various nanoparticles onto polyolefin-based separators has garnered significant research attention due to its cost-effectiveness and potential to improve separator performance. These nanocomposite separators enhance mechanical strength and thermal stability, facilitating seamless migration of lithium ions by improving electrolyte wettability and uptake. This, in turn, boosts ion conductivity, ultimately elevating battery performance and safety.

Among various nanoparticles for coating separators, ceramic nanoparticles, including SiO_2, Al_2O_3, TiO_2, CeO_2, and ZrO_2 have been employed to develop composite separators. SiO_2-based composite separators, in particular, have been extensively studied due to their cost-effectiveness, high polarity, and thermal stability (Mong, 2021; Mun, 2021). Wang (2015), for instance, developed polyetheimide/SiO_2 (PEI/SiO_2)-modified PE separators using a simple and environmentally friendly self-assembly process (Fig. 7.10A). The construction of an ultrathin PEI/SiO_2 layer on the PE surface has demonstrated improvements in the electrolyte wetting, electrolyte uptake, thermal stability, ionic conductivity, and lithium-ion transference number of the separator. Furthermore, the application of the coated separator in LIB cells (Li/$LiCoO_2$) has shown enhanced performance. Similar advancements were observed by Kim and coworkers using a dip-coating approach with multi-scale nanoporous SiO_2 nanoparticles. The macro- and mesopores of SiO_2 collaboratively enhance the electrolyte absorption and ionic conductivity of the coated separator (Fig. 7.10B) (Kim, 2015).

Al_2O_3 particles are known to enhance not only the thermal stability of coated separators but also the wettability of electrolytes and ion conductivity, owing to their high thermal stability and strong affinity with electrolytes (Cai et al., 2019). While there has been extensive research on separators coated solely with Al_2O_3 nanoparticles, Shin (2023) have recently

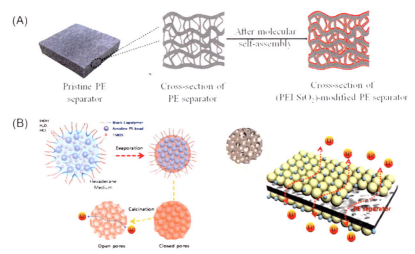

Figure 7.10 (A) Schematic illustrations of the structure of PE separator after PEI/SiO$_2$ modification. (B) The preparation of multiscale nanoporous SiO$_2$ particles along with the coated separator with these particles. *Permission from (A) Wang, Z., Guo, F., Chen, C., Shi, L., Yuan, S., Sun, L., & Zhu, J. (2015). Self-Assembly of PEI/SiO2 on Polyethylene Separators for Li-Ion Batteries with Enhanced Rate Capability. ACS Applied Materials & Interfaces, 7(5), 3314. (B) Kim, Y. B., Tran-Phu, T., Kim, M., Jung, D.-W., Yi, G.-R., & Park, J. H. (2015). Facilitated Ion Diffusion in Multiscale Porous Particles: Application in Battery Separators. ACS Applied Materials & Interfaces, 7(8), 4511.*

investigated separators coated with a combination of Al$_2$O$_3$ nanoparticles and nanocellulose (NC). While the Al$_2$O$_3$ coating alone improved the thermal stability and electrolyte wettability of the coated separator, the addition of NC in the coating further enhanced ion conductivity through increased electrolyte wettability and uptake. This enhancement is attributed to the abundant hydrophilic functional groups of NC. Additionally, NC serves as a spacer between Al$_2$O$_3$ particles, further improving the porosity of the coated separator (Fig. 7.11A and B).

ZrO$_2$, with excellent chemical and thermal stability, enhances ion conductivity and Li-ion transference number when coated on separators (Kim, 2014). Similarly, TiO$_2$ nanoparticles, known for high hydrophilicity, chemical stability, and superior thermal stability, have been applied as coating materials (Zhang, 2013). In addition, materials such as AlOOH (Xiao, 2022; Yang, 2017), CeO$_2$ (Luo, 2017), NiO (Prasanna, 2014), Mg(OH)$_2$ (Yeon, 2015), and Ca$_3$(PO$_4$)$_2$ (Xie, 2022) along with nonceramic BN nanostructures and functional nanocarbons, also exhibit excellent comprehensive properties, indicating the potential for their use as composite battery separators.

Nanoelectrochemistry in next generation lithium batteries 241

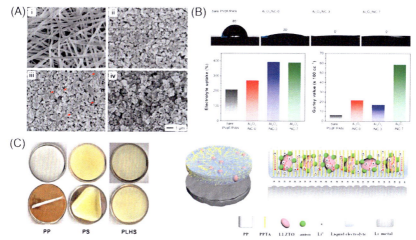

Figure 7.11 (A) SEM images of bare and Al$_2$O$_3$/NC-coated separators: (i) bare, (ii) Al$_2$O$_3$/NC-0, (iii) Al$_2$O$_3$/NC-3, and (iv) Al$_2$O$_3$/NC-7 separators. (B) Results showing electrolyte wettability, electrolyte uptake, and Gurley value of bare and Al$_2$O$_3$/NC-coated separators. (C) Thermal shrinkage photographs before and after 150°C for 1 h of PP, PS, and PLHS separators. (D) Schematic illustration of the electrochemical deposition behavior of the Li metal anode using PLHS. *Permission from (B) Shin, D.-M., Son, H., Park, K. U., Choi, J., Suk, J., Kang, E. S., Kim, D.-W., & Kim, D. Y. (2023). In Coatings, Vol. 13. (D) Mao, Y., Sun, W., Qiao, Y., Liu, X., Xu, C., Fang, L., Hou, W., Wang, Z., & Sun, K. (2021). A high strength hybrid separator with fast ionic conductor for dendrite-free lithium metal batteries. Chemical Engineering Journal, 416, 129119.*

Recent studies have also explored separators coated with solid electrolyte nanoparticles. For instance, Sun and coworkers designed a composite separator coated with Li$_{6.75}$La$_3$Zr$_{1.75}$Ta$_{0.25}$O$_{12}$ (LLZTO) nanoparticles on a PP separator (Mao, 2021). This composite separator exhibited excellent mechanical strength and thermal stability, with the LLZTO coating layer uniformly distributing lithium ions and suppressing the formation of Li dendrites (Fig. 7.11C and D).

7.5.3 Nanofibrous separators

As mentioned earlier, various studies have explored nanocomposite separators to address the limitations of polyolefin-based separators. Despite these efforts, challenges such as weak bonding between organic and inorganic materials, as well as intricate production processes, remain unresolved. Recent research has directed attention toward advanced separators utilizing nonpolyolefin fibrous materials instead of conventional polyolefin-based separators. This shift explores alternative approaches to conventional separators, with the overarching goal of enhancing overall battery performance and safety (Xing, 2022).

Nanofibrous membrane separators are produced through methods like electrospinning (Li, 2018a; Li, 2021), vacuum filtration (Hao, 2016; Wang, 2021), wet-laid (Zhu, 2012), and melt-blown (Zhang, 2019) using nonpolyolefin materials such as polyacrylonitrile (PAN) (Zhu, 2017), polyimide (PI) (Kong, 2017; Li, 2021), poly(ether ether ketone) (PEEK) (Li, 2018a; Li, 2018b), polyetherimide (PEI) (Kong, 2018), aramid (Patel, 2020), and cellulose (Lv, 2021; Xie, 2019). For instance, Li (2018a) fabricated nanofibrous separators via electrospinning, employing thermally stable PEEK and fluorinated PEEK (FPEEK) with superior wettability (Fig. 7.12A). Both separators

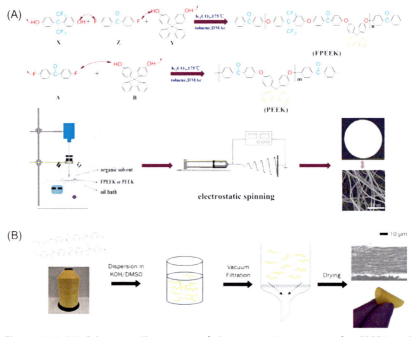

Figure 7.12 (A) Schematic illustrations of the preparation process for FPEEK and PEEK electrospun separators. (B) The formation of ANF separator by vacuum filtration and freeze-drying. *Permission from (A) Li, H., Zhang, B., Liu, W., Lin, B., Ou, Q., Wang, H., Fang, M., Liu, D., Neelakandan, S., & Wang, L. (2018). Effects of an electrospun fluorinated poly(ether ether ketone) separator on the enhanced safety and electrochemical properties of lithium ion batteries. Electrochimica Acta, 290, 150. (B) Patel, A., Wilcox, K., Li, Z., George, I., Juneja, R., Lollar, C., Lazar, S., Grunlan, J., Tenhaeff, W. E., & Lutkenhaus, J. L. (2020). High Modulus, Thermally Stable, and Self-Extinguishing Aramid Nanofiber Separators. ACS Applied Materials & Interfaces, 12(23), 25756.*

exhibited excellent mechanical properties and thermal stability with cells using FPEEK separators demonstrating outstanding electrochemical performance. Patel (2020) reported the fabrication of an aramid nanofiber (ANF) separator through vacuum filtration and freeze-drying, showing excellent thermal stability and mechanical strength (Fig. 7.12B).

In the realm of nanofibrous separators, the introduction of nanoparticles into polymers, from a nanoscale design perspective, is also employed to enhance properties such as thermal stability, mechanical strength, and wettability. In this context, nanofillers such as SiO_2, Al_2O_3, TiO_2, CeO_2, and ZrO_2 nanoparticles, commonly used for coating polyolefin-based separators, are utilized. Illustrative techniques involve blending nanoparticles with polymer solutions (Deng, 2020; Yanilmaz, 2017) or polymer nanofibers (Huang, 2019), along with coating the surface of polymer nanofibrous separators with nanoparticles (Fu, 2021; Xiao, 2020).

References

An, W., Gao, B., Mei, S., Xiang, B., Fu, J., Wang, L., Zhang, Q., Chu, P. K., & Huo, K. (2019). Scalable synthesis of ant-nest-like bulk porous silicon for high-performance lithium-ion battery anodes. *Nature Communications, 10*(1), 1447.

Ardhi, R. E. A., Liu, G., Tran, M. X., Hudaya, C., Kim, J. Y., Yu, H., & Lee, J. K. (2018). Self-Relaxant Superelastic Matrix Derived from C60 Incorporated Sn Nanoparticles for Ultra-High-Performance Li-Ion Batteries. *ACS Nano, 12*(6), 5588.

Armand, M., & Tarascon, J. M. (2008). Building better batteries. *Nature, 451*(7179), 652.

Babiker, D. M. D., Usha, Z. R., Wan, C., Hassaan, M. M. E., Chen, X., & Li, L. (2023). Recent progress of composite polyethylene separators for lithium/sodium batteries. *Journal of Power Sources, 564*, 232853.

Bae, J., Li, Y., Zhang, J., Zhou, X., Zhao, F., Shi, Y., Goodenough, J. B., & Yu, G. (2018). A 3D Nanostructured Hydrogel-Framework-Derived High-Performance Composite Polymer Lithium-Ion Electrolyte. *Angewandte Chemie International Edition, 57*(8), 2096.

Cai, H., Yang, G., Meng, Z., Yin, X., Zhang, H., & Tang, H. (2019). *In Polymers, 11*.

Chan, C. K., Peng, H., Liu, G., McIlwrath, K., Zhang, X. F., Huggins, R. A., & Cui, Y. (2008). High-performance lithium battery anodes using silicon nanowires. *Nature Nanotechnology, 3*(1), 31.

Chen, F., Han, J., Kong, D., Yuan, Y., Xiao, J., Wu, S., Tang, D.-M., Deng, Y., Lv, W., Lu, J., et al. (2021). 1000 Wh L − 1 lithium-ion batteries enabled by crosslink-shrunk tough carbon encapsulated silicon microparticle anodes. *National Science Review, 8*(9), nwab012.

Chen, R., Zhao, T., Zhang, X., Li, L., & Wu, F. (2016). Advanced cathode materials for lithium-ion batteries using nanoarchitectonics. *Nanoscale horizons, 1*(6), 423.

Chen, S., Zheng, J., Mei, D., Han, K. S., Engelhard, M. H., Zhao, W., Xu, W., Liu, J., & Zhang, J.-G. (2018). High-Voltage Lithium-Metal Batteries Enabled by Localized High-Concentration Electrolytes. *Advanced Materials, 30*(21), 1706102.

Cheng, H., Shapter, J. G., Li, Y., & Gao, G. (2021). Recent progress of advanced anode materials of lithium-ion batteries. *Journal of Energy Chemistry, 57*, 451.

Chintapalli, M., Chen, X. C., Thelen, J. L., Teran, A. A., Wang, X., Garetz, B. A., & Balsara, N. P. (2014). Effect of Grain Size on the Ionic Conductivity of a Block Copolymer Electrolyte. *Macromolecules*, *47*(15), 5424.

Deng, N., Wang, L., Liu, Y., Zhong, C., Kang, W., & Cheng, B. (2020). Functionalized polar Octa(γ-chloropropyl) polyhedral oligomeric silsesquioxane assisted polyimide nanofiber composite membrane with excellent ionic conductivity and wetting mechanical strength towards enhanced lithium-ion battery. *Composites Science and Technology*, *192*, 108080.

Ding, M. S., von Cresce, A., & Xu, K. (2017). Conductivity, Viscosity, and Their Correlation of a Super-Concentrated Aqueous Electrolyte. *The Journal of Physical Chemistry C*, *121*(4), 2149.

Ding, P., Wu, L., Lin, Z., Lou, C., Tang, M., Guo, X., Guo, H., Wang, Y., & Yu, H. (2023). Molecular Self-Assembled Ether-Based Polyrotaxane Solid Electrolyte for Lithium Metal Batteries. *Journal of the American Chemical Society*, *145*(3), 1548.

Dong, H., & Koenig, G. M. (2020). A review on synthesis and engineering of crystal precursors produced via coprecipitation for multicomponent lithium-ion battery cathode materials. *CrystEngComm*, *22*(9), 1514.

Efaw, C. M., Wu, Q., Gao, N., Zhang, Y., Zhu, H., Gering, K., Hurley, M. F., Xiong, H., Hu, E., Cao, X., et al. (2023). Localized high-concentration electrolytes get more localized through micelle-like structures. *Nature Materials*, *22*(12), 1531.

Feng, W., Zhao, Y., & Xia, Y. (2024). Solid Interfaces for the Garnet Electrolytes. *Advanced Materials*, 2306111, n/a (n/a).

Franco Gonzalez, A., Yang, N.-H., & Liu, R.-S. (2017). Silicon Anode Design for Lithium-Ion Batteries: Progress and Perspectives. *The Journal of Physical Chemistry C*, *121*(50), 27775.

Fu, Q., Zhang, W., Muhammad, I. P., Chen, X., Zeng, Y., Wang, B., & Zhang, S. (2021). Coaxially electrospun PAN/HCNFs@PVDF/UiO-66 composite separator with high strength and thermal stability for lithium-ion battery. *Microporous and Mesoporous Materials*, *311*, 110724.

Goriparti, S., Miele, E., De Angelis, F., Di Fabrizio, E., Proietti Zaccaria, R., & Capiglia, C. (2014). Review on recent progress of nanostructured anode materials for Li-ion batteries. *Journal of Power Sources*, *257*, 421.

Guo, K., Wang, J., Shi, Z., Wang, Y., Xie, X., & Xue, Z. (2023). One-Step In Situ Polymerization: A Facile Design Strategy for Block Copolymer Electrolytes. *Angewandte Chemie International Edition*, *62*(9), e202213606.

Hailu Mengesha, T., Lemma Beshahwured, S., Wu, Y.-S., Wu, S.-H., Jose, R., & Yang, C.-C. (2023). A polydopamine-modified garnet−based polymer-in-ceramic hybrid solid electrolyte membrane for high-safety lithium metal batteries. *Chemical Engineering Journal*, *452*, 139340.

Hao, X., Zhu, J., Jiang, X., Wu, H., Qiao, J., Sun, W., Wang, Z., & Sun, K. (2016). Ultrastrong Polyoxyzole Nanofiber Membranes for Dendrite-Proof and Heat-Resistant Battery Separators. *Nano Letters*, *16*(5), 2981.

He, R., Deng, K., Mo, D., Guan, X., Hu, Y., Yang, K., Yan, Z., & Xie, H. (2024). Active Diluent-Anion Synergy Strategy Regulating Nonflammable Electrolytes for High-Efficiency Li Metal Batteries. *Angewandte Chemie International Edition*, n/a (n/a), e202317176.

Huang, C., Ji, H., Guo, B., Luo, L., Xu, W., Li, J., & Xu, J. (2019). Composite nanofiber membranes of bacterial cellulose/halloysite nanotubes as lithium ion battery separators. *Cellulose*, *26*(11), 6669.

Huang, Z., Pang, W., Liang, P., Jin, Z., Grundish, N., Li, Y., & Wang, C.-A. (2019). A dopamine modified Li6.4La3Zr1.4Ta0.6O12/PEO solid-state electrolyte: enhanced thermal and electrochemical properties. *Journal of Materials Chemistry A*, *7*(27), 16425.

Jain, R., Lakhnot, A. S., Bhimani, K., Sharma, S., Mahajani, V., Panchal, R. A., Kamble, M., Han, F., Wang, C., & Koratkar, N. (2022). Nanostructuring versus microstructuring in battery electrodes. *Nature Reviews Materials*, 7(9), 736.

Je, M., Han, D.-Y., Ryu, J., & Park, S. (2023). Constructing Pure Si Anodes for Advanced Lithium Batteries. *Accounts of Chemical Research*, 56(16), 2213.

Julien, C., Mauger, A., Zaghib, K., & Groult, H. (2016). Optimization of layered cathode materials for lithium-ion batteries. *Materials*, 9(7), 595.

Jung, S.-K., Hwang, I., Chang, D., Park, K.-Y., Kim, S. J., Seong, W. M., Eum, D., Park, J., Kim, B., Kim, J., et al. (2020). Nanoscale Phenomena in Lithium-Ion Batteries. *Chemical Reviews*, 120(14), 6684.

Kang, S., Wang, C., Chen, J., Meng, T., & Jiaqiang, E. (2023). Progress on solvo/hydrothermal synthesis and optimization of the cathode materials of lithium-ion battery. *Journal of Energy Storage*, 67, 107515.

Kim, J., Jeong, K.-J., Kim, K., Son, C. Y., & Park, M. J. (2022). Enhanced Electrochemical Properties of Block Copolymer Electrolytes with Blended End-Functionalized Homopolymers. *Macromolecules*, 55(6), 2028.

Kim, K. J., Kwon, H. K., Park, M.-S., Yim, T., Yu, J.-S., & Kim, Y.-J. (2014). Ceramic composite separators coated with moisturized ZrO2 nanoparticles for improving the electrochemical performance and thermal stability of lithium ion batteries. *Physical Chemistry Chemical Physics*, 16(20), 9337.

Kim, P. J. (2021). In *Nanomaterials, Vol. 11*.

Kim, Y. B., Tran-Phu, T., Kim, M., Jung, D.-W., Yi, G.-R., & Park, J. H. (2015). Facilitated Ion Diffusion in Multiscale Porous Particles: Application in Battery Separators. *ACS Applied Materials & Interfaces*, 7(8), 4511.

Kong, L., Liu, B., Ding, J., Yan, X., Tian, G., Qi, S., & Wu, D. (2018). Robust polyetherimide fibrous membrane with crosslinked topographies fabricated via in-situ micro-melting and its application as superior Lithium-ion battery separator with shutdown function. *Journal of Membrane Science*, 549, 244.

Kong, L., Yuan, L., Liu, B., Tian, G., Qi, S., & Wu, D. (2017). Crosslinked Polyimide Nanofiber Membrane Prepared via Ammonia Pretreatment and Its Application as a Superior Thermally Stable Separator for Li-Ion Batteries. *Journal of The Electrochemical Society*, 164(6), A1328.

Kravchyk, K., Protesescu, L., Bodnarchuk, M. I., Krumeich, F., Yarema, M., Walter, M., Guntlin, C., & Kovalenko, M. V. (2013). Monodisperse and Inorganically Capped Sn and Sn/SnO2 Nanocrystals for High-Performance Li-Ion Battery Anodes. *Journal of the American Chemical Society*, 135(11), 4199.

Kurzweil, P. (2015). Lithium battery energy storage: State of the art including lithium−air and lithium−sulfur systems. *Electrochemical energy storage for renewable sources and grid balancing*, 269.

Lee, H., Yanilmaz, M., Toprakci, O., Fu, K., & Zhang, X. (2014). A review of recent developments in membrane separators for rechargeable lithium-ion batteries. *Energy & Environmental Science*, 7(12), 3857.

Lee, S. H., Yu, S.-H., Lee, J. E., Jin, A., Lee, D. J., Lee, N., Jo, H., Shin, K., Ahn, T.-Y., Kim, Y.-W., et al. (2013). Self-Assembled Fe3O4 Nanoparticle Clusters as High-Performance Anodes for Lithium Ion Batteries via Geometric Confinement. *Nano Letters*, 13(9), 4249.

Li, D., Wang, H., Luo, L., Zhu, J., Li, J., Liu, P., Yu, Y., & Jiang, M. (2021). Electrospun Separator Based on Sulfonated Polyoxadiazole with Outstanding Thermal Stability and Electrochemical Properties for Lithium-Ion Batteries. *ACS Applied Energy Materials*, 4(1), 879.

Li, H., Zhang, B., Lin, B., Yang, Y., Zhao, Y., & Wang, L. (2018). Electrospun Poly (ether ether ketone) Nanofibrous Separator with Superior Performance for Lithium-Ion Batteries. *Journal of The Electrochemical Society*, 165(5), A939.

Li, H., Zhang, B., Liu, W., Lin, B., Ou, Q., Wang, H., Fang, M., Liu, D., Neelakandan, S., & Wang, L. (2018). Effects of an electrospun fluorinated poly(ether ether ketone) separator on the enhanced safety and electrochemical properties of lithium ion batteries. *Electrochimica Acta, 290*, 150.

Li, M., Sheng, L., Xu, R., Yang, Y., Bai, Y., Song, S., Liu, G., Wang, T., Huang, X., & He, J. (2021). Enhanced the mechanical strength of polyimide (PI) nanofiber separator via PAALi binder for lithium ion battery. *Composites Communications, 24*, 100607.

Li, S., Wang, J., Ji, F., Wang, M., Hu, Z., Huo, S., Zhang, S., Cheng, H., & Zhang, Y. (2024). Surface modification strategies for an improved interfacial compatibility between LLZO and a polymer substrate for applications in high-performance solid-state Li-metal batteries. *Journal of Power Sources, 592*, 233969.

Li, Y., Yan, K., Lee, H.-W., Lu, Z., Liu, N., & Cui, Y. (2016). Growth of conformal graphene cages on micrometre-sized silicon particles as stable battery anodes. *Nature Energy, 1*(2), 15029.

Liang, S., Cheng, Y.-J., Zhu, J., Xia, Y., & Müller-Buschbaum, P. (2020). A Chronicle Review of Nonsilicon (Sn, Sb, Ge)-Based Lithium/Sodium-Ion Battery Alloying Anodes. *Small. Methods, 4*(8), 2000218.

Liu, D., & Cao, G. (2010). Engineering nanostructured electrodes and fabrication of film electrodes for efficient lithium ion intercalation. *Energy & Environmental Science, 3*(9), 1218.

Liu, F., & Chuan, X. (2021). Recent developments in natural mineral-based separators for lithium-ion batteries. *RSC Advances, 11*(27), 16633.

Liu, H., Wu, Y., Rahm, E., Holze, R., & Wu, H. (2004). Cathode materials for lithium ion batteries prepared by sol-gel methods. *Journal of Solid State Electrochemistry, 8*, 450.

Liu, N., Lu, Z., Zhao, J., McDowell, M. T., Lee, H.-W., Zhao, W., & Cui, Y. (2014). A pomegranate-inspired nanoscale design for large-volume-change lithium battery anodes. *Nature Nanotechnology, 9*(3), 187.

Liu, N., Wu, H., McDowell, M. T., Yao, Y., Wang, C., & Cui, Y. (2012). A Yolk-Shell Design for Stabilized and Scalable Li-Ion Battery Alloy Anodes. *Nano Letters, 12*(6), 3315.

Liu, P., Vajo, J. J., Wang, J. S., Li, W., & Liu, J. (2012). Thermodynamics and Kinetics of the Li/FeF3 Reaction by Electrochemical Analysis. *The Journal of Physical Chemistry C, 116*(10), 6467.

Liu, S., Liu, W., Ba, D., Zhao, Y., Ye, Y., Li, Y., & Liu, J. (2023). Filler-Integrated Composite Polymer Electrolyte for Solid-State Lithium Batteries. *Advanced Materials, 35*(2), 2110423.

Liu, X. H., Zhong, L., Huang, S., Mao, S. X., Zhu, T., & Huang, J. Y. (2012). Size-Dependent Fracture of Silicon Nanoparticles During Lithiation. *ACS Nano, 6*(2), 1522.

Lu, J., Chen, Z., Ma, Z., Pan, F., Curtiss, L. A., & Amine, K. (2016). The role of nanotechnology in the development of battery materials for electric vehicles. *Nature nanotechnology, 11*(12), 1031.

Luo, X., Liao, Y., Zhu, Y., Li, M., Chen, F., Huang, Q., & Li, W. (2017). Investigation of nano-CeO2 contents on the properties of polymer ceramic separator for high voltage lithium ion batteries. *Journal of Power Sources, 348*, 229.

Lv, D., Chai, J., Wang, P., Zhu, L., Liu, C., Nie, S., Li, B., & Cui, G. (2021). Pure cellulose lithium-ion battery separator with tunable pore size and improved working stability by cellulose nanofibrils. *Carbohydrate Polymers, 251*, 116975.

Maier, J. (2005). Nanoionics: ion transport and electrochemical storage in confined systems. *Nature Materials, 4*(11), 805.

Majdi, H. S., Latipov, Z. A., Borisov, V., Yuryevna, N. O., Kadhim, M. M., Suksatan, W., Khlewee, I. H., & Kianfar, E. (2021). Nano and Battery Anode: A Review. *Nanoscale Research Letters, 16*(1), 177.

Mao, Y., Sun, W., Qiao, Y., Liu, X., Xu, C., Fang, L., Hou, W., Wang, Z., & Sun, K. (2021). A high strength hybrid separator with fast ionic conductor for dendrite-free lithium metal batteries. *Chemical Engineering Journal, 416*, 129119.

Melodia, D., Bhadra, A., Lee, K., Kuchel, R., Kundu, D., Corrigan, N., & Boyer, C. (2023). 3D Printed Solid Polymer Electrolytes with Bicontinuous Nanoscopic Domains for Ionic Liquid Conduction and Energy Storage. *Small, 19*(50), 2206639.

Mo, R., Rooney, D., Sun, K., & Yang, H. Y. (2017). 3D nitrogen-doped graphene foam with encapsulated germanium/nitrogen-doped graphene yolk-shell nanoarchitecture for high-performance flexible Li-ion battery. *Nature Communications, 8*(1), 13949.

Mong, A. L., Shi, Q. X., Jeon, H., Ye, Y. S., Xie, X. L., & Kim, D. (2021). Tough and Flexible, Super Ion-Conductive Electrolyte Membranes for Lithium-Based Secondary Battery Applications. *Advanced Functional Materials, 31*(12), 2008586.

Mun, J., Yim, T., Gap Kwon, Y., & Jae Kim, K. (2021). Self-assembled nano-silica-embedded polyethylene separator with outstanding physicochemical and thermal properties for advanced sodium ion batteries. *Chemical Engineering Journal, 405*, 125844.

Nunes-Pereira, J., Costa, C. M., & Lanceros-Méndez, S. (2015). Polymer composites and blends for battery separators: State of the art, challenges and future trends. *Journal of Power Sources, 281*, 378.

Nzereogu, P. U., Omah, A. D., Ezema, F. I., Iwuoha, E. I., & Nwanya, A. C. (2022). Anode materials for lithium-ion batteries: A review. *Applied Surface Science Advances, 9*, 100233.

Orendorff, C. J. (2012). The Role of Separators in Lithium-Ion Cell Safety. *The Electrochemical Society Interface, 21*(2), 61.

Pan, H., Zhang, S., Chen, J., Gao, M., Liu, Y., Zhu, T., & Jiang, Y. (2018). Li-and Mn-rich layered oxide cathode materials for lithium-ion batteries: a review from fundamentals to research progress and applications. *Molecular systems design & engineering, 3*(5), 748.

Panday, A., Mullin, S., Gomez, E. D., Wanakule, N., Chen, V. L., Hexemer, A., Pople, J., & Balsara, N. P. (2009). Effect of Molecular Weight and Salt Concentration on Conductivity of Block Copolymer Electrolytes. *Macromolecules, 42*(13), 4632.

Park, C.-M., Kim, J.-H., Kim, H., & Sohn, H.-J. (2010). Li-alloy based anode materials for Li secondary batteries. *Chemical Society Reviews, 39*(8), 3115.

Patel, A., Wilcox, K., Li, Z., George, I., Juneja, R., Lollar, C., Lazar, S., Grunlan, J., Tenhaeff, W. E., & Lutkenhaus, J. L. (2020). High Modulus, Thermally Stable, and Self-Extinguishing Aramid Nanofiber Separators. *ACS Applied Materials & Interfaces, 12* (23), 25756.

Poizot, P., Laruelle, S., Grugeon, S., Dupont, L., & Tarascon, J. (2000). Nano-sized transition-metal oxides as negative-electrode materials for lithium-ion batteries. *Nature, 407* (6803), 496.

Prasanna, K., & Lee, C. W. (2013). Physical, thermal, and electrochemical characterization of stretched polyethylene separators for application in lithium-ion batteries. *Journal of Solid State Electrochemistry, 17*(5), 1377.

Prasanna, K., Subburaj, T., Lee, W. J., & Lee, C. W. (2014). Polyethylene separator: stretched and coated with porous nickel oxide nanoparticles for enhancement of its efficiency in Li-ion batteries. *Electrochimica Acta, 137*, 273.

Qi, W., Shapter, J. G., Wu, Q., Yin, T., Gao, G., & Cui, D. (2017). Nanostructured anode materials for lithium-ion batteries: principle, recent progress and future perspectives. *Journal of Materials Chemistry A, 5*(37), 19521.

Qiu, Y. (2023). Nanotechnology Applications in Cathode and Anode Materials of Li-Ion Battery. *Highlights in Science, Engineering and Technology, 58*, 379.

Ryu, J., Chen, T., Bok, T., Song, G., Ma, J., Hwang, C., Luo, L., Song, H.-K., Cho, J., Wang, C., et al. (2018). Mechanical mismatch-driven rippling in carbon-coated silicon sheets for stress-resilient battery anodes. *Nature Communications, 9*(1), 2924.

Ryu, J., Hong, D., Choi, S., & Park, S. (2016). Synthesis of Ultrathin Si Nanosheets from Natural Clays for Lithium-Ion Battery Anodes. *ACS Nano, 10*(2), 2843.

Schipper, F., Nayak, P. K., Erickson, E. M., Amalraj, S. F., Srur-Lavi, O., Penki, T. R., Talianker, M., Grinblat, J., Sclar, H., & Breuer, O. (2017). Study of cathode materials for lithium-ion batteries: Recent progress and new challenges. *Inorganics, 5*(2), 32.

Seo, J., Lee, G.-H., Hur, J., Sung, M.-C., Seo, J.-H., & Kim, D.-W. (2021). Mechanically Interlocked Polymer Electrolyte with Built-In Fast Molecular Shuttles for All-Solid-State Lithium Batteries. *Advanced Energy Materials, 11*(44), 2102583.

Shin, D.-M., Son, H., Park, K. U., Choi, J., Suk, J., Kang, E. S., Kim, D.-W., & Kim, D. Y. (2023). *In Coatings, Vol. 13.*

Sun, L., Liu, Y., Shao, R., Wu, J., Jiang, R., & Jin, Z. (2022). Recent progress and future perspective on practical silicon anode-based lithium ion batteries. *Energy Storage Materials, 46*, 482.

Sun, Y., Liu, N., & Cui, Y. (2016). Promises and challenges of nanomaterials for lithium-based rechargeable batteries. *Nature Energy, 1*(7), 16071.

Sung, J., Kim, N., Ma, J., Lee, J. H., Joo, S. H., Lee, T., Chae, S., Yoon, M., Lee, Y., Hwang, J., et al. (2021). Subnano-sized silicon anode via crystal growth inhibition mechanism and its application in a prototype battery pack. *Nature Energy, 6*(12), 1164.

Wang, D., Yu, Y., He, H., Wang, J., Zhou, W., & Abruña, H. D. (2015). Template-Free Synthesis of Hollow-Structured Co3O4 Nanoparticles as High-Performance Anodes for Lithium-Ion Batteries. *ACS Nano, 9*(2), 1775.

Wang, H., Cui, L.-F., Yang, Y., Sanchez Casalongue, H., Robinson, J. T., Liang, Y., Cui, Y., & Dai, H. (2010). Mn3O4 − Graphene Hybrid as a High-Capacity Anode Material for Lithium Ion Batteries. *Journal of the American Chemical Society, 132*(40), 13978.

Wang, J., Yamada, Y., Sodeyama, K., Chiang, C. H., Tateyama, Y., & Yamada, A. (2016). Superconcentrated electrolytes for a high-voltage lithium-ion battery. *Nature Communications, 7*(1), 12032.

Wang, M., Wang, C., Fan, Z., Wu, G., Liu, L., & Huang, Y. (2021). Aramid nanofiber-based porous membrane for suppressing dendrite growth of metal-ion batteries with enhanced electrochemistry performance. *Chemical Engineering Journal, 426*, 131924.

Wang, P., Liu, J.-H., Cui, W., Li, X., Li, Z., Wan, Y., Zhang, J., & Long, Y.-Z. (2023). Electrospinning techniques for inorganic−organic composite electrolytes of all-solid-state lithium metal batteries: a brief review. *Journal of Materials Chemistry A, 11*(31), 16539.

Wang, Y., Li, H., He, P., Hosono, E., & Zhou, H. (2010). Nano active materials for lithium-ion batteries. *Nanoscale, 2*(8), 1294.

Wang, Z., Guo, F., Chen, C., Shi, L., Yuan, S., Sun, L., & Zhu, J. (2015). Self-Assembly of PEI/SiO2 on Polyethylene Separators for Li-Ion Batteries with Enhanced Rate Capability. *ACS Applied Materials & Interfaces, 7*(5), 3314.

Wu, F., Maier, J., & Yu, Y. (2020). Guidelines and trends for next-generation rechargeable lithium and lithium-ion batteries. *Chemical Society Reviews, 49*(5), 1569.

Xiao, W., Song, J., Huang, L., Yang, Z., & Qiao, Q. (2020). PVA-ZrO2 multilayer composite separator with enhanced electrolyte property and mechanical strength for lithium-ion batteries. *Ceramics International, 46*(18, Part A), 29212.

Xiao, Y., Fu, A., Zou, Y., Huang, L., Wang, H., Su, Y., & Zheng, J. (2022). High safety lithium-ion battery enabled by a thermal-induced shutdown separator. *Chemical Engineering Journal, 438*, 135550.

Xie, W., Liu, W., Dang, Y., Tang, A., Deng, T., & Qiu, W. (2019). Investigation on electrolyte-immersed properties of lithium-ion battery cellulose separator through multi-scale method. *Journal of Power Sources, 417*, 150.

Xie, X., Sheng, L., Xu, R., Gao, X., Yang, L., Gao, Y., Bai, Y., Liu, G., Dong, H., Fan, X., et al. (2022). In situ mineralized Ca3(PO4)2 inorganic coating modified polyethylene separator for high-performance lithium-ion batteries. *Journal of Electroanalytical Chemistry*, *920*, 116570.

Xing, J., Li, J., Fan, W., Zhao, T., Chen, X., Li, H., Cui, Y., Wei, Z., & Zhao, Y. (2022). A review on nanofibrous separators towards enhanced mechanical properties for lithium-ion batteries. *Composites Part B: Engineering*, *243*, 110105.

Xu, B., Qian, D., Wang, Z., & Meng, Y. S. (2012). Recent progress in cathode materials research for advanced lithium ion batteries. *Materials Science and Engineering: R: Reports*, *73*(5-6), 51.

Xu, K. (2014). Electrolytes and Interphases in Li-Ion Batteries and Beyond. *Chemical Reviews*, *114*(23), 11503.

Xu, Y., Wang, K., Zhang, X., Ma, Y., Peng, Q., Gong, Y., Yi, S., Guo, H., Zhang, X., Sun, X., et al. (2023). Improved Li-Ion Conduction and (Electro)Chemical Stability at Garnet-Polymer Interface through Metal-Nitrogen Bonding. *Advanced Energy Materials*, *13*(14), 2204377.

Yang, C., Tong, H., Luo, C., Yuan, S., Chen, G., & Yang, Y. (2017). Boehmite particle coating modified microporous polyethylene membrane: A promising separator for lithium ion batteries. *Journal of Power Sources*, *348*, 80.

Yang, X., Liu, J., Pei, N., Chen, Z., Li, R., Fu, L., Zhang, P., & Zhao, J. (2023). The Critical Role of Fillers in Composite Polymer Electrolytes for Lithium Battery. *Nano-Micro Letters*, *15*(1), 74.

Yanilmaz, M., Zhu, J., Lu, Y., Ge, Y., & Zhang, X. (2017). High-strength, thermally stable nylon 6,6 composite nanofiber separators for lithium-ion batteries. *Journal of Materials Science*, *52*(9), 5232.

Yeon, D., Lee, Y., Ryou, M.-H., & Lee, Y. M. (2015). New flame-retardant composite separators based on metal hydroxides for lithium-ion batteries. *Electrochimica Acta*, *157*, 282.

Yu, S.-H., Feng, X., Zhang, N., Seok, J., & Abruña, H. D. (2018). Understanding Conversion-Type Electrodes for Lithium Rechargeable Batteries. *Accounts of Chemical Research*, *51*(2), 273.

Zhang, H., Huang, X., Noonan, O., Zhou, L., & Yu, C. (2017). Tailored Yolk–Shell Sn@C Nanoboxes for High-Performance Lithium Storage. *Advanced Functional Materials*, *27*(8), 1606023.

Zhang, H., Zhen, Q., Liu, Y., Liu, R., & Zhang, Y. (2019). Branched polyethylene glycol/polypropylene micro-nanofiber nonwovens for fast liquid planar transmission. *Journal of Engineered Fibers and Fabrics*, *14*, 1558925019850798.

Zhang, L., Al-Mamun, M., Wang, L., Dou, Y., Qu, L., Dou, S. X., Liu, H. K., & Zhao, H. (2022). The typical structural evolution of silicon anode. *Cell Reports Physical Science*, *3*(4), 100811.

Zhang, R.-X., Braeken, L., Luis, P., Wang, X.-L., & Van der Bruggen, B. (2013). Novel binding procedure of TiO2 nanoparticles to thin film composite membranes via self-polymerized polydopamine. *Journal of Membrane Science*, *437*, 179.

Zhang, X., Porras-Gutierrez, A.-G., Mauger, A., Groult, H., & Julien, C. M. (2017). Nanotechnology of positive electrodes for Li-ion batteries. *Inorganics*, *5*(2), 25.

Zhang, X., Qiu, X., Kong, D., Zhou, L., Li, Z., Li, X., & Zhi, L. (2017). Silicene Flowers: A Dual Stabilized Silicon Building Block for High-Performance Lithium Battery Anodes. *ACS Nano*, *11*(7), 7476.

Zhang, X., Wang, D., Qiu, X., Ma, Y., Kong, D., Müllen, K., Li, X., & Zhi, L. (2020). Stable high-capacity and high-rate silicon-based lithium battery anodes upon two-dimensional covalent encapsulation. *Nature Communications*, *11*(1), 3826.

Zhao, Y., Fan, L., Xiao, B., Cai, S., Chai, J., Liu, X., Liu, J., & Liu, Z. (2023). Preparing 3D Perovskite Li0.33La0.557TiO3 Nanotubes Framework Via Facile

Coaxial Electro-Spinning Towards Reinforced Solid Polymer Electrolyte. *ENERGY & ENVIRONMENTAL MATERIALS*, 6(4), e12636.

Zheng, J., & Hu, Y.-Y. (2018). New Insights into the Compositional Dependence of Li-Ion Transport in Polymer–Ceramic Composite Electrolytes. *ACS Applied Materials & Interfaces*, 10(4), 4113.

Zheng, J., Tang, M., & Hu, Y.-Y. (2016). Lithium Ion Pathway within Li7La3Zr2O12-Polyethylene Oxide Composite Electrolytes. *Angewandte Chemie International Edition*, 55(40), 12538.

Zhu, G., Chao, D., Xu, W., Wu, M., & Zhang, H. (2021). Microscale Silicon-Based Anodes: Fundamental Understanding and Industrial Prospects for Practical High-Energy Lithium-Ion Batteries. *ACS Nano*, 15(10), 15567.

Zhu, M., Xu, G., Yu, M., Liu, Y., & Xiao, R. (2012). Preparation, properties, and application of polypropylene micro/nanofiber membranes. *Polymers for Advanced Technologies*, 23(2), 247.

Zhu, Y., Yin, M., Liu, H., Na, B., Lv, R., Wang, B., & Huang, Y. (2017). Modification and characterization of electrospun poly (vinylidene fluoride)/poly (acrylonitrile) blend separator membranes. *Composites Part B: Engineering*, 112, 31.

Zou, Z., Hu, Z., & Pu, H. (2023). Lithium-ion battery separators based-on nanolayer co-extrusion prepared polypropylene nanobelts reinforced cellulose. *Journal of Membrane Science*, 666, 121120.

CHAPTER 8

Coordination materials for supercapacitors

Diab Khalafallah[1,2], Mohamed S. Abdel-Latif[3], Mohamed A Ibrahim[4] and Qinfang Zhang[1]

[1]School of Materials Science and Engineering, Yancheng Institute of Technology, Yancheng, P.R. China
[2]Mechanical Design and Materials Department, Faculty of Energy Engineering, Aswan University, Aswan, Egypt
[3]Engineering Physics and Mathematics Department, Faculty of Engineering, Tanta University, Tanta, Egypt
[4]Faculty of Engineering, Aswan University, Aswan, Egypt

8.1 Introduction

Currently, nonrenewable and sporadic-type energy resources are highly undesirable due to the increasing energy demands and environmental concerns. Therefore, not only the production of green energy from renewable systems but also the storage has become an important direction of reaction in recent years. With significant technological advancements and intensive scientific progress, the rapid development and emergence of sustainable and ecofriendly energy storage devices have boosted to replace the traditional polluting energy resources and satisfy the energy/power requirements and long-lasting features (González et al., 2016; Vlad et al., 2015). Rechargeable batteries and electrochemical supercapacitors (ESCs) as potential energy storage technologies have attracted noticeable interest. These advanced systems possess inherent advantages (e.g., safety operations and environmental friendliness) and can supply electricity for various day-to-day configurations such as portable electronic devices, electric vehicles, and stationary energy-storing appliances. Nevertheless, the widespread usage of batteries is hindered by their low power density, extended charging durations, and limited lifespan (Fleischmann et al., 2020; Wei et al., 2011). Moreover, the energy-storing mechanism of batteries depends on chemical reactions within the electrode configuration, allowing slow ion transport rate and limited power density. In this context, ESCs are being innovated because of their excellent charging rate, good ability to promote capacitance and mechanical flexibility, high power density output, and long cyclability/calendar life (Muzaffar et al., 2019).

Electrochemistry and Photo-Electrochemistry of Nanomaterials
DOI: https://doi.org/10.1016/B978-0-443-18600-4.00009-0
© 2025 Elsevier Inc. All rights are reserved, including those for text and data mining, AI training, and similar technologies.

The charge storage capability of such potential power resources correlates strongly with the adopted electrode materials. In general, the electrode for ESCs can be divided into three groups depending on their intrinsic charge storage mechanism, namely electrical double layer, pseudocapacitive, and battery type. In the case of conventional electrical double layer electrodes based on state-of-the-art carbonaceous materials, charges are electrostatically stored via the physical adsorption of ions at the electrode/electrolyte interface under a non-Faradaic process (Khalafallah, Zhang, et al., 2024; Muzaffar et al., 2019). Pseudocapacitive materials existed in the picture with quite different electrochemical characteristics, where the current signal is neither bulk Faradaic (like batteries) nor artlessly capacitive. These compounds experience rapid and reversible surface redox reactions at the electrode-exposed surface through the adsorption or intercalation of ions, thereby affording an efficient path to acquire simultaneous high energy/power densities (Hong et al., 2024). Similar to double-layer capacitive compounds, pseudocapacitive materials can reveal the desired rate capability, but they differ from well-known battery materials as the kinetics of the redox reaction are very quick and not limited by diffusion. Battery-type electrode materials can store charges via a diffusion-controlled Faradaic redox reaction (Wu et al., 2021). They are different from the electrical double layer-like compounds in terms of the bulk redox processes, which can guide the phase change of electrode active components during electrochemical reactions. Besides, the potential of the battery-type electrode remains unchanged during the charge−discharge processes, therefore revealing well-defined redox signatures in the cyclic voltammetry (CV) responses and obvious flat plateau in the galvanostatic charge−discharge (GCD) profiles similar to batteries (Chodankar et al., 2020; Khalafallah et al., 2023; Wu et al., 2021).

The criteria for identifying the charge storage mechanism and information related to electrochemical reactions can be realized based on the empirical description "Log $i(V) = \log a + b \log v$"), where V and v denote the potential and sweep rate, respectively. Both a and b are adjustable parameters. The slope of the linear relationship of the logarithmic peak current response (i) versus the logarithmic sweep rate (v) represents the value of b. The pseudocapacitive electrodes always show a b-value of ≈ 1.0, signifying contributions from rapid surface redox reactions and near-surface activity (Khalafallah, Huang, et al., 2024; Liu et al., 2018; Saleem et al., 2016). In contrast, diffusion-controlled Faradaic processes take place in bulk with a b-value of ≈ 0.5, verifying a typical battery-type behavior. With the *b*-value ranging from 0.5 to 1.0, a transition region between the pseudocapacitive-like material and the battery-type material may be formed. The smaller *b*-value demonstrates a larger contribution from the diffusion-controlled intercalation process, whereas increasing the *b*-value indicates a more capacitive contribution.

As the charge storage mechanism in supercapacitors (SCs) is either a surface-capacitive phenomenon or a diffusion-controlled process, the rational engineering of nanostructured electrodes plays a considerable role in exploring promising supercapacitor devices with satisfactory electrochemical matrices (Khalafallah, Zhi, et al., 2022). Hierarchical nanostructured materials with various morphologies (nanoparticles, rods, sheets, branches, etc.) can boost the electrochemical performance by expanding the surface area, maximizing the number of accessible sites, and reducing the ion transport distances. Electrode materials with smart morphological, compositional and structural advantages have generally been thought of as the key to offering large exposed active surfaces, regulated electronic/chemical configurations of sites, and rich immigration passages together with amplified energy storage ability (Khalafallah, Huang, et al., 2022; Shi et al., 2022). Accordingly, enormous research efforts have been devoted to achieving efficient electrochemical performance through several ways such as multicomponent incorporation, internal electronic optimization, and coupling effect modification. Of particular, novel heterointerfacing, core−shell nanostructuring, and defect engineering are other feasible approaches to modify the charge transfer kinetics in the surface or the bulk, realize significant synergistic effect, and improve the electrical conductivity, as well as tunable electronic properties of electrode materials (Khalafallah et al., 2023).

We here aim to discuss the intrinsic cost-efficient coordination materials with novel structural/compositional configurations for high-performance ESCs. This chapter presents a brief overview of ESCs and the representative types of ESCs including electrical double layer capacitors (EDLCs), pseudocapacitors (PCs), and hybrid supercapacitors (HSCs) and outlines the dominant charge storage mechanisms. Afterward, the state-of-the-art strategies for manipulating the energy storage capability of electrode materials are comprehensively reviewed with numerous suitable studies. Several emerging methods such as hybridization, defect engineering, elemental doping, heterointerface engineering, and surface functionalization are covered.

8.2 An overview of electrochemical SCs

A supercapacitor device is a capacitor that can store a significant amount of energy at high power density while functioning at lower voltages (Chodankar et al., 2020). A typical schematic illustration of the supercapacitor system is shown in Fig. 8.1. SCs have a structure similar to that of capacitors, with two electrodes made of highly porous materials and immersed in an electrolyte. A dielectric membrane separates the two electrodes, allowing ions to pass through. When an external electric field is applied, positive and negative charges accumulate on the surfaces, forming layers on each electrode. Opposite charges attract each other, causing ions in the electrolyte solution to diffuse across the separator and enter the pores of the electrode materials (Chodankar et al., 2020; Khalafallah et al., 2023). The electrodes are designed to prevent ion recombination, creating a double layer behavior of charges at each electrode. SCs require porous materials with larger specific surface areas to achieve higher energy density and capacitance based on their energy storage mechanisms. SCs follow the

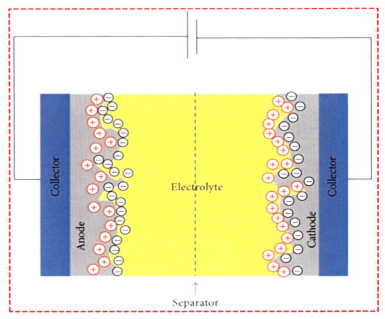

Figure 8.1 Representative diagram of a supercapacitor system. *Reprinted with permission from Saleem, A. M., Desmaris, V., & Enoksson, P. (2016). Performance enhancement of carbon nanomaterials for supercapacitors.* Journal of Nanomaterials, *1537269, 17 pages. Copyright 2016, Hindawi Publishing Corporation.*

same basic principles as capacitors but use high surface area electrodes and a thin membrane as a dielectric layer, resulting in a significant increase of capacitance and energy. Additionally, they have a low equivalent series resistance (ESR) value similar to conventional capacitors, allowing them to maintain a high power density (Liu et al., 2018; Saleem et al., 2016).

Despite the effectiveness of SCs in storing intermittent energy, the energy density of these devices is limited by their specific capacitance/capacity and working potential as expressed by Eq. (8.1) (Khalafallah, Huang, et al., 2024):

$$E = \frac{1}{2} \, CV^2, \tag{8.1}$$

where capacitance and cell voltage are defined as C and V, respectively. The amount of energy that can be stored by a supercapacitor system depends on the width of the voltage window and the area as well as the thickness of the double-layer of ions at the surface. For increasing the energy density, there are two main pathways: developing electrode materials with higher capacitance and electrolytes with a broader potential window to increase the cell voltage. Although much research has been dedicated for developing new electrode materials, commercialization has proved challenging due to factors such as low capacitance, weak conductivity, limited surface area, limited energy density, poor cycle life, and complex synthesis processes (Noori et al., 2019). Therefore, it is crucial to identify practical and sustainable materials and techniques for supercapacitor energy storage technologies.

Classification of SCs can be made based on details of the construction and manufacturing process. The fundamental principle behind their operation is the energy storage, which divides them into three main groups depending on the energy storage method. SCs can be classified into three main types: EDLCs (Fig. 8.2A), PCs (Fig. 8.2B), and HSCs (Fig. 8.2C), which differ in their energy storage mechanism (Chodankar et al., 2020; Khalafallah et al., 2023; Noori et al., 2019; Şahin et al. 2022). Principally, in two ways, SCs can hold charges: EDLCs rely on charge storage between the electrolyte and electrodes, while PCs involve reversible and fast Faradaic redox reactions to enhance the capacitance of the supercapacitor. HSCs store charges by combining capacitive carbon electrodes with pseudocapacitive components (Khalafallah, Li, et al., 2021).

Moreover, based on electrodes, SCs can be classified into symmetrical and asymmetrical ones. Electrodes are the most vital parts of SCs, as they

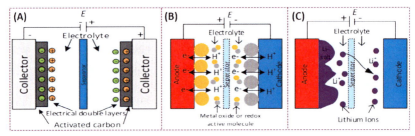

Figure 8.2 Charge storage mechanism in (A) EDLC-, (B) PC-, and (C) HSC-based energy storage devices. Reprinted with permission from Şahin, M. E., Blaabjerg, F., & Sangwongwanich, A. (2022). A comprehensive review on supercapacitor applications and developments. Energies, 15, 674. https://doi.org/10.3390/en15030674. Copyright 2022, MDPI, Basel, Switzerland.

store the charge at the interface between the electrode and electrolyte. Symmetric SCs have electrodes made of the same material. These are often used for EDLCs and some pseudocapacitors. Although the design and manufacturing of symmetric SCs is simple, their energy storage capability may be limited by the voltage window. Asymmetric supercapacitors (ASCs) use different electrode materials for the positive and negative electrodes, allowing for a wider voltage window and resulting in higher energy storage capability. HSCs combine the features of EDLCs and PCs, commonly known as ASCs (Fleischmann et al., 2020; Khalafallah, Miao, et al., 2021; Majumdar et al., 2019).

Based on construction design, ESCs can be constructed in the form of planar (flat), cylindrical, or rectangular models. Thin-film techniques are utilized to produce planar SCs, where electrodes and separators are deposited on a substrate. A new trend in the manufacturing process is the use of flexible SCs. The integration of electronics, flexibility, and weight reduction are among the potential advantages they offer. Electrode materials, such as carbon nanotube yarns, are twisted or aligned into fibers to create fiber-shaped SCs. These SCs can be used in flexible applications or woven into textiles (Yadav, 2023).

EDLCs are a specific kind of supercapacitor that stores charge in electrostatic or non–Faradic way by creating a double layer at the interface of the electrode and electrolyte (Chodankar et al., 2020; Khalafallah et al., 2023; Liu et al., 2018). Strong interactions between ions and molecules at this interface are crucial for the formation of this electrical double layer. The storage of charge in EDLCs depends on the reversibility of ion adsorption onto the surface of the electrode, which is usually composed of carbon. This means that the charge storage in EDLCs is primarily a surface phenomenon, unlike bulk reactions that occur in batteries. Therefore, the available/exposed surface area, thickness of the electrical double layer, electrolyte concentration, size as well as the shape of cations and anions in the electrolyte, diffusion rate of charged ions, electrode size and shape, and surface morphology of the electrode material all play a crucial role in the complex process of forming the electrical double layer (Khalafallah, Quan, et al., 2021; Khalafallah, Zhi, et al., 2021). Furthermore, unlike batteries, the capacitance values of EDLCs do not change with the surface potential and concentration as they do not have any redox behavior.

EDLCs consist of carbon-based positive and negative electrode, a separator, and an electrolyte. The primary types of carbon-based electrodes used in EDLCs are activated carbon (AC), carbon nanotubes (CNTs), and graphene. EDLCs can be distinguished not only by their arrangement but also by their electrochemical behavior, as they have a tribular-shaped charge−discharge profile, and their GCD profiles have a rectangular-like shape as shown in Fig. 8.3A (González et al., 2016). Based on their composition and electrochemical behavior, EDLCs have a relatively low capacitance value, high power density, good durability and cyclability in several thousands of cycles. Although EDLCs have their drawbacks, they are currently the most developed type of SCs dominating the market, despite the possibility of using a variety of materials and other devices' compositions. Carbon in all its forms is currently the most researched and often used electrode material in commercial EDLCs.

Taslim et al. developed an electrode material for EDLC applications made of biomass-based AC derived from Moringa oleifera leaves. The material was prepared through a consolidated carbon disk binder-free design as shown in Fig. 8.3B, which underwent a series of treatments, including pre-carbonization, chemical impregnation, integrated pyrolysis, and physical activation. The carbon disks were optimized at physical activation temperatures of 650°C, 750°C, and 850°C to achieve a high specific capacitance of 307 F g^{-1} at a current density of 1 A g^{-1} in an aqueous electrolyte solution of 1 M H_2SO_4 (Taslim et al., 2022). Hussein El-Shafei et al. (2023) reported another recent study focusing on flexibility and developed a graphene-like material using the laser carbonization of PAN nanofibers to create interdigitated micro-SCs as shown in Fig. 8.3C. The authors first produced a sheet of PAN nanofibers and then used the CO_2 laser engraving machine to convert them into laser-carbonized nanofibers and draw the pattern of the micro-supercapacitor simultaneously.

258 Electrochemistry and Photo-Electrochemistry of Nanomaterials

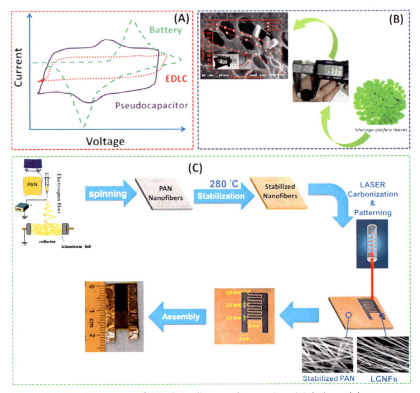

Figure 8.3 (A) CV curves of EDLC (red), pseudocapacitor (violet), and battery-type (green) configurations. (B) Porous carbon electrode material derived from Moringa oleifera leaves. (C) Schematic illustration of the laser-induced carbonization of the carbon nanofiber-based free-standing electrode. *(A) Reprinted with permission from González A., Goikolea, E., Barrena, J. A., & Mysyk, R. (2016). Renewable and Sustainable Energy Reviews, 58, 1189–1206. Copyright 2016, Elsevier. (B) Reprinted with permission from Taslim, R., Apriwandi, A., & Taer, E. (2022). ACS Omega, 7, 41, 36489–36502. Copyright 2022, American Chemical Society. (C) Reprinted with permission from Hussein El-Shafei, M., Abdel-Latif, M. S., Hessein, A., & Abd El-Moneim, A. (2023). FlatChem, 42, 100570. Copyright 2023, Elsevier.*

The resulting micro-supercapacitor exhibited high energy and power densities along with excellent electrochemical stability and mechanical cyclability (Pan et al., 2019).

8.2.1 Pseudocapacitors

Pseudocapacitance is a term used to describe the occurrence of fast and reversible Faradic reactions as an intermediate electron transfer takes place in electrode materials. This results in a capacitance that is different from the electrical double layer behavior and is not electrostatic in origin. The term "pseudo" is used to distinguish this capacitance from electrostatic capacitance and it arises due to an electrochemical charge transfer process that is limited by the finite amount of the active material or available surface (Chodankar et al., 2020). Pseudocapacitance can be caused by various processes such as the redox reaction of adsorbed ions from the electrolytes onto the electrode surface, doping and de-doping of conductive polymer composites, or variable oxidation state of the transition metal present in the electrode material. PCs have the same composition as EDLCs, but they use different materials for the electrode, such as metal oxides/hydroxides, chalcogenides, and their composites. Due to the redox behavior of these materials, PCs have higher specific capacitances than EDLCs, but they have a narrower potential window and poor cycle life due to the irreversible nature of redox reactions (Chodankar et al., 2020; Khalafallah et al., 2023; Liu et al., 2018). To improve the charge storage capabilities of PCs, factors such as material porosity, conductivity, particle size, electrode surface area, electrolyte concentration, and cell design should to be optimized.

Transition metal-based electrodes have been extensively studied as electrode materials for PCs. Compounds that are made up of oxygen atoms bonded to transition metals are known as transition metal oxides. These materials have high theoretical specific capacitances, making them promising candidates for supercapacitor electrodes. They are utilized for their semiconducting properties and catalytic activities. There are various types of transition metal oxides, including monoxides such as TiO and dioxides like MnO_2, as well as ternary oxides such as SrV_2O_6 and perovskite structures like $LaNiO_3$.

Ruthenium oxide (RuO_2) is a well-known material with favorable properties, such as high theoretical specific capacitance (1400$-$2000 F g^{-1}), excellent rate capacity, large voltage window, highly reversible redox reactions, high conductivity, and long cycle life (Majumdar et al., 2019). Recently, researchers focused on the studies of combining RuO_2 with other elements. According to Raja's group, a capacitor composed of RuO_2 particles adorned onto a phosphate-doped reduced graphene oxide (RuO_2-P-rGO) showed a specific capacitance of 606 F g^{-1} and remained stable for up to 2000 test cycles (Raja et al., 2022). The proposed RuO_2 and RuO_2-P-rGO obtained good capacitive performances (Fig. 8.4A and B). Yadav prepared an efficient electrode by spraying a mixture of Co-doped

260 Electrochemistry and Photo-Electrochemistry of Nanomaterials

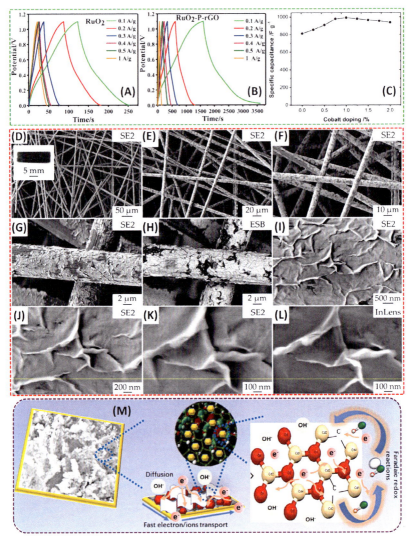

Figure 8.4 GCD curves at different current densities of (A) RuO$_2$ and (B) RuO$_2$-P-rGO. (C) Specific capacitance as a function of Co content. (D–L) Scanning electron microscope (SEM) images of the as-fabricated NiO/CP material. Specific capacitance as a function of the Co content. (M) A schematic diagram illustrating the reasons behind the superior performance of the proposed HSC cell. *(A, B) Reprinted with permission from Raja A., Son, N., Swaminathan, M., Kang, M. (2022). Journal of the Taiwan Institute of Chemical Engineers, 138, 104471. Copyright 2022, Elsevier. (C) Reprinted with permission from Yadav C. R A.A., Electrochimica Acta 437 (2023) 141521. Copyright 2023, Elsevier. (D–L) Reprinted with permission from Simonenko T. L., N. P. Simonenko, P. Y. Gorobtsov, E. P. Simonenko, N. T. Kuznetsov, Materials. 16 (2023) 5208. Copyright 2023, MDPI, Basel, Switzerland. (M) Reprinted with permission from Sajjad M., M. Z. U. Shah, F. Mahmood, M. S. Javed, R. Maryam, F. Ahmad, A. Shah, R. Hussain, A. M. Toufiq, Z. Mao, S. Rahman, Journal of Alloys and Compounds. 938 (2023) 168462. Copyright 2023, Elsevier.*

RuO_2 in an aqueous organic solvent. As a result, the specific capacitance reached 1158 F g^{-1} at 0.5 A g^{-1}, compared to 893 F g^{-1} for the non-doped RuO_2. As seen in Fig. 8.4C, the incorporated Co could positively impact the specific capacitance. Specifically, the 1.0 mol% Co-doped RuO_2 electrode yielded outstanding long-term cyclic stability and preserved 94% of the initial capacitance after 1000 charge—discharge cycles (Yadav, 2023).

Although RuO_2 electrode material has several benefits, including high conductivity, durability, and stability, its usage is restricted because of its expensive cost and toxic nature. Transition metal oxides have become popular for their potential use as pseudocapacitor electrode materials. Examples include MnO_2, NiO, and Co_3O_4. These candidates stand out as promising alternatives to RuO_2 due to their low cost, environmental friendliness, and high theoretical capacitance. Among these, MnO_2 can effectively work in mild aqueous electrolytes like Na_2SO_4, NaCl, and KCl, whereas RuO_2 requires stronger acid or base electrolytes, making it less versatile compared to MnO_2 (Wei et al., 2011). However, most transition metal oxides are limited by insufficient ion diffusion, weak structural stability, and poor conductivity. As a result, the electrochemical utilization is low and they have poor cycling life (Bounor et al., 2023). In a recent study, Khalid et al. (2023) have developed a method to enhance the electrochemical performance of MnO_2 for next-generation SCs. The Ag-doping helped to modify the band structure of MnO_2 and improve its specific conductivity. Additionally, sandwiching Ag-MnO_2 nanowires between the voids of MXene sheets and dispersing them over the sheet's surface led to the formation of a heterostructured composite with excellent conductivity, high surface area, and structural openings. These features enabled the Ag-MnO_2/MXene@NF electrode to exhibit a capacitance of 1188 F g^{-1} @ 1 A g^{-1}, which is higher than both the MnO_2@NF electrode and the Ag-MnO_2@NF electrode (643 and 795 F g^{-1}, respectively). The electrode also demonstrated an impressive rate capability (85.8% @9 A g^{-1}) and a high cyclic activity of 96.4% after 6000 tests (Khalid et al., 2023). Recent studies have been focused on creating flexible SCs on substrates. Simonenko et al. (2023) investigated the formation of a cellular hierarchically organized NiO film on a carbon paper (CP) substrate under hydrothermal conditions to develop a flexible NiO/CP composite electrode as illustrated in Fig. 8.4D—L. The resulting electrode exhibited an exceptional cycling stability (95% capacitance retention during 2000 cycles at 5 A g^{-1}) and a good specific capacitance value $(207 \text{ F g}^{-1}$ at $0.5 \text{ A g}^{-1})$ when tested using a three-electrode scheme (Simonenko et al., 2023).

8.2.2 HSCs

HSCs are a type of energy storage device that combines both electric double layer and pseudocapacitive materials to store charges efficiently. They work by using a thin polymeric sheet infused with electrolytes to enable the integration of both capacitive and redox materials into a single cell. This implies that they can deliver improved specific energy and power outcomes without compromising on cycle life while exhibiting enhanced charge storage performance over a broad potential range (Chodankar et al., 2020; Khalafallah et al., 2023; Liu et al., 2018). To achieve higher energy and power densities by HSCs, it is essential to design appropriate electrode materials. The literature suggests that conductive carbon-based materials such as graphene, CNTs, and AC are often used as capacitive materials, while metal oxides, hydroxides, metal chalcogenides, phosphides, polymers, and their composites are used as pseudocapacitive electrodes (Chodankar et al., 2020; Khalafallah et al., 2023; Liu et al., 2018). There are three types of HSCs: asymmetric, composite, and rechargeable batteries. The asymmetric hybrid supercapacitor has a capacitive electrode and a faradaic electrode, while a composite hybrid supercapacitor combines the properties of metal oxides and carbon to create a composite with high conductivity, cyclic stability, and specific capacitance (Muzaffar et al., 2019). Metal-ion capacitors have emerged recently. These capacitors have metal ions in their anodes, which change the potential of the electrode and can increase the voltage window of the cell. The most common types of metal-ion capacitors are Li-ion capacitors and Na-ion capacitors, while Al-ion capacitors have also been developed successfully (Khalafallah, Zhang, et al., 2024). SCs that combine a high-capacity battery-type electrode with a high-rate capacitive electrode are known as rechargeable battery type. Wider cell voltage windows and larger capacities, and therefore higher energy densities, can be achieved by replacing one capacitive electrode of the typical symmetrical supercapacitor with a battery electrode (Khalafallah, Zhang, et al., 2024). A hybrid supercapacitor positive electrode made of CdO-rGO nanocomposites with varying rGO concentrations was developed by Rahman's group. The composite with 25% rGO concentration demonstrated excellent specific capacitance and rate performance $(1195 \text{ F g}^{-1}$ at $1 \text{ A g}^{-1})$.

Furthermore, the HSC device demonstrated an exceptional cyclic stability, retaining 88% of its initial capacitance over 7000 charge−discharge cycles at a high current density of 10 A g^{-1} with remarkable coulombic efficiency. The smart compositional/morphological/structural features favoured the enhancement in the electrode's performance (Fig. 8.4M) (Sajjad et al., 2023).

8.3 Cost-efficient coordination materials for electrochemical SCs

For the wide utilization of ESCs, electrode material should be inexpensive and abundant. According to the nature of the electrode material, there are three major classes of coordination compounds that have been extensively researched for ESCs: (1) carbon-based materials (e.g., graphene, AC, tubular carbon nanofibers, CNTs, carbon aerogels, hierarchical porous carbon), (2) Faradaic conductive polymers (e.g., polyaniline, polypyrrole (PPy), polythiophene, poly (3,4-ethylene dioxythiophene), etc.), and (3) transition metal-based materials (e.g., earth-abundant transition oxides, hydroxides, chalcogenides, phosphides, nitrides, and borides). Carbon nanomaterials with formidable specific surface area, excellent chemical/mechanical/thermal stability, large power density, and tunable electrical conductivity have been found to serve as electrode materials for SCs (Chodankar et al., 2020; Khalafallah et al., 2023; Liu et al., 2018). The intrinsic rectangular shape of the CV curves and symmetrical GCD spectra of carbon-based materials make them proper capacitive electrodes, while transition metals and conducting polymers serve as pseudocapacitive or battery-type materials. The factors influencing the electrochemical properties of nanostructured carbon electrodes are their specific surface area, pore architecture, and surface functionality. However, their specific capacitance is much lower than those of conducting polymers and transition metal-based Faradaic electrode materials.

Conducting polymers have attained considerable importance for applications in ESCs owing to their high conductivity, distinct redox behavior, relatively high specific capacitance, good mechanical characteristics, accelerated charge/ion penetration kinetics, and low ESR. Especially, the n/p type polymer configurations have a notable potential for high energy/power densities. Nevertheless, the intensive swelling and shrinking of polymeric material volume during the charge−discharge process due to ion insertion/deinsertion may accelerate the material structure deterioration, resulting in poor cycling performance. To surmount this limitation, a polymeric layer is coated on a metal or carbon support, constructing a binary or ternary hybrid with a high capacitance and improved long-term cyclic stability (Fleischmann et al., 2020; González et al., 2016; Muzaffar et al., 2019; Wei et al., 2011). In principle, pseudocapacitive and battery-type materials can provide larger capacitance/capacity and higher energy density than electric double-layer capacitive materials, therefore stimulating the development of sustainable energy storage devices. Transition metal oxides/hydroxides have been applied as electrode materials for SCs because of their large theoretical capacitances; however, the applications of transition metal oxides/hydroxides are largely inhibited owing to their weak electrical conductivity and unsatisfactory rate capability. On the other hand, a large number of advanced nanomaterials have been explored to meet the requirements of practical utilization such as chalcogenides (e.g., sulfides, selenides, and tellurides), phosphides, nitrides, and borides of first-row transition metals owing to their well-defined electrochemical redox characteristics, multiple reactive equivalents of variably valent metal ions, favorable kinetic process of Faradaic reactions, and higher conductivity in comparison with their oxide/hydroxide counterparts (Hong et al., 2024; Wu et al., 2021). Transition metal phosphides that comprise one or more transition metal elements with the phosphorus (P) atom hold a potential as effective electrode materials for SCs with good electrochemical performance and metalloid properties. The advantages of phosphides originate from the lower electronegativity and larger atomic radius of P atoms compared with the oxygen, revealing different physicochemical characteristics.

Additionally, transition metal chalcogenides have also been widely designed as a unique class of electrode materials with smart hierarchical morphologies, superior electrochemical properties, rich species of electrochemical redox reactions, acquiring higher charge storage capability and rate performance. Recently, Ni- and Co-based boride with typical metal–metal, metal–B, and B–B coordination environments have piqued the focus of the scientific community. These materials also displayed short path lengths for ion diffusion, hence promoting the utilization efficiency of active sites. On the other hand, metal boride-based electrodes can undergo severe structural degradation and dissolution during electrochemical reactions due to chemical reactions with electrolyte components, inducing the loss of active material and performance deterioration (Khalafallah, Zhi, et al., 2022; Shi et al., 2022). To solve this problem, support materials are often integrated with metal borides to strengthen their durability. Support components (e.g., carbonaceous substrates, metal oxides or conducting polymers) protect metal borides from dissolution by physically immobilizing or confining them. Furthermore, transition metal nitrides with various chemical valence states, strong metal–nitrogen bonds and reasonably high capacity are viable candidates for ESCs. However, some nitrides (e.g., Mo_2N, WN, TiN, and NbN) suffer from limited cycling life under specific circumstances and low capacity issues. Compositing transition metal nitrides with other metallic or nonmetallic components can expected to enhance the electrochemical efficiency.

8.4 Manipulating the energy storage capability of electrodes

To configure an efficient electrode material with satisfactory performance for practical applications, it is necessary to apply modification approaches to effectively manipulate the intrinsic electrochemical properties. Besides, specific surface area, intrinsic surface chemistry, accessible active sites, and electronic configuration also exhibit significant influence on the electrochemical performance. Rational modification strategies such as hybridization, elemental doping, heterointerface engineering, surface functionalization, defect engineering, and morphological optimization can significantly regulate the physicochemical properties, increasing the active site exposure and maximizing the energy storage ability.

8.4.1 Elemental doping

Heteroatomic implanting (e.g., single-element doping and multi-element co-doping) by substituting atoms in the support material with foreign elements or inserting foreign elements into interstitial voids, has been considered to be a highly feasible pathway to enhance the conductivity and increase the electrochemically active sites for redox processes, thus boosting reaction kinetics (Khalafallah, Quan, et al., 2021). The cationic and anionic doping effect strongly correlates with the intrinsic properties of dopants and interfacial interactions between dopants and their coordination atoms. Heteroatom dopants can boost the adsorption kinetics of ions, which benefits the volumetric capacitance. For example, Huang et al. (2018) explored binder-free Al-modified hierarchical Co-sulfide anchored on vertically oriented Ni nanotubes (H−Ni@Al−Co−S) as a flexible and efficient cathode for high-performance ASCs. The core—shell nanocomposite electrode was rationally constructed and directly grown on a conductive carbon cloth (CC) substrate as schematically depicted in Fig. 8.5A (Huang et al., 2018). The open-arrayed architectures are favorable to facilitate the electron/ion transportation kinetics within the active electrode material, thereby boosting electrochemical properties. The abundant interspaces improved the electrolyte ion accessibility, reduced ion diffusion distances, and strengthened the redox reactions (Fig. 8.5B and C), thus simultaneously reinforcing the energy storage capability. Consequently, the as-yielded core—shell structured CC/H−Ni@Al−Co−S positive electrode exhibited enhanced electrochemical properties in a 2.0 M KOH electrolyte solution, delivering a remarkable specific capacitance of 2434 F g^{-1} at a current density of 1 A g^{-1} and a superior rate capability with 72.3% capacitance retention at the higher current density of 100 A g^{-1} (Fig. 8.5D and E). Obtained results demonstrated that a maximum specific capacitance and better electric conductivity were acquired with an appropriate amount of the Al (2 at.%), whereas increasing the Al content resulted in a lower specific capacitance. This might be ascribed to the inhibition of the Faradaic redox reactions with the excessive Al content. Furthermore, the all-solid-state ASC device integrating the CC/H-Ni@Al-Co-S as a cathode and multilayered graphene/CNTs film as the anode could attain a high energy density of 65.7 Wh kg^{-1} at a power density of 765.3 W kg^{-1}. The ASC full device provided 90.6% of its original capacitance even after consecutive 10,000 GCD times, confirming the prominent cyclic stability (Fig. 8.5F and G). The device also displayed high flexibility under various bending conditions (Fig. 8.5H). Similarly, Au-doped porous MnO_2 (Kang et al., 2013), defect-rich Cu-doped MnO_2 nanowires (Cu_xMnO_2) (Wu et al., 2023), Al-$NiCo_2O_4$ NSW (Chen et al., 2022), Zn-doped MnO_2 (ZnMO) (Wu et al., 2022), MnO_2-Ag_3/GO (Mane et al., 2021), 0.3% Ag-MnO_2 HMS (Yi et al., 2020), and Al-doped $CuCo_2O_4$ nanowire arrays (Cheng et al., 2020) were designed with tunable intrinsic electron configurations and regulated capacitive performances.

Apart from the metallic doping, incorporating nonmetallic nitrogen (N), sulfur (S), boron (B), and P species into the carbon framework can efficiently modulate the electronic properties of neighboring carbon atoms because of different electronegativity. These species can create surface functional groups and optimize the electronic distribution, favoring the synergistic enhancement in the charge storage performance by forming

Coordination materials for supercapacitors 267

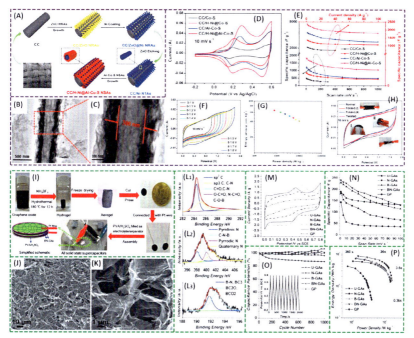

Figure 8.5 (A) Representative scheme for the synthesis of hierarchical CC/H-Ni@Al-Co-S core−shell nanostructures, (B and C) transition electron microscope (TEM) images of Ni nanotube@Al-doped Co-S nanosheets, (D) CV curves of electrodes at scan rate of 10 mV s^{-1}, (E) specific capacitance of electrodes derived from the collected CV and GCD results, (F) CV curves of the as-assembled ASC device recorded with different working potential windows, (G) Ragone plot, and (H) CV profiles of flexible ASC cell under different bending conditions. (I) Construction procedure of BN-GAs-based all-solid-state supercapacitor system via a combined hydrothermal approach and freeze-drying process. (J) Low- and (K) high-magnification SEM images of the BN-GAs-based electrode. High-resolution XPS spectrun of C 1s (L1), N 1s (L2), and B 1s (L3) core levels. (M) CV curves of different electrodes obtained at 5 mV s^{-1}. (N) Specific capacitance and rate performance of electrodes as a function of applied scan rate, (O) cycling stability of electrodes. *Inset-O* illustrates the GCD profile of BN-GAs at 2 A g^{-1}, and (P) Ragone plots of as-fabricated all-solid-state SCs based on various proposed materials. *(A-H) Reprinted with permission from Huang, J., Wei, J., Xiao, Y., Xu, Y., Xiao, Y., Wang, Y., Tan, L., Yuan, K., & Chen, Y. (2018). When Al-doped cobalt sulfide nanosheets meet nickel nanotube arrays: A highly efficient and stable cathode for asymmetric supercapacitors. ACS Nano, 12, 3030. Copyright 2018, American Chemical Society. (I-P) Reprinted with permission from Wu, Z., Winter, A., Chen, L., Sun, Y., Turchanin, A., Feng, X., Mullen, K. (2012). Three-dimensional nitrogen and boron Co-doped graphene for high-performance all-solid-state supercapacitors. Advanced Materials, 24, 5130. Copyright 2012, WILEY-VCH.*

many structural defects. In this regard, Wu et al. achieved the synthesis of N,B-codoped well-interconnected ultrathin graphene nanosheet aerogels (GAs) through a hydrothermal reaction (Fig. 8.5I−K) (Wu et al., 2012). As illustrated in Fig. 8.5L1−L3, the implanting of heteroatomic N/B species into the carbon matrix was verified by X-ray photoelectron spectroscopy (XPS) analyses. The tailored monolithic GAs with a 3D interconnected network and prominent macroporous architecture could accelerate the electron transport and ion diffusion in the bulk electrode. Besides, the introduced N/B species into the carbon framework improved the charge transfer between adjacent carbon atoms and supplied extra pseudocapacitance contribution. Given this, the as-obtained BN-GAs with strong synergetic effects of N/B codoping realized a high capacitance of 239 F g^{-1} at a low scan rate of 5 mV s^{-1} and 132 F g^{-1} at a higher scan rate of 100 mV s^{-1} in a 1.0 M aqueous H_2SO_4 electrolyte with a three-electrode configuration (Fig. 8.5M−O). The as-designed all-solid-state supercapacitor cell employing the BN-GAs as both the positive and negative electrodes with the polyvinyl alcohol (PVA)/H_2SO_4 gel as a solid electrolyte attained good energy/power outcomes (≈ 8.65 Wh kg^{-1}/ ≈ 1600 W kg^{-1}) (Fig. 8.5P). Accordingly, the functions of heteroatomic doping are follows: (1) cause a robust electronic interaction as the dopants can serve as an electron donor or acceptor, thereby optimizing the band gap of the electrode material and achieving good conductivity and (2) modify the surface chemical states of the electrode material, therefore heteroatomic dopants facilitating the charge penetration at the electrode/ electrode interface.

8.4.2 Surface functionalization

It is generally recognized that Faradaic redox reactions usually take place at the exposed surface. Thus, the capacitive properties of active components are closely related to the reactivity at the electrode/electrolyte interface and the number of accessible active sites (Chodankar et al., 2020). Taking these into consideration, researchers have focused on coordinating ionic groups with particular functional characteristics to the electrode material surface. This can manipulate the number of electroactive sites, maintain the reactivity of reaction centers, and boost the charge transfer rate. Surface functional groups adjust the electronic configuration of the host material. According to previous studies, surface modification and optimization enrich the active sites and further promote the charge storage capacity. Zhai et al. functionalized the surface of Co_3O_4 with phosphate ions (denoted as PCO nanosheet arrays) to stimulate the high chemical reactivity for quick and effective Faradaic redox reactions. $NaH_2PO_2 \cdot H_2O$ salt as the P source was investigated to form a phosphate radical under the thermal annealing treatment in the Ar atmosphere (Fig. 8.6A) (Zhai et al., 2017). The structural and surface characterizations validated the coexistence of $(H_2PO_4)^-$ and $(PO_3)^-$ species within the rationally engineered PCO material (Fig. 8.6B–D). The authors elucidated substantially enhanced pseudocapacitive behavior and high rate capability for the functionalized PCO nanosheet arrays electrode compared with the pristine Co_3O_4 nanosheet arrays electrode (Fig. 8.6E and F). The phosphate ion functionalization and the unique porous architecture could enable plentiful active sites for redox reactions. The PO_3^- species strengthened the redox reactions at the interface by increasing the iconicity of Co–O bonding and modified the effectiveness of the Co-involved Faradaic process. Side by side, the $H_2PO_4^-$ species promoted the electron transfer process. Impressively, the PCO electrode achieved remarkable gravimetric capacitance (1716 F g^{-1} at 5 mV s^{-1}), exceptional rate capability (1016 F g^{-1} at 100 mV s^{-1}), and notable cycle performance (85% capacitance retention after 10,000 CV cycles), outperforming those of pristine Co_3O_4 electrode (Fig. 8.6E–G). More importantly, the established ASC device utilizing the PCO cathode and 3D porous graphene gel (3DPG) anode could reach a stable operational potential window of 1.5 V (Fig. 8.6H), revealing an excellent energy density of 71.58 Wh kg^{-1} and an average power density of 1500 W kg^{-1} (Fig. 8.6I).

Surface functionalization can be also performed by immobilizing conducting polymers such as PPy and polyaniline (PANI) as a nanosized thin layer onto the surface of the support material. This can lead to a high electronic conductivity of electrodes. For instance, highly conducting PPy was firmly anchored on well-aligned CoO nanowire arrays grown on a 3D nickel foam (NF) skeleton (Zhou et al., 2013). As can be seen, obvious hierarchical structure of the CoO@PPy nanowire arrays could be obtained, in which the surface of CoO nanowires was decorated with the PPy film without changing the ordered nanoarchitecture (Fig. 8.6J and J1). Moreover, the PPy not only regulated the electron transport kinetics but also participated in the redox reaction to store energy (Fig. 8.6K and L). The thus obtained CoO@PPy hybrid electrode showed a lower charge transfer resistance and larger charger storage capacitance

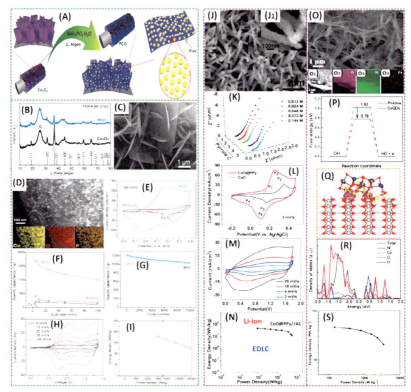

Figure 8.6 (A) Synthesis procedure of the PCO nanosheet arrays electrode, (B) X-ray diffraction (XRD) patterns of as-fabricated samples, (C) SEM image of porous PCO framework, (D) typical high-angle annular dark-field scanning TEM image and related energy-dispersive X-ray spectrometry (EDS) mapping profiles of Co, O, and P elements throughout the regulated PCO material. (E) CV spectra of electrodes at 100 mV s^{-1}, (F) calculated specific capacitance of electrodes as a function of applied scan rate, (G) cycle stability of the PCO electrode analyzed at 50 mV s^{-1} for 10,000 CV cycles, (H) CV signals of the as-assembled PCO//3D porous graphene gel (3DPG) ASC device at different scan rates, and (I) Ragone plot of the PCO//3DPG ASC system. (J–J1) High-magnification SEM portraits of the CoO@PPy nanowires hybrid electrode, (K) Nyquist responses of the CoO@PPy hybrid electrode constructed with various PPy concentrations, (L) CV curves of optimized CoO@PPy and pure CoO electrodes, (M) CV signals of the as-designed CoO@PPy//AC ASC device collected at different scan rates, and (N) Ragone plot of the CoO@PPy//AC device compared with conventional EDLC system and Li-ion battery cell. (O) SEM image of 2D/0D Ni(OH)$_2$-CoQDs nanohybrid heterostructures, (O1–O4) cross-sectional SEM image and related EDS mapping spectra of O, Ni, and Co components within the Ni(OH)$_2$-CoQDs nanostructures, (P) adsorption free energies of OH$^-$ over pure Ni(OH)$_2$ and CoQDs coupled Ni(OH)$_2$ (110) surface, (Q) the charge density difference of CoQDs coupled Ni(OH)$_2$ (110) surface, and (R) the total and partial DOS of CoQDs coupled Ni(OH)$_2$ (110) surface. The

$(2223\ \text{F g}^{-1}$ at 1 mA cm^{-2}) nearly twice that of the individual CoO electrode $(1212\ \text{F g}^{-1}$ at 1 mA cm^{-2}). The 4 cm^2 ASC device approached an expanded cell voltage of 1.8 V in a 3.0 M NaOH electrolyte, delivering a high energy density of 43.5 Wh kg^{-1} at a power density of 87.5 W kg^{-1}. These findings were much superior to that of a conventional EDLC system with the same power scale and even comparable to a traditional Li-ion battery cell (Fig. 8.6M and N).

In addition to organic polymers, inorganic quantum dots can be used in a similar procedure to modify the electrode's active material surface. Gong's group designed two-dimensional (2D) Ni(OH)$_2$ nanosheet arrays reinforced with zero-dimensional (0D) semiconductinge Co(OH)$_2$ quantum dots (CoQDs) through a hydrothermal reaction and proper room temperature immersing route, enabling large surface area, unique porosity, and wide exposure to electrolyte ions (Shi et al., 2018). Significantly, the 2D/0D Ni(OH)$_2$-CoQDs nanohybrid heterostructures (Fig. 8.6O and O1−O4) could reach an excellent capacitance (3244 F g^{-1} at 5 mA cm^{-2}) with a remarkable rate activity (2047 F g^{-1} at 50 mA cm^{-2}). The density functional theory (DFT) calculation results elucidated an electron accumulation at the interface of Ni(OH)$_2$ (110) and CoQDs. The Bader charge analysis indicated that the encapsulated CoQDs donated electrons to the neighboring Ni(OH)$_2$ (110) surface. The analyzed total and partial density of states (DOS) revealed that the Co $3d$ orbitals improved the DOS near the Fermi level and enhanced the conduction band, thus modifying the OH$^-$ binding. The Co atoms supplied additional electrons to the Ni(OH)$_2$ slab, thereby manipulating the electrical properties of the Ni(OH)$_2$ (110) surface and decreasing the adsorption-free energy of OH$^-$ (Fig. 8.6P−Q). Notably, the sandwiched ASC system-based Ni(OH)$_2$-CoQDs nanocomposite revealed 46 Wh kg^{-1} at 141 W kg^{-1} (Fig. 8.6S).

---

yellow and purple colors indicate charge depletion and accumulation, respectively. The H, O, Ni, and Co atoms are presented by light pink, red, light blue, and blue spheres, respectively. (S) Specific energy density versus the specific power density of the established ASC cell-based Ni(OH)$_2$-CoQDs positive electrode. *(A-I) Reprinted with permission from Zhai T., Wan, L., Sun, S., Chen, Q., Sun, J., Xia, Q., Xia, H. (2017). Advanced Materials, 29, 1604167. Copyright 2017, WILEY-VCH. (J-N) Reprinted with permission from Zhou C., Zhang, Y., Li, Y., Liu, J. (2013). Nano Letters, 13, 2078−2085. Copyright 2013, American Chemical Society. (O-S) Reprinted with permission from Shi, D., Zhang, L., Zhang, N., Zhang, Y., Yu, Z., Gong, H. (2018). Nanoscale, 10, 10554. Copyright 2013, The Royal Society of Chemistry.*

8.4.3 Hybridization with carbons

Although the enhanced conductivity and large specific surface area of carbon-based electrodes, their specific capacitance is highly limited by the restriction of the electrical double layer mechanism. To promote the energy-storing capability of carbon-based electrodes, tremendous efforts have been focused on engineering hybrid nanostructures by combining carbon components and pseudocapacitive nanomaterials, especially earth-abundant transition metals (Khalafallah, Miao, et al., 2021). The attractive synergistic effects of conductive carbonaceous matrix and nonconductive or semiconducting pseudocapacitive component can significantly strengthen the enhancement of state-of-the-art SCs. The robust coupling between the carbonaceous nanoframe and metallic nanomaterial improves the efficiency of electron movement and reinforces the structural stability during the electrochemical reaction, therefore ameliorating the rate performance and boosting the cycling stability of resulting supercapacitor electrodes (Hong et al., 2024; Khalafallah, Zhang, et al., 2024). Impressed by the above discussion, Yang et al. proposed a skeleton CNT sponge with a 3D porous network as a compressible substrate to support $NiCo_2S_4$ nanoparticles through a solvothermal method. The resultant $NiCo_2S_4/$ CNT sponge electrode with an enhanced hydrophilic character could increase the contact area between the underlying substrate and electrochemically active ions, thus affording more active sites for redox reactions. The $NiCo_2S_4/CNT$ positive electrode exhibited a good discharge capacitance of $1110 \, F \, g^{-1}$ at $1 \, A \, g^{-1}$ and possessed an exceptional capacitance retention of 81% at a higher current density of $20 \, A \, g^{-1}$. To establish an ASC system with an appropriate voltage window and a high energy density, the Fe_2O_3/CNT sponge anode electrode was explored to match the $NiCo_2S_4/CNT$ sponge cathode electrode. The utilization of Fe_2O_3/CNT as a negative electrode in the voltage window of -1.0 to $0.0 \, V$ enabled a high discharge capacitance of $330 \, F \, g^{-1}$ at $1 \, A \, g^{-1}$. Furthermore, the constructed $NiCo_2S_4/CNT//Fe_2O_3/CNT$ ASC cell with a working voltage window of $1.7 \, V$ displayed suitable energy/power ($41.6 \, Wh \, kg^{-1}/ 800 \, W \, kg^{-1}$) outputs in a $3.0 \, M$ KOH electrolyte (Yang et al., 2021).

Transition metal sulfides are novel battery-type positive electrode materials owing to their supreme redox feasibility, excellent pseudocapacitive charge-storing ability, and tunable electronic conductivity. The multireactive centers in metal sulfides with variable valent metal ions promote the redox reactions, resulting in larger charge storage behaviors. However, heterostructurization with carbon can improve the bulk diffusivity and further tune the intrinsic conductivity, hence inducing excellent electro-functionality and yielding unique charge storage capability (Khalafallah, Zhi, et al., 2021). In this regard, controllable growth strategies were adopted to fabricate diffuse-porous nanostructure-shaped $Co_9S_8-NiCo_2S_4$/defective rGO ($Co_9S_8-NiCo_2S_4$/D-rGO) with a large assembly of interconnected nanoparticles as a battery-type positive electrode and flaky FeS/nitrogen-doped D-rGO (FeS/ND-rGO) with copious mesopores and macropores as the negative electrode to strengthen the energy storing efficiency of hybrid pseudocapacitor systems (Sonia & Meher, 2022). The nanostructured $Co_9S_8-NiCo_2S_4$/D-rGO hybrid showed a wide range of defects and disorders in the embedded graphene network, proper surface wettability, multireactive Co/Ni equivalents, and strong heterostructurization. The electroredox employment of $Co_9S_8-NiCo_2S_4$/D-rGO positive electrode exhibited integrated redox/pseudocapacitive charge storage pathways, high-rate capacitance efficacy, and diffuse-porous architecture-directed modulated redox kinetics. As a positive electrode material, the multicomponent $Co_9S_8-NiCo_2S_4$/D-rGO revealed a high discharge capacitance of 2610 and 1751 F g^{-1} at the applied charge$-$discharge current densities of 1 and 24 A g^{-1}, respectively, approaching a superior rate capability of $\sim 68\%$. Further, the FeS/ND-rGO nanocomposite showed a high pseudocapacitive charge storage efficacy under a wide negative potential window of 1.2 V, achieving profound kinetic reversibility, desired charge transfer resistance, modified rate efficiency, and enlarged negative potential window . The thus obtained FeS/ND-rGO negative electrode presented a high capacitance of 393 F g^{-1} at 50 mV s^{-1}. Accordingly, the electrochemical behavior of the FeS/ND-rGO negative electrode truly fitted the compatibility with the $Co_9S_8-NiCo_2S_4$/D-rGO positive electrode. The all-solid-state hybrid pseudocapacitor system-based $Co_9S_8-NiCo_2S_4$/D-rGO‖FeS/ND-rGO configuration in the PVA-KOH solid electrolyte could operate stably at a broad voltage window of 1.9 V and demonstrate diffusion-controlled charge storage physiognomies. Such a full device realized maximum energy/power densities (~ 91 Wh $kg^{-1}/\sim 5221$ W kg^{-1}) and superior long-term service stability with 96.9% capacitance preservation even after 11,000 charge$-$discharge time at 12 mA cm^{-2}. The regulated performance of such a pseudocapacitor device was attributed to the integration of electromicrostructurally compatible sulfide/carbon-based electrodes with rich redox-active Co, Ni, and Fe ions. Additionally, the ion-buffering pool mechanism of materials' bulk and intensified charge transfer ability of D-rGO and ND-rGO matrices contributed significantly to the overall charge storage activity.

8.4.4 Heterointerface engineering

Interface engineering of multiple electroactive components is another efficient route to significantly adjust the physicochemical properties. The heterointerfaces in nanostructured core−shell architectures usually reveal modified internal electronic characteristics and synergistic behavior compared to single-component electrodes, which accelerate electron transport kinetics and boost the synergistic behavior of the whole electrode material (Khalafallah et al., 2023). The utilization of hybrid compounds to harvest the advantages of single components can greatly regulate the electrochemical performance of the whole. Functionally, multicomponent hybrid electrodes, which incorporate a blend of different components, can offer the following aspects: (1) the integrated active components bring in the coupling effect to rearrange electrons, enabling quicker electron transfer rate and richer redox reactions; (2) the synergism may be brought in to take advantage of each component, achieving lower charge transfer resistance and smaller material resistance. The synergetic effects with complementary properties have been proven to extensively improve the electrochemical properties of ESCs (Muzaffar et al., 2019; Wei et al., 2011). Bian and coworkers tailored an elastic 3D metallic conductive platform along cotton yarns by confining a thin Au layer into cotton fibers. Hence, the redox reaction kinetics and utilization efficiency of active components could be intensified due to the fast electron transport feature. In such a pseudocapacitive system, hierarchical core−shell paired $NiCo_2S_4$@Ni-Co LDH nanotube arrays were directly grown on the Au nanoparticle layer-coated cotton yarn fiber ($NiCo_2S_4$@Ni-Co LDH/ ACY) through hydrothermal approach, anion−exchange reaction, and electrolytic deposition process (Fig. 8.7A−C) (Wang et al., 2019). In this design, the arrayed $NiCo_2S_4$ nanotube afforded two functions: (1) the conductive bridges created by the vertically aligned $NiCo_2S_4$ nanotubes and the coated Au film greatly promoted the electron transportation kinetics and (2) the as-grown $NiCo_2S_4$ nanotubes served as a conductive scaffold to load the Ni-Co LDH nanosheets with a large active surface area and additional pseudocapacitance. Besides, the hierarchically porous architectures could reduce the ion diffusion path. The $NiCo_2S_4$@Ni-Co

Coordination materials for supercapacitors 275

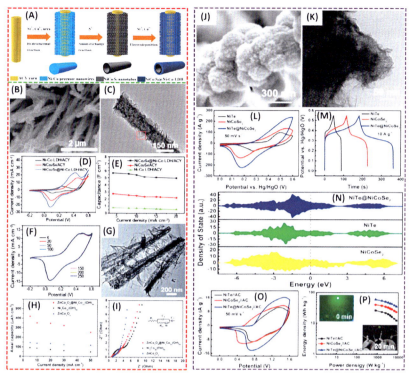

Figure 8.7 (A) Representative diagram for the fabrication strategy of NiCo$_2$S$_4$@Ni–Co LDH/ACY electrode, (B) SEM image, and (C) TEM image of as-grown NiCo$_2$S$_4$@Ni–Co LDH/ACY core–shell nanostructures. (D) CV plots of electrodes at 5 mV s^{-1}, (E) areal capacitance values at various areal current densities, and (F) CV curves over 250 bending cycles at 5 mV s^{-1}. (G) TEM image of as-obtained ZnCo$_2$O$_4$@Ni$_x$Co$_{2x}$(OH)$_{6x}$ core–shell nanostructures, (H) the discharge areal capacity versus the current density, and (I) Nyquist plots of proposed samples. (J) SEM and (K) TEM observations of the core–shell-coupled NiTe@NiCoSe$_2$ nanohybrid. (L) Comparative CV curves of as-synthesized electrodes, (M) GCD plots of electrodes, (N) calculated DOS of the NiTe@NiCoSe$_2$ nanocomposite, (O) comparative CV plots of different assembled ASC devices, and (P) Ragone plots of constructed full cells. *Inset-P* shows an optical picture of two series connected NiTe@NiCoSe$_2$//AC ASC systems with a green light-emitting diode. *(A-F) Reprinted with permission from Wang, Y.-F., Wang, H.-T., Yang, S.-Y., Yue, Y., & Bian, S.-W. (2019). ACS Applied Materials & Interfaces, 11, 33, 30384–30390. Copyright 2019, American Chemical Society. (G-I) Reprinted with permission from Fu W., Wang, Y., Han, W., Zhang, Z., Zha, H., & Xie, E. (2016). Journal of Materials Chemistry A, 4, 173. Copyright 2016, The Royal Society of Chemistry. (J-P) Reprinted with permission from Ye, B., Xiao, S., Cao, X., Chen, J., Zhou, A., Zhao, Q., Huang, W., & Wang, J. (2021). Journal of Power Sources 506, 2021, 230056. Copyright 2021, Elsevier.*

LDH/ACY electrode displayed remarkable electrochemical properties in terms of specific capacitance (5680 mF cm^{-2} at 2 mA cm^{-2}), proper rate ability (87% of its initial capacitance even at 20 mA cm^{-2}), and desired mechanical flexibility (Fig. 8.7D$-$F). The two-ply symmetric yarn supercapacitor device merging the NiCo$_2$S$_4$@Ni-Co LDH/ACY electrode acquired an areal energy density of 3.5 μWh cm^{-2}. The hierarchical hollow nanotubes could offer numerous growth sites for anchoring active materials than solid nanotube scaffolding, ensuring more capacitance. The highly porous Ni-Co LDH nanosheet network could expose abundant redox active sites, yielding a large capacitance.

Additionally, different work was also presented to accelerate the ion transfer at the electrode/electrolyte interface by manufacturing hybrid components. Fu et al. (2016) constructed a vertically oriented core$-$branch hetero-nanostructures on NF scaffold by coating Ni$_x$Co$_{2x}$(OH)$_{6x}$ on the surface of ZnCo$_2$O$_4$ nanowires. Consequently, the areal capacity increased by up to 419.1 μAh cm^{-2} at 5 mA cm^{-2}, which was four times that of the Ni$_x$Co$_{2x}$(OH)$_{6x}$ and three times as large as that of the ZnCo$_2$O$_4$. The enhancement in the capacitive behavior could be ascribed to the formation of core$-$branch hetero-architectures with mixed phases and the coexistence of rich active species. The integrated Ni$_x$Co$_{2x}$(OH)$_{6x}$ and ZnCo$_2$O$_4$ components allowed lower bulk resistance (0.75 Ω) and charge transfer resistance (1.12 Ω) than those of single materials, implying rapid electron transfer and flexible ion diffusion (Fig. 8.7G$-$I).

The aggregation of electroactive components can be well-restricted through the construction of a composite compound with hierarchical architectures. These hierarchical nanostructures can offer sufficient spaces to attenuate the side effects caused by the volume expansion of the electrode, particularly when the loading of electroactive materials and particle size increase. This is attractive for ESCs to yield a high energy density by engineering high mass loads. A core$-$shell-shaped NiTe@NiCoSe$_2$ nanocomposite with 3D hierarchical architectures was designed by encapsulating the large theoretical capacity NiCoSe$_2$ shell on NiTe nanorods core via a simple hydrothermal reaction and electrodeposition process (Fig. 8.7J and K) (Ye et al., 2021). The NiTe solid backbone with more metallic character induced a fluent electron transfer property. The self-standing NiTe@NiCoSe$_2$ core$-$shell heterostructures with favorable synergistic effects approached remarkable electrochemical performances with an excellent specific capacity of 560.6 mAh g^{-1} (Fig. 8.7L and M).

The porous $NiCoSe_2$ shell not only supplied tremendous channels for the migration of ions, enlarged the surface area for OH^- adsorption but also reinforced the durability of the electrode during electrochemical redox reactions and long-term charge–discharge operations. The integrated NiTe enhanced the electric/ionic conductivities, resulting in a splendid energy storage performance. Based on the DFT calculation results, the $NiTe@NiCoSe_2$ hybrid illustrated a gapless DOS near the Fermi level, verifying that both NiTe and $NiCoSe_2$ components were metallic conductors. Besides, the synergetic behavior at the heterojunction between the NiTe and $NiCoSe_2$ indicated tunable capability of electron transfer and better conductivity (Fig. 8.7N). The developed $NiTe@NiCoSe_2//AC$ ASC configuration obtained high energy/power densities (59.8 Wh kg^{-1}/ 800 W kg^{-1}) with enhanced capacitance retention at 2 A g^{-1} (96.6% retention after 10,000 cycles) (Fig. 8.7O and P).

Furthermore, the distribution of electrons across different cations or anions can be adjusted through multicomponent hybridization, demonstrating richer redox reactions. Lin et al. (2019) achieved the synthesis of P-rich CoP/ metal-rich NiCoP heterojunction with nanotadpoles-like morphologies and plenty of exposed interfaces through a solid-state phase conversion route. The XPS and X-ray absorption spectroscopy (XAS) analyses demonstrated that engineering a heterojunction with many disorders including coordination environment and electronic structure at the heterointerface of CoP/NiCoP could dramatically intensify the electrochemical properties and induce the electron redistribution. Moreover, the XPS spectra revealed a shift in the binding energies of Co atoms due to the change in their corresponding chemical states, thereby indicating a more obvious electron transport along the CoP/NiCoP heterojunction. The $P^{\delta-}$ for the P 2p region of the CoP/NiCoP was negatively shifted compared to that of the single component, implying dense electrons around the P center. The tight connection between CoP and NiCoP not only generated strong electron interactions and rich redox reactions but also facilitated the kinetics of Faradaic reactions. The modulated surface electronic structure and corresponding electron distribution might grant many preferable sites for electrochemical redox reactions.

Core–shell-structured $MnCo-LDH@Ni(OH)_2$ heterojunction with core–branch architectures was synthesized through a two-step hydrothermal approach (Liu et al., 2017). The diffusion-favored architectures and hybrid components play a considerable role in strengthening the pseudocapacitive performance. The polyvalent Mn species extremely promoted the rich redox reactions. The conductive core and pseudocapacitive shell could afford a remarkable synergism. As a result, the nanostructured $MnCo-LDH@Ni(OH)_2$ heterojunction showed a high specific capacitance of 2320 F g^{-1} at 3 A g^{-1}. Similarly, Ning et al. (2014) developed $Co_3O_4@NiAl-LDH$ core—shell hetero-nanostructures by using a two-step hydrothermal strategy. The lattice distortion at the the interface reinforced the electronic interactions between Co_3O_4 and $NiAl-LDH$ components. The formation of $Co(OH)_2$ due to the binding Co in the core and OH^- in the shell confirmed the robust chemical bonding between the core and shell, therefore boosting the charge storage and transport capability. Accordingly, the regulated Faradaic redox reaction activity was acquired, benefiting from the novel core—branch hetero-nanostructures and enhanced charge-storing characteristics.

As addressed by the above discussions, hetero-nanostructures with complex hierarchical architectures have become a hotspot in the field sustainable electrochemical energy storage. However, the impacts of heterogeneous interfaces are often difficult to measure due to the concomitant disturbance of the structure itself. To deeply achieve and explore a more accurate understanding of heterointerfaces, adopting high-resolution interface characterization techniques and implementing an easy structure represent an effective way. From a perspective of manipulating the capacitive properties, these conclusions can be drawn. The deployment of multiactive hybrid components aims to realize the influence of "the whole is better than the sum of the parts." Though the compositions of diverse electroactive components have already been explored, confusion still appears regarding the mechanism of the capacitance improvement arising from multiphase. More profound studies and efforts are urgently necessary. It is suggested that the prospering DFT simulation approach may pave a new direction to attain a deep understanding based on ion absorption—desorption and electron orbitals of materials.

8.4.5 Defect engineering

Several supercapacitive electrode materials still suffer from some inherent shortcomings. Typically, earth-abundant transition metal oxides reveal limited intrinsic conductivity (Fleischmann et al., 2020; Wei et al., 2011). On the other hand, 2D graphene experiences restricted mass transfer because of the blocking effect by the restacking, which is induced by strong van der Waals forces. To overcome these issues, defect engineering has been widely investigated and gradually established as a highly efficient method to reinforce capacitive properties (Khalafallah, Huang, et al., 2024). In general, defects involve vacancies and porous architectures. To tailor defects, previous works utilized the following techniques such as plasma etching, reducing agents, sacrificial agents, and phase conversion methods. Even though these strategies are proper for a wide range of cost-efficient materials, the preferable strategy should be adopted based on the specific situation.

Defect engineering has emerged to boost the electrochemical properties of earth-abundant transition metals-based energy storage configurations. The electrochemical activity and stability are strongly related to defect-dominated crystal and external environments (Khalafallah et al., 2023). Thus, optimizing a defect engineering recipe presents a direct path to advance the electrochemical behavior of electrode materials. For example, Zuo et al. (2015) fabricated a binder-free graphene hydrogel with large specific surface area of up to 835 m^2 g^{-1}, smart hierarchical architecture, and enhanced powder conductivity of 400 S m^{-1} by using the sodium iodide as a sacrificial agent, facilitating the formation of highly defective and porous nanostructures. The in-plane pores could serve as channels for reducing the mass cross-plane diffusion distance, meanwhile, interlayer pores serve as ion-buffering reservoirs, inhibiting the interlayer restacking and decreasing the ion response time. Obtained findings confirmed the existence of hierarchical pore architecture involving interlayer and in-plane pores, which could minimize the length of diffusion channels, leading to fast mass diffusion with low resistance. The as-developed graphene hydrogel revealed a specific capacitance of 169 F g^{-1} and a power density of 7.5 kW kg^{-1} in a 6.0 M KOH solution.

Apart from the investigation of sacrificial agents, plasma etching can be utilized to create defective nanostructures. In this context, a versatile recipe involving H-plasma and O-plasma etching were used to achieve the birnessite-MnO_2 with enriched lattice oxygen vacancies (LOV-MnO_2) (Fig. 8.8A and B), thereby delivering an improved electrochemical energy storage behavior (Cui et al., 2021). More importantly, the occurrence of rich lattice oxygen vacancies could induce the unsaturated sites of Mn^{3+}, imparting some metallic features to MnO_2 and increasing its electronic conductivity and corresponding charge transport capability. The theoretical DFT analyses proved the modified ion intercalation/deintercalation behaviors and diffusion kinetics of LOV-MnO_2 due to a smaller energy

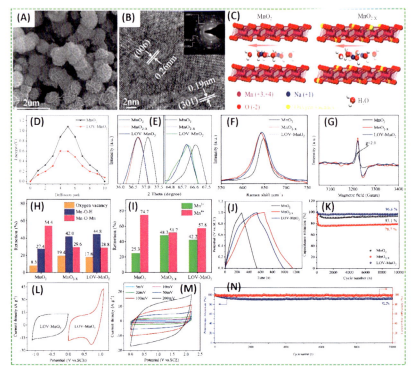

Figure 8.8 (A) SEM and (B) HR-TEM images of the LOV-MnO$_2$-based material. (C) Ion diffusion kinetics in non-defective MnO$_2$ (left) and defect-rich LOV-MnO$_2$ (right), (D) the analyzed ion diffusion energy barrier of non-defective MnO$_2$ and defect-rich LOV-MnO$_2$ materials, (E) XRD signals near the (0 0 6) and (1 1 9) diffraction planes, (F) Raman spectra, (G) electron spin resonance responses of different samples, (H) the fraction of oxygen vacancy, lattice oxygen and surface oxygen derived from XPS measurements, (I) content of Mn^{3+} and Mn^{4+} in prepared samples, (J) GCD curves of as-prepared materials at 1 A g^{-1}, (K) long-term cyclic performance at 20 A g^{-1} for 10,000 times, (L) CV profiles of LOV-MnO$_2$ as the positive and negative potential ranges at 50 mV s^{-1}, (M) CV curves at various sweep rate of the configured LOV-MnO$_2$//LOV-MnO$_2$ device, and (N) long-term durability test and related coulombic efficiency of the LOV-MnO$_2$//LOV-MnO$_2$ symmetric supercapacitor device over 10,000 cycles at 20 A g^{-1}. *Reprinted with permission from Cui, P., Zhang, Y., Cao, Z., Liu, Y., Sun, Z., Cheng, S., Wu, Y., Fu, J., & Xie, E. (2021). Plasma-assisted lattice oxygen vacancies engineering recipe for high-performing supercapacitors in a model of birnessite-MnO2. Chemical Engineering Journal, 412, 128676. Copyright 2021, Elsevier.*

barrier ($\sim$ 0.6 eV) (Fig. 8.8C and D). Briefly, the lattice oxygen vacancies in the as-synthesized LOV-MnO$_2$ would extensively speed up the diffusion of solvated Na$^+$ ions during the charge—discharge process, enabling favorable enhancement in the pseudocapacitive reactions of MnO$_2$. The local structural defects after the H-plasma and O-plasma exposure were monitored by XRD, Raman spectra and electron spin resonance tools (Fig. 8.8E−G). The changes in electronic states after plasma treatment were also studied by the XPS analysis, which demonstrated an increase in the valence state of Mn after the O-plasma treatment. Moreover, the ratio of Mn-O-Mn was significantly intensified after the H-plasma treatment, suggesting the construction of abundant oxygen vacancies. The oxygen vacancy ratio of MnO_{2-X} was greater than that of LOV-MnO_2 because the O-plasma treatment was filled in some surface oxygen vacancies. The fraction of Mn^{3+} in MnO_{2-X} and LOV-MnO_2 was increased due to the loss of oxygen, which induced a charge imbalance. Besides, the Mn^{4+} valence state was reduced to Mn^{3+} (Fig. 8.8H and I). The copresence of different valence states of Mn favored the enhancement of the electrochemical performance. The LOV-MnO_2-based electrode presented a capacitance of 445 F g^{-1} at 1 A g^{-1} with 96.6% capacitance retention after successive 10,000 charge−discharge cycles (Fig. 8.8J and K). The configured LOV-MnO_2//LOV-MnO_2 symmetric supercapacitor cell with an operational voltage window of 2.2 V realized a notable electrochemical energy storage performance in terms of energy/power densities and long-term lifespan behavior (Fig. 8.8L−N).

Furthermore, reducing agents can be used to create vacancies. Zhai et al. (2014) designed free-standing MnO_2 nanorods on CC with dense oxygen vacancies through a hydrogenation treatment. TEM observations evidenced the existence of the Mn_3O_4 lattice owing to the reduction process by the H_2. Some micropores were also formed, which could hasten the mass transfer. The generated oxygen vacancies induced the copresence of Mn^{2+}, Mn^{3+}, and Mn^{4+} states, which enhanced the ionic and electronic properties. The Mn^{3+} in MnO_2 supplied additional negative charges to the adjacent O atoms, thereby further boosting the charge storage behavior. Strikingly, a good capacitance of 306 F g^{-1} at 10 mV s^{-1} and a lower charge transfer resistance were achieved.

Combining these findings, it can be concluded that both experimental and theoretical studies have displayed that vacancies not only tune the charge transfer kinetics by inserting charge carriers into the electrode material but also raise the number of active sites for Faradaic redox reactions. Defect engineering has been extensively researched in the field of SCs. Nevertheless, more efforts to discover other functional types of defects such as cation or hybrid defects are desirable to synthesize a wide domain of electrode materials. Emerging in situ characterization devices can provide more understanding of the defect evolution. Such a smart exploration may pave an advanced window for the defect engineering.

8.5 Conclusions and perspectives

ESCs with high power density, long lifespan, and short charging time duration are one of the key energy storage technologies in portable and wearable electronics. The research in this field is principally focusing on strengthening the energy output without compromising the specific power and cyclic performance. The optimal potential of nanostructured coordination materials, particularly extrinsic hybrid pseudocapacitive and battery-type materials, has not yet been yielded. The energy storage matrices such as specific capacitance, rate capability, and long-operation cyclic stability should be further regulated and suitable balance needs to be adjusted. In this context, several advanced approaches aiming at manipulating the energy-storing characteristic, exposing rich electrochemically active sites, and diminishing the ion diffusion lengths and electron transport paths for supercapacitor applications through hybridization, elemental doping, heterointerfacing, defect engineering, and surface functionalization were discussed. However, a more in-depth understanding of the charge storage pathways and surface/bulk reaction kinetics requires to be better illustrated. It is well elaborated that despite the significant achievements, innovative strategies are still desirable to enlarge the operational potential window of ASCs to reinforce the energy density. Therefore, various important points should be considered regarding the field of pseudocapacitive and hybrid electrode materials.

1. The development of low-cost ternary hybrid or multicomponent structures is a feasible path for efficient electrode design since the positive features of all components can be harvested. A better understanding of the surface chemical interactions between integrated components is necessary to tune the synergistic effects for regulating charge transport and storage. This knowledge is lacking for multicomponent nanohybrid materials. In situ measurements and theoretical simulation tools such as Raman spectroscopy, XRD, X-ray absorption spectroscopy (XAS), XPS, and electrochemical mass spectrometry may clarify the mechanism of energy storage systems.

2. The rational engineering of nanoscale electrode materials for SCs has many limitations related to reproducibility, controllable morphologies, and optimized composition on a large scale. Conventional methods-

based annealing treatment of precursors can raise safety and environmental concerns. High temperature and high-pressure growth reactions make the process more complex. Thus, cost-effective, ecofriendly, and industrially applicable techniques are highly desirable.

3. The electrolyte is a critical issue of ESCs, affording the ionic conductivity and charge reparation; the stable working voltage window of the electrolyte greatly influences the overall voltage window of ESCs and the corresponding energy density. Reaching a wider potential window with high ionic conductivity is desirable to extremely enhance the efficiency of ASCs.

References

Bounor, B., Seenath, J. S., Patnaik, S. G., Bourrier, D., Tran, C. C. H., Esvan, J., Weingarten, L., Descamps-Mandine, A., Rochefort, D., Guay, D., & Pech, D. (2023). Low-cost micro-supercapacitors using porous Ni/MnO_2 entangled pillars and Na-based ionic liquids. *Energy Storage Materials, 63*, 102986.

Chen, X., Song, L., Zeng, M., Tong, L., Zhang, C., Xie, K., & Wang, Y. (2022). Regulation of morphology and electronic configuration of $NiCo_2O_4$ by aluminum doping for high performance supercapacitors. *Journal of Colloid and Interface Science, 610*, 70−79.

Cheng, L., Xu, M., Zhang, Q., Li, G., Zhao, K., Chen, J., & Lou, Y. (2020). When Al^{3+} meets $CuCo_2O_4$ nanowire arrays: an enhanced positive electrode for energy storage devices. *Journal of Alloys and Compounds, 834*, 155001.

Chodankar, N. R., Pham, H. D., Nanjundan, A. K., Fernando, J. F. S., Jayaramulu, K., Golberg, D., Han, Y.-K., & Dubal, D. P. (2020). True meaning of pseudocapacitors and their performance metrics: asymmetric versus hybrid supercapacitors. *Small (Weinheim an der Bergstrasse, Germany), 16*, 2002806.

Cui, P., Zhang, Y., Cao, Z., Liu, Y., Sun, Z., Cheng, S., Wu, Y., Fu, J., & Xie, E. (2021). Plasma-assisted lattice oxygen vacancies engineering recipe for high-performing supercapacitors in a model of birnessite-MnO_2. *Chemical Engineering Journal., 412*, 128676.

Fleischmann, S., Mitchell, J. B., Wang, R., Zhan, C., Jiang, D., Presser, V., & Augustyn, V. (2020). Pseudocapacitance: from fundamental understanding to high power energy storage materials. *Chemical Reviews, 120*, 6738−6782.

Fu, W., Wang, Y., Han, W., Zhang, Z., Zha, H., & Xie, E. (2016). Construction of hierarchical $ZnCo_2O_4@Ni_xCo_{2x}$ $(OH)_{6x}$ core/shell nanowire arrays for high-performance supercapacitors. *Journal of Materials Chemistry A, 4*, 173−182.

González, A., Goikolea, E., Barrena, J. A., & Mysyk, R. (2016). Review on supercapacitors: Technologies and materials. *Renewable and Sustainable Energy Reviews, 58*, 1189−1206.

Hong, H., Tu, H., Jiang, L., Du, Y., & Wong, C. (2024). Advances in fabric-based supercapacitors and batteries: Harnessing textiles for next-generation energy storage. *Journal of Energy Storage, 75*, 109561.

Huang, J., Wei, J., Xiao, Y., Xu, Y., Xiao, Y., Wang, Y., Tan, L., Yuan, K., & Chen, Y. (2018). When Al-doped cobalt sulfide nanosheets meet nickel nanotube arrays: a highly efficient and stable cathode for asymmetric supercapacitors. *ACS Nano, 12*, 3030−3041.

Hussein El-Shafei, M., Abdel-Latif, M. S., Hessein, A., & Abd El-Moneim, A. (2023). Laser-induced carbonization of carbon nanofibers free-standing electrodes for flexible interdigitated micro-supercapacitors. *FlatChem*, *42*, 100570.

Kang, J., Hirata, A., Kang, L., Zhang, X., Hou, Y., Chen, L., Li, C., Fujita, T., Akagi, K., & Chen, M. (2013). Enhanced supercapacitor performance of MnO2 by atomic doping. *Angew Chem. Int. Ed.*, *52*, 1664.

Khalafallah, D., Huang, W., Wunn, M., Zhi, M., & Hong, Z. (2022). Promoting the energy storage capability via selenium-enriched nickel bismuth selenide/graphite composites as the positive and negative electrodes. *Journal of Energy Storage*, *45*, 103716.

Khalafallah, D., Huang, W., Zhi, M., & Hong, Z. (2024). Synergistic Tuning of Nickel Cobalt Selenide@Nickel Telluride Core−Shell Heteroarchitectures for Boosting Overall Urea Electrooxidation and Electrochemical Supercapattery. *Energy and Environmental Materials*, *7*, e12528.

Khalafallah, D., Li, X., Zhi, M., & Hong, Z. (2021). Nanostructuring Nickel−Zinc−Boron/graphitic carbon nitride as the positive electrode and BiVO$_4$-immobilized nitrogen-doped defective carbon as the negative electrode for asymmetric capacitors. *ACS Applied Nano Materials.*, *4*, 14258−14273.

Khalafallah, D., Miao, J., Zhi, M., & Hong, Z. (2021). Structuring graphene quantum dots anchored CuO for high-performance hybrid supercapacitors. *Journal of the Taiwan Institute of Chemical Engineers*, *122*, 168−175.

Khalafallah, D., Qiao, F., Liu, C., Wang, J., Zhang, Y., Wang, J., Zhang, Q., & Notten, P. H. L. (2023). Heterostructured transition metal chalcogenides with strategic heterointerfaces for electrochemical energy conversion/storage. *Coordination Chemistry Reviews*, *496*, 215405.

Khalafallah, D., Quan, X., Ouyang, C., Zhi, M., & Hong, Z. (2021). Heteroatoms doped porous carbon derived from waste potato peel for supercapacitors. *Renewable Energy*, *170*, 60−71.

Khalafallah, D., Zhang, Y., Dai, H., Liu, C., & Qinfang, Z. (2024). Manipulating the overall capacitance of hierarchical porous carbons via structure-and pore-tailoring approach. *Carbon*, *227*, 119250.

Khalafallah, D., Zhi, M., & Hong, Z. (2021). Rational engineering of hierarchical mesoporous Cu$_x$Fe$_y$Se battery-type electrodes for asymmetric hybrid supercapacitors. *Ceramics International*, *47*, 29081−29090.

Khalafallah, D., Zhi, M., & Hong, Z. (2022). Bi-Fe chalcogenides anchored carbon matrix and structured core−shell Bi-Fe-P@ Ni-P nanoarchitectures with appealing performances for supercapacitors. *Journal of Colloid and Interface Science*, *606*, 1352−1363.

Khalid, M. U., Katubi, K. M., Zulfiqar, S., Alrowaili, Z. A., Aadil, M., Al-Buriahi, M. S., Shahid, M., & Warsi, M. F. (2023). Boosting the electrochemical activities of MnO$_2$ for next-generation supercapacitor application: adaptation of multiple approaches. *Fuel.*, *343*, 127946.

Lin, Y., Sun, K., Liu, S., Chen, X., Cheng, Y., Cheong, W., Chen, Z., Zheng, L., Zhang, J., Li, X., Pan, Y., & Chen, C. (2019). Construction of CoP/NiCoP nanotadpoles heterojunction interface for wide pH hydrogen evolution electrocatalysis and supercapacitor. *Advanced Energy Materials.*, *9*, 1901213.

Liu, J., Wang, J., Xu, C., Jiang, H., Li, C., Zhang, L., Lin, J., & Shen, Z. X. (2018). Advanced energy storage devices: basic principles, analytical methods, and rational materials design. *Advanced Science*, *5*, 1700322.

Liu, S., Lee, S., Patil, U., Shackery, I., Kang, S., Zhang, K., Park, J., Chung, K., & Jun, S. (2017). Hierarchical MnCo-layered double hydroxides@ Ni(OH)$_2$ core−shell heterostructures as advanced electrodes for supercapacitors. *Journal of Materials Chemistry A*, *5*, 1043−1049.

Majumdar, D., Maiyalagan, T., & Jiang, Z. (2019). Recent progress in ruthenium oxide-based composites for supercapacitor applications. *ChemElectroChem, 6*, 4343−4372.

Mane, V. J., Kale, S. B., Ubale, S. B., Lokhande, V. C., & Lokhande, C. D. (2021). Enhanced specific energy of silver-doped MnO_2/graphene oxide electrodes as facile fabrication symmetric supercapacitor device. *Materials Today Chemistry, 20*, 100473.

Muzaffar, A., Ahamed, M. B., Deshmukh, K., & Thirumalai, J. (2019). A review on recent advances in hybrid supercapacitors: Design, fabrication and applications. *Renewable and Sustainable Energy Reviews, 101*, 123−145.

Ning, F., Shao, M., Zhang, C., Xu, S., Wei, M., & Duan, X. (2014). $Co_3O_4@$ layered double hydroxide core/shell hierarchical nanowire arrays for enhanced supercapacitance performance. *Nano Energy, 7*, 134−142.

Noori, A., El-Kady, M. F., Rahmanifar, M. S., Kaner, R. B., & Mousavi, M. F. (2019). Towards establishing standard performance metrics for batteries, supercapacitors and beyond. *Chemical Society Reviews, 48*, 1272−1341.

Pan, Z., Yang, J., Zhang, Q., Liu, M., Hu, Y., Kou, Z., Liu, N., Yang, X., Ding, X., Chen, H., Li, J., Zhang, K., Qiu, Y., Li, Q., Wang, J., & Zhang, Y. (2019). All-solid-state fiber supercapacitors with ultrahigh volumetric energy density and outstanding flexibility. *Advanced Energy Materials., 9*, 1802753.

Raja, A., Son, N., Swaminathan, M., & Kang, M. (2022). Electrochemical behavior of heteroatom doped on reduced graphene oxide with RuO_2 for HER, OER, and supercapacitor applications. *Journal of the Taiwan Institute of Chemical Engineers., 138*, 104471.

Şahin, M., Blaabjerg, F., & Sangwongwanich, A. (2022). A comprehensive review on supercapacitor applications and developments. *Energies, 15*, 674.

Sajjad, M., Shah, M. Z. U., Mahmood, F., Javed, M. S., Maryam, R., Ahmad, F., Shah, A., Hussain, R., Toufiq, A. M., Mao, Z., & Rahman, S. (2023). CdO nanocubes decorated on rGO sheets as novel high conductivity positive electrode material for hybrid supercapacitor. *Journal of Alloys and Compounds., 938*, 168462.

Saleem, A. M., Desmaris, V., & Enoksson, P. (2016). Performance enhancement of carbon nanomaterials for supercapacitors. *Journal of Nanomaterials., 2016*, 1537269.

Shi, D., Zhang, L., Zhang, N., Zhang, Y., Yu, Z., & Gong, H. (2018). Boosted electrochemical properties from the surface engineering of ultrathin interlaced $Ni(OH)_2$ nanosheets with $Co(OH)_2$ quantum dot modification. *Nanoscale, 10*, 10554−10563.

Shi, Y., Zhu, B., Guo, X., Li, W., Ma, W., Wu, X., & Pang, H. (2022). MOF-derived metal sulfides for electrochemical energy applications. *Energy Storage Materials, 51*, 840−872.

Simonenko, T. L., Simonenko, N. P., Gorobtsov, P. Y., Simonenko, E. P., & Kuznetsov, N. T. (2023). Hydrothermal synthesis of a cellular NiO film on carbon paper as a promising way to obtain a hierarchically organized electrode for a flexible supercapacitor. *Materials., 16*, 5208.

Sonia, Y. K., & Meher, S. K. (2022). Electrostructural compatibility of battery-type diffuse-porous Co_9S_8−$NiCo_2S_4$/defective reduced graphene oxide and flaky FeS/nitrogen-doped defective reduced graphene oxide for ultra-high-performance all-solid-state hybrid pseudocapacitors. *ACS Applied Energy Materials., 5*(11), 13672−13691.

Taslim, R., Apriwandi, A., & Taer, E. (2022). Novel moringa oleifera leaves 3D porous carbon-based electrode material as a high-performance EDLC supercapacitor. *ACS Omega, 7*(41), 36489−36502.

Vlad, A., Singh, N., Galande, C., & Ajayan, P. M. (2015). Design considerations for unconventional electrochemical energy storage architectures. *Advanced Energy Materials., 5*, 1402115.

Wang, Y.-F., Wang, H.-T., Yang, S.-Y., Yue, Y., & Bian, S.-W. (2019). Hierarchical $NiCo_2S_4@$nickel−cobalt layered double hydroxide nanotube arrays on metallic cotton yarns for flexible supercapacitors. *ACS Applied Materials & Interfaces*, *11*(33), 30384−30390.

Wei, W., Cui, X., Chen, W., & Ivey, D. G. (2011). Manganese oxide-based materials as electrochemical supercapacitor electrodes. *Chemical Society Reviews*, *40*, 1697−1721.

Wu, J., Faiz, Y., Hussain, S., Faiz, F., Zarshad, N., Ur Rahman, A., Masood, M. A., Deng, Y., Pan, X., & Ahmad, M. (2023). Defects rich-Cu-doped MnO_2 nanowires as an efficient and durable electrode for high performance aqueous supercapacitors. *Electrochimica Acta*, *443*, 141927.

Wu, J., Raza, W., Wang, P., Hussain, A., Ding, Y., Yu, J., Wu, Y., & Zhao, J. (2022). Zn-doped MnO_2 ultrathin nanosheets with rich defects for high performance aqueous supercapacitors. *Electrochimica Acta*, *418*, 140339.

Wu, N., Bai, X., Pan, D., Dong, B., Wei, R., Naik, N., Patil, R. R., & Guo, Z. (2021). Recent advances of asymmetric supercapacitors. *Advanced Materials Interfaces.*, *8*, 2001710.

Wu, Z., Winter, A., Chen, L., Sun, Y., Turchanin, A., Feng, X., & Mullen, K. (2012). Three-dimensional nitrogen and boron co-doped graphene for high-performance all-solid-state supercapacitors. *Advanced Materials.*, *24*, 5130−5135.

Yadav, C. R. A. A. (2023). Spray-deposited cobalt-doped RuO_2 electrodes for high-performance supercapacitors. *Electrochimica Acta*, *437*, 141521.

Yang, X., He, X., Li, Q., Sun, J., Lei, Z., & Liu, Z.-H. (2021). 3D hierarchical $NiCo_2S_4$ nanoparticles/carbon nanotube sponge cathode for highly compressible asymmetric supercapacitors. *Energy & Fuels: An American Chemical Society Journal*, *35*(4), 3449−3458.

Ye, B., Xiao, S., Cao, X., Chen, J., Zhou, A., Zhao, Q., Huang, W., & Wang, J. (2021). Interface engineering for enhancing performance of additive-free $NiTe@NiCoSe_2$ core/shell nanostructure for asymmetric supercapacitors. *Journal of Power Sources*, *506*, 230056.

Yi, H., Gao, A., Pang, X., Ao, Z., Shu, D., Deng, S., Yi, F., He, C., Zhou, X., & Zhu, Z. (2020). Preparation of single-atom Ag-decorated MnO_2 hollow microspheres by redox etching method for high-performance solid-state asymmetric supercapacitors. *ACS Applied Energy Materials.*, *3*(10), 10192−10201.

Zhai, T., Wan, L., Sun, S., Chen, Q., Sun, J., Xia, Q., & Xia, H. (2017). Phosphate ion functionalized Co_3O_4 ultrathin nanosheets with greatly improved surface reactivity for high performance pseudocapacitors. *Advanced Materials.*, *29*, 1604167.

Zhai, T., Xie, S., Yu, M., Fang, P., Liang, C., Lu, X., & Tong, Y. (2014). Oxygen vacancies enhancing capacitive properties of MnO_2 nanorods for wearable asymmetric supercapacitors. *Nano Energy*, *8*, 255−263.

Zhou, C., Zhang, Y., Li, Y., & Liu, J. (2013). Construction of high-capacitance 3D $CoO@$polypyrrole nanowire array electrode for aqueous asymmetric supercapacitor. *Nano Letters*, *13*, 2078−2085.

Zuo, Z., Kim, T., Kholmanov, I., Li, H., Chou, H., & Li, Y. (2015). Ultra-light hierarchical graphene electrode for binder-free supercapacitors and lithium-ion battery anodes. *Small (Weinheim an der Bergstrasse, Germany)*, *11*, 4922−4930.

CHAPTER 9

Coordination materials for metal−sulfur batteries

Dominika Capková[1,2,3] and Miroslav Almáši[4]

[1]Department of Physical Chemistry, Faculty of Sciences, Pavol Jozef Šafárik University in Košice, Košice, Slovak Republic
[2]Department of Chemical Sciences, Bernal Institute, University of Limerick, Limerick, Ireland
[3]Department of Electrical and Electronic Technology, Faculty of Electrical Engineering and Communication, Brno University of Technology, Brno, Czech Republic
[4]Department of Inorganic Chemistry, Faculty of Sciences, Pavol Jozef Šafárik University in Košice, Košice, Slovak Republic

9.1 Introduction

Energy is currently one of the most important human needs and is also necessary for the development of technology and human civilization (Huang et al., 2021). The constant consumption of nonrenewable natural energy sources, which are primarily represented by fossil fuels, represents a serious problem in terms of their sustainability due to their depletion. On the other hand, using these fuels, especially in transport, causes global issues in the form of atmospheric warming, contamination of water sources, and soil pollution (Ehigiamusoe & Dogan, 2022; Mujtaba et al., 2022). Nonrenewable energy sources thus directly affect, influence, and, above all, pollute the geosphere, hydrosphere, and atmosphere of our planet. For the stated reasons, it is necessary and desirable to use green energy sources to alleviate the ecological crisis and to avert it in the future. It is desirable to use renewable energy sources that the surrounding nature offers us in the form of solar, wind, geothermal energy, or energy generated in hydropower plants. However, the mentioned forms of obtaining energy have their limitations because they cannot be used continuously due to atmospheric conditions. The intermittent nature of solar and wind energy generation, irregular rainfall, and the state of water reservoirs in the case of hydroelectric power plants require electrical energy storage systems for efficient and continuous energy supply. Rechargeable batteries have been shown as promising energy storage devices in recent decades. In general, batteries convert chemical energy into electrical energy and inversely for efficient energy storage, which facilitates the use

Electrochemistry and Photo-Electrochemistry of Nanomaterials
DOI: https://doi.org/10.1016/B978-0-443-18600-4.00010-7
© 2025 Elsevier Inc. All rights reserved, including those for text and data mining, AI training, and similar technologies.

of clean energy. Currently, rechargeable lithium ion batteries (Li-ion), which have dominated the market since the 1990s, are the most commercially used (Shahjalal et al., 2022). Although continuous Li-ion battery research is progressing, and batteries are achieving energy densities approaching their theoretical values, they are no longer sufficient for current energy consumption demands (Camargos et al., 2022; Roy et al., 2022). The energy densities of commercially used Li-ion batteries range from 200 to 400 Wh/kg, 500 to 1300 Wh/L, and 140 to 180 mAh/g depending on the battery chemistry [lower values are for LiFePO$_4$ (LFP) cathode material and higher capacities for LiNi$_{0.8}$Co$_{0.15}$Al$_{0.05}$O$_2$ (NCA) or LiNi$_x$Mn$_y$Co$_{1-x-y}$O$_2$ (NMC)] (Xue et al., 2017). For electrical devices to be used effectively, it is necessary to use a more significant number of batteries and charge them more often, which leads to stress on the material and its gradual degradation. From an ecological point of view, this creates battery waste, the elimination of which is financially demanding and, if handled incorrectly, burdens the environment. Therefore, it is currently necessary to develop new batteries with higher theoretical capacities and get closer to them. Examples include advanced metal−sulfur batteries, Li−S, Na−S, K−S, and multivalent M−S batteries (Mg, Ca, Al), which are covered in the present chapter. Basic electrochemical and physical properties such as standard metal electrode potential, gravimetric and volumetric energy densities (comparison in Fig. 9.1), Giggs free energy, and metal radius are summarized in Table 9.1 (Yang et al., 2019; Ye & Li, 2022). Moreover, in this chapter, various additives

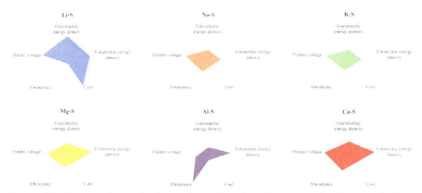

Figure 9.1 A comparison of volumetric and gravimetric energy density, metal abundance, output voltage, and cost of Li, Na, K, Mg, Al, and Ca anodes used in metal−S batteries.

Table 9.1 Theoretical parameters of metal−S batteries (Yang et al., 2019; Ye & Li, 2022).

Battery type	Standard metal electrode potential (vs SHE) (V)	Gravimetric energy density (Wh/kg)	Volumetric energy density (Wh/L)	Gibbs free energy (ΔrG/kJ/mol)	Metal ion radius (pm)
Li−S	−3.045	2612	2955	−432.57	76
Na−S	−2.710	1270	1545	−357.77	102
K−S	−2.925	916	952	−362.72	138
Mg−S	−2.356	1685	3221	−341.44	72
Al−S	−1.676	1319	2981	−713.20	54
Ca−S	−2.840	1838	3202	−477.40	100

in the form of carbonaceous materials, metal−organic frameworks, MXenes, organic polymers, and ecomaterials to M−S batteries with the intention to improve the electrochemical properties of the batteries are also presented and discussed.

9.2 Mechanism of metal−sulfur batteries

The electrochemical reduction of sulfur in metal−sulfur (metal−S) batteries is not based on an intercalation/deintercalation process like lithium-ion (Li-ion) batteries but on a conversion reaction mechanism where intermediates are formed during the charging/discharging (Sun et al., 2020; Yan et al., 2019). The metal−S batteries are constructed with a metal anode and sulfur cathode separated by a separator saturated by an organic electrolyte. During discharging, the ring cyclo-octasulfur (S_8) scaffold is opened and reduced with metal ions to form a series of intermediates, sulfide anions (see Fig. 9.2). First, higher polysulfides are formed (S_n^{2-}, $n \geq 4$) in region I, and further reduction leads to the production of lower polysulfides (S_n^{2-}, $n < 4$) in regions II and III (Hu et al., 2023). The progressive reduction of sulfur to sulfide ions and its lower states can be expressed by the following reactions (Camargos et al., 2022; Ehigiamusoe & Dogan, 2022; Huang et al., 2021; Mujtaba et al., 2022; Shahjalal et al., 2022):

$$\tfrac{1}{2}S_8 + e^- \leftrightarrows \tfrac{1}{2}S_8^{2-}$$

$$\tfrac{3}{2}S_8^{2-} + e^- \leftrightarrows 2S_6^{2-}$$

$$S_6^{2-} + e^- \leftrightarrows \tfrac{3}{2}S_4^{2-}$$

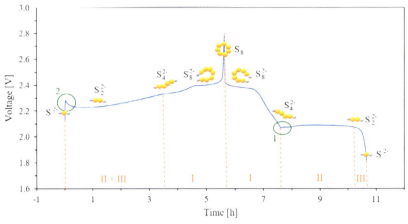

Figure 9.2 A typical charge and discharge profiles of metal–S batteries and the associated phase changes of sulfur.

$$\tfrac{1}{2}S_4^{2-} + e^- \leftrightarrows S_2^{2-}$$
$$\tfrac{1}{2}S_2^{2-} + e^- \leftrightarrows S^{2-}$$

Most sulfur (S$_8$) is in a solid state at the beginning of discharging, the intermediates are in a liquid phase, and the final discharge products are in a solid phase (Kumaresan et al., 2008). A typical charge and discharge profiles of metal–S batteries are illustrated in Fig. 9.2. One can see two charge and discharge plateaus, which agree with the multistep reaction of sulfur. The high-voltage plateau (region I) is attributed to the reduction of sulfur to higher polysulfides. At the end of the high-voltage plateau is the highest electrolyte viscosity and polarization represented by rapid voltage drop circled as point 1 in Fig. 9.2. Furthermore, the subsequent reduction to lower polysulfides and the formation of the low-voltage plateau (regions II and III) is observed. However, the process described is more complex, and the coexistence of different polysulfides in region I is present in the electrolyte, and the reaction in region II between S_2^{2-} and S^{2-} is described as a solid-solid reaction.

During charging, the long low-voltage plateau corresponding to the oxidation of lower polysulfides is observed first. Circled point 2 in Fig. 9.2 represents reduced polarization due to S_2^{2-} and S^{2-} dissolution. The oxidation process of higher polysulfides to sulfur may be seen as a high-voltage plateau. Considering the overall reduction of sulfur S$_8$ to S^{2-}, the theoretical capacity of M–S batteries is 1675 mAh/g (Capkova et al., 2020, 2021; Zhu et al., 2019).

9.3 Challenges of metal–sulfur batteries

The working mechanism of metal–S batteries is accompanied by several challenges, which hinder their practical application. One of the problems is the insulating nature of sulfur and the discharge products (Wang et al., 2013). The main challenge of metal–S batteries is the dissolution of polysulfides and their shuttle between the electrodes (Mikhaylik & Akridge, 2004). The shuttle effect may be described as follows: The higher polysulfides may dissolve in the organic electrolyte, and due to the concentration gradient, they may migrate to the metal anode, where they are reduced to lower polysulfides. The formed lower polysulfides may migrate back to the cathode side as an impact of the concentration gradient. Polysulfide shuttle may result in an endless charging of the metal–S battery (Capkova, Almasi, et al., 2022; Capkova, Knap, et al., 2022). Multivalent-metal polysulfides (e.g, MgS_n, AlS_n, and CaS_n) are less soluble in ether-based electrolytes than LiS_n, NaS_n, and KS_n (Cohn et al., 2015; Guo et al., 2019; Zhao-Karger et al., 2015). Due to hindered solubility of multivalent-metal polysulfides, the charging time of these batteries is not significantly prolonged due to polysulfide shuttle as Li–S batteries (Yang, Yang, et al., 2022; Zhang, 2013). Another challenge of metal–S batteries is the passivation of the cathode by uncontrollable aggregation of insulating discharge products. The passivation layer grows more significantly during long-term operation (Lei et al., 2020). The densities between sulfur and final discharge products (metal–sulfides) are very different, which results in significant volume changes (Capkova et al., 2023). Based on the metal anode, the volume changes are different, for example, up to 80% for Li–S, around 170% for Na–S, and approximately 300% for K–S (Cheng et al., 2022; Hu et al., 2023). The stated issues result in low sulfur utilization, loss of active material, fast capacity decay, limited rate capability, low Coulombic efficiency, high self-discharge rate, and increased cell resistance (Lang et al., 2022; Zhou et al., 2022). A well-defined, structured, and effective additive must be added to counteract the negative effects on cell properties (Han et al., 2021; Yan et al., 2016; Yu & Manthiram, 2020).

9.4 Requirements for host materials

Experimental and theoretical research focuses on the design of various host materials for sulfur that can capture polysulfides effectively

(Fan & Huang, 2022; Mačák et al., 2020; Mačák, Jasso, et al., 2021; Mačák, Kazda, et al., 2021). The investigation of materials concerns physical confinement and chemical bonding (see Fig. 9.3) (Li, Wang, et al., 2018; Peng & Zhang, 2015; Zeng et al., 2022). The idea of physical confinement is to spatially capture polysulfides inside the pores of host materials. Chemical bonding is based on the strong interaction of polysulfides and host material, for example, Lewis acid−base, covalent binding, polar−polar, and redox interactions (Li, Xi, et al., 2018; Pang et al., 2015).

The suppression of polysulfide shuttle is possible by application of host material with a large surface area, capturing, and confinement of sulfur and polysulfides (Tang & Hou, 2020). The affinity for polysulfides and the ability to provide conversion of sulfur to the discharge products may improve the polysulfide trapping and conversion kinetics (Xu et al., 2021). The low electronic conductivity of sulfur may be resolved by applying highly conductive host material (Xia et al., 2021). Furthermore, the host material should be thermally, dynamically, and structurally stable (Zhang et al., 2021). Whereas the electrode is in direct contact with the electrolyte, the material requires strong corrosion resistance. Due to the significant volume changes during cycling, a material that provides

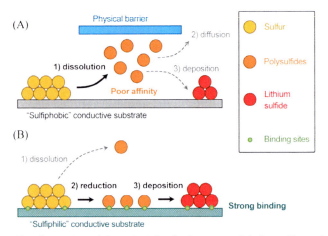

Figure 9.3 The illustration of the principles for host materials for sulfur cathodes: (A) physical confinement—capturing of sulfur by an external barrier and (B) chemical bonding—application of polar adsorbent as a substrate. *Reprinted with permission from Peng, H. J. & Zhang, Q. (2015). Designing host materials for sulfur cathodes: From physical confinement to surface chemistry. Angewandte Chemie—International Edition, 54(38), 11018−11020.*

a sizeable spatial expansion is essential. In addition, the host material should be lightweight and easily synthesized on a large scale for practical applications. Based on the demands mentioned above, various materials have been investigated as hosts for sulfur in metal−S batteries.

9.5 5 Lithium−sulfur batteries

The increasing demand for high-energy-density batteries may be satisfied by the application of lithium metal as an anode due to its high theoretical capacity (3860 mAh/g) and low electrode potential (−3.04 V vs the standard hydrogen electrode—SHE) (Wang, Liu, et al., 2019). The first prototype of the Li−S battery was introduced in 1962 by Danuta Herbert and Juliusz Ulam (Danuta & Juliusz, 1957). The operating potential of the Li−S battery is 2.1 V versus Li/Li^+, representing a theoretical energy density of ~ 2500 Wh/kg and ~ 2800 Wh/L (Kang et al., 2021). Among all metal−S batteries, Li−S batteries are the most widely investigated due to their high storage capacity and are considered as the next-generation energy-storage technology (Chung & Manthiram, 2019).

Except for the aforementioned challenges, the Li−S batteries suffer from the growth of lithium dendrites, which may lead to safety hazards due to the increased potential of the short circuit and Li deposition/striping process (Xu et al., 2014; Zhu et al., 2019). In order to improve the performance of Li−S batteries, different materials were applied in cathodes as support for sulfur, including carbonaceous materials, metal−organic frameworks, MXenes, or polymers.

9.5.1 Carbonaceous materials

Porous carbon materials have been investigated as host materials for sulfur due to their large surface area, which is beneficial for the encapsulation of sulfur and high sulfur loading. These materials are well electronically conductive and mechanically stable (Brückner et al., 2014; Kazda et al., 2021).

Graphite is a naturally occurring allotrope of carbon and has a layered 2D structure, and carbon atoms form hexagonal layers stacked together by van der Waals interactions to create a final 3D supramolecular motif (Tatar & Rabii, 1982). The application of graphite in sulfur cathode material was one of the first studies and patents in Li−S batteries (Cheng, 2000). Based on this patent, Sion Power Corporation constructed prismatic cells containing sulfur/graphite composite (Mikhaylik & Akridge, 2003).

The presented battery cells exceed 200 Wh/kg and have a capacity retention of 80% at −40°C compared to room-temperature cycle performance.

A single layer (2D) of hexagonal carbon net represents graphene's structure, which has improved conductivity and material strength compared to graphite (Evers & Nazar, 2013). Sulfur/graphene composite with low-defect graphene sheets and sulfur loading of 73% exhibits a capacity of 615 mAh/g after 100 cycles at 1 C (Tianquan Lin et al., 2013). The synthesis of graphene material and defects in the structure affect the material properties. Graphene-containing defects can rearrange the uniformity of the charge distribution and enhance the polysulfide confinement (Kamisan et al., 2022). Another approach to improve the electrochemical properties of graphene is to functionalize the sheet with heteroatoms or functional groups. The electrode with nitrogen-doped graphene showed long cycle life of 2000 cycles at 2 C with a capacity retention of 44% (see Fig. 9.4) (Qiu et al., 2014). Cobalt atoms embedded in nitrogen-doped graphene improved capacity and delivered

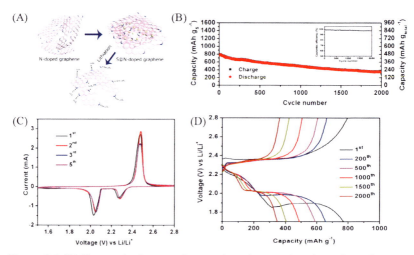

Figure 9.4 (A) The synthesis route for creating nitrogen-doped graphene for trapping sulfur and polysulfides. (B) Cycling performance of the sulfur-based electrode with nitrogen-doped graphene despite cycling at 2 C over 2000 cycles. (C) Cyclic voltammograms of the Li−S cell containing nitrogen-doped graphene in the cathode material. (D) Charge/discharge profiles despite cycling at 2 C at different cycles. *Reprinted with permission from Qiu, Y., Li, W., Zhao, W., Li, G., Hou, Y., Liu, M., Zhou, L., Ye, F., Li, H., Wei, Z., Yang, S., Duan, W., Ye, Y., Guo, J., & Zhang, Y. (2014). High-rate, ultralong cycle-life lithium/sulfur batteries enabled by nitrogen-doped graphene. Nano Letters, 14(8), 4821−4827.*

a gravimetric capacity of 1210 mAh/g at 0.2 C with a capacity fading rate of 0.029% per cycle over 100 cycles (Du et al., 2019). Pyrrole modification of graphene improved the sulfur loading to $\sim 6.2\,mg/cm^2$ compared to $\sim 4\,mg/cm^2$ for graphene without modification (Zhang et al., 2017). The pyrrole-graphene cathode exhibits a discharge capacity of 985.8 mAh/g at 0.5 C with a capacity retention of 81% after 100 cycles. Combining graphene with other carbon materials is an effective way to synthesize efficient support for sulfur. Wen et al. (2020) fabricated composite material formed of nitrogen-doped graphene and carbon nanotubes by ultrasonic spray method. The discharge capacity at 0.05 C was more than 1500 mAh/g, and the capacity retention after 500 cycles at 2 C corresponded to the value of 77%.

Oxidized graphite, graphene oxide (GO), is intensively studied in cathode materials as a host for sulfur or additive in the cathode material due to its high electronic conductivity and adsorbing properties (Wang et al., 2011). Chabu et al. (2019) presented a composite of GO and sulfur where the specific capacity at 0.2 C was around 600 mAh/g. However, the electrode material containing GO and polyvinyl pyrrolidone (PVP) showed improved cycle performance, and the discharge capacity at 0.2 C reached the value of 1301 mAh/g. The oxygen in GO may be removed by chemical or thermal treatment to achieve reduced graphene oxide (rGO). Chen et al. (2023) proposed composite rGO/VO_2 synthesized by hydrothermal reaction for sulfur utilization in Li$-$S batteries. The proposed composite achieved 1093.8 mAh/g at 0.1 C, and after 1000 cycles at 1 C, the remaining capacity was 395.8 mAh/g with a decay rate per cycle of 0.016%. Self-assembled rGO fibers with polypyrrole (PPy) and sulfur were investigated as a cathode in Li$-$S batteries (see Fig. 9.4) (Huang et al., 2022). The prepared cathodes showed excellent mechanical stability, good conductivity, and high sulfur utilization. The capacity retention after 200 cycles at 0.1 A/g was 81.9%, and capacity at high-rate of 5 A/g reached the value of 523 mAh/g.

Curled graphene sheets formed carbon nanotubes (CNTs) depending on the number of the graphene layers may be formed single-walled (SWCNTs) and multiwalled carbon nanotubes (MWCNTs) (Razzaq et al., 2019; Zheng et al., 2019). A composite cathode material fabricated with CNTs showed high sulfur loading of 90% prepared by the ball milling method (Cheng et al., 2014). The Li$-$S battery cell containing the presented CNT electrode showed an increase in energy density. CNTs combined with metal boride remarkably improved sulfur loading to 15.28 mg/cm^2 (Li et al., 2023). The capacity retention after 120 cycles at a current density of 21.8 mA/cm^2 of this electrode with ultrahigh sulfur loading was 86.96%. The electrode material with sulfur loading of 3.8 mg/cm^2 showed a capacity retention of 81.58% after 100 cycles at 1 C.

Isooctahedral C$_{60}$, fullerene, represents another type of carbon material where the graphene structure forms a sphere. Fullerenes possess remarkable physicochemical properties, such as the ability to accept electrons and high charge mobility (Bao et al., 2022). Feng et al. (2019) applied fullerene as a host for sulfur in Li−S batteries, and after 200 cycles at 0.2 C, they observed a specific capacity of 721 mAh/g. Functionalization of fullerene with cyanide can form covalent bonds with sulfur and polysulfides to improve their confinement. Numerical results showed that fullerene modified by cyanide ligands could enhance the interaction of polysulfides and host material thanks to the strong ionic bond between them (Ramezanitaghartapeh et al., 2022).

Amorphous carbon is able to immobilize the diffusion of polysulfides and shows stronger adsorption of polysulfides than graphene (Jeon et al., 2021). Acetylene black represents a type of amorphous carbon that is well electrically conductive but has a smaller surface area. Kwon et al. (2021) compared acetylene black and active carbon as a host for sulfur in Li−S batteries. Despite a larger surface area and pore volume of activated carbon than acetylene black, the improved cycle performance was observed for the composite with acetylene black due to higher electrical conductivity, which seems to be more dominant than the material's porosity. Another research group studied the influence of amorphous carbons, carbon Super P and Ketjen black, in sulfur cathode on the electrochemical properties of Li−S batteries (Capková et al., 2019). It was found that the larger surface area of Ketjen black is not as beneficial as the higher conductivity of carbon Super P, which resulted in enhanced capacity and cycle stability of the electrode with carbon Super P. The specific discharge capacity at 0.2 C of the electrode with carbon Super P was around 1058 mAh/g, and for Ketjen black was 886 mAh/g. Moreover, the cycling stability of the electrode with carbon Super P was highly stable, and the capacity retention after 60 cycles was 97.7% compared to 71% for the electrode containing Ketjen black.

In addition, carbon materials may be prepared by the carbonization process of biological materials or biomass (Yang, Wang et al., 2022). Kazda et al. (2018) annealed *Spongia officinalis* sea sponge in order to obtain a 3D structured carbon matrix for encapsulation of sulfur in Li−S batteries. The initial discharge capacity at 0.2 C was around 650 mAh/g, and after 80 cycles rate performance test, the capacity retention was 85%. Moreover, a 3D structured electrode showed a high sulfur loading of about 5 mg/cm^2 and very high capacity per centimeter squared up to 3.1 mAh/cm^2. A summary of the electrochemical performance of the used carbon materials in Li−S batteries is listed in Table 9.2.

9.5.2 Metal−organic frameworks (MOFs)

Metal−organic frameworks (MOFs) represent a novel class of highly porous materials with a large surface area and adjustable pore size. The tunable materials' morphology may be by controlling the ratio of raw materials, reaction time, and reaction conditions such as temperature and pH (Almáši et al., 2020; Almáši et al., 2021; Almáši, 2021). MOFs are hybrid inorganic−organic materials assembled by metal ions or clusters connected by organic likers to form a final robust framework with permanent porosity (Almáši, 2022; Cui et al., 2023). Metal ions, various organic ligands, or postsynthetic modification may modify their properties and resulting application properties (Almáši, 2013; Garg, 2022; Zelenka, 2022, 2024). The presence of metal compounds can improve polysulfide trapping through strong chemical interaction (Xu et al., 2018). The electrochemical properties of MOF material may be improved by the carbonization process (Capkova, Almasi, et al., 2022; Capkova, Knap, et al., 2022). Remarkably, their morphology can be almost preserved after the carbonization process in an inert atmosphere, and electron contacts should increase (Chen et al., 2019; Yu & Lin, 2022).

Demir-Cakan et al. (2011) introduced the application of MOF material in Li−S battery cells. They proposed MIL-100(Cr) (MIL—Material Institute Lavoisier) as a host for sulfur in a cathode material. MIL-100(Cr) is a mesoporous material consisting of Cr(III) ions and benzene-1,3,5-dicarboxylate ligands published by Férey et al. (2004). The specific discharge capacity at 0.1 C was around 1100 mAh/g, and after 50 cycles, the capacity retention was approximately 40%. Since then, many enhancements have been applied to improve the Li−S cell's cycle performance with the application of MOF material in the cathode. As other MIL materials that have been studied as hosts for sulfur in Li−S batteries are MIL-88(Fe) (Benítez et al., 2020), MIL-101(Cr) (Gao, Feng et al., 2020; Zhao, Zhao et al., 2014), and MIL-101(Fe)-NH$_2$ (Capková et al., 2020). MIL 88(Fe) delivered a specific capacity of about 400 mAh/g at 0.5 C

Table 9.2 Summary of performance parameters for carbonaceous materials used in Li−S batteries.

Host material	S content (wt %)	Specific capacity (mAh/g)	Retained capacity (mAh/g)	C-rate	Cycle number	Capacity decay rate (%)	Ref.
G	73	860	615	1 C	100	0.285	Lin et al. (2013)
N−G	60	788	347	2 C	2000	0.028	Qiu et al. (2014)
Co/N−G	90	1210	1173	0.2 C	100	0.029	Du et al. (2019)
Pyrrole/G	80	986	798	0.5 C	100	0.190	Zhang et al. (2017)
N−G/CNTs	80	1103	849	2 C	500	0.046	Wen et al. (2020)
GO	92	275	227	1 C	160	0.109	Chabu et al. (2019)
GO/PVP	92	705	609	1 C	115	0.118	Chabu et al. (2019)
VO_2/rGO	75	476	396	1 C	1000	0.017	Chen et al. (2023)
PPy/rGO	71	1041	853	0.1 A/g	200	0.091	(Huang et al., 2022)
CNTs	90	737	420	0.1 C	90	0.478	Cheng et al. (2014)
MBoride/ CNTs	75	941	768	1 C	100	0.184	Li et al. (2023)
C_{60}	70	792	721	0.2 C	200	0.045	Feng et al. (2019)
Acetylene black	70	920	360	0.5 C	100	0.609	Kwon et al. (2021)
Super P	60	1058	1033	0.2 C	60	0.039	Capková et al. (2019)
Ketjen black	60	886	628	0.2 C	60	0.485	Capková et al. (2019)
Sea sponge	60	650	537	0.2 C	80	0.217	Kazda et al. (2018)

with a capacity retention of 50% after 1000 cycles (Benítez et al., 2020). MIL-101(Cr), composed of Cr(III) ions and terephthalate ligands (BDC), was studied as support for sulfur by Zhao, Wang, et al. (2014) and Gao, Feng et al. (2020). Zhao, Wang, et al. (2014) reported that the MIL-101 (Cr)/S composite reached the initial discharge capacity of 715 mAh/g at 0.1 C with about 80% capacity retention after 100 cycles. However, Gao, Feng, et al. (2020) obtained an initial discharge capacity of around 1400 mAh/g at 0.1 C with the same MOF. The discharge capacity decreased very fast in the first cycles, and after 200 cycles, the remained capacity was 630 mAh/g representing capacity retention of around 45%. Nevertheless, MIL-101(Fe)-NH$_2$, composed of Fe(III) ions and 2-aminoterephthalate linker (BDC-NH$_2$), showed stable cycle performance with an initial discharge capacity of 775.2 Ah/g at 0.2 C (Capková et al., 2020). The specific discharge capacity at 0.5 C was 705 mAh/g, and after 200 cycles, the capacity decreased to 476 mAh/g, representing a capacity retention of 67.5%. The self-discharge rate of the fully charged cell after 48 hours of idling time represents 25.8%, representing the capacity of a high voltage plateau where higher polysulfides are formed. The capacity after the self-discharge test was recovered and reached the value of 95% of the capacity in the cycle before storage.

One of the most widely used MOF materials is HKUST-1 (Hong Kong University of Science and Technology), composed of benzene-1,3,5-tricarboxylate linker (BTC) and Cu(II) ions (Chui et al., 1999). de Haro et al. (2021) applied composite material containing HKUST-1 as a cathode in Li—S batteries. The specific capacity at 0.1 C was 620 mAh/g, although, during the first cycles, capacity decreased significantly, and in the second cycle, the capacity was around 400 mAh/g. After 100 cycles at 0.1 C, the capacity reached the value of 200 mAh/g, representing a capacity retention of 32%.

Another frequently applied MOF material belongs to a ZIF group (zeolitic imidazole framework) (Kukkar et al., 2021). Using ZIF-67 in sulfur cathode material delivered an initial discharge capacity of around 800 mAh/g at 0.1 C (Liu et al., 2022). The discharge capacity of the presented electrode was high even at 5 C; it was around 400 mAh/g. Moreover, after 500 cycles at 0.5 C, the capacity retention was about 70%. The electrochemical properties of the presented electrode material were improved by graphene oxide. Higher discharge capacities than ZIF-67 were observed for ZIF-8, where the specific capacity at 0.1 C was around 1000 mAh/g (Wu et al., 2021). However, the discharge capacity at 3 C significantly decreased to about 100 mAh/g. Long-term galvanostatic cycling at 0.5 C showed capacity retention of around 56%. The application of tannic acid and CNTs enhanced the electrode containing ZIF-8. The sulfur cathode material with hollow ZIF-8 with CNTs reached the initial discharge capacity of almost 1400 mAh/g at 0.1 C, although the capacity decay was more significant in the first cycles. The capacity decay per cycle was 0.12% during 500 cycles at 0.5 C. The process of synthesis of ZIF-8, the electrode material, and its electrochemical properties are depicted in Figs. 9.5 and 9.6.

As a last example of the MOF compound without further modification can be mentioned UPJS-15 with the chemical composition [Sr_2(MTA)

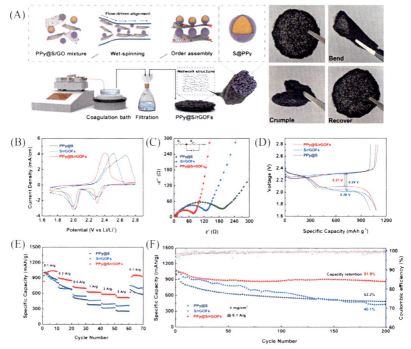

Figure 9.5 (A) The process of assembly of the sulfur-based electrode containing rGO fibers and photos of different deformations. (B) Cyclic voltammogram curves of the prepared electrode for Li−S batteries. (C) Electrochemical impedance spectra and used equivalent circuit. (D) Charge/discharge profiles for the prepared electrodes. (E) Galvanostatic cycling at current rates from 0.1 A/g to 5 A/g. (F) Long-term galvanostatic cycling at a current density of 0.1 A/g. *Reprinted with permission from Huang, L., Guan, T., Su, H., Zhong, Y., Cao, F., Zhang, Y., Xia, X., Wang, X., Bao, N., & Tu, J. (2022). Synergistic interfacial bonding in reduced graphene oxide fiber cathodes containing polypyrrole@sulfur nanospheres for flexible energy storage. Angewandte Chemie International Edition, 61(44).*

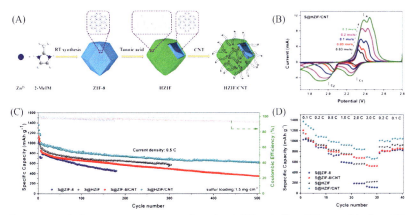

Figure 9.6 (A) The preparation route of synthesis ZIF-8, hollow ZIF-8, and hollow ZIF-8 with CNTs. (B) Cyclic voltammograms of the electrode with hollow ZIF-8 and CNTs at different scan rates. (C) Long-term galvanostatic cycling at 0.5 C of other electrodes containing ZIF-8. (D) Rate performance at different C-rates of the electrodes containing ZIF-8. *Reprinted with permission from Wu, Z., Wang, L., Chen, S., Zhu, X., Deng, Q., Wang, J., Zeng, Z., & Deng, S. (2021). Facile and low-temperature strategy to prepare hollow ZIF-8/CNT polyhedrons as high-performance lithium-sulfur cathodes.* Chemical Engineering Journal, 404.

(H_2O)] · H_2O · 4DMF [MTA—(methanetetrayltetrakis(benzene-4,1diyl)tetrakis(aza))tetrakis(methan-1-yl-1-yliden)tetrabenzoate, DMF—N,N-dimethylformamide] (Király et al., 2023). The initial discharge capacity was 337 mAh/g at 0.2 C and decreased continuously to 235 mAh/g after 100 charge/discharge cycles. The capacity retention after 100 cycles was 69.8%, corresponding to a fading rate of 0.3% per cycle. Coulombic efficiency during the cycling procedure at 0.2 C reached around 81.8%.

Porphyrin-based MOF with gallium(III) ions and its postsynthetically modified forms were proposed as a host for sulfur in Li−S batteries by Király et al. (2022). They analyzed GaTCPP [TCPP = 4,4′,4′′,4′′′-(5,10,15,20-porphyrintetrayl)-tetrabenzoate] in its activated state and two forms of GaTCPP doped by additional ions [Co(II) and Ni(II)]. The highest discharge capacity at 0.2 C was observed for the electrode containing doped GaTCPP by nickel(II) ions and reached the value of 733.8 mAh/g. Additional cobalt(II) ions resulted in a decrease in capacity compared to pristine GaTCPP. However, the capacity retention after 200 cycles at 0.5 C was 86.1% for pristine GaTCPP, 78.5% for GaTCPP(Ni), and 71.4% for GaTCPP(Co). Based on the stated results, it was proved that additional ions may improve the electrochemical properties of prepared MOF materials, although it has to be done very carefully as it may decrease the material stability. The best cycle performance for long-term galvanostatic cycling was observed for the electrode with pristine GaTCPP.

The electrochemical properties of MOF material may be enhanced by the carbonization process as presented in Capková et al. (2022). In this paper, MOF-76 with gadolinium(III) ions in the activated and carbonized state was proposed as a matrix for sulfur. The specific discharge capacity at 0.2 C for the cathode material with activated MOF-76(Gd) or carbonized MOF-76(Gd) was 684.4 mAh/g and 763.3 mAh/g, respectively. The capacity retention after 200 cycles at 0.5 C was significantly higher for the carbonized MOF-76(Gd), 92.8%, compared to 80.9% for its activated form. The capacity fading rate per cycle for the carbonized MOF-76(Gd) was only 0.036%. The carbonization process may improve cycle performance and stability, although it requires an additional step in the electrode material preparation. The MOF materials used as a host for sulfur in Li−S batteries are summarized in Table 9.3.

9.5.3 MXenes

2D metal carbides or nitrides represent a new family of materials called MXene phases. MXenes are highly conductive, and their surface is active, which is beneficial for bonding polysulfides via M−sulfur interactions. They are prepared by selectively etching the A layers laminated in aqueous HF from MAX phases (Zhao et al., 2021). Their composition is $M_{n+1}AX_n$, where M is the early transition metal (V, Ti, Nb, and Zr), A represents the IIIA and IVA elements (Al, Si, Ga, etc.), X is carbon and/or nitrogen, and $n = 1, 2,$ or 3 (Barsoum, 2013). The exfoliated sheets are created by these layers, and they may be delaminated by solvation with donor solvents (Yu et al., 2022). The A atoms are selectively removed after the etching process, and some surface terminations are bonded. The results are MXenes with a general formula of $M_{n+1}X_nT_x$, where T is the surface termination, such as -O, -OH, -F, and -Cl. The surface terminations are intended by etching compounds and may be modified by posttreatment (Naguib et al., 2011). Similar to MOFs, MXenes provide a strong chemical interaction between metal atoms and polysulfides that can reduce the influence of the shuttle effect, improve sulfur utilization, enhance redox kinetics, and have a large surface area (Iqbal et al., 2022; Wang, Yu, et al., 2019).

The first introduction of MXenes in Li−S batteries was by Liang et al. (2015). They demonstrated the application of titanium carbide (Ti_2C) as

Table 9.3 Summary of electrochemical performance of MOF materials applied in Li–S batteries.

Host material	S content (wt %)	Specific capacity (mAh/g)	Retained capacity (mAh/g)	C-rate	Cycle number	Capacity decay rate (%)	Ref.
MIL-100(Cr)	48	1100	440	0.1 C	50	1.200	Demir-Cakan et al. (2011)
MIL-88(Fe)	40	400	200	0.5 C	1000	0.050	Benítez et al. (2020)
MIL-101(Cr)	59	715	695	0.1 C	100	0.028	Zhao, Wang, et al. (2014)
MIL-101(Cr)	49	1400	630	0.1 C	200	0.275	Gao, Feng, et al. (2020)
MIL-101(Fe)-NH$_2$	60	705	476	0.5 C	200	0.162	Capková et al. (2020)
HKUST-1	65	620	200	0.1 C	100	0.677	de Haro et al. (2021)
ZIF-67	63	590	410	0.5 C	500	0.061	Liu et al. (2022)
ZIF-67/GO	63	800	672	0.5 C	500	0.032	Liu et al. (2022)
ZIF-8	70	750	460	0.5 C	180	0.215	Wu et al. (2021)
ZIF-8/CNTs	73	870	520	0.5 C	500	0.120	Wu et al. (2021)
UPJS-15	60	337	235	0.2 C	100	0.303	Király et al. (2023)
GaTCPP	60	601	517	0.5 C	200	0.069	Király et al. (2022)
GaTCPP(Co)	60	524	374	0.5 C	200	0.143	Király et al. (2022)
GaTCPP(Ni)	60	610	479	0.5 C	200	0.108	Király et al. (2022)
MOF-76(Gd)(AC)	60	621	502	0.5 C	200	0.096	Capková et al. (2022)
MOF-76(Gd)(C)	60	658	610	0.5 C	200	0.036	Capková et al. (2022)

AC, Activated; C, Carbonized.

a host for sulfur in cathode material for Li−S batteries. The specific capacity with 70% of sulfur was close to 1400 mAh/g at 0.2 C, and after 100 cycles, the capacity reached the value of around 1050 mAh/g. The significant decrease in capacity was in the first cycles then the capacity stabilizes. Moreover, the discharge capacity was high (1000 mAh/g) even at a high C-rate of 1 C, indicating improvement in the electrochemical performance and effectiveness of the cathode material in Li−S batteries. Zhang, Zhou et al. (2020) prepared and applied $Ti_3C_2T_x$ in sulfur cathode material. The preparation of sulfur composite with MXene was performed by two different methods, ball-milling and hydrothermal, and the results were compared. The hydrothermal method significantly improved the electrochemical performance of the Li−S battery cell. For the electrode prepared by ball-milling, the specific discharge capacity at 0.2 C was 840 mAh/g and decreased to almost 400 mAh/g after 100 cycles. However, the preparation of the cathode material by hydrothermal method improved the discharge capacity to 1477 mAh/g, and after 100 cycles, the capacity decreased to 1213 mAh/g.

The electrochemical performance of carbon materials can be improved by chemical modifications; therefore, it is expected that the properties of MXenes after further modification will be enhanced (Bao et al., 2018). Single-atom zinc-doped Ti_3AlC_2 layers improved the redox kinetics of polysulfides and accelerated the formation of discharge products on the widely exposed 2D surface (Zhang, Wang, et al., 2020). The reversible capacity of 1136 mAh/g was achieved at 0.2 C, and even at a high C-rate of 6 C, the discharge capacity was stable at around 517 mAh/g (see Fig. 9.7). After 90 cycles of the rate-capability test, the long-term cycling at 1 C was performed. The specific capacity at 1 C was 706 mAh/g with a capacity decay rate of 0.03% per cycle after 400 cycles. An overview of applied MXenes in cathode materials in Li−S batteries is listed in Table 9.4.

Nevertheless, the electrical conductivity of nitride MXenes is expected to be higher than carbide MXenes, nitride MXenes are less studied than carbide MXenes (Urbankowski et al., 2017). The potential application of V_2N/V_2NT_2 as the support for sulfur in Li−S batteries by density functional theory (DFT) calculations was explored by Fan et al. (2021). They showed that both nitride monolayers possessed an excellent metallic character. Due to the strong interaction between V and S/Li, bare V_2N cannot be directly applied in Li−S batteries. Another research group investigated Ti_2NO_2 and Ti_2NF_2 as the sulfur cathode host for Li−S batteries (Lin et al., 2019). Both nitrides exhibit moderate adsorption energies toward polysulfides. The anchoring mechanism between mentioned

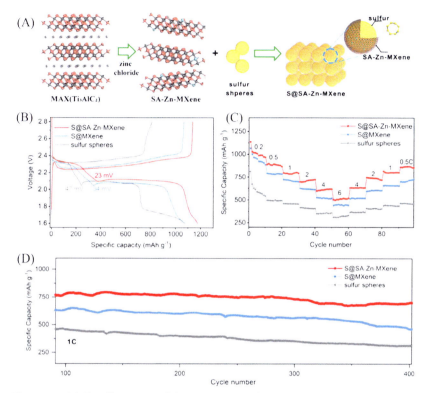

Figure 9.7 (A) The illustration of the preparation of single atom zinc MXene (SA-Zn-MXene) and its coating by sulfur to prepare S@SA-Zn-MXene. (B) Charge–discharge profiles of S@SA-Zn-MXene, S@MXene, and sulfur spheres. (C) Galvanostatic cycling at different C-rates from 0.2 C to 6 C, and (D) cycling stabilities at 1 C of S@SA-Zn-MXene, S@MXene, and sulfur spheres. *Reprinted with permission from Zhang, D., Wang, S., Hu, R., Gu, J., Cui, Y., Li, B., Chen, W., Liu, C., Shang, J., & Yang, S. (2020). Catalytic conversion of polysulfides on single atom zinc implanted MXene toward high-rate lithium–sulfur batteries.* Advanced Functional Materials, 30(30).

nitrides and polysulfides is different. The interactions between polysulfides and Ti_2NO_2 are dominated by chemical interaction, and Ti_2NF_2 is mainly represented by van der Waals interaction.

9.5.4 Polymeric materials

The most widely used noncarbonaceous materials in sulfur cathode in Li–S batteries are polymers due to their abundance, chemical stability, lightweight, and multiple structures (Huang et al., 2019). Polymeric materials represent long-chain structures with different functional groups.

Table 9.4 Summary of electrochemical properties of MXenes in cathode materials in Li–S batteries.

Host material	S content (wt %)	Specific capacity (mAh/g)	Retained capacity (mAh/g)	C-rate	Cycle number	Capacity decay rate (%)	Ref.
Ti_2C	70	1400	1050	0.2 C	100	0.250	Liang et al. (2015)
$Ti_3C_2T_x$—Ball mill	67	840	399	0.2 C	100	0.525	Fan Zhang, Zhou, et al. (2020)
$Ti_3C_2T_x$— Hydrothermal	67	1477	1213	0.2 C	100	0.179	Fan Zhang, Zhou, et al. (2020)
SA-Zn- Ti_3AlC_2	89	760	706	1 C	400	0.030	Zhang, Wang, et al. (2020)
Ti_3AlC_2	89	630	467	1 C	400	0.065	Zhang et al. (2020)

Their structures may accommodate well to volumetric sulfur changes during the Li−S battery's cycling. The functional groups in the polymeric structure can improve the interaction between sulfur species and polymer material (Arias et al., 2018; Manthiram et al., 2013).

The most widely used conductive polymer in Li−S batteries is polypyrrole (PPy). Wang et al. (2006) prepared the first composite material S-PPy, where the initial discharge capacity was 1280 mAh/g at 50 mA/g. Afterward, various polymerization processes were implemented to improve the electrochemical properties of PPy, such as surface modification or doping. Ball milling of sulfur and PPy improved the electrode material's electrochemical properties, resulting in an enhanced capacity and material stability (Xin et al., 2017). The specific discharge capacity at 1 C was 1085 mAh/g, and after 100 cycles, the capacity retention reached the value of 56.87%. The interstitial structure of sulfur-PPy composite reduced the polarization and improved the rate performance of the Li−S battery. Another example of a PPy-containing composite is a core−shell complex containing sulfur and porous hollow carbon aerogel (PHCA), which showed a discharge capacity of 650 mAh/g after 200 cycles at 1 C (Gao, Huang, et al., 2020). The core−shell structure hinders the dissolution of polysulfides and their shuttle between the electrodes.

Another highly stable, available, and easily preparable polymeric material is polyaniline (PANI) (Cheng & Wang, 2014). Encapsulation of sulfur into PANI nanotubes was proposed by Xiao et al. (2012). Sulfur/PANI nanotube composite was prepared via an in situ vulcanization process. After 100 cycles, the electrode retains a discharge capacity of 837 mAh/g at 0.1 C. Moreover, the highest discharge capacity at 1 C reached the value of 511 mAh/g, and the capacity retention after 500 cycles was around 84.54%. A further approach was the preparation of a yolk−shell nanoarchitecture composed of sulfur and PANI (Zhou et al., 2013). The yolk−shell structure showed a discharge capacity of 765 mAh/g after 200 cycles at 0.2 C, corresponding to a capacity retention of 69.5%. At a higher current density of 0.5 C, the capacity retention after 200 cycles was 68.3%, and the capacity reached the value of around 628 mAh/g. Furthermore, an example of PANI application is that $CoFe_2O_4$ nanotubes were decorated by PANI, resulting in $CoFe_2O_4$@PANI nanotubes being used as a host for sulfur in Li−S batteries (Gu et al., 2021). The initial discharge capacity during long-term cycling at 2 C was around 920 mAh/g, and after 500 cycles, the capacity decreased to 432 mAh/g. The electrode with $CoFe_2O_4$@PANI nanotubes showed stable cycle performance with a capacity decay rate per cycle of 0.065% and a Coulombic efficiency of around 97.9%.

The third form of the most well-known polymers applied in Li−S batteries is poly(3,4-ethylenedioxythiophene) (PEDOT). Chen et al. (2013) reported a core−shell structure where sulfur nanoparticles were coated with conductive polymer PEDOT. The sulfur encapsulation into PEDOT hindered the dissolution of polysulfides, resulting in improved cycling stability. The discharge capacity after 50 cycles at a current density of 400 mA/g was 930 mAh/g, corresponding to a capacity retention of 83.26%. Sulfur with a coating of PEDOT prepared by dielectric barrier discharge (DBD) plasma technology was reported by Shafique et al. (2021). PEDOT coating by DBD plasma treatment is low cost and safe; in addition, the surface conductivity of the resulting material increased. The discharge capacity after 100 cycles at 0.1 C was around 503 mAh/g with a Coulombic efficiency of around 99.4%. The coating composed of PEDOT in combination with poly(styrenesulfonate) (PSS) effectively enhances the low current rate (Zeng et al., 2021). Although the discharge capacity decreases faster in the first 30 cycles, and after 100 cycles at 0.1 C, the remained capacity was 296 mAh/g. Li et al. (2013) investigated the influence of PPy, PANI, and PEDOT polymers in sulfur cathode material on the cycling stability of Li−S batteries. The polymers were coated onto hollow sulfur nanospheres via a facile polymerization process. PEDOT showed the most remarkable improvement in rate performance, cycling stability, and cycle life. All prepared electrodes with polymer materials were charged/discharged for 500 cycles at 0.5 C, and the capacity retention for PPy, PANI, and PEDOT was 74%, 65%, and 86%, respectively (see Fig. 9.3). After 500 cycles at 0.5 C for the electrode containing PEDOT, the discharge capacity was around 780 mAh/g. The overview of the electrochemical characterization of applied polymeric materials in Li−S batteries is presented in Table 9.5 (Fig. 9.8).

9.6 Other metal−sulfur battery systems

Regardless of the progress made in Li−S batteries, challenges still need to be addressed to make this technology commercially available. Using Li−metal anodes suffers from the formation of dendrites during cell operation (He et al., 2018). Most of the other metal−S systems, except Li−S, suffer from poor kinetics and fast capacity decay due to larger ionic radii and higher charge density. The drawback of these metals is their higher

Table 9.5 Overview of the electrochemical characterization of polymers used in cathode material in Li−S battery cells.

Host material	S content (wt %)	Specific capacity (mAh/g)	Retained capacity (mAh/g)	C-rate	Cycle number	Capacity decay rate (%)	Ref.
PPy	50	1280	630	0.05 A/g	20	2.539	Wang et al. (2006)
PPy	60	1085	617	1 C	100	0.431	Xin et al. (2017)
PPy	77	1201	726	0.5 C	500	0.079	Li et al. (2013)
PHCA/PPy	70	876	650	1 C	200	0.129	Gao, Huang, et al. (2020)
PANI	62	511	432	1 C	500	0.031	Xiao et al. (2012)
PANI	58	920	628	2 C	500	0.159	Zhou et al. (2013)
PANI	74	1140	516	0.5 C	500	0.110	Li et al. (2013)
$CoFe_2O_4$@PANI	75	864	583	2 C	500	0.065	Gu et al. (2021)
PEDOT	72	1117	930	0.4 A/g	50	0.335	Chen et al. (2013)
PEDOT	78	1165	780	0.5 C	500	0.066	Li et al. (2013)
PEDOT (DBD-plasma)	66	690	503	0.1 C	100	0.271	Shafique et al. (2021)
PEDOT/PSS	75	827	296	0.1 C	100	0.642	Zeng et al. (2021)

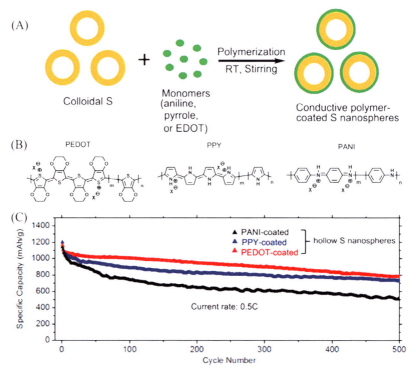

Figure 9.8 (A) The schematic illustration of PPy, PANI, and PEDOT polymer coatings preparation on sulfur nanospheres (RT = room temperature). (B) Chemical structures of applied polymeric materials. (C) Long-term cycling performance at 0.5 C for 500 cycles of the electrodes with different polymer coatings on cathodes in Li–S batteries. *Reprinted with permission from Li, W., Zhang, Q., Zheng, G., She, Z. W., Yao, H., & Cui, Y. (2013). Understanding the role of different conductive polymers in improving the nanostructured sulfur cathode performance. Nano Letters, 13(11), 5534–5540.*

reactivity with water and electrolytes (Muldoon et al., 2014). Therefore, it would be prudent to explore metal anodes for battery systems that can interface with high-capacity sulfur cathodes and are more abundant in the Earth's crust than lithium. The research of nonlithium metal anodes for metal−S batteries is in the early stage and hindered by several challenges. The optimization of high sulfur loading, content, and electrolyte needs to be addressed (Slater et al., 2013).

9.6.1 Sodium−sulfur batteries

The sixth most abundant element on the Earth's crust is sodium; due to this reason, sodium is inexpensive. The reduction potential of sodium is

-2.71 V (vs SHE), which makes sodium less reducing compared to lithium (-3.04 V). On the other hand, the reactivity of sodium is higher compared to lithium. In the 1960s, similarly to Li$-$S batteries, the first Na$-$S battery was introduced by Kummer and Weber (1967). The described Na$-$S battery operated at a high temperature of 300°C to 350°C, and the battery components, sulfur, sodium, and solid beta-alumina electrolyte, were in a molten state during battery operation. However, molten sodium, sulfur, and polysulfides are highly corrosive, so alloy steels were used in these cells. Their extensive application is limited by their high cost and operating temperature (Hueso et al., 2013). Thus, scientists have focused on discovering and developing room-temperature Na$-$S batteries. In addition to the fundamental problem of metal$-$S batteries, Na$-$S batteries suffer from the complicated reaction mechanisms of sodium and sulfur, which causes low active material utilization.

In order to improve the electrochemical performance of room-temperature Na$-$S batteries, several developments have been introduced. Carbon nanostructures are widely studied as a sulfur host for sulfur hosts. N-doped carbon nanosheets (CN) were used for sulfur encapsulation in Na$-$S by Wang et al. (2020). The initial discharge capacity at 0.1 A/g was 1253 mAh/g, and after 110 cycles, the capacity retention was around 26.34%. The multichannel carbon fibers (MCCFs) have a porous structure and good electrical conductivity; therefore, they were applied as a matrix for sulfur in Na$-$S batteries (Ye et al., 2021). The discharge capacity after 100 cycles at 0.1 A/g was 347 mAh/g representing a capacity retention of 45%. The electrochemical properties of MCCFs were improved by the modification with TiN-TiO$_2$. The discharge capacity of the TiN-TiO$_2$ heterostructures incorporated in MCCFs electrode material reached after 100 cycles at 0.1 A/g, the value of 640 mAh/g with a capacity retention of 49% (see Fig. 9.4). Activated ultramicroporous coffee carbon (AUCC) as a host for sulfur in Na$-$S batteries achieved a reversible specific capacity of 1492 mAh/g at 0.1 C (Guo et al., 2020). The capacity retention after 700 cycles at 0.5 C was 90%. The reactivity of sulfur and inhibition of the shuttle effect are expected to increase with the application of inherent polarization sulfur hosts. Zhang et al. (2019) reported transition-metal (Fe, Cu, and Ni) nanoclusters loaded onto hollow carbon nanospheres (HC) as a polarized host for sulfur. Fe nanoclusters enhance conductivity and activity and assist in the immobilization of sulfur. The reversible discharge capacity of 394 mAh/g at 0.1 A/g after 1000 cycles was observed for the electrodes with Fe-HC, and the capacity retention was about 38.5%.

The electrodes with Cu and Ni nanoclusters showed a reversible capacity of 263 and 201 mAh/g, representing capacity retention of 27.8% and 25.7%, respectively (Fig. 9.9).

MOFs or their derivatives are widely used as a support for sulfur also in Na−S batteries. Xiao et al. (2021) reported polydopamine-coated Co (II)-MOF with composition $Co(OBA)(DMBY)$ [4,4′-dimethyl-2,2′-bipyridyl (DMBY), 4,4′-oxybisbenzoic acid (OBA)] as host composite for sulfur storage in Na−S batteries (Xiao et al., 2021). They analyzed electrode materials based on MOF with different contents of sulfur 10:10, 10:9, 10:6, and 10:4 weight ratios (composite:S). The best cycle performance showed the electrode with ∼50% of sulfur, and the electrodes showed a reversible capacity of 403 mAh/g at a current rate of 1 A/g after 1000 cycles. The electrochemical properties of various materials used in cathode in Na−S batteries are summarized in Table 9.6.

9.6.2 Potassium−sulfur batteries

The abundance of potassium is higher compared to lithium; however, the reduction potential of potassium is slightly less reducing than lithium (-2.93 V for K and -3.04 V for Li vs SHE). The large ionic radius of potassium ions results in the difficult formation of K_2S in K−S batteries (Hwang et al., 2018). Moreover, the reactivity of potassium is even higher than that of sodium, which further aggravates the anode problems, so the K−S battery design must be proposed very carefully.

The design principles for suppressing polysulfide shuttle in cathode materials for K−S batteries are similar to those for Li−S batteries. The first room-temperature K−S battery was proposed in 2014, where mesoporous carbon CMK-3 with PANI was presented as a matrix for sulfur in K−S batteries (Fig. 9.10) (Zhao, Hu, et al., 2014). The initial discharge capacity at a current density of 0.05 A/g was 513 mAh/g, and after 50 cycles, the capacity retention was around 40%. Pyrolized polyacrylonitrile (PAN) was used as a positive electrode material for sulfur storage in K−S batteries (Liu et al., 2018). PAN-sulfur nanocomposite showed a specific capacity of 270 mAh/g at 0.5 C, and after 100 cycles, the capacity retention was around 54.5%. MOFs may also be applied in K−S batteries to enhance the S-based cathode's cycle performance. Ge et al. (2020) proposed zeolitic imidazolate framework-67 (ZIF-67) with catalytic Co−N bonds as self-assembled templates for the synthesis of N-doped nanoclusters of

Coordination materials for metal−sulfur batteries 313

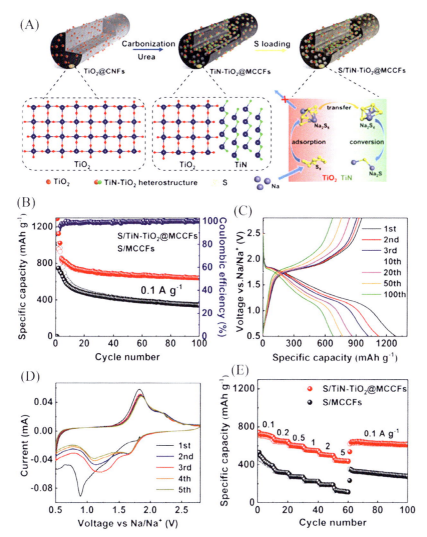

Figure 9.9 (A) The illustration of the design principle of the TiN-TiO$_2$ heterostructure. (B) Long-term cycle performance test for S/TiN-TiO$_2$@MCCFs and S/MCCFs cathodes in Na−S batteries at 0.1 A/g. (C) Charge and discharge profiles for the Na−S battery with S/TiN-TiO$_2$@MCCFs electrode at 0.1 A/g. (D) Cyclic voltammogram for the Na−S battery with S/TiN-TiO$_2$@MCCFs cathode. (E) Rate performance test for S/TiN-TiO$_2$@MCCFs and S/MCCFs cathodes in Na−S batteries. *Reprinted with permission from Ye, X., Ruan, J., Pang, Y., Yang, J., Liu, Y., Huang, Y., & Zheng, S. (2021). Enabling a stable room-temperature sodium−sulfur battery cathode by building heterostructures in multichannel carbon fibers. ACS Nano, 15(3), 5639−5648.*

Table 9.6 Summary of the electrochemical performance of various materials used in cathode material in Na—S battery cells.

Host material	S content (wt %)	Specific capacity (mAh/g)	Retained capacity (mAh/g)	C-rate	Cycle number	Capacity decay rate (%)	Ref.
CN	53	1253	330	0.1 A/g	110	0.670	Wang et al. (2020)
MCCFs	52	780	347	0.1 A/g	100	0.555	Ye et al. (2021)
TiN-TiO$_2$/ MCCFs	57	1308	640	0.1 A/g	100	0.511	Ye et al. (2021)
AUCC	40	920	830	0.5 C	700	0.014	Guo et al. (2020)
Fe-HC	40	1023	394	0.1 A/g	1000	0.062	Zhang et al. (2019)
Cu-HC	35	945	263	0.1 A/g	1000	0.072	Zhang et al. (2019)
Ni-HC	30	783	201	0.1 A/g	1000	0.074	Zhang et al. (2019)
Co-MOF	50	625	403	1 A/g	1000	0.036	Xiao et al. (2021)

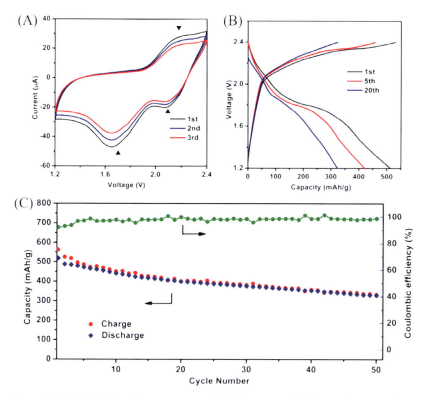

Figure 9.10 (A) Cyclic voltammogram of K−S battery with S/PANI/CMK-3 cathode material at a scan rate of 0.1 mV/s. (B) Charge and discharge profiles for the K−S battery with PANI/CMK-3 sulfur cathode and (C) its cycling performance at 0.05 A/g. *Reprinted with permission from Zhao, Q., Hu, Y., Zhang, K., & Chen, J. (2014). Potassium−sulfur batteries: A new member of room-temperature rechargeable metal−sulfur batteries. Inorganic Chemistry, 53(17), 9000−9005.*

Co-encrusted hollow hierarchical porous carbon polyhedron. The electrodes based on ZIF-67 showed a reversible discharge capacity of 355 mAh/g at 0.2 A/g after 150 cycles. The performance parameters of the presented electrode materials for K−S batteries are summarized in Table 9.7.

9.6.3 Multivalent metal−sulfur batteries

Multivalent metal−S (metal = Mg, Al, and Ca) batteries offer high theoretical capacity, an abundance of metal elements, and cost-effectiveness. Their redox potentials are less reducing compared to lithium (Mg—2.36 V, Al—1.68 V, Ca—2.84 V, and Li 3.04 V vs SHE) (Yang et al., 2019).

Table 9.7 Summary of performance parameters of applied materials in K−S batteries.

Host material	S content (wt %)	Specific capacity (mAh/g)	Retained capacity (mAh/g)	C-rate	Cycle number	Capacity decay rate (%)	Ref.
PANI/CMK-3	41	513	202	0.05 A/g	50	1.212	Zhao, Hu, et al. (2014)
PAN	38	270	147	0.5 C	100	0.456	Liu et al. (2018)
ZIF-67	63	595	355	0.2 A/g	150	0.267	Ge et al. (2020)

Metalic Ca is more reactive than Li, but Al and Mg are stable in the air. Presented multivalent metals are environmentally friendly, abundant, and low cost. Ions of mentioned metals transfer more than one electron, which may result in a high capacity and energy density (Liang et al., 2020). In addition to characteristic issues for metal−S batteries, multivalent metal−S batteries, especially Al−S, suffer from strong solvation of multivalent ions and high internal charge density. These properties cause slow diffusion and a high reaction energy barrier, leading to battery polarization and poor electrochemical kinetics (Yang, Yang, et al., 2022).

Sulfur cathodes for the advancement of multivalent metal−S batteries include particularly carbon materials containing elemental sulfur. Yu and Manthiram (2016) reported an Mg−S battery where a carbon nanofiber (CNF) matrix was filled with active sulfur material. In addition, the CNF coating was applied on the glass fiber separator to improve cyclability. The discharge capacity at 0.02 C was approaching 1200 mAh/g, and the cycling stability after 20 cycles was around 63%. Nitrogen-doped graphene (N−G) showed in Li−S batteries improved cycle performance, so the same impact is expected in Mg−S batteries. The specific discharge capacity at 0.01 C was around 700 mAh/g, although rapid capacity decay was observed in the first cycles (Li et al., 2016). Zhou et al. (2018) proposed ZIF-67 derivative framework as a host for sulfur in Mg−S batteries where the initial discharge capacity at 0.1 C was around 850 mAh/g (Fig. 9.11A). The cycle performance of the MOF-based electrode was highly stable, and after 200 cycles, the capacity retention was around 53%.

The activated carbon cloth (ACC) was proposed as a sulfur host in Al−S batteries by Gao et al. (2016). The presented Al−S battery cell

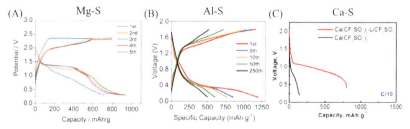

Figure 9.11 Galvanostatic charge/discharge curves of (A) Mg−S battery with S/ZIF-67 cathode at 0.1 C, (B) Al−S battery system with S/HKUST-1 cathode at 1 A/g, and (C) discharge profiles for the Ca−S battery with S/CNF cathode at 0.1°C. *(A) Reprinted with permission from Zhou et al. (2018). (B) Reprinted with permission from Guo et al. (2019). (C) Reprinted with permission from Yu et al. (2019).*

Table 9.8 Overview of the electrochemical characteristics of applied materials in multivalent metal−sulfur (Mg−S, Al−S, and Ca−S) batteries.

	Host material	S content (wt %)	Specific capacity (mAh/g)	Retained capacity (mAh/g)	C-rate	Cycle number	Capacity decay rate (%)	Ref.
Mg−S	CNF	–	1200	750	0.02 C	20	1.875	Yu and Manthiram (2016)
	N−G	50	700	70	0.01 C	20	4.500	Li et al. (2016)
	ZIF-67	60	850	450	0.1 C	250	0.188	Zhou et al. (2018)
Al−S	ACC	58	1300	1000	0.05 A/g	20	1.154	Gao et al. (2016)
	HKUST-1	34	1200	500	1 A/g	500	0.117	Guo et al. (2019)
Ca−S	CMK-3	–	600	–	∼0.3 C	–	–	See et al. (2013)
	CNF	–	800	300	0.1 C	20	3.125	Yu et al. (2019)

showed a discharge capacity of 1000 mAh/g after 20 cycles at 0.05 A/g. The carbonized MOF HKUST-1 was applied for anchoring polysulfides in Al−S batteries (Guo et al., 2019). The proposed electrode material showed highly stable cycle performance, and after 500 cycles at 1 A/g, the capacity was around 500 mAh/g representing a capacity retention of around 41.7% (Fig. 9.11B).

See et al. (2013) reported the Ca−S battery containing CMK-3 in sulfur cathode material. The presented electrode material in the Ca−S battery showed a discharge capacity of 600 mAh/g. The research group of Manthiram proposed also for Ca−S batteries CNF as support for sulfur in the cathode material (Yu et al., 2019). The specific discharge capacity of the Ca−S battery at 0.1 C was around 800 mAh/g, and the capacity fading rate per cycle reached 3.1% despite 20 cycles (Fig. 9.11C).

The summary of the presented performance parameter of applied materials in multivalent metal−sulfur (Mg−S, Al−S, and Ca−S) batteries is presented in Table 9.8.

9.7 Conclusion

A typical battery usually consists of a cathode, an anode, a separator, and an electrolyte. Considerable progress has been made in improving the performance of rechargeable batteries by modifying the electrode and electrolyte compositions. Most of the work focuses on designing new electrode materials with different nanostructures that aid charge transfer, exploring electrolyte synthesis and additives to build a stable electrode−electrolyte interface. The metal−sulfur (M−S) batteries are one of the most prominent candidates for the next generation of energy storage systems to alleviate the energy crisis. However, for the successful commercialization of M−S batteries, it is necessary to increase the usability of active materials, extend the lifetime, and improve the energy density.

In the present chapter, we focused on a different type of metal−sulfur batteries (M−S, where M = Li, Na, K, Mg, Ca, and Al) and a variety of materials, including carbonaceous materials, metal−organic frameworks, MXenes, organic polymers, and carbonized natural products, for example, biomass and coffee as additives for polysulfide immobilization. Currently, the most studied M−S batteries are Li−S. Depending on the additive and C-rate, they can reach values of initial discharge capacity 275−1210 mAh/g for carbonaceous materials, 400−1400 mAh/g for metal−organic frameworks, 337−1477 mAh/g for MXenes, and 511−1280 mAh/g for polymeric compounds. Scientists still face the challenge of complex research to achieve the maximum theoretical capacity of 1675 mAh/g for Li—S batteries and to reduce the capacity decay rate. Currently less studied, but no less important, are other types of M—S batteries, which are still waiting for their more detailed examination. To date, M—S batteries are able to reach initial discharge capacities of 625—1308 mAh/g for Na—S, 270—595 mAh/g for K—S, and 600—1300 mAh/g for multivalent M—S (M = Mg, Ca, Al) batteries depending on the additive used.

Acknowledgments

The authors are grateful for financial support from the Slovak Research and Development Agency No. APVV-20—0138, No. SK-CZ-RD-21—0068, COST Action - Organic Batteries and specific graduate research of the Brno University of Technology no. FEKT-S-23-8286. DC acknowledges the EU Horizon 2023 research and innovation program under the Marie Sklodowska-Curie Postdoctoral Fellowship Grant no. 101152715 (SALSA project). M.A. acknowledges the EU NextGenerationEU through the Recovery and Resilience Plan for Slovakia under the project No. 09I03-03-V03-00034.

References

Almáši, M.Zeleňák, V., Gyepes, R., Zukal, A., & Čejka, J. (2013). Synthesis, characterization and sorption properties of zinc(II) metal—organic framework containing methanetetrabenzoate ligand. *Colloids and Surfaces. A, Physicochemical and Engineering Aspects, 437*, 101—107. Available from https://doi.org/10.1016/j.colsurfa.2012.11.067.

Almáši, M. (2021). A review on state of art and perspectives of Metal-Organic frameworks (MOFs) in the fight against coronavirus SARS-CoV-2. *Journal of Coordination Chemistry, 74*(13), 2111—2127. Available from https://doi.org/10.1080/00958972.2021.1965130.

Almáši, M. (2022). *Current development in MOFs for hydrogen storage. Metal-organic framework-based nanomaterials for energy conversion and storage* (pp. 631—661). Slovakia: Elsevier. Available from https://www.sciencedirect.com/book/9780323911795, https://doi.org/10.1016/B978-0-323-91179-5.00020-6.

Almáši, M., Király, N., Zeleňák, V., Vilková, M., & Bourrelly, S. (2021). Zinc(ii) and cadmium(ii) amorphous metal—organic frameworks (aMOFs): study of activation process and high-pressure adsorption of greenhouse gases. *RSC Advances, 11*(33), 20137—20150. Available from https://doi.org/10.1039/d1ra02938j.

Almáši, M., Zeleňák, V., Gyepes, R., Zauška, Ľ., & Bourrelly, S. (2020). A series of four novel alkaline earth metal—organic frameworks constructed of Ca(ii), Sr(ii), Ba(ii) ions and tetrahedral MTB linker: structural diversity, stability study and low/high-pressure gas adsorption properties. *RSC Advances, 10*(54), 32323—32334. Available from https://doi.org/10.1039/d0ra05145d.

Arias, A. N., Tesio, A. Y., & Flexer, V. (2018). Review—non-carbonaceous materials as cathodes for lithium-sulfur batteries. *Journal of the Electrochemical Society, 165*(1), A6119—A6135. Available from https://doi.org/10.1149/2.0181801jes, http://jes.ecsdl.org/content/by/year.

Bao, L., Xu, T., Guo, K., Huang, W., & Lu, X. (2022). Supramolecular engineering of crystalline fullerene micro-/nano-architectures. *Advanced Materials*, *34*(52). Available from https://doi.org/10.1002/adma.202200189.

Bao, W., Liu, L., Wang, C., Choi, S., Wang, D., & Wang, G. (2018). Facile synthesis of crumpled nitrogen-doped MXene nanosheets as a new sulfur host for lithium−sulfur batteries. *Advanced Energy Materials*, *8*(13). Available from https://doi.org/10.1002/aenm.201702485.

Barsoum, M. W. (2013). *MAX phases: Properties of machinable ternary carbides and nitrides* (pp. 1−421). United States: Wiley. Available from http://onlinelibrary.wiley.com/book/10.1002/9783527654581, https://doi.org/10.1002/9783527654581.

Benítez, A., Amaro-Gahete, J., Esquivel, D., Romero-Salguero, F. J., Morales, J., & Caballero, Á. (2020). MIL-88A metal-organic framework as a stable sulfur-host cathode for long-cycle Li-S batteries. *Nanomaterials*, *10*(3). Available from https://doi.org/10.3390/nano10030424.

Brückner, J., Thieme, S., Grossmann, H. T., Dörfler, S., Althues, H., & Kaskel, S. (2014). Lithium-sulfur batteries: Influence of C-rate, amount of electrolyte and sulfur loading on cycle performance. *Journal of Power Sources*, *268*, 82−87. Available from https://doi.org/10.1016/j.jpowsour.2014.05.143.

Camargos, P. H., dos Santos, P. H. J., dos Santos, I. R., Ribeiro, G. S., & Caetano, R. E. (2022). Perspectives on Li-ion battery categories for electric vehicle applications: A review of state of the art. *International Journal of Energy Research*, *46*(13), 19258−19268. Available from https://doi.org/10.1002/er.7993, http://onlinelibrary.wiley.com/journal/10.1002/(ISSN)1099-114X.

Capkova, D., Almasi, M., Macko, J., Kiraly, N., Cech, O., Cudek, P., Fedorkova, A. S., Knap, V., & Kazda, T. (2022). Activated and carbonized metal-organic frameworks for improved cycle performance of cathode material in lithium-sulphur batteries. *Journal of Physics: Conference Series*, *2382*(1). Available from https://doi.org/10.1088/1742-6596/2382/1/012010.

Capková, D., Almáši, M., Kazda, T., Čech, O., Király, N., Čudek, P., Straková Fedorková, A., & Hornebecq, V. (2020). Metal-organic framework MIL-101(Fe)−NH2 as an efficient host for sulphur storage in long-cycle Li−S batteries. *Electrochimica Acta*, *354*. Available from https://doi.org/10.1016/j.electacta.2020.136640.

Capkova, D., Kazda, T., Petruš, O., Macko, J., Jasso, K., Baskevich, A., Shembel, E., & Strakova Fedorkova, A. (2021). Pyrite as a low-cost additive in sulfur cathode material for stable cycle performance. *ECS Transactions*, *105*(1), 191−198. Available from https://doi.org/10.1149/10501.0191ecst.

Capková, D., Kazda, T., Straková Fedorková, A., Čudek, P., & Oriňaková, R. (2019). Carbon materials as the matrices for sulfur in Li-S batteries. *ECS Transactions*, *95*(1), 19−26. Available from https://doi.org/10.1149/09501.0019ecst.

Capková, D., Kazda, T., Čech, O., Király, N., Zelenka, T., Čudek, P., Sharma, A., Hornebecq, V., Straková Fedorková, A., & Almáši, M. (2022). Influence of metal-organic framework MOF-76(Gd) activation/carbonization on the cycle performance stability in Li-S battery. *Journal of Energy Storage*, *51*. Available from https://doi.org/10.1016/j.est.2022.104419.

Capkova, D., Kazda, T., Čudek, P., & Strakova Fedorkova, A. (2020). Binder influence on electrochemical properties of Li-S batteries. *ECS Transactions*, *99*(1), 161−167. Available from https://doi.org/10.1149/09901.0161ecst.

Capkova, D., Knap, V., Fedorkova, A. S., & Stroe, D. I. (2022). Analysis of 3.4 Ah lithium-sulfur pouch cells by electrochemical impedance spectroscopy. *Journal of Energy Chemistry*, *72*, 318−325. Available from https://doi.org/10.1016/j.jechem.2022.05.026, elsevier.com/journals/journal-of-energy-chemistry/2095-4956.

Capkova, D., Knap, V., Fedorkova, A. S., & Stroe, D. I. (2023). Investigation of the temperature and DOD effect on the performance-degradation behavior of lithium—sulfur pouch cells during calendar aging. *Applied Energ*, *332*. Available from https://doi.org/10.1016/j.apenergy.2022.120543, https://www.journals.elsevier.com/applied-energy.

Chabu, J. M., Zeng, K., Chen, W., Mustapha, A., Li, Y., & Liu, Y. N. (2019). A novel graphene oxide-wrapped sulfur composites cathode with ultra-high sulfur content for lithium—sulfur battery. *Applied Surface Science*, *493*, 533—540. Available from https://doi.org/10.1016/j.apsusc.2019.07.061, http://www.journals.elsevier.com/applied-surface-science/.

Chen, B., Wei, J., Li, X., Ji, Y., Liang, D., & Chen, T. (2023). Vanadium dioxide plates reduced graphene oxide as sulfur cathodes for efficient polysulfides trap in long-life lithium-sulfur batteries. *Journal of Colloid and Interface Science*, *629*, 1003—1011. Available from https://doi.org/10.1016/j.jcis.2022.09.028.

Chen, H., Dong, W., Ge, J., Wang, C., Wu, X., Lu, W., & Chen, L. (2013). Ultrafine sulfur nanoparticles in conducting polymer shell as cathode materials for high performance lithium/sulfur batteries. *Scientific Reports*, *3*(1). Available from https://doi.org/10.1038/srep01910.

Chen, H., Shen, K., Tan, Y., & Li, Y. (2019). Multishell hollow metal/nitrogen/carbon dodecahedrons with precisely controlled architectures and synergistically enhanced catalytic properties. *ACS Nano*, *13*(7), 7800—7810. Available from https://doi.org/10.1021/acsnano.9b01953.

Cheng, Sulfur—containing cathode. (2000).

Cheng, H., & Wang, S. (2014). Recent progress in polymer/sulphur composites as cathodes for rechargeable lithium—sulphur batteries. *J. Mater. Chem. A.*, *2*(34), 13783—13794. Available from https://doi.org/10.1039/C4TA02821J.

Cheng, M., Yan, R., Yang, Z., Tao, X., Ma, T., Cao, S., Ran, F., Li, S., Yang, W., & Cheng, C. (2022). Polysulfide catalytic materials for fast-kinetic metal—sulfur batteries: principles and active centers. *Advanced Science*, *9*(2). Available from https://doi.org/10.1002/advs.202102217, http://onlinelibrary.wiley.com/journal/10.1002/(ISSN)2198-3844.

Cheng, X. B., Huang, J. Q., Zhang, Q., Peng, H. J., Zhao, M. Q., & Wei, F. (2014). Aligned carbon nanotube/sulfur composite cathodes with high sulfur content for lithium-sulfur batteries. *Nano Energy*, *4*, 65—72. Available from https://doi.org/10.1016/j.nanoen.2013.12.013.

Chui, S. S. Y., Lo, S. M. F., Charmant, J. P. H., Orpen, A. G., & Williams, I. D. (1999). A chemically functionalizable nanoporous material [Cu3(TMA)2 (H2O)3](n). *Science*, *283* (5405), 1148—1150. Available from https://doi.org/10.1126/science.283.5405.1148.

Chung, S. -H., & Manthiram, A. (2019). Current status and future prospects of metal—sulfur batteries. *Advanced Materials*, *31*(27). Available from https://doi.org/10.1002/adma.201901125.

Cohn, G., Ma, L., & Archer, L. A. (2015). A novel non-aqueous aluminum sulfur battery. *Journal of Power Sources*, *283*, 416—422. Available from https://doi.org/10.1016/j.jpowsour.2015.02.131.

Cui, H., Yan, X., Liu, B., Zhao, X., Zhang, X., Zhao, X., Tong, X., Wang, Y., & Xing, Y. (2023). Flower-like spherical Ni-benzimidazole derived Ni-NiO-C complexed with carbon nanotubes as electrocatalysts for lithium-sulfur battery. *Journal of Alloys and Compounds*, *931*. Available from https://doi.org/10.1016/j.jallcom.2022.167402.

Danuta, H., & Juliusz, U. (1957). *Electric dry cells and storage batteries*, US3043896A Patient.

de Haro, J., Benítez, A., Caballero, Á., & Morales, J. (2021). Revisiting the HKUST-1/S composite as an electrode for Li-S batteries: Inherent problems that hinder its performance. *European Journal of Inorganic Chemistry*, *2021*(2), 177—185. Available from https://doi.org/10.1002/ejic.202000837, http://onlinelibrary.wiley.com/journal/10.1002/(ISSN)1099-0682c.

Demir-Cakan, R., Morcrette, M., Nouar, F., Davoisne, C., Devic, T., Gonbeau, D., Dominko, R., Serre, C., Férey, G., & Tarascon, J. M. (2011). Cathode composites for Li-S batteries via the use of oxygenated porous architectures. *Journal of the American Chemical Society*, *133*(40), 16154−16160. Available from https://doi.org/10.1021/ja2062659, http://pubs.acs.org/journal/jacsat.

Du, Z., Chen, X., Hu, W., Chuang, C., Xie, S., Hu, A., Yan, W., Kong, X., Wu, X., Ji, H., & Wan, L. J. (2019). Cobalt in nitrogen-doped graphene as single-atom catalyst for high-sulfur content lithium-sulfur batteries. *Journal of the American Chemical Society*, *141*(9), 3977−3985. Available from https://doi.org/10.1021/jacs.8b12973, http://pubs.acs.org/journal/jacsat.

Ehigiamusoe, K. U., & Dogan, E. (2022). The role of interaction effect between renewable energy consumption and real income in carbon emissions: Evidence from low-income countries. *Renewable and Sustainable Energy Reviews*, *154*. Available from https://doi.org/10.1016/j.rser.2021.111883.

Evers, S., & Nazar, L. F. (2013). New approaches for high energy density lithium-sulfur battery cathodes. *Accounts of Chemical Research*, *46*(5), 1135−1143. Available from https://doi.org/10.1021/ar3001348.

Fan, K., & Huang, H. (2022). Two-dimensional host materials for lithium-sulfur batteries: A review and perspective. *Energy Storage Materials*, *50*, 696−717. Available from https://doi.org/10.1016/j.ensm.2022.06.009.

Fan, K., Ying, Y., Luo, X., & Huang, H. (2021). Nitride MXenes as sulfur hosts for thermodynamic and kinetic suppression of polysulfide shuttling: a computational study. *Journal of Materials Chemistry A*, *9*(45), 25391−25398. Available from https://doi.org/10.1039/d1ta06759a.

Feng, X., Huang, X., Ma, Y., Song, G., & Li, H. (2019). New structural carbons via industrial gas explosion for hybrid cathodes in Li−S batteries. *ACS Sustainable Chemistry & Engineering*, *7*(15), 12948−12954. Available from https://doi.org/10.1021/acssuschemeng.9b01951.

Férey, G., Serre, C., Mellot-Draznieks, C., Millange, F., Surblé, S., Dutour, J., & Margiolaki, I. (2004). A hybrid solid with giant pores prepared by a combination of targeted chemistry, simulation, and powder diffraction. *Angewandte Chemie International Edition*, *43*(46), 6296−6301. Available from https://doi.org/10.1002/anie.200460592.

Gao, G., Feng, W., Su, W., Wang, S., Chen, L., Li, M., & Song, C. (2020). Preparation and modification of MIL-101(Cr) metal organic framework and its application in lithium-sulfur batteries. *International Journal of Electrochemical Science*, *15*(2), 1426−1436. Available from https://doi.org/10.20964/2020.02.26.

Gao, T., Li, X., Wang, X., Hu, J., Han, F., Fan, X., Suo, L., Pearse, A. J., Lee, S. B., Rubloff, G. W., Gaskell, K. J., Noked, M., & Wang, C. (2016). A rechargeable Al/S battery with an ionic-liquid electrolyte. *Angewandte Chemie—International Edition*, *55* (34), 9898−9901. Available from https://doi.org/10.1002/anie.201603531, http://onlinelibrary.wiley.com/journal/10.1002/(ISSN)1521-3773.

Gao, X., Huang, Y., Zhang, Z., Batool, S., Li, X., & Li, T. (2020). Porous hollow carbon aerogel-assembled core@polypyrrole nanoparticle shell as an efficient sulfur host through a tunable molecular self-assembly method for rechargeable lithium/sulfur batteries. *ACS Sustainable Chemistry & Engineering*, *8*(42), 15822−15833. Available from https://doi.org/10.1021/acssuschemeng.0c02456.

Garg, A.Almáši, M., Bednarčík, J., Sharma, R., Rao, V. S., Panchal, P., Jain, A., & Sharma, A. (2022). Gd(III) metal-organic framework as an effective humidity sensor and its hydrogen adsorption properties. *Chemosphere*, *305*, 135467. Available from https://doi.org/10.1016/j.chemosphere.2022.135467.

Ge, X., Di, H., Wang, P., Miao, X., Zhang, P., Wang, H., Ma, J., & Yin, L. (2020). Metal−organic framework-derived nitrogen-doped cobalt nanocluster inlaid porous carbon as high-efficiency catalyst for advanced potassium−sulfur batteries. *ACS Nano*, *14*(11), 16022−16035. Available from https://doi.org/10.1021/acsnano.0c07658.

Gu, L. L., Wang, C., Qiu, S. Y., Wang, K. X., Gao, X. T., Zuo, P. J., Sun, K. N., & Zhu, X. D. (2021). Cobalt-iron oxide nanotubes decorated with polyaniline as advanced cathode hosts for Li-S batteries. *Electrochimica Acta*, *390*. Available from https://doi.org/10.1016/j.electacta.2021.138873, http://www.journals.elsevier.com/electrochimica-acta/.

Guo, Q., Li, S., Liu, X., Lu, H., Chang, X., Zhang, H., Zhu, X., Xia, Q., Yan, C., & Xia, H. (2020). Ultrastable sodium−sulfur batteries without polysulfides formation using slit ultramicropore carbon carrier. *Advanced Science*, *7*(11). Available from https://doi.org/10.1002/advs.201903246.

Guo, Y., Jin, H., Qi, Z., Hu, Z., Ji, H., & Wan, L. J. (2019). Carbonized-MOF as a sulfur host for aluminum−sulfur batteries with enhanced capacity and cycling life. *Advanced Functional Materials*, *29*(7). Available from https://doi.org/10.1002/adfm.201807676, http://onlinelibrary.wiley.com/journal/10.1002/(ISSN)1616-3028.

Han, X., Cai, J., Wang, X., Liu, Y., Zhou, H., & Meng, X. (2021). Understanding effects of conductive additives in lithium-sulfur batteries. *Materials Today Communications*, *26*. Available from https://doi.org/10.1016/j.mtcomm.2020.101934.

He, Y., Chang, Z., Wu, S., Qiao, Y., Bai, S., Jiang, K., He, P., & Zhou, H. (2018). Simultaneously inhibiting lithium dendrites growth and polysulfides shuttle by a flexible MOF-based membrane in Li−S batteries. *Advanced Energy Materials*, *8*(34). Available from https://doi.org/10.1002/aenm.201802130.

Hu, X., Huang, T., Zhang, G., Lin, S., Chen, R., Chung, L.-H., & He, J. (2023). Metal-organic framework-based catalysts for lithium-sulfur batteries. *Coordination Chemistry Reviews*, *475*. Available from https://doi.org/10.1016/j.ccr.2022.214879.

Huang, J. Q., Chong, W. G., Zhang, B., & Ma, X. (2021). Advances in multi-functional flexible interlayers for Li−S batteries and metal-based batteries. *Materials Today Communications*, *28*. Available from https://doi.org/10.1016/j.mtcomm.2021.102566, http://www.journals.elsevier.com/materials-today-communications/.

Huang, L., Guan, T., Su, H., Zhong, Y., Cao, F., Zhang, Y., Xia, X., Wang, X., Bao, N., & Tu, J. (2022). Synergistic interfacial bonding in reduced graphene oxide fiber cathodes containing polypyrrole@sulfur nanospheres for flexible energy storage. *Angewandte Chemie International Edition*, *61*(44). Available from https://doi.org/10.1002/anie.202212151.

Huang, S., Guan, R., Wang, S., Xiao, M., Han, D., Sun, L., & Meng, Y. (2019). Polymers for high performance Li-S batteries: Material selection and structure design. *Progress in Polymer Science*, *89*, 19−60. Available from https://doi.org/10.1016/j.progpolymsci.2018.09.005.

Hueso, K. B., Armand, M., & Rojo, T. (2013). High temperature sodium batteries: Status, challenges and future trends. *Energy & Environmental Science*, *6*(3). Available from https://doi.org/10.1039/c3ee24086j.

Hwang, J. Y., Kim, H. M., Yoon, C. S., & Sun, Y. K. (2018). Toward high-safety potassium-sulfur batteries using a potassium polysulfide catholyte and metal-free anode. *ACS Energy Letters*, *3*(3), 540−541. Available from https://doi.org/10.1021/acsenergylett.8b00037, http://pubs.acs.org/journal/aelccp.

Iqbal, N., Ghani, U., Liao, W., He, X., Lu, Y., Wang, Z., & Li, T. (2022). Synergistically engineered 2D MXenes for metal-ion/Li−S batteries: Progress and outlook. *Materials Today Advances*, *16*. Available from https://doi.org/10.1016/j.mtadv.2022.100303.

Jeon, T., Lee, Y. C., Hwang, J. Y., Choi, B. C., Lee, S., & Jung, S. C. (2021). Strong lithium-polysulfide anchoring effect of amorphous carbon for lithium−sulfur batteries. *Current Applied Physics*, *22*, 94−103. Available from https://doi.org/10.1016/j.cap.2020.11.004, http://www.elsevier.com/.

Kamisan, A. I., Tunku Kudin, T. I., Kamisan, A. S., Che Omar, A. F., Mohamad Taib, M. F., Hassan, O. H., Ali, A. M. M., & Yahya, M. Z. A. (2022). Recent advances on graphene-based materials as cathode materials in lithium-sulfur batteries. *International Journal of Hydrogen Energy*, *47*(13), 8630−8657. Available from https://doi.org/10.1016/j.ijhydene.2021.12.166, http://www.journals.elsevier.com/international-journal-of-hydrogen-energy/.

Kang, H. J., Park, J. W., Hwang, H. J., Kim, H., Jang, K. S., Ji, X., Kim, H. J., Im, W. B., & Jun, Y. S. (2021). Electrocatalytic and stoichiometric reactivity of 2D layered siloxene for high-energy-dense lithium−sulfur batteries. *Carbon Energy*, *3*(6), 976−990. Available from https://doi.org/10.1002/cey2.152, onlinelibrary.wiley.com/journal/26379368.

Kazda, T., Capková, D., Jaššo, K., Fedorková Straková, A., Shembel, E., Markevich, A., & Sedlaříková, M. (2021). Carrageenan as an ecological alternative of polyvinylidene difluoride binder for Li-S batteries. *Materials*, *14*(19). Available from https://doi.org/10.3390/ma14195578.

Kazda, T., Čudek, P., Vondrák, J., Sedlaříková, M., Tichý, J., Slávik, M., Fafilek, G., & Čech, O. (2018). Lithium-sulphur batteries based on biological 3D structures. *Journal of Solid State Electrochemistry*, *22*(2), 537−546. Available from https://doi.org/10.1007/s10008-017-3791-0.

Király, N., Capková, D., Almáši, M., Kazda, T., Čech, O., Čudek, P., Fedorková, A. S., Lisnichuk, M., Meynen, V., & Zeleňák, V. (2022). Post-synthetically modified metal-porphyrin framework GaTCPP for carbon dioxide adsorption and energy storage in Li-S batteries. *RSC Advances*, *12*(37), 23989−24002. Available from https://doi.org/10.1039/d2ra03301a, http://pubs.rsc.org/en/journals/journal/ra.

Király, N., Capková, D., Gyepes, R., Vargová, N., Kazda, T., Bednarčík, J., Yudina, D., Zelenka, T., Čudek, P., Zeleňák, V., Sharma, A., Meynen, V., Hornebecq, V., Straková Fedorková, A., & Almáši, M. (2023). Sr(II) and Ba(II) alkaline earth metal−organic frameworks (AE-MOFs) for selective gas adsorption, energy storage, and environmental application. *Nanomaterials*, *13*(2). Available from https://doi.org/10.3390/nano13020234.

Kukkar, P., Kim, K.-H., Kukkar, D., & Singh, P. (2021). Recent advances in the synthesis techniques for zeolitic imidazolate frameworks and their sensing applications. *Coordination Chemistry Reviews*, *446*. Available from https://doi.org/10.1016/j.ccr.2021.214109.

Kumaresan, K., Mikhaylik, Y., & White, R. E. (2008). A Mathematical model for a lithium−sulfur cell. *Journal of The Electrochemical Society*, *155*(8). Available from https://doi.org/10.1149/1.2937304.

Kummer, J.T., & Weber, N. (1967). A sodium-sulfur secondary battery. SAE Technical Papers, SAE International, United States. http://papers.sae.org/, 10.4271/670179.

Kwon, S., Song, H., Çakmakçı, N., & Jeong, Y. (2021). A practical approach to design sulfur host material for lithium-sulfur batteries based on electrical conductivity and pore structure. *Materials Today Communications*, *27*. Available from https://doi.org/10.1016/j.mtcomm.2021.102309.

Lang, S., Yu, S. H., Feng, X., Krumov, M. R., & Abruña, H. D. (2022). Understanding the lithium−sulfur battery redox reactions via operando confocal Raman microscopy. *Nature Communications*, *13*(1). Available from https://doi.org/10.1038/s41467-022-32139-w, http://www.nature.com/ncomms/index.html.

Lei, T., Hu, Y., Chen, W., Lv, W., Jiao, Y., Wang, X., Lv, X., Yan, Y., Huang, J., Chu, J., Yan, C., Wu, C., Wang, X., He, W., & Xiong, J. (2020). Genetic engineering of porous sulfur species with molecular target prevents host passivation in lithium sulfur batteries. *Energy Storage Materials*, *26*, 65−72. Available from https://doi.org/10.1016/j.ensm.2019.12.036.

Li, C., Xi, Z., Guo, D., Chen, X., & Yin, L. (2018). Chemical immobilization effect on lithium polysulfides for lithium-sulfur batteries. *Small*, *14*(4). Available from https://doi.org/10.1002/smll.201701986.

Li, S. Y., Wang, W. P., Duan, H., & Guo, Y. G. (2018). Recent progress on confinement of polysulfides through physical and chemical methods. *Journal of Energy Chemistry, 27* (6), 1555−1565. Available from https://doi.org/10.1016/j.jechem.2018.04.014. Available from:, elsevier.com/journals/journal-of-energy-chemistry/2095-4956.

Li, W., Cheng, S., Wang, J., Qiu, Y., Zheng, Z., Lin, H., Nanda, S., Ma, Q., Xu, Y., Ye, F., Liu, M., Zhou, L., & Zhang, Y. (2016). Synthesis, crystal structure, and electrochemical properties of a simple magnesium electrolyte for magnesium/sulfur batteries. *Angewandte Chemie, 128*(22), 6516−6520. Available from https://doi.org/10.1002/ange.201600256.

Li, W., Zhang, Q., Zheng, G., Seh, Z. W., Yao, H., & Cui, Y. (2013). Understanding the role of different conductive polymers in improving the nanostructured sulfur cathode performance. *Nano Letters, 13*(11), 5534−5540. Available from https://doi.org/10.1021/nl403130h.

Li, Z., Zeng, Q., Yu, Y., Liu, Y., Chen, A., Guan, J., Wang, H., Liu, W., Liu, X., Liu, X., & Zhang, L. (2023). Application of transition metal boride nanosheet as sulfur host in high loading Li-S batteries. *Chemical Engineering Journal, 452.* Available from https://doi.org/10.1016/j.cej.2022.139366.

Liang, X., Garsuch, A., & Nazar, L. F. (2015). Sulfur cathodes based on conductive MXene nanosheets for high-performance lithium−sulfur batteries. *Angewandte Chemie, 127*(13), 3979−3983. Available from https://doi.org/10.1002/ange.201410174.

Liang, Y., Dong, H., Aurbach, D., & Yao, Y. (2020). Current status and future directions of multivalent metal-ion batteries. *Nature Energy, 5*(9), 646−656. Available from https://doi.org/10.1038/s41560-020-0655-0.

Lin, H., Yang, D. D., Lou, N., Zhu, S. G., & Li, H. Z. (2019). Functionalized titanium nitride-based MXenes as promising host materials for lithium-sulfur batteries: A first principles study. *Ceramics International, 45*(2), 1588−1594. Available from https://doi.org/10.1016/j.ceramint.2018.10.033.

Lin, T., Tang, Y., Wang, Y., Bi, H., Liu, Z., Huang, F., Xie, X., & Jiang, M. (2013). Scotch-tape-like exfoliation of graphite assisted with elemental sulfur and graphene−sulfur composites for high-performance lithium-sulfur batteries. *Energy & Environmental Science, 6*(4). Available from https://doi.org/10.1039/c3ee24324a.

Liu, Y., Liu, X., Qiu, W., Song, Y., Yang, J., & Zhang, Y. (2022). Synthesis and electrochemical analysis of S/ZIF-67@rGO composite cathodes for lithium-sulfur batteries. *Journal of Nanoparticle Research, 24*(7). Available from https://doi.org/10.1007/s11051-022-05519-y.

Liu, Y., Wang, W., Wang, J., Zhang, Y., Zhu, Y., Chen, Y., Fu, L., & Wu, Y. (2018). Sulfur nanocomposite as a positive electrode material for rechargeable potassium−sulfur batteries. *Chemical Communications, 54*(18), 2288−2291. Available from https://doi.org/10.1039/C7CC09913D.

Manthiram, A., Fu, Y., & Su, Y. S. (2013). Challenges and prospects of lithium-sulfur batteries. *Accounts of Chemical Research, 46*(5), 1125−1134. Available from https://doi.org/10.1021/ar300179v.

Mačák, M., Jasso, K., Vyroubal, P., Kazda, T., & Cudek, P. (2021). Numerical investigation of cathode structure influence on electrochemical behavior of lithium-sulfur battery. *ECS Transactions, 105*(1), 617−625. Available from https://doi.org/10.1149/10501.0617ecst.

Mačák, M., Kazda, T., Jasso, K., & Vyroubal, P. (2021). Equivalent circuit modelling of Li-S batteries. *ECS Transactions, 105*(1), 609−616. Available from https://doi.org/10.1149/10501.0609ecst.

Mačák, M., Vyroubal, P., Kazda, T., & Jašso, K. (2020). Numerical investigation of lithium-sulfur batteries by cyclic voltammetry. *Journal of Energy Storage, 27.* Available from https://doi.org/10.1016/j.est.2019.101158.

Mikhaylik, Y. V., & Akridge, J. R. (2003). Low temperature performance of Li/S batteries. *Journal of the Electrochemical Society, 150*(3), A306−A311. Available from https://doi.org/10.1149/1.1545452.

Mikhaylik, Y. V., & Akridge, J. R. (2004). Polysulfide shuttle study in the Li/S battery system. *Journal of the Electrochemical Society, 151*(11), A1969−A1976. Available from https://doi.org/10.1149/1.1806394.

Mujtaba, A., Jena, P. K., Bekun, F. V., & Sahu, P. K. (2022). Symmetric and asymmetric impact of economic growth, capital formation, renewable and non-renewable energy consumption on environment in OECD countries. *Renewable and Sustainable Energy Reviews, 160*. Available from https://doi.org/10.1016/j.rser.2022.112300, https://www.journals.elsevier.com/renewable-and-sustainable-energy-reviews.

Muldoon, J., Bucur, C. B., & Gregory, T. (2014). Quest for nonaqueous multivalent secondary batteries: Magnesium and beyond. *Chemical Reviews, 114*(23), 11683−11720. Available from https://doi.org/10.1021/cr500049y, http://pubs.acs.org/journal/chreay.

Naguib, M., Kurtoglu, M., Presser, V., Lu, J., Niu, J., Heon, M., Hultman, L., Gogotsi, Y., & Barsoum, M. W. (2011). Two-dimensional nanocrystals produced by exfoliation of Ti 3AlC 2. *Advanced Materials, 23*(37), 4248−4253. Available from https://doi.org/10.1002/adma.201102306.

Pang, Q., Liang, X., Kwok, C. Y., & Nazar, L. F. (2015). Review-the importance of chemical interactions between sulfur host materials and lithium polysulfides for advanced lithium-sulfur batteries. *Journal of the Electrochemical Society, 162*(14), A2567−A2576. Available from https://doi.org/10.1149/2.0171514jes, http://jes.ecsdl.org/content/by/year.

Peng, H. J., & Zhang, Q. (2015). Designing host materials for sulfur cathodes: From physical confinement to surface chemistry. *Angewandte Chemie—International Edition, 54*(38), 11018−11020. Available from https://doi.org/10.1002/anie.201505444, http://onlinelibrary.wiley.com/journal/10.1002/(ISSN)1521-3773.

Qiu, Y., Li, W., Zhao, W., Li, G., Hou, Y., Liu, M., Zhou, L., Ye, F., Li, H., Wei, Z., Yang, S., Duan, W., Ye, Y., Guo, J., & Zhang, Y. (2014). High-rate, ultralong cycle-life lithium/sulfur batteries enabled by nitrogen-doped graphene. *Nano Letters, 14*(8), 4821−4827. Available from https://doi.org/10.1021/nl5020475.

Ramezanitaghartapeh, M., Achazi, A. J., Soltani, A., Miró, P., Kaghazchi, P., Mahon, P. J., Hollenkamp, A. F., & Musameh, M. (2022). Sustainable cyanide-C60 fullerene cathode to suppress the lithium polysulfides in a lithium-sulfur battery. *Sustainable Materials and Technologies, 32*. Available from https://doi.org/10.1016/j.susmat.2022.e00403, https://www.journals.elsevier.com/sustainable-materials-and-technologies.

Razzaq, A. A., Yao, Y., Shah, R., Qi, P., Miao, L., Chen, M., Zhao, X., Peng, Y., & Deng, Z. (2019). High-performance lithium sulfur batteries enabled by a synergy between sulfur and carbon nanotubes. *Energy Storage Materials, 16*, 194−202. Available from https://doi.org/10.1016/j.ensm.2018.05.006.

Roy, K., Banerjee, A., & Ogale, S. (2022). Search for new anode materials for high performance Li-ion batteries. *ACS Applied Materials & Interfaces, 14*(18), 20326−20348. Available from https://doi.org/10.1021/acsami.1c25262.

See, K. A., Gerbec, J. A., Jun, Y. S., Wudl, F., Stucky, G. D., & Seshadri, R. (2013). A high capacity calcium primary cell based on the Ca-S system. *Advanced Energy Materials, 3*(8), 1056−1061. Available from https://doi.org/10.1002/aenm.201300160.

Shafique, A., Rangasamy, V. S., Vanhulsel, A., Safari, M., Gross, S., Adriaensens, P., Van Bael, M. K., Hardy, A., & Sallard, S. (2021). Dielectric barrier discharge (DBD) plasma coating of sulfur for mitigation of capacity fade in lithium-sulfur batteries. *ACS Applied Materials and Interfaces, 13*(24), 28072−28089. Available from https://doi.org/10.1021/acsami.1c04069, http://pubs.acs.org/journal/aamick.

Shahjalal, M., Roy, P. K., Shams, T., Fly, A., Chowdhury, J. I., Ahmed, M. R., & Liu, K. (2022). A review on second-life of Li-ion batteries: Prospects, challenges, and issues. *Energy*, *241*. Available from https://doi.org/10.1016/j.energy.2021.122881, https://www.journals.elsevier.com/energy.

Slater, M. D., Kim, D., Lee, E., & Johnson, C. S. (2013). Sodium-ion batteries. *Advanced Functional Materials*, *23*(8), 947−958. Available from https://doi.org/10.1002/adfm.201200691, http://onlinelibrary.wiley.com/journal/10.1002/(ISSN)1616-3028.

Sun, Z., Vijay, S., Heenen, H. H., Eng, A. Y. S., Tu, W., Zhao, Y., Koh, S. W., Gao, P., Seh, Z. W., Chan, K., & Li, H. (2020). Catalytic polysulfide conversion and physiochemical confinement for lithium−sulfur batteries. *Advanced Energy Materials*, *10* (22). Available from https://doi.org/10.1002/aenm.201904010, http://onlinelibrary. wiley.com/journal/10.1002/(ISSN)1614-6840.

Tang, T., & Hou, Y. (2020). Chemical confinement and utility of lithium polysulfides in lithium sulfur batteries. *Small Methods*, *4*(6). Available from https://doi.org/10.1002/smtd.201900001. Available from:, onlinelibrary.wiley.com/journal/23669608.

Tatar, R. C., & Rabii, S. (1982). Electronic properties of graphite: A unified theoretical study. *Physical Review B*, *25*(6), 4126−4141. Available from https://doi.org/10.1103/physrevb.25.4126.

Urbankowski, P., Anasori, B., Hantanasirisakul, K., Yang, L., Zhang, L., Haines, B., May, S. J., Billinge, S. J. L., & Gogotsi, Y. (2017). 2D molybdenum and vanadium nitrides synthesized by ammoniation of 2D transition metal carbides (MXenes). *Nanoscale*, *9* (45), 17722−17730. Available from https://doi.org/10.1039/c7nr06721f, http://www.rsc.org/publishing/journals/NR/Index.asp.

Wang, H., Liu, Y., Li, Y., & Cui, Y. (2019). Lithium metal anode materials design: Interphase and host. *Electrochemical Energy Reviews*, *2*(4), 509−517. Available from https://doi.org/10.1007/s41918-019-00054-2.

Wang, H., Yang, Y., Liang, Y., Robinson, J. T., Li, Y., Jackson, A., Cui, Y., & Dai, H. (2011). Graphene-wrapped sulfur particles as a rechargeable lithium-sulfur battery cathode material with high capacity and cycling stability. *Nano Letters*, *11*(7), 2644−2647. Available from https://doi.org/10.1021/nl200658a.

Wang, J., Chen, J., Konstantinov, K., Zhao, L., Ng, S. H., Wang, G. X., Guo, Z. P., & Liu, H. K. (2006). Sulphur-polypyrrole composite positive electrode materials for rechargeable lithium batteries. *Electrochimica Acta*, *51*(22), 4634−4638. Available from https://doi.org/10.1016/j.electacta.2005.12.046.

Wang, L., Zhang, T., Yang, S., Cheng, F., Liang, J., & Chen, J. (2013). A quantum-chemical study on the discharge reaction mechanism of lithium-sulfur batteries. *Journal of Energy Chemistry*, *22*(1), 72−77. Available from https://doi.org/10.1016/s2095-4956(13)60009-1.

Wang, N., Wang, Y., Bai, Z., Fang, Z., Zhang, X., Xu, Z., Ding, Y., Xu, X., Du, Y., Dou, S., & Yu, G. (2020). High-performance room-temperature sodium−sulfur battery enabled by electrocatalytic sodium polysulfides full conversion. *Energy & Environmental Science*, *13*(2), 562−570. Available from https://doi.org/10.1039/c9ee03251g.

Wang, Z., Yu, K., Feng, Y., Qi, R., Ren, J., & Zhu, Z. (2019). VO 2 (p)-V 2 C (MXene) grid structure as a lithium polysulfide catalytic host for high-performance Li−S battery. *ACS Applied Materials & Interfaces*, *11*(47), 44282−44292. Available from https://doi.org/10.1021/acsami.9b15586.

Wen, X., Xiang, K., Zhu, Y., Xiao, L., Liao, H., Chen, W., Chen, X., & Chen, H. (2020). 3D hierarchical nitrogen-doped graphene/CNTs microspheres as a sulfur host for high-performance lithium-sulfur batteries. *Journal of Alloys and Compounds*, *815*. Available from https://doi.org/10.1016/j.jallcom.2019.152350.

Wu, Z., Wang, L., Chen, S., Zhu, X., Deng, Q., Wang, J., Zeng, Z., & Deng, S. (2021). Facile and low-temperature strategy to prepare hollow ZIF-8/CNT polyhedrons as high-performance lithium-sulfur cathodes. *Chemical Engineering Journal, 404*. Available from https://doi.org/10.1016/j.cej.2020.126579.

Xia, G., Ye, J., Zheng, Z., Li, X., Chen, C., & Hu, C. (2021). Catalytic FeP decorated carbon black as a multifunctional conducting additive for high-performance lithium-sulfur batteries. *Carbon, 172*, 96−105. Available from https://doi.org/10.1016/j.carbon.2020.09.094.

Xiao, F., Wang, H., Yao, T., Zhao, X., Yang, X., Yu, D. Y. W., & Rogach, A. L. (2021). MOF-derived CoS2/N-doped carbon composite to induce short-chain sulfur molecule generation for enhanced sodium-sulfur battery performance. *ACS Applied Materials and Interfaces, 13*(15), 18010−18020. Available from https://doi.org/10.1021/acsami.1c02301, http://pubs.acs.org/journal/aamick.

Xiao, L., Cao, Y., Xiao, J., Schwenzer, B., Engelhard, M. H., Saraf, L. V., Nie, Z., Exarhos, G. J., & Liu, J. (2012). A soft approach to encapsulate sulfur: Polyaniline nanotubes for lithium-sulfur batteries with long cycle life. *Advanced Materials, 24*(9), 1176−1181. Available from https://doi.org/10.1002/adma.201103392.

Xin, P., Jin, B., Li, H., Lang, X., Yang, C., Gao, W., Zhu, Y., Zhang, W., Dou, S., & Jiang, Q. (2017). Facile synthesis of sulfur−polypyrrole as cathodes for lithium−sulfur batteries. *ChemElectroChem, 4*(1), 115−121. Available from https://doi.org/10.1002/celc.201600479.

Xu, H., Hu, R., Zhang, Y., Yan, H., Zhu, Q., Shang, J., Yang, S., & Li, B. (2021). Nano high-entropy alloy with strong affinity driving fast polysulfide conversion towards stable lithium sulfur batteries. *Energy Storage Materials, 43*, 212−220. Available from https://doi.org/10.1016/j.ensm.2021.09.003.

Xu, J., Lawson, T., Fan, H., Su, D., & Wang, G. (2018). Updated metal compounds (MOFs, −S, −OH, −N, −C) used as cathode materials for lithium−sulfur batteries. *Advanced Energy Materials, 8*(10). Available from https://doi.org/10.1002/aenm.201702607.

Xu, W., Wang, J., Ding, F., Chen, X., Nasybulin, E., Zhang, Y., & Zhang, J.-G. (2014). Lithium metal anodes for rechargeable batteries. *Energy Environ. Sci., 7*(2), 513−537. Available from https://doi.org/10.1039/c3ee40795k.

Xue, W., Miao, L., Qie, L., Wang, C., Li, S., Wang, J., & Li, J. (2017). Gravimetric and volumetric energy densities of lithium-sulfur batteries. *Current Opinion in Electrochemistry, 6*(1), 92−99. Available from https://doi.org/10.1016/j.coelec.2017.10.007.

Yan, J., Liu, X., & Li, B. (2016). Capacity fade analysis of sulfur cathodes in lithium−sulfur batteries. *Advanced Science, 3*(12). Available from https://doi.org/10.1002/advs.201600101.

Yan, Y., Cheng, C., Zhang, L., Li, Y., & Lu, J. (2019). Deciphering the reaction mechanism of lithium−sulfur batteries by in situ/operando synchrotron-based characterization techniques. *Advanced Energy Materials, 9*(18). Available from https://doi.org/10.1002/aenm.201900148.

Yang, H., Li, H., Li, J., Sun, Z., He, K., Cheng, H. -M., & Li, F. (2019). The rechargeable aluminum battery: Opportunities and challenges. *Angewandte Chemie International Edition, 58*(35), 11978−11996. Available from https://doi.org/10.1002/anie.201814031.

Yang, J., Wang, G., Paula Teixeira, A., Goulart Silva, G., Hansen, Z., Jamal M Jamal, M., Mathew, K., Xiong, J., Zhou, T., Mackowiak, M., Dan Fleming, P., & Wu, Q. (2022). A biomass-based cathode for long-life lithium-sulfur batteries. *Electrochemistry Communications, 140*. Available from https://doi.org/10.1016/j.elecom.2022.107325.

Yang, Y., Yang, H., Wang, X., Bai, Y., & Wu, C. (2022). Multivalent metal−sulfur batteries for green and cost-effective energy storage: Current status and challenges. *Journal of Energy Chemistry, 64*, 144−165. Available from https://doi.org/10.1016/j.jechem.2021.04.054.

Ye, H., & Li, Y. (2022). Room-temperature metal−sulfur batteries: What can we learn from lithium−sulfur? *InfoMat*, *4*(5). Available from https://doi.org/10.1002/inf2.12291.

Ye, X., Ruan, J., Pang, Y., Yang, J., Liu, Y., Huang, Y., & Zheng, S. (2021). Enabling a stable room-temperature sodium−sulfur battery cathode by building heterostructures in multichannel carbon fibers. *ACS Nano*, *15*(3), 5639−5648. Available from https://doi.org/10.1021/acsnano.1c00804.

Yu, D., & Lin, F. (2022). MOF derived Co-N-C/CNT crosslinked nets as enhanced carriers for Li-S batteries. *Chemical Physics Letters*, *789*. Available from https://doi.org/10.1016/j.cplett.2021.139327.

Yu, X., Boyer, M. J., Hwang, G. S., & Manthiram, A. (2019). Toward a reversible calcium-sulfur battery with a lithium-ion mediation approach. *Advanced Energy Materials*, *9*(14). Available from https://doi.org/10.1002/aenm.201803794, http://onlinelibrary.wiley.com/journal/10.1002/(ISSN)1614-6840.

Yu, X., & Manthiram, A. (2016). Performance enhancement and mechanistic studies of magnesium−sulfur cells with an advanced cathode structure. *ACS Energy Letters*, *1*(2), 431−437. Available from https://doi.org/10.1021/acsenergylett.6b00213.

Yu, X., & Manthiram, A. (2020). A progress report on metal−sulfur batteries. *Advanced Functional Materials*, *30*(39). Available from https://doi.org/10.1002/adfm.202004084.

Yu, X., Yang, Y., Si, L., Cai, J., Lu, X., & Sun, Z. (2022). V4C3TX MXene: First-principles computational and separator modification study on immobilization and catalytic conversion of polysulfide in Li-S batteries. *Journal of Colloid and Interface Science*, *627*, 992−1002. Available from https://doi.org/10.1016/j.jcis.2022.07.082, http://www.elsevier.com/inca/publications/store/6/2/2/8/6/1/index.htt.

Zelenka, T.Simanova, K., Saini, R., Zelenkova, G., Nehra, S. P., Sharma, A., & Almasi, M. (2022). Carbon dioxide and hydrogen adsorption study on surface-modified HKUST-1 with diamine/triamine. *Scientific Reports*, *12*(1), 17366. Available from https://doi.org/10.1038/s41598-022-22273-2.

Zelenka, T.Baláž, M., Férová, M., Diko, P., Bednarčík, J., Királyová, A., Zauška, Ľ., Bureš, R., Sharda, P., Király, N., Badač, A., Vyhlídalová, J., Želinská, M., & Almáši, M. (2024). The influence of HKUST-1 and MOF-76 hand grinding/mechanical activation on stability, particle size, textural properties and carbon dioxide sorption. *Scientific reports*, *14*(1), 15386. Available from https://doi.org/10.1038/s41598-024-66432-z.

Zeng, S., Peng, J., Liang, X., Wu, X., Zheng, H., Zhong, H., Guo, T., Luo, S., Hong, J., Li, Y., Wu, Q., & Xu, W. (2022). Combined physical confinement and chemical adsorption on co-doped hollow TiO 2 for long-term cycle lithium−sulfur batteries. *Nanoscale*, *14*(26), 9401−9408. Available from https://doi.org/10.1039/d2nr01815b.

Zeng, S. Z., Yao, Y., Tu, J., Tian, Z., Zou, J., Zeng, X., Zhu, H., Kong, L. B., & Han, P. (2021). Improving the low-rate stability of lithium-sulfur battery through the coating of conductive polymer. *IONICS*, *27*(9), 3887−3893. Available from https://doi.org/10.1007/s11581-021-04164-0, http://www.springerlink.com/content/120106/.

Zhang, B. -W., Sheng, T., Wang, Y. -X., Chou, S., Davey, K., Dou, S. -X., & Qiao, S. -Z. (2019). Long-life room-temperature sodium−sulfur batteries by virtue of transition-metal-nanocluster−sulfur interactions. *Angewandte Chemie*, *131*(5), 1498−1502. Available from https://doi.org/10.1002/ange.201811080.

Zhang, D., Wang, S., Hu, R., Gu, J., Cui, Y., Li, B., Chen, W., Liu, C., Shang, J., & Yang, S. (2020). Catalytic conversion of polysulfides on single atom zinc implanted MXene toward high-rate lithium−sulfur batteries. *Advanced Functional Materials*, *30* (30). Available from https://doi.org/10.1002/adfm.202002471.

Zhang, F., Zhou, Y., Zhang, Y., Li, D., & Huang, Z. (2020). Facile synthesis of sulfur@titanium carbide Mxene as high performance cathode for lithium-sulfur batteries. *Nanophotonics*, *9*(7), 2025−2032. Available from https://doi.org/10.1515/nanoph-2019-0568.

Zhang, H., Wang, S., Wang, Y., Huang, B., Dai, Y., & Wei, W. (2021). Borophosphene: A potential anchoring material for lithium-sulfur batteries. *Applied Surface Science, 562.* Available from https://doi.org/10.1016/j.apsusc.2021.150157, http://www.journals.elsevier.com/applied-surface-science/.

Zhang, K., Xie, K., Yuan, K., Lu, W., Hu, S., Wei, W., Bai, M., & Shen, C. (2017). Enabling effective polysulfide trapping and high sulfur loading via a pyrrole modified graphene foam host for advanced lithium—sulfur batteries. *Journal of Materials Chemistry A, 5*(16), 7309—7315. Available from https://doi.org/10.1039/C7TA00445A.

Zhang, S. S. (2013). Liquid electrolyte lithium/sulfur battery: Fundamental chemistry, problems, and solutions. *Journal of Power Sources, 231,* 153—162. Available from https://doi.org/10.1016/j.jpowsour.2012.12.102.

Zhao, Q., Hu, Y., Zhang, K., & Chen, J. (2014). Potassium—sulfur batteries: A new member of room-temperature rechargeable metal—sulfur batteries. *Inorganic Chemistry, 53*(17), 9000—9005. Available from https://doi.org/10.1021/ic500919e.

Zhao, Q., Zhu, Q., Liu, Y., & Xu, B. (2021). Status and prospects of MXene-based lithium—sulfur batteries. *Advanced Functional Materials, 31*(21). Available from https://doi.org/10.1002/adfm.202100457.

Zhao, Z., Wang, S., Liang, R., Li, Z., Shi, Z., & Chen, G. (2014). Graphene-wrapped chromium-MOF(MIL-101)/sulfur composite for performance improvement of high-rate rechargeable Li—S batteries. *J. Mater. Chem. A., 2*(33), 13509—13512. Available from https://doi.org/10.1039/C4TA01241K.

Zhao-Karger, Z., Zhao, X., Wang, D., Diemant, T., Behm, R. J., & Fichtner, M. (2015). Performance improvement of magnesium sulfur batteries with modified non-nucleophilic electrolytes. *Advanced Energy Materials, 5*(3). Available from https://doi.org/10.1002/aenm.201401155, http://onlinelibrary.wiley.com/journal/10.1002/(ISSN)1614-6840.

Zheng, M., Chi, Y., Hu, Q., Tang, H., Jiang, X., Zhang, L., Zhang, S., Pang, H., & Xu, Q. (2019). Carbon nanotube-based materials for lithium—sulfur batteries. *Journal of Materials Chemistry A, 7*(29), 17204—17241. Available from https://doi.org/10.1039/c9ta05347f.

Zhou, C., Wang, X., Zhao, R., Li, L., Li, J., & Wu, F. (2022). MOF-808-derived Ce-doped ZrOF composite as an efficient polysulfide inhibitor for advanced lithium-sulfur batteries. *Journal of Alloys and Compounds, 924.* Available from https://doi.org/10.1016/j.jallcom.2022.166486.

Zhou, W., Yu, Y., Chen, H., Disalvo, F. J., & Abruña, H. D. (2013). Yolk-shell structure of polyaniline-coated sulfur for lithium-sulfur batteries. *Journal of the American Chemical Society, 135*(44), 16736—16743. Available from https://doi.org/10.1021/ja409508q.

Zhou, X., Tian, J., Hu, J., & Li, C. (2018). High rate magnesium—sulfur battery with improved cyclability based on metal—organic framework derivative carbon host. *Advanced Materials, 30*(7). Available from https://doi.org/10.1002/adma.201704166.

Zhu, K., Wang, C., Chi, Z., Ke, F., Yang, Y., Wang, A., Wang, W., & Miao, L. (2019). How far away are lithium-sulfur batteries from commercialization? *Frontiers in Energy Research, 7.* Available from https://doi.org/10.3389/fenrg.2019.00123, http://www.frontiersin.org/Energy_Research/.

CHAPTER 10

Nano/photoelectrochemistry for environmental applications

Zahraa Alqallaf[1], Hamda Bukhatir[1], Fayne D'Souza[1], Abdullah Ali[1], Amani Al-Othman[1] and Muhammad Tawalbeh[2,3]

[1]Department of Chemical and Biological Engineering, American University of Sharjah, Sharjah, United Arab Emirates
[2]Sustainable and Renewable Energy Engineering Department, University of Sharjah, Sharjah, United Arab Emirates
[3]Sustainable Energy & Power Systems Research Centre, RISE, University of Sharjah, Sharjah, United Arab Emirates

10.1 Introduction

In 2015, the United Nations presented its global Sustainability Development Goals (SDGs). These goals comprise 17 overarching goals and 244 targets that are collectively geared toward improving the triple bottom line of sustainability around the world. The environmental aspect of sustainability is one of the focal areas of this agenda and encompasses 169 of the 244 targets (Stafford-Smith et al., 2017; Tawalbeh et al., 2023). Based on this, global fixation, research, and development, across the various engineering disciplines, have focused on the development of new technologies and operational methods that can achieve a greater degree of environmental sustainability. One emerging trend, in this domain of ecofriendly engineering, is the potential of nanoelectrochemistry in the water treatment sector for the purpose of detoxification. Electrochemistry can be defined as a discipline that deals with the physical and chemical alterations in chemical energy storage and conversions (Bueno & Gabrielli, 2009). As elaborated by Bueno and Gabrielli (2009), electrochemistry is divided into two classes that include interfacial electrochemistry and bulk electrochemistry. Interfacial electrochemistry pertains to heterogeneous systems that address subjects relating to the nature of the electrode—electrolyte interphase considering the thermodynamic and kinetic properties of the reactions occurring within that interphase. Bulk electrochemistry, on the other hand, pertains to homogenous systems that address the ion—solvent and ion—ion interactions, ionic conductivity and mobility, and the activity coefficient.

Electrochemistry and Photo-Electrochemistry of Nanomaterials
DOI: https://doi.org/10.1016/B978-0-443-18600-4.00011-9
© 2025 Elsevier Inc. All rights are reserved, including those for text and data mining, AI training, and similar technologies.

333

The incorporation of nanoscience in electrochemistry has ensured the development and furtherance of this exciting field. The inclusion of the prefix "nano" denotes the conduction of these electrochemical processes in the nanoscale ranging from 1 to 100 nm. Nanomaterials such as carbon nanotubes, nanowires, quantum dots, and graphene could be amalgamated with electrochemical materials to enhance their activity and efficiency.

Besides this, one emerging subdiscipline in the more panoptic domain of electrochemistry is that of photoelectrochemistry. As explained by Yu et al. (2020), there has been a growing interest in the integration of photoredox catalysis and synthetic electrochemistry with the objective of optimizing organic synthesis operations. This is achieved via the introduction of a multitude of chemical transformation processes that were thought to be impossible in the past. Both photochemistry and electrochemistry share the common feature of utilizing electrons as reagents for the generation of open-shell radical intermediates. This, in turn, enables many transformations to be translated from electrochemistry to photoredox catalysis and vice versa. Nonetheless, so far, there has been minimal research and development attention directed toward the integration of photochemistry and electrochemistry in the domain of organic synthesis. This constructive merger of the two disciplines in the aforementioned context could generate a scenario in which the flaws of each respective technique can be perfectly compensated for by the core competency of the other. Ultimately, this can aid in the development of novel reaction pathways that were otherwise unachievable, via the utilization of each individual method separately as demonstrated in Fig. 10.1 (Ding et al., 2021). As highlighted earlier, the integration of nanotechnology into their merger has the potential to further proliferate the possible applications of this novel methodology. For example, nanowires, which are one-dimensional nanostructures, possess several unique features for photochemical applications that aid in the improvement of performance across various metrics compared to the bulk alternative (Deng et al., 2019). This innovation of nano/photoelectrochemistry carries potential applications across several fields such as sustainable battery development, wastewater treatment, and organic material preservative waste amongst many others. This chapter will specifically focus on the use of nano/photoelectrochemistry for different environmental applications.

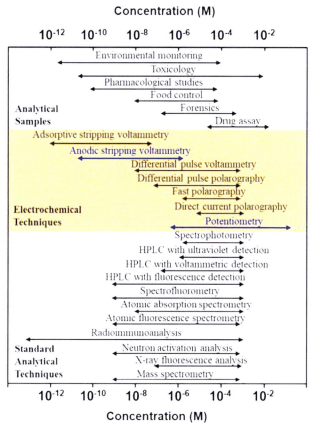

Figure 10.1 Heavy metals detection techniques (Ding et al., 2021).

10.2 Nanomaterials used in environmental applications

10.2.1 Carbon nanomaterials

Carbon nanomaterials have triggered great interest from the academic and industrial sectors due to their innovative applications. Environmentally, carbon nanomaterials offer high affinity, enhanced capacity, and precise selectivity for adsorption purposes. They could be integrated into applications, including water treatment and nanosensors. In the case of water treatment, various carbon nanomaterials, such as the fullerenes from the carbon nanotubes and the graphene family, have demonstrated promising adsorption capabilities. They have been extensively studied for adsorption purposes and other industrial applications due to their structural variety,

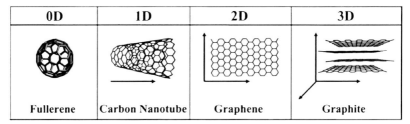

Figure 10.2 Visual representation of carbon nanomaterials structures (Bergmann & Machado, 2015).

large surface area, high chemical stability, and most importantly potential regeneration and reusability (Bergmann & Machado, 2015; Tawalbeh et al., 2022). Fig. 10.2 comparatively depicts the different allotropes of carbon: fullerenes, carbon nanotubes, graphene, and graphite. Fullerenes are characterized by their spherical or ellipsoid structural arrangement of numerous carbons that have high thermal and chemical stabilities and technically great electron acceptors (Bergmann & Machado, 2015). Fullerenes were found to have great sorption capacity for the removal of organic pollutants from solutions. Carbon nanotubes (CNT) are characterized by large pore size, large surface area, and high aspect ratio (Li et al., 2004). Therefore, they were found to be effective for various adsorbates, including organic and inorganic pollutants from water. Graphene family nanomaterials, specifically graphene oxide and graphene nanosheets, have been shown to offer great potential for the removal of toxic contaminants, such as metal ions, dyes, drugs, and pesticides from aquatic systems (Upadhyay et al., 2014; Wang et al., 2013).

10.2.2 Nanocatalysts

The basis for the design of chemical processes investigates the minimization or prevention of pollution at its source before integrating a pollution control system. Therefore, optimization could begin with the development and implementation of improved catalysts. Compared to conventional solid heterogeneous catalysts, nanocatalysts are excellent candidates for process sustainability due to their high surface area. They are characterized by high catalytic activity, selectivity, and stability (Narayan et al., 2019; Ndolomingo et al., 2020). There are a variety of materials lying below nanocatalysts, including magnetic nanocatalysts, nanomixed metal oxides, and core-shell nanocatalysts. Because of their exceptional activity, economic effectiveness, high selectivity, and efficient recovery, magnetic nanocatalysts have found numerous useful applications. For example, nanocrystalline zeolite could be used to catalyze various industrial reactions making the process environmentally friendly (Halalsheh et al., 2022; Khan et al., 2018). To elaborate, the energy consumption would be reduced as oxidation is triggered by visible light. Moreover, wasteful secondary photoreactions are eliminated by visible light and therefore only low-energy reaction paths are triggered, leading to higher yields of the desired product. Among nanomixed metal oxides, nano-TiO_2 is used in the production of photodegradable polymers, therefore, reducing the released toxic byproducts during polymer incineration. For example, a previously conducted study discussed by Khan et al. (2018) explained the suppression of dioxin emission for polyvinylchloride incineration when nano-TiO_2 was integrated into the process. The TiO_2 is low in cost, stable physically and chemically, nontoxic, and has a high efficiency to adsorb several heavy metals with the ability to be regenerated and reused (Al Bsoul et al., 2019).

10.2.3 Modified nanomaterials

The modification of existing materials into nanostructured materials contributes to various applications. The modification is achieved through various routes, mainly by manipulating the material at a nanoscale, such as coating. For example, nanofertilizers are used a substitute for conventional polymer coated fertilizers that are often quite expensive because of high energy requirements during their synthesis. Nanofertilizers were found to be more effective on account of their high surface area to volume ratios (Dambale, 2019; Jakhar et al., 2022). Such nanostructured formulations offer the reduction of lost fertilizer nutrients in the soil. The encapsulation of nanoparticles within fertilizers is achieved through the encapsulation of nutrients inside nanoporous materials, coating a thin film on the fertilizer, or simply delivering particles of nanoscale dimensions (Khan et al., 2018). Another example of modified nanomaterials is the nanoparticles mediated gene transfer that offers a potential green and ecofriendly alternative to conventional pesticides that negatively impact the environment. Such modification is achieved through the formulation of nanocapsules with polymers, nanoemulsions or nanoparticles supported products (metals, metal oxides, and nanoclays) (Khan et al., 2018). Other promising materials are the ones employed in nanomaterial-based biosensors (Mohamed et al., 2021). They offer to enhance the agriculture sector by remote

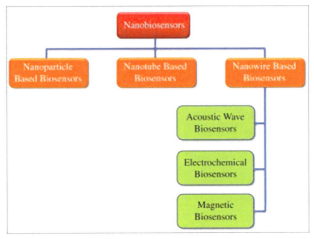

Figure 10.3 Types of nanobiosensors (Bose et al., 2022).

sensing and precisely measuring pollutants' concentration (Pal et al., 2022). Besides that, nanomaterial-based biosensors are used in the detection of heavy metals that pose grave environmental concerns when introduced into the food chain (Bose et al., 2022). Fig. 10.3 illustrates the different types of nanomaterials, such as nanotubes, nanowires, and quantum dots, that could be integrated into biosensors for heavy metal detection (Bose et al., 2022). As a result, this would create biosensors with unique electrical, thermal, fluidic, and magnetic properties. The nanobiosensor is manufactured through a technique called nanofabrication. Nanofabrication incorporates different processes that ultimately merge the conventional chemical and biological sensors in one device resulting in a biosensing process that is considerably easier, quicker, and economical.

10.3 Environmental applications of nanoelectrochemistry

10.3.1 Water splitting

The decomposition of water into hydrogen and oxygen is defined as water splitting. There are numerous pathways of water splitting that are broadly classified under electrolytic, thermal, and photonic headings (Dincer & Acar, 2017). The diversity of the electrolytic processes primarily stems from the type of electrolyte and the separating material, usually membrane, employed. Some of the common electrolytes and membranes, include solid oxide materials, alkaline solutions, proton exchange membranes (PEM), and anion exchange membranes (Mohammed et al., 2019; Tahir & Batool, 2022). The efficiency and cost of the electrolytic methods hinge on the fuel source used to produce electricity that stimulates the redox reaction. The use of electricity renders this method of production $4-10$ times more expensive since the efficiency of H_2 production is 75% (Dincer & Acar, 2015). Thermal methods, on the other hand, work on the principle of utilizing heat to disassociate the water molecules. Temperature ranges dictate the extent of disassociation; a 64% efficiency is achieved for a temperature of 3000°C (Tahir & Batool, 2022). Finally, the photonic processes wield the light energy to drive the splitting process. Examples include photoelectrochemical (PEC) and photobiological (Parthasarathy & Narayanan, 2014). In this chapter, the PEC process will be discussed.

10.3.1.1 Photoelectrochemical water splitting for solar hydrogen production

Hydrogen is popularly considered to be a sustainable fuel source when compared to other fossil fuels because of its ability to burn without harmful emissions. Currently, hydrogen is produced from steam reforming processes (Ahmad et al., 2015). Production through these methods is unsustainable for the environment. Therefore, the production and implementation of hydrogen through water splitting is hitherto the most sustainable method. Amongst the various water splitting, PEC water splitting is effective because of its low cost since it utilizes the absorption of sunlight for the PE to split water into oxygen and hydrogen (Fajrina & Tahir, 2019). Sunlight is an abundant and clean source of energy. Establishing PEC systems that have a zero-carbon footprint are of immense environmental importance. There are myriad of PEC setups that are implemented for hydrogen production. The PEC water splitting using TiO_2 will be discussed in the following sections.

10.3.1.2 Basic fundaments of water splitting

As highlighted previously, the production of hydrogen via water splitting is given special attention because it employs a renewable energy source (Almomani et al., 2023). In the splitting process, semiconductors are utilized because of their ability to absorb photons that are converted into electrochemical energy. Specifically, the semiconductor photocatalysts used can be single-phase photocatalysts or poly phase photocatalysts. The

$TiO_2 \xrightarrow{h\nu} e^-_{TiO_2} + h^+_{TiO_2}$ (1)

$H_2O + 2h^+ \rightarrow 2H^+ + 1/2 O_2$ (2)

$2H^+ + 2e^- \rightarrow H_2$ (3)

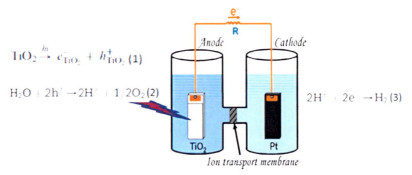

Figure 10.4 Overall water splitting reactions (Ahmad et al., 2015).

TiO_2 is a commonly used photocatalyst in the process for hydrogen production via a redox reaction (Ahmad et al., 2015). As demonstrated in Fig. 10.4, there are three main functional steps to produce hydrogen (Ahmad et al., 2015). The first step is the formation of electron hole pairs when light falls on the anode. Step 2 involves the oxidation of water at the anode to give O_2 and H^+, while the last stage would be the reduction of H^+ at the cathode to form H_2. The basic setup of a PEC device consists of electrodes and an aqueous electrolyte solution. The two electrodes are called counter electrode and photoelectrode. Within the experiment, the PEC absorbs solar energy through the photoelectrode, which is in contact with the electrolyte solution; eventually, producing hydrogen and oxygen upon reaction completion. The uncertainty of the product being either hydrogen or oxygen depends on the type of semiconductor deployed as the electrode within the PEC system (Narayanan et al., 2019). For the photoelectrode, two types of semiconductors are used: n-type semiconductor and p-type semiconductor. When a p-type semiconductor is used, hydrogen is produced at that electrode while oxygen is produced at the counter electrode. Similarly, when the n-type semiconductor is used, hydrogen is produced at the counter electrode while oxygen is produced at the photoelectrode (Narayanan et al., 2019). Carbon nanomaterials can also be employed as photocatalysts for photocatalytic water splitting (Syed et al., 2019). These nanomaterials are carbon nanotubes (CNTs), graphitic carbon nitride, and graphene (Singla et al., 2021). The porous structure of carbon allows organic or inorganic functional groups to be incorporated easily that aids the photocatalytic behavior of carbon.

10.3.1.3 Mechanism of photo electrocatalytic water splitting for H₂ production

Initially, photocatalysis occurs with irradiation of light with energy larger than the band gap of the photocatalyst (Narayanan et al., 2019). Once enough amount of energy is attained by the photocatalyst, an electron transition takes place that generates electron hole pairs, as shown in Eq. (1). After the first step, holes are left in the valence band because of charge separation and excitation of electrons from the valence band to conduction band (Narayanan et al., 2019). The holes and electrons formed will then take place in oxidation and reduction reactions. Water undergoes the oxidation reaction with the holes formed in the previous step, according to Eq. (2). From the oxidation reaction, the H^+ ion is formed on the decomposition of water with the holes. The next step involves the reduction of the H^+ ions into hydrogen, as shown in Eq. (3) (Wang et al., 2020). Fig. 10.5 illustrates a diagrammatic representation of the mechanism explained above (Fajrina & Tahir, 2019).

$$\text{Catalyst} \rightarrow e^- + h^+$$

$$2H_2O + 2h^+ \rightarrow O_2 + 4H^+$$

$$2e^- + 2H^+ \rightarrow H_2$$

10.3.2 Reduction of CO₂ to CH₃OH

The gradual escalation of CO_2 emissions has led to detrimental environmental repercussions over the years. The CO_2 is the primary greenhouse

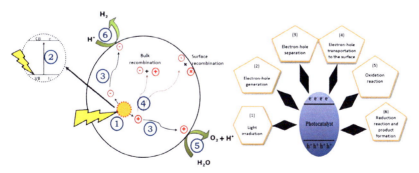

Figure 10.5 Mechanism of photo electrocatalytic water splitting . (Fajrina and Tahir, 2019).

gas present in the atmosphere, and the continuous CO_2 emissions have aggravated global warming and climate change and are expected to worsen in the coming years (Tawalbeh, Al-Ismaily, et al., 2021). Massive research has been conducted to decrease these emissions, capture, and utilize them (Alami et al., 2020). Therefore, establishing methods to reduce CO_2 emissions from the source or convert it to nonhazardous compounds that can be used as energy sources is imperative (Alami et al., 2021). There have been recognized methods such as photocatalytic reduction of CO_2 into hydrocarbons like CH_3OH (Almomani et al., 2019; Dai et al., 2015; Duangkaew et al., 2020; Lais et al., 2018). The general reduction of CO_2 to CH_3OH consists of three steps. First is illuminating the photocatalyst using energy of light, which is larger than the band gap energy of the semiconductor, to evolve electron hole pairs. Second is the generated electrons and holes will be initiated to migrate toward the surface of the semiconductor. Third is the reduction of CO_2 to CH_3OH using the electrons generated in the first step (Bharath et al., 2021). The TiO_2/RGO with Ag nanoparticles was utilized for the photo electrocatalytic reduction in (Bharath et al., 2021). The photo electrocatalyst consists of Pt-modified TiO_2 nanotubes and Pt/RGO. The TiO_2/RGO modified with Ag nanoparticles is used as cathode while Pt/RGO are used as the anode. The TiO_2 is modified with Ag nanoparticles using hydrothermal methods that is followed by microwave irradiation for better CO_2 reduction efficiency. These electrodes are placed within the electrolyte solution. The polymer Nafion was used as a separator in a homemade H-type reactor. Finally, the entire process was actualized in a 1 M KOH aqueous solution, which is used as an electrolyte, under UV light for a duration of 2 hours at a potential of -0.5 V at the cathode. The yield of the produced fuels is a function of potential at the electrodes. Thus, photo electrocatalysis is an attractive approach for the conversion of solar energy into chemical energy courtesy its green and safe nature.

10.3.2.1 Mechanism of photoelectrochemical reduction of CO_2 to CH_3OH

The mechanism of CO_2 reduction to CH_3OH involves both reduction and oxidation reactions at the electrodes. As hinted in the previous section, a proton exchange membrane, Nafion, separates the anodic and the cathodic chambers (Bharath et al., 2021). A platinum sheet is used as the anode, at which, oxidation of water takes place that forms H^+ ions. When light falls on the photocathode, which is Ag modified with $TiO_2/$

RGO, electrons and holes are generated. On formation of the electron and holes, reduction of CO_2 occurs. The Ag acts as separator to separate the electrons formed from the conduction band of TiO_2.

10.3.3 Photoelectrochemical TiO_2 catalyst on pollutant degradation

Water bodies are exponentially becoming polluted due to the industries disposing waste, often toxic, into rivers causing negative environmental effects (Chang et al., 2015). These contaminants often contain pollutants with low biodegradability; therefore, natural treatment processes are not capable of eliminating them. Industries that produce dyes, cosmetics, plastics, and various other pollutants tend to discard their waste into water bodies. The released dyes and cosmetic pollutants could cause carcinogenic and mutagenic effects to aquatic and human life. Dye pollutants are toxic and can cause eutrophication (Berradi et al., 2019). Therefore, elimination or reduction of these pollutants is important for a healthy aquatic life and in turn the environment. A potential solution to mitigate this escalating concern is the PEC oxidation of water using TiO_2 (Chang et al., 2015). As highlighted in the preceding sections, the TiO_2 photocatalyst semiconductor is commonly employed for reduction and oxidation reactions on its surface. For water purification, electrochemical oxidation of titanium in dimethyl sulfoxide is implemented. Undoped TiO_2, exhibits weak response for solar energy utilization. Doping TiO_2 with carbon, nitrogen, and sulfur not only increases response to UV light but also advances photocatalytic activity under visible light, consequently, making it favorable, particularly for low-priced applications (Piątkowska et al., 2021). The experiment conducted by Chang et al. (2015) used RhB as an organic pollutant to assess the response of non-porous TiO_2. This was examined under visible light, solar light, and UV light at an electrode potential of 1 V. It was observed the organic pollutant, RhB, could be more efficiently oxidized using TiO_2 under UV light. One drawback associated with unmodified TiO_2 is its limited application due to its small photocatalytic wavelength (Dong et al., 2015). Another limitation is the low adsorption capacity of TiO_2 toward some organic pollutants. Hydrophobic organic pollutants are unable to adsorb on the surface of TiO_2 (Dong et al., 2015). The low adsorption of these pollutants results in reduced degradation of the pollutant.

Integrating TiO_2 with carbon has exhibited advanced adsorption capabilities; therefore, improving pollutant degradation. The TiO_2 with compositions of different carbon forms such as graphene, activated carbon and carbon nanotubes show greater performance in degradation of organic pollutants (Dong et al., 2015). Graphene has high thermal and electrical conductivity, and good mobility of charge carriers (Gosling et al., 2021; Nauman Javed et al., 2022). Additionally, the high surface area and adsorptive capacity of activated carbon enables the organic pollutants in water to be adsorbed at high efficiencies (Martín-Lara et al., 2020; Tawalbeh et al., 2005). Finally, the large surface area of carbon nanotubes improves the degradation of these organic pollutants (Alves et al., 2022). Hence, the integration of carbon-based nanomaterials into the TiO_2 structure enhances their photocatalytic performance ultimately improving pollutant degradation.

10.3.4 Heavy metals detection

Given their toxic and nonbiodegradable nature, heavy metals are one of the most challenging pollutants to remove from the environment (Shams Jalbani et al., 2021). Such characteristics lead to their bioaccumulation in the ecological systems. Hence, the determination of heavy metals level is of critical importance for environmental quality. If heavy metals enter the environment, they could interfere with the food chain and bio magnify inside animals and human body through the consumption of contaminated bodies. Since there does not exist a natural mechanism for the removal of heavy metals from the body and the environment, even traces of heavy metals can damage human health and the environment. The present standard analytical methods to quantify the levels of heavy metals include cold vapor atomic fluorescence, atomic absorption, and emission spectroscopies (Ding et al., 2021). Such techniques are applicable over multiple samples and offer high accuracy if done in centralized facilities. However, such procedures require complex infrastructures, manpower, and precise sample storage with only trained personnel able to obtain accurate results. Because of the increased cost of sampling procedures available that require accreditation, problems caused by heavy metal pollution are most present in countries with low domestic gross. Fig. 10.1 further elaborates on the techniques used to monitor the levels of heavy metals in different sample arrays indicating that there is no specific universal technique but rather based on concentration of samples examined (Ding et al., 2021). This is vital especially for heavy metals that need to be quantified at trace levels and are likely to be ingested or come in close contact with humans.

A simple, cheap, and fast method for direct monitoring and one that does not require sample pretreatment is potentiometry. Potentiometric sensors work on a wide concentration range (between 10^{-7} and 10^{-1} M). Hence, they could ascertain heavy metal ions in various samples, such as in the industrial wastewater and its sludge (Agoro et al., 2020; Buaisha et al., 2020; Ding et al., 2021; Gupta, 2005; Qasem et al., 2021).

The use of paper as a sample medium in heavy metal sensors has been thoroughly researched in recent years. Electrochemical sensing devices that integrated a paper medium yielded fast response time and high sensitivity (Ding et al., 2021). The cellulose, a major part of the paper, contains multiple carboxyl and hydroxyl groups (Tawalbeh, Rajangam, et al., 2021). Those negatively charged sites provide the area for heavy metals adsorption. In a studied exothermic process as the concentration of the ions increase the adsorption potential also increases. Additionally, the impact of the ion interactions with the paper medium has been studied to determine the preferred properties of the paper. It was discovered that the integrated paper should be hydrophilic, porous, light, and flexible.

The utilization of paper exclusively as a substrate for the sensor has also been investigated (Krikstolaityte et al., 2020). Rather than the electrochemical ion measurement that is performed in milliliters samples scale, wicking the sample through the paper provide the chance to perform the quantifying procedure on microliter volume samples. An important advantage of low-volume sampling is the ability to reduce the consumption of buffers and modification of the electrodes (Ding et al., 2021). Thus, this could increase the lifetime of the reagents that are used in the operation.

Fig. 10.6 shows the mechanism of potentiometry and voltammetry techniques where paper is integrated as a microliter sampling medium (Ding et al., 2021). The driving force is capillary pushing the solution flow

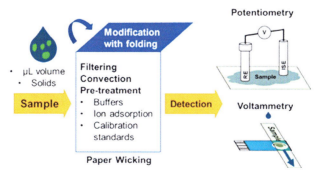

Figure 10.6 Potentiometry and voltammetry with integrated paper medium (Ding et al., 2021).

into the porous of the paper where in interdiffusion mixing takes place (Ding et al., 2021). Electrodes, whether they were glassy carbon or screen-printed, are ion selective and are passed against the surface of the paper for readings. Such a procedure is done under static or dynamic states of the sample. Another critical advantage of using paper as the sample medium is their ability to sample on uneven surfaces as well as disposability that therefore can lower contamination and possible biofouling between different measurements. This also allows the paper to transport the sample around the device making additional tests to be performed simultaneously.

10.3.5 Algal toxins detection and decontamination

To ensure that people have access to safe and clean water, various state of the art technologies has long been designed and the emerging ones are under continuous study. The purification and detection of algal toxins, in aquatic life, is a major issue that is under in-depth investigation. Algal toxins are of a critical concern as they present a threat to human health and ecological system. If found in drinking water, they can trigger tumors and liver cancer under long exposure times. The World Health Organization (WHO) and several agencies as of 2021 have established a minimum threshold for the concentration of algal toxins in drinking water of 1 μg/L (Bilibana et al., 2022). One example of a concerning algal toxin is the MC-LR, consisting of two leucine-arginine (LR) and made of five sustained amino acids. MC-LR is among the various MCs congeners that are deeply studied pollutants in drinking water (Bilibana et al., 2022). Existing detection methods of algal toxins include high-performance liquid chromatography, protein phosphate inhibition assay, and enzyme-linked immunosorbent assay. Such methods are costly given its complex preparation requirement and the time consuming nature of the process. Hence, cheap, portable, and timesaving materials for algal detection and purification should be employed. The sensor in use can be aided with a molecular recognition element (MRE) and thus provide both quality and quantity readings of algal toxins in water. One of the MRE that can accomplish this job is aptamer. Aptamers are nucleic acids that can recognize predetermined algal toxins with precision and high affinity using a technique called Systematic Evolution of Ligands by Exponential Enrichment (SELEX) (Xu et al., 2021). Such a technique has been previously tested in which aptamers were employed against various other environmental pollutants (Piątkowska et al., 2021). Courtesy their unique structure, aptamers have high stability, even under severe conditions. The versatility in structure also allows them to interact with targeted molecules.

Aptamers are used for both detection and decontamination owing to their recognition capacity. Such materials are compatible with hybrid nanomaterials that in turn enrich the structure by providing unique physical and chemical properties. A hybrid nanomaterial is referred to nanomaterials that are made from mixture of organic nanoparticles or inorganic nanoparticles molecules (Ananikov, 2019). Such mixtures propose novel properties advantages of exhibiting large surface area to volume ratios, biocompatibility, and unique shapes, therefore, making them a special candidate for sensing and decontamination techniques. The type of nanomaterial molecules can be chosen from various materials. Most commonly inorganic nanoparticles, such as TiO_2, molybdenum disulfide sheets, quantum dots are used. Aptamer hybrid nanomaterial conjugates can selectively capture specific targets rendering them ideal for sensing and decontamination procedures.

Presently, numerous aptamers have been reported that are capable of rapid detection of algal toxins and, more importantly, efficient algal toxin decontamination in aquatic systems. For sensing applications, the functionalization of aptamer surface is of critical importance, the high surface to volume ratio translates into high avidity against the targets (Dincer & Acar, 2015). Such high avidity ultimately promotes better detection sensitivity of the sensor.

10.3.6 The use of photocatalysts for detoxification

Photocatalysts are materials that can change the rate of reaction without having direct involvement in the chemical transformation process utilizing light to stimulate the process. The main difference between photocatalysts and traditional thermal catalysts is that, while the latter involves the activation by the use of heat, the former relies on photons of light energy (Mohan et al., 2020). The process of photocatalysis comprises two separate processes that are mineralization and degradation of pollutants. Mineralization involves the complete destruction of pollutants into carbon dioxide, water, and other inorganic ions. Degradation, on the other hand, involves the breaking down or decomposition of pollutants into smaller byproducts (Dai et al., 2008). Recently, the emergence of nanotechnology has introduced a new domain of nanocatalysts. The use of nanophotocatalysts in industrial applications, particularly in wastewater purification, is becoming more prevalent due to the ability of these materials to improve the reactivity of conventional catalysts given their relatively larger surface area to volume ratio and shape-dependent characteristics. These nanosized photocatalysts exhibit a unique response relative to bulk material due to their distinctive quantum and surface properties, which have been found to aid in increasing their mechanical, magnetic, chemical, and electrical reactivity (Yaqoob et al., 2020). In the context of wastewater treatment, studies have shown that the use of nanophotocatalysts can improve the oxidation due to their ability to effectively produce oxidizing species at the surface of the material which allows for the degradation of pollutants (Gómez-Pastora et al., 2017). Experimental research has investigated many nanomaterials that can be used in this photocatalysis process, the most common being SiO_2, ZnO, TiO_2, and Al_2O_3. The TiO_2, however, is by far the most common and significant metal oxide that is used for photocatalysis. The prevalent use of this material is due to its low cost, chemical stability, availability, and toxic-free characteristics. The TiO_2 exists in nature in three states that include brookite, rutile, and anatase. Anatase is the most used form in the process of nanophotocatalysis (Zhang et al., 2014). This information thus provides a general overview on the use of nanophotocatalyst in the detoxification of wastewater.

In line with their beneficial characteristics, an experiment was carried out to evaluate the efficacy of TiO_2 in the degradation of endotoxin as well as its antibacterial properties (Sunada et al., 1998). The process of TiO_2 film preparation was carried out using titanium isopropoxide solution annealed at 500°C. The antibacterial effect of the titanium oxide film was evaluated based on the activity of $E.$ $coli$ cells through the calculation of its survival ratio from the number of viable cells, while the concentration of the toxin was determined using a limulus test. The experimental results indicated that the use of TiO_2 in conjunction with exposure to ultraviolet radiation caused both a detoxification impact on the endotoxins and an antibacterial effect on the $E.$ $coli$. The bactericidal effect, in particular, has been found to be caused by the deactivation of the viability of the bacteria as well as the direct destruction of the bacterial cells. In a more practical context, Abdollahizadeh et al. (2021) carried out an experimental study to investigate the impact of nanophotocatalyst in the removal of pharmaceutical containments from water. The removal of pharmaceutical toxins from aqueous solutions can be particularly challenging and exhibit hazardous characteristics in the water solution due to their highly stable nature that makes their degradation difficult (Fanourakis et al., 2020). The experiment conducted by Fanourakis et al. (2020) involved the fabrication of a novel staggered AgX (Br, I)/CoCrNO₃LDH nanophotocatalyst, which were evaluated in terms of their efficacy for their removal of antibiotic containment in a simulated natural sunlight environment. The analysis of the study involved the use of X-ray diffraction (XRD), field emission scanning electron microscope (FESEM), transmission electron microscopy (TEM), and atomic force microscopy (AFM). The results of the study indicated that the use of 1 g/L of the photocatalysts in a weighted ratio of 75:25 leads to the removal of the constituent of the toxins to the degree outlined in Table 10.1 (Fanourakis et al., 2020). This information thus highlights the efficacy in the use of nanophotocatalyst in a practical or industrial setting.

10.3.7 Photochemical enzymatic systems in wastewater

Various physiochemical processes, such as nanofiltration, photolysis, adsorption, advanced oxidization, and reverse osmosis, have been historically practiced in wastewater treatment (Al Bsoul et al., 2021; Al-Bsoul et al., 2020; Tawalbeh et al., 2018). While these processes have exhibited parametric success in a laboratory setting, they nonetheless have their respective economic and environmental limitations for large-scale industrial or municipal wastewater treatment. In response to this limitation, extensive research and development efforts have been allocated toward the development of artificial enzymes or synthetic model enzymes that have robust molecular components. These research efforts thus led to the introduction of photochemical enzyme models. The development of these models is based on the exploitation of the excited state properties and molecular kinetics of low-molecular weight catalysts precursors that enable the creation of light-controlled electronic, energetic, and structural

Table 10.1 % Rates of removal for different antibiotics using 1 g/L of the photocatalysts in a weighted ratio of 75:25 (Fanourakis et al., 2020).

Contaminant	Rate of removal (%)
Tetracycline	95.1
Ofloxacin	81.8
Levofloxacin	81.8
Ciprofloxacin	60.3

changes at a substrate-binding level which can aid in the replacement of the selective and efficient chemistry of biocatalysts (Unuofin et al., 2019).

The existence of a mixture of effluents with heterogeneous physiochemical properties in wastewater poses a challenge. These effluents are typically degradation resistant and can have detrimental effect on the vitality of aquatic life. Unuofin et al. (2019) explained that the enzyme system of laccases possesses catalytic properties that makes it suitable for various environmental cleaning applications. In terms of waste-water treatment, laccase-catalyzed xenobiotic process can be executed in various ways that include the use of crude-free laccases, the use of purified free laccases and

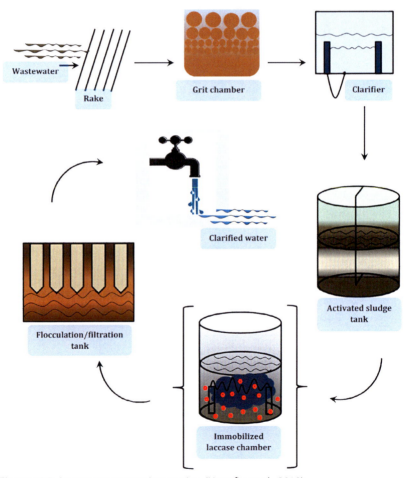

Figure 10.7 Laccase treatment integration (Unuofin et al., 2019).

the use of immobilized crude laccases. To evaluate the efficacy of this enzymatic model in wastewater treatment, Gomaa and Momtaz (2015) carried out an experimental study on wastewater in a textile factory. The findings of the study indicated that the application of laccase individually led to the achievement of a 71% decolorization and further improvement in the process outcomes was achieved using a laccase-cooper compound (Gomaa & Momtaz, 2015). Fig. 10.7 depicts a wastewater treatment schematic that shows how laccase treatment can be integrated (Unuofin et al., 2019).

10.4 Conclusions

The integration of photoelectrochemistry into environmental applications has garnered special attention in the past few decades. Various environmental issues, including pollution treatment, detection of chemicals, and process optimization, could be greatly improved using nanomaterials. Nanomaterials have been advertised as effective substances for the purpose of purification and detection of toxic pollutants. They exhibit novel properties and can function as adsorbents and catalysts for the removal of heavy metals, SO_2, and CO, among many other environmental pollutants. The primary reasons for their excellent performance are their high chemical and thermal stability, high surface area, reactivity, and, most importantly, regeneration ability. Furthermore, depending upon the application desired, they can be effortlessly designed to suit different environments and modified for targeted species. Moreover, their utilization in various environmental applications tackled in the chapter has demonstrated promising results compared to conventional methods. Finally, the integration of nanomaterials in photoelectrochemistry, as a subbranch of electrochemistry, has resulted in enhanced processes or the development of new ones with the advantages of high efficiency and sustainability.

References

Abdollahizadeh, Z., Haghighi, M., & Shabani, M. (2021). Photocatalytic removal of pharmaceutical contaminants from aqueous effluents using staggered AgX(Br, I)/CoCrNO3LDH plasmon nanophotocatalysts under simulated solar-light. *Separation and Purification Technology.*, *278*. Available from https://doi.org/10.1016/j.seppur.2021.119574.

Agoro, M. A., Adeniji, A. O., Adefisoye, M. A., & Okoh, O. O. (2020). Heavy metals in wastewater and sewage sludge from selected municipal treatment plants in eastern cape province, south africa. *Water (Switzerland).*, *12*(10). Available from https://doi.org/10.3390/w12102746, http://www.mdpi.com/journal/water.

Ahmad, H., Kamarudin, S. K., Minggu, L. J., & Kassim, M. (2015). Hydrogen from photo-catalytic water splitting process: A review. *Renewable and Sustainable Energy Reviews.*, *43*, 599−610. Available from https://doi.org/10.1016/j.rser.2014.10.101.

Al Bsoul, A., Hailat, M., Abdelhay, A., Tawalbeh, M., Al-Othman, A., Al-kharabsheh, I. ' N., & Al-Taani, A. A. (2021). Efficient removal of phenol compounds from water environment using Ziziphus leaves adsorbent. *Science of The Total Environment.*, *761*. Available from https://doi.org/10.1016/j.scitotenv.2020.143229.

Al Bsoul, A., Hailat, M., Abdelhay, A., Tawalbeh, M., Jum'h, I., & Bani-Melhem, K. (2019). Treatment of olive mill effluent by adsorption on titanium oxide nanoparticles. *Science of the Total Environment.*, *688*, 1327−1334. Available from https://doi.org/10.1016/j.scitotenv.2019.06.381, http://www.elsevier.com/locate/scitotenv.

Alami, A. H., Hawili, A. A., Tawalbeh, M., Hasan, R., Al Mahmoud, L., Chibib, S., Mahmood, A., Aokal, K., & Rattanapanya, P. (2020). Materials and logistics for carbon dioxide capture, storage and utilization. *Science of The Total Environment.*, *717*. Available from https://doi.org/10.1016/j.scitotenv.2020.137221.

Alami, A. H., Tawalbeh, M., Alasad, S., Ali, M., Alshamsi, M., & Aljaghoub, H. (2021). Cultivation of Nannochloropsis algae for simultaneous biomass applications and carbon dioxide capture. *Energy Sources, Part A: Recovery, Utilization and Environmental Effects.* Available from https://doi.org/10.1080/15567036.2021.1933267, http://www.tandf.co.uk/journals/titles/15567036.asp.

Al-Bsoul, A., Al-Shannag, M., Tawalbeh, M., Al-Taani, A. A., Lafi, W. K., Al-Othman, A., & Alsheyab, M. (2020). Optimal conditions for olive mill wastewater treatment using ultrasound and advanced oxidation processes. *Science of The Total Environment.*, *700*. Available from https://doi.org/10.1016/j.scitotenv.2019.134576.

Almomani, F., Al-Rababah, A., Tawalbeh, M., & Al-Othman, A. (2023). A comprehensive review of hydrogen generation by water splitting using 2D nanomaterials: Photo vs electro-catalysis. *Fuel.*, *332*. Available from https://doi.org/10.1016/j.fuel.2022.125905.

Almomani, F., Bhosale, R., Khraisheh, M., Kumar, A., & Tawalbeh, M. (2019). Photocatalytic conversion of O_2 and H_2O to useful fuels by nanostructured composite catalysis. *Applied Surface Science*, *483*, 363−372. Available from https://doi.org/10.1016/j.apsusc.2019.03.304, http://www.journals.elsevier.com/applied-surface-science/.

Alves D.Cd. S., B.S. de Farias, C. Breslin, L.Ad.A. Pinto, T.R.S.A. Cadaval, Carbon nanotube-based materials for environmental remediation processes. In D. Giannakoudakis, L. Meili, & I. Anastopoulos (Eds.), Advanced materials for sustainable environmental remediation: Terrestrial and aquatic environments. Elsevier, Brazil (2022), 475−513, Available from https://www.sciencedirect.com/book/9780323904858, https://doi.org/10.1016/B978-0-323-90485-8.00017-5.

Ananikov, V. P. (2019). Organic−inorganic hybrid nanomaterials. *Nanomaterials.*, *9*(9). Available from https://doi.org/10.3390/nano9091197.

Bergmann C.P., F.M. Machado, Carbon nanomaterials as adsorbents for environmental and biological applications. In C. P. Bergmann & F. M. Machado (Eds.), Carbon nanostructures. Springer International Publishing, (2015), Available from https://doi.org/10.1007/978-3-319-18875-1.

Berradi, M., Hsissou, R., Khudhair, M., Assouag, M., Cherkaoui, O., El Bachiri, A., & El Harfi, A. (2019). Textile finishing dyes and their impact on aquatic environs. *Heliyon.*, *5*(11). Available from https://doi.org/10.1016/j.heliyon.2019.e02711.

Bharath, G., Prakash, J., Rambabu, K., Venkatasubbu, G. D., Kumar, A., Lee, S., Theerthagiri, J., Choi, M. Y., & Banat, F. (2021). Synthesis of TiO_2/RGO with plasmonic Ag nanoparticles for highly efficient photoelectrocatalytic reduction of CO_2 to methanol toward the removal of an organic pollutant from the atmosphere. *Environmental Pollution.*, *281.* Available from https://doi.org/10.1016/j. envpol.2021.116990.

Bilibana, M. P., Citartan, M., Fuku, X., Jijana, A. N., Mathumba, P., & Iwuoha, E. (2022). Aptamers functionalized hybrid nanomaterials for algal toxins detection and decontamination in aquatic system: Current progress, opportunities, and challenges. *Ecotoxicology and Environmental Safety, 232.* Available from https://doi.org/10.1016/j. ecoenv.2022.113249, http://www.elsevier.com/inca/publications/store/6/2/2/8/1/9/index.htt.

Bose S., S. Maity, A. Sarkar, Nano-materials as biosensor for heavy metal detection. In K. Pal, A. Sarkar, P. Sarkar, N. Bandara, & V. Jegatheesan (Eds.), Food, medical, and environmental applications of nanomaterials. Elsevier, India (2022), 493−526, Available from https://www.sciencedirect.com/book/9780128228586, https://doi. org/10.1016/B978-0-12-822858-6.00018-2.

Buaisha, M., Balku, S., & Özalp-Yaman, Ş. (2020). Heavy metal removal investigation in conventional activated sludge systems. *Civil Engineering Journal (Iran).*, *6*(3), 470−477. Available from https://doi.org/10.28991/cej-2020-03091484, https://www.civile-journal.org/index.php/cej/article/view/1990/pdf.

Bueno, P. R., & Gabrielli, C. (2009). *Electrochemistry, nanomaterials, and nanostructures* (pp. 81−149). Springer Science and Business Media LLC. Available from http://doi. org/10.1007/978-0-387-49323-7_3.

Chang, X., Thind, S. S., Tian, M., Hossain, M. M., & Chen, A. (2015). Significant enhancement of the photoelectrochemical activity of nanoporous TiO_2 for environmental applications. *Electrochimica Acta, 173,* 728−735. Available from https://doi.org/10.1016/j. electacta.2015.05.122, http://www.journals.elsevier.com/electrochimica-acta/.

Dai, K., Peng, T., Chen, H., Zhang, R., & Zhang, Y. (2008). Photocatalytic degradation and mineralization of commercial methamidophos in aqueous titania suspension. *Environmental Science and Technology.*, *42*(5), 1505−1510. Available from https://doi. org/10.1021/es702268p.

Dai, W., Xu, H., Yu, J., Hu, X., Luo, X., Tu, X., & Yang, L. (2015). Photocatalytic reduction of CO_2 into methanol and ethanol over conducting polymers modified Bi_2WO_6 microspheres under visible light. *Applied Surface Science, 356,* 173−180. Available from https://doi.org/10.1016/j.apsusc.2015.08.059, http://www.journals. elsevier.com/applied-surface-science/.

Deng, J., Su, Y., Liu, D., Yang, P., Liu, B., & Liu, C. (2019). Nanowire photoelectrochemistry. *Chemical Reviews, 119*(15), 9221−9259. Available from https://doi.org/ 10.1021/acs.chemrev.9b00232, http://pubs.acs.org/journal/chreay.

Dincer, I., & Acar, C. (2015). A review on clean energy solutions for better sustainability. *International Journal of Energy Research.*, *39*(5), 585−606. Available from https://doi.org/ 10.1002/er.3329, http://onlinelibrary.wiley.com/journal/10.1002/(ISSN)1099-114X.

Dincer, I., & Acar, C. (2017). Innovation in hydrogen production. *International Journal of Hydrogen Energy, 42*(22), 14843−14864. Available from https://doi.org/10.1016/j. ijhydene.2017.04.107, http://www.journals.elsevier.com/international-journal-of-hydrogen-energy/.

Ding, R., Cheong, Y. H., Ahamed, A., & Lisak, G. (2021). Heavy metals detection with paper-based electrochemical sensors. *Analytical Chemistry, 93*(4), 1880−1888. Available from https://doi.org/10.1021/acs.analchem.0c04247, http://pubs.acs.org/journal/ancham.

Dong, H., Zeng, G., Tang, L., Fan, C., Zhang, C., He, X., & He, Y. (2015). An overview on limitations of TiO_2-based particles for photocatalytic degradation of organic pollutants and the corresponding countermeasures. *Water Research, 79,* 128−146. Available from https://doi.org/10.1016/j.watres.2015.04.038, http://www.elsevier. com/locate/watres.

Duangkaew, N., Suparee, N., Athikaphan, P., Neramittagapong, A., & Neramittagapong, S. (2020). Photocatalytic reduction of CO_2 into methanol over $Cu-Ni-TiO_2$ supported on $SiO_2-Al_2O_3$ catalyst. *Energy Reports.*, *6*, 1157−1161. Available from https://doi.org/10.1016/j.egyr.2020.11.060, http://www.journals.elsevier.com/energy-reports/.

Fajrina, N., & Tahir, M. (2019). A critical review in strategies to improve photocatalytic water splitting towards hydrogen production. *International Journal of Hydrogen Energy*, *44*(2), 540−577. Available from https://doi.org/10.1016/j.ijhydene.2018.10.200.

Fajrina, N., & Tahir, M. (2019). A critical review in strategies to improve photocatalytic water splitting towards hydrogen production. *International Journal of Hydrogen Energy*, *44*(2), 540−577. Available from https://doi.org/10.1016/j.ijhydene.2018.10.200, http://www.journals.elsevier.com/international-journal-of-hydrogen-energy/.

Fanourakis, S. K., Peña-Bahamonde, J., Bandara, P. C., & Rodrigues, D. F. (2020). Nano-based adsorbent and photocatalyst use for pharmaceutical contaminant removal during indirect potable water reuse. *Npj Clean Water.*, *3*(1). Available from https://doi.org/10.1038/s41545-019-0048-8, https://www.nature.com/npjcleanwater/.

Gomaa, O. M., & Momtaz, O. A. (2015). Copper induction and differential expression of laccase in *Aspergillus flavus*. *Brazilian Journal of Microbiology.*, *46*(1), 285−292. Available from https://doi.org/10.1590/S1517-838246120120118, http://www.scielo.br/pdf/bjm/v46n1/1517-8382-bjm-46-01-0285.pdf.

Gómez-Pastora, J., Dominguez, S., Bringas, E., Rivero, M. J., Ortiz, I., & Dionysiou, D. D. (2017). Review and perspectives on the use of magnetic nanophotocatalysts (MNPCs) in water treatment. *Chemical Engineering Journal.*, *310*, 407−427. Available from https://doi.org/10.1016/j.cej.2016.04.140, http://www.elsevier.com/inca/publications/store/6/0/1/2/7/3/index.htt.

Gosling, J. H., Makarovsky, O., Wang, F., Cottam, N. D., Greenaway, M. T., Patanè, A., Wildman, R. D., Tuck, C. J., Turyanska, L., & Fromhold, T. M. (2021). Universal mobility characteristics of graphene originating from charge scattering by ionised impurities. *Communications Physics*, *4*(1). Available from https://doi.org/10.1038/s42005-021-00518-2, nature.com/commsphys/.

Gupta, V. K. (2005). Potentiometric sensors for heavy metals—An overview. *Chimia*, *59*(5). Available from https://doi.org/10.2533/000942905777676614.

Halalsheh, N., Alshboul, O., Shehadeh, A., Al Mamlook, R. E., Al-Othman, A., Tawalbeh, M., Almuflih, A. S., & Papelis, C. (2022). Breakthrough curves prediction of selenite adsorption on chemically modified zeolite using boosted decision tree algorithms for water treatment applications. *Water.*, *14*(16). Available from https://doi.org/10.3390/w14162519.

Jakhar, A. M., Aziz, I., Kaleri, A. R., Hasnain, M., Haider, G., Ma, J., & Abideen, Z. (2022). Nano-fertilizers: A sustainable technology for improving crop nutrition and food security. *NanoImpact.*, *27*. Available from https://doi.org/10.1016/j.impact.2022.100411.

Khan, S. B., Asiri, A. M., & Akhtar, K. (2018). *Nanomaterials for environmental applications and their fascinating attributes* (2). Bentham Science Publishers. Available from http://doi.org/10.2174/97816810864531180201.

Krikstolaityte, V., Ding, R., Xia, E. C. H., & Lisak, G. (2020). Paper as sampling substrates and all-integrating platforms in potentiometric ion determination. *TrAC Trends in Analytical Chemistry.*, *133*. Available from https://doi.org/10.1016/j.trac.2020.116070.

Lais, A., Gondal, M. A., Dastageer, M. A., & Al-Adel, F. F. (2018). Experimental parameters affecting the photocatalytic reduction performance of CO_2 to methanol: A review. *International Journal of Energy Research.*, *42*(6), 2031−2049. Available from https://doi.org/10.1002/er.3965, http://onlinelibrary.wiley.com/journal/10.1002/(ISSN)1099-114X.

Li, Z., Pan, Z., & Dai, S. (2004). Nitrogen adsorption characterization of aligned multi-walled carbon nanotubes and their acid modification. *Journal of Colloid and Interface Science*, 277(1), 35−42. Available from https://doi.org/10.1016/j.jcis.2004.05.024.

Martín-Lara, M. A., Calero, M., Ronda, A., Iáñez-Rodríguez, I., & Escudero, C. (2020). Adsorptive behavior of an activated carbon for bisphenol A removal in single and binary (bisphenol A-heavy metal) solutions. *Water (Switzerland).*, 12(8). Available from https://doi.org/10.3390/W12082150, https://res.mdpi.com/d_attachment/water/water-12-02150/article_deploy/water-12-02150-v2.pdf.

Mohamed, O., Al-Othman, A., Al-Nashash, H., Tawalbeh, M., Almomani, F., & Rezakazemi, M. (2021). Fabrication of titanium dioxide nanomaterial for implantable highly flexible composite bioelectrode for biosensing applications. *Chemosphere*, 273. Available from https://doi.org/10.1016/j.chemosphere.2021.129680.

Mohammed, H., Al-Othman, A., Nancarrow, P., Tawalbeh, M., & El Haj Assad, M. (2019). Direct hydrocarbon fuel cells: A promising technology for improving energy efficiency. *Energy.*, 172, 207−219. Available from https://doi.org/10.1016/j.energy.2019.01.105, http://www.elsevier.com/inca/publications/store/4/8/3/.

Mohan, A., Ulmer, U., Hurtado, L., Loh, J., Li, Y. F., Tountas, A. A., Krevert, C., Chan, C., Liang, Y., Brodersen, P., Sain, M. M., & Ozin, G. A. (2020). Hybrid photo- and thermal catalyst system for continuous CO_2 reduction. *ACS Applied Materials and Interfaces.*, 12(30), 33613−33620. Available from https://doi.org/10.1021/acsami.0c06232, http://pubs.acs.org/journal/aamick.

Narayan, N., Meiyazhagan, A., & Vajtai, R. (2019). Metal nanoparticles as green catalysts. *Materials.*, 12(21). Available from https://doi.org/10.3390/ma12213602, https://res.mdpi.com/d_attachment/materials/materials-12-03602/article_deploy/materials-12-03602.pdf.

Narayanan H., B. Viswanathan, K.R. Krishnamurthy, H. Nair, Hydrogen from photo-electrocatalytic water splitting. In F. Calise, M. D. D'Accadia, M. Santarelli, A. Lanzini, & D. Ferrero (Eds.), Solar hydrogen production: Processes, systems and technologies. Elsevier, India (2019), 419−486, Available from http://www.sciencedirect.com/science/book/9780128148532, https://doi.org/10.1016/B978-0-12-814853-2.00012-6.

Nauman Javed, R. M., Al-Othman, A., Tawalbeh, M., & Olabi, A. G. (2022). Recent developments in graphene and graphene oxide materials for polymer electrolyte membrane fuel cells applications. *Renewable and Sustainable Energy Reviews.*, 168. Available from https://doi.org/10.1016/j.rser.2022.112836, https://www.journals.elsevier.com/renewable-and-sustainable-energy-reviews.

Ndolomingo, M. J., Bingwa, N., & Meijboom, R. (2020). Review of supported metal nanoparticles: Synthesis methodologies, advantages and application as catalysts. *Journal of Materials Science*, 55(15), 6195−6241. Available from https://doi.org/10.1007/s10853-020-04415-x, http://www.springer.com/journal/10853.

Nitumoni, M., Dambale, A., & Montrishna, R. (2019). Nutrient use efficiency through Nano fertilizers. *International Journal of Chemical Studies*, 7(3), 2839−2842.

Pal K., A. Sarkar, P. Sarkar, N. Bandara, V. Jegatheesan, Food, medical, and environmental applications of nanomaterials. Elsevier, India (2022), 1−554, Available from https://www.sciencedirect.com/book/9780128228586, https://doi.org/10.1016/C2019-0-04943-6.

Parthasarathy, P., & Narayanan, K. S. (2014). Hydrogen production from steam gasification of biomass: Influence of process parameters on hydrogen yield—A review. *Renewable Energy.*, 66, 570−579. Available from https://doi.org/10.1016/j.renene.2013.12.025, http://www.journals.elsevier.com/renewable-and-sustainable-energy-reviews/.

Piątkowska, A., Janus, M., Szymański, K., & Mozia, S. (2021). C-,n-and s-doped tio2 photo. *catalysts: A review. Catalysts.*, 11(1), 1−56. Available from https://doi.org/10.3390/catal11010144, https://www.mdpi.com/2073-4344/11/1/144/pdf.

Qasem, N. A. A., Mohammed, R. H., & Lawal, D. U. (2021). Removal of heavy metal ions from wastewater: A comprehensive and critical review. *Npj Clean Water.*, *4*(1). Available from https://doi.org/10.1038/s41545-021-00127-0, https://www.nature.com/npjcleanwater/.

Shams Jalbani, N., Solangi, A. R., Memon, S., Junejo, R., Bhatti, A. A., Yola, M. L., Tawalbeh, M., & Karimi-Maleh, H. (2021). Synthesis of new functionalized Calix[4]arene modified silica resin for the adsorption of metal ions: Equilibrium, thermodynamic and kinetic modeling studies. *Journal of Molecular Liquids.*, *339*. Available from https://doi.org/10.1016/j.molliq.2021.116741.

Singla, S., Sharma, S., Basu, S., Shetti, N. P., & Aminabhavi, T. M. (2021). Photocatalytic water splitting hydrogen production via environmental benign carbon based nanomaterials. *International Journal of Hydrogen Energy*, *46*(68), 33696−33717. Available from https://doi.org/10.1016/j.ijhydene.2021.07.187, http://www.journals.elsevier.com/international-journal-of-hydrogen-energy/.

Stafford-Smith, M., Griggs, D., Gaffney, O., Ullah, F., Reyers, B., Kanie, N., Stigson, B., Shrivastava, P., Leach, M., & O'Connell, D. (2017). Integration: The key to implementing the sustainable development goals. *Sustainability Science*, *12*(6), 911−919. Available from https://doi.org/10.1007/s11625-016-0383-3, http://www.springer.com/east/home?SGWID = 5-102-70-144940151-0&changeHeader = true&SHORT CUT = http://www.springer.com/journal/11625.

Sunada, K., Kikuchi, Y., Hashimoto, K., & Fujishima, A. (1998). Bactericidal and detoxification effects of TiO_2 thin film photocatalysts. *Environmental Science and Technology.*, *32*(5), 726−728. Available from https://doi.org/10.1021/es970860o.

Syed, N., Huang, J., Feng, Y., Wang, X., & Cao, L. (2019). Carbon-based nanomaterials via heterojunction serving as photocatalyst. *Frontiers in Chemistry*, *7*. Available from https://doi.org/10.3389/fchem.2019.00713, http://journal.frontiersin.org/journal/chemistry.

Tahir M.B., A. Batool, Recent development in sustainable technologies for clean hydrogen evolution: Current scenario and future perspectives. In K. Y. Cheong & A. Apblett (Eds.), Sustainable materials and green processing for energy conversion. Elsevier, Pakistan (2022), 97−130, Available from https://www.sciencedirect.com/book/9780128228388, https://doi.org/10.1016/B978-0-12-822838-8.00008-9.

Tawalbeh, M., Aljaghoub, H., Alami, A. H., & Olabi, A. G. (2023). Selection criteria of cooling technologies for sustainable greenhouses: A comprehensive review. *Thermal Science and Engineering Progress.*, *38*. Available from https://doi.org/10.1016/j.tsep.2023.101666, https://www.journals.elsevier.com/thermal-science-and-engineering-progress.

Tawalbeh, M., Allawzi, M. A., & Kandah, M. I. (2005). Production of activated carbon from jojoba seed residue by chemical activation residue using a static bed reactor. *Journal of Applied Sciences.*, *5*(3), 482−487. Available from https://doi.org/10.3923/jas.2005.482.487.

Tawalbeh, M., Al-Ismaily, M., Kruczek, B., & Tezel, F. H. (2021). Modeling the transport of CO_2, N_2, and their binary mixtures through highly permeable silicalite-1 membranes using Maxwell − Stefan equations. *Chemosphere*, *263*. Available from https://doi.org/10.1016/j.chemosphere.2020.127935, http://www.elsevier.com/locate/chemosphere.

Tawalbeh, M., Al Mojjly, A., Al-Othman, A., & Hilal, N. (2018). Membrane separation as a pre-treatment process for oily saline water. *Desalination.*, *447*, 182−202. Available from https://doi.org/10.1016/j.desal.2018.07.029, https://www.journals.elsevier.com/desalination.

Tawalbeh, M., Muhammad Nauman Javed, R., Al-Othman, A., & Almomani, F. (2022). The novel advancements of nanomaterials in biofuel cells with a focus on electrodes' applications. *Fuel.*, *322*. Available from https://doi.org/10.1016/j.fuel.2022.124237, http://www.journals.elsevier.com/fuel/.

Tawalbeh, M., Rajangam, A. S., Salameh, T., Al-Othman, A., & Alkasrawi, M. (2021). Characterization of paper mill sludge as a renewable feedstock for sustainable hydrogen and biofuels production. *International Journal of Hydrogen Energy*, *46*(6), 4761−4775. Available from https://doi.org/10.1016/j.ijhydene.2020.02.166, http://www.journals.elsevier.com/international-journal-of-hydrogen-energy/.

Unuofin, J. O., Okoh, A. I., & Nwodo, U. U. (2019). Aptitude of oxidative enzymes for treatment of wastewater pollutants: A laccase perspective. *Molecules (Basel, Switzerland)*, *24*(11). Available from https://doi.org/10.3390/molecules24112064, https://www.mdpi.com/1420-3049/24/11/2064/pdf.

Upadhyay, R. K., Soin, N., & Roy, S. S. (2014). Role of graphene/metal oxide composites as photocatalysts, adsorbents and disinfectants in water treatment: a review. *RSC Advances.*, *4*(8), 3823−3851. Available from https://doi.org/10.1039/c3ra45013a.

Wang, S., Ding, Z., Chang, X., Xu, J., & Wang, D. H. (2020). Modified nano-TiO_2 based composites for environmental photocatalytic applications. *Catalysts.*, *10*(7), 1−38. Available from https://doi.org/10.3390/catal10070759, https://www.mdpi.com/2073-4344/10/7/759/pdf.

Wang, Y., Liang, S., Chen, B., Guo, F., Yu, S., & Tang, Y. (2013). Synergistic removal of Pb(II), Cd(II) and humic acid by Fe3O4@mesoporous silica-graphene oxide composites. *PLoS One*, *8*(6). Available from https://doi.org/10.1371/journal.pone.0065634China, http://www.plosone.org/article/fetchObject.action?uri = info%3Adoi%2F10.1371%2Fjournal.pone.0065634&representation = PDF.

Xu, Y., Jiang, X., Zhou, Y., Ma, M., Wang, M., & Ying, B. (2021). Systematic evolution of ligands by exponential enrichment technologies and aptamer-based applications: Recent progress and challenges in precision medicine of infectious diseases. *Frontiers in Bioengineering and Biotechnology.*, *9*. Available from https://doi.org/10.3389/fbioe.2021.704077, http://journal.frontiersin.org/journal/bioengineering-and-biotechnology#archive.

Yaqoob, A. A., Parveen, T., Umar, K., & Ibrahim, M. N. M. (2020). Role of nanomaterials in the treatment of wastewater: A review. *Water.*, *12*(2). Available from https://doi.org/10.3390/w12020495.

Yu, Y., Guo, P., Zhong, J.-S., Yuan, Y., & Ye, K.-Y. (2020). Merging photochemistry with electrochemistry in organic synthesis. *Organic Chemistry Frontiers.*, *7*(1), 131−135. Available from https://doi.org/10.1039/c9qo01193e.

Zhang, J., Zhou, P., Liu, J., & Yu, J. (2014). New understanding of the difference of photocatalytic activity among anatase, rutile and brookite TiO_2. *Physical Chemistry Chemical Physics.*, *16*(38), 20382−20386. Available from https://doi.org/10.1039/c4cp02201g.

CHAPTER 11

Nanomaterials for electrochemical chlorine evolution reaction in the Chlor-alkali process

Waseem Ahmad[1], Kaidi Zhang[1], Yu Zou[1], Liang Wang[1], Mengyang Dong[1], Huai Qin Fu[1], Huajie Yin[2], Yonggang Jin[3], Porun Liu[1] and Huijun Zhao[1]

[1]Centre for Catalysis and Clean Energy, Gold Coast Campus, Griffith University, Gold Coast, QLD, Australia
[2]Key Laboratory of Materials Physics, Centre for Environmental and Energy Nanomaterials, Anhui Key Laboratory of Nanomaterials and Nanotechnology, CAS Center for Excellence in Nanoscience, Institute of Solid State Physics, Hefei Institutes of Physical Science, Chinese Academy of Sciences, Hefei, Anhui, China
[3]Commonwealth Scientific and Industrial Research Organization (CSIRO) Mineral Resources, Pullenvale, QLD, Australia

11.1 Introduction

Chlorine (Cl_2) gas is a vital commodity chemical crucial to many chemical industries such as water treatment, a wide variety of organic solvents, as a disinfectant in chemical products, and in the production of commercial synthetic polymer polyvinyl chloride (PVC) (Lakshmanan & Murugesan, 2014). Chlorine gas is commercially produced by the Chlor-alkali process on an industrial scale using a dimensionally stable anode (DSA) made of RuO_2/TiO_2 and the reaction is called the Chlorine evolution reaction (CER). CER is an important reaction as the global chlorine demands most depend on the electrolysis of brine. The global demand for chlorine reached 88 million tons per year in 2020 and its commercial production mostly relies on the Chlor-alkali process (Wang et al, 2021a). CER is the main energy-consuming reaction in the Chlor-alkali process, which occurs at anode requiring a high potential along with the hydrogen evolution reaction (HER) which occurs at the cathode (Chen et al., 2013). Anodic CER is a highly energy-consuming reaction and requires a highly reactive and selective electrocatalyst to facilitate the CER with minimum overpotential.

Electrochemistry and Photo-Electrochemistry of Nanomaterials
DOI: https://doi.org/10.1016/B978-0-443-18600-4.00012-0
© 2025 Elsevier Inc. All rights are reserved, including those for text and data mining, AI training, and similar technologies.

The Chlor-alkali industry process has been developed in the 19th century and has been the only commercial process for chlorine gas manufacturing on industrial scale ever since (Crook & Mousavi, 2016). The Chlor-alkali process consists of an electrochemical cell in which Cl_2 gas is produced at the anode and sodium hydroxide (NaOH) is produced in the catholyte. In an industrial electrochemical cell, both the anode and cathode are separated from each other. There are three main reaction systems adopted in the Chlor-alkali process, namely, the diaphragm cell, mercury cell, and membrane cell. The separation of anodic and cathodic compartments constitutes the main difference between the diaphragm cell and the membrane cell. In a membrane cell, a selective cation membrane while in a diaphragm, a microporous diaphragm separates the anodic and cathodic compartments. In a mercury cell, mercury amalgam is formed at the cathode which is later converted to hydrogen gas and sodium hydroxide in a separation reactor (Karlsson & Cornell, 2016).

The mercury-cell process, also known as the Castner—Kellner process, was the first process to be developed in the 1980s in which mercury acted as a cathode by floating on the saturated brine solution forming an amalgam. Amalgam would be continuously removed and made to decompose into hydrogen, mercury, and NaOH. Mercury was continuously recycled into the system (Lakshmanan & Murugesan, 2014). The mercury cell process has been discontinued due to mercury poisoning causing severe environmental problems (Crook & Mousavi, 2016). In the diaphragm cell process, both the anode and cathode compartments were separated where Cl_2 is produced at the anode and H_2 and NaOH at the cathode side. The diaphragm cell process and membrane cell process are very similar, with the only difference being that a cation-selective membrane is used in membrane cell while microporous diaphragm is used in the diaphragm process.

At present, membrane cell technology is the most popular technology in the Chlor-alkali process because of its high efficiency, environmental benefits, and high product purity (Brinkmann et al., 2014). The Chlor-alkali process is an energy-intensive process mainly because the CER reaction has high reaction overpotential. The cost for the process has been estimated as 2100 kWh per ton of NaOH while for every ton of NaOH roughly 886 kg of Cl_2 gas is being produced (Karlsson & Cornell, 2016). The Chlor-alkali process on average consumes about 150TWh every year (Karlsson & Cornell, 2016). Even a small improvement in the energy efficiency of the Chlor-alkali process may significantly reduce world energy consumption. The main area where these energy improvements can be made is in the development of more efficient electrocatalysts for CER.

Historically, dimensionally stable anode (DSA) had been used in the Chlor-alkali process which is made of mixed metal oxides (MMOs) of ruthenium(IV) oxide (RuO_2) and titanium dioxide (TiO_2) deposited on Ti substrate. RuO_2 is found to have the most active sites for CER; however, it is not very stable at higher potential. To improve its stability, other mixed oxides such as TiO_2 and IrO_2 are added (Hansen et al., 2010). The discovery and use of DSA have been termed the most successful breakthrough in the field of electrochemistry in the last 50 years and thus making the Chlor-alkali process the most successful commercial electrochemical process (Trasatti, 2000). In this book chapter, we will briefly discuss the technical aspects of the Chlor-alkali process, reaction mechanism of CER, understanding the competition between OER and CER and finally, the future perspectives of the Chlor-alkali process in terms of facing future challenges like climate change and increasing global energy demands.

11.2 Chlor-alkali process

The Chlor-alkali process uses brine water to produce chlorine gas and sodium hydroxide by the following overall reaction (Sohrabnejad-Eskan et al., 2017):

$$2NaCl + 2H_2O \rightarrow Cl_2 + H_2 + 2NaOH \tag{11.1}$$

In the Chlor-alkali process, anodic and cathodic compartments are separated from each other. Chlorine gas is formed at the DSA electrode (anolyte) as per the following reduction reaction:

$$2Cl^- \rightarrow Cl_2 + 2e^-, \quad E° = 1.35 \ V \ vs \ SHE \tag{11.2}$$

While at the cathode (catholyte) side following reactions take place:

$$2H^+ + 2e^- \rightarrow H_2 \tag{11.3}$$

$$2Na^+ + 2OH^- \rightarrow 2NaOH \tag{11.4}$$

As already discussed, membrane cell technology is the most popular technology in the Chlor-alkali process and other cells are converting their process to membrane cells. One of the main advantages of membrane cell technology is that its power consumption is the lowest compared to other technologies (Lakshmanan & Murugesan, 2014). The commercial membrane process operates at cell voltage between 2.4 and 2.7 V with a current density between 1.5 and 7 kA/m^2 and NaCl concentration in the anolyte is 200 g/dm^3 with pH in the range of 2−4 (Brinkmann et al., 2014). The overall process operates at 90°C and the concentration of NaOH is kept at 32 wt% (Karlsson & Cornell, 2016). In membrane cells, DSAs are used as an electrode for CER, which comprises RuO$_2$ and TiO$_2$. Membrane, due to its property of selective permeability, suits best for the Chlor-alkali process.

A schematic diagram of the membrane cell is shown in Fig. 11.1. The anode and cathode compartments are separated by a double-layered ion-exchange membrane made from perfluoro-sulfonic acid. As illustrated in the figure, A pretreated saturated brine solution is injected into the anode compartment while demineralized water is channeled to the cathode compartment. Cl$_2$ gas is liberated at the anode and H$_2$ gas is formed at the cathode by the electrolysis of water. The Na$^+$ ions that are produced in the anolyte pass through the ion-exchange membrane to react with OH$^−$ ions that are produced during the electrolysis of water forming NaOH. Water is continuously fed into the cathode compartment to keep the concentration of NaOH at 32 wt% (Karlsson & Cornell, 2016). Membrane cell technology is developing fast as further research is going on to make the process more energy efficient with the development of new membranes. In understanding the Chlor-alkali process, it is important

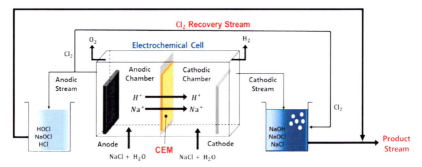

Figure 11.1 Schematic diagram of membrane electrolytic cell (Kim et al., 2021). Taken from Kim, S.K., Shin, D.-M., & Rhim, J.W. (2021). Designing a high-efficiency hypochlorite ion generation system by combining cation exchange membrane aided electrolysis with chlorine gas recovery stream. Journal of Membrane Science, 630, 119318.

to know the reaction mechanism of CER and how Cl_2 gas is formed on the electrode surface loaded with catalyst.

11.2.1 Reaction mechanism of CER

Until now, the actual reaction mechanism of CER has been poorly understood on the atomic scale level. The reaction mechanism along with the chemical nature of the electrocatalyst used plays an important role in the mechanisms of CER. It is best to model the reaction mechanism of CER on a well-defined structure like that of RuO_2. As per the ab initio thermodynamic calculations, three probable mechanistic reaction pathways of CER on RuO_2 have been devised, that is, Volmer–Tafel (VT), Volmer–Heyrovsky (VH), and Krishtalik. These reaction pathways are given below where the "*" represents the active sites on the catalyst (Hansen et al., 2010):

A. Volmer–Tafel pathway

$$2Cl^- + 2^* \rightarrow 2Cl^* + 2e^-$$

$$Cl^* + Cl^* \rightarrow 2^* + Cl_2$$

B. Volmer–Heyrovsky pathway

$$2Cl^- + {}^* \rightarrow Cl^* + e^- + Cl^-$$

$$Cl^* + Cl^- \rightarrow {}^* + Cl_2 + e^-$$

C. Krishatalik pathway

$$2Cl^- + {}^* \rightarrow Cl^* + e^- + Cl^-$$

$$Cl^* + Cl^- \rightarrow Cl^{*+} + e^- + Cl^-$$

$$Cl^{*+} + Cl^- \rightarrow {}^* + Cl_2$$

CER is a two-electron-transfer reaction. Adsorbed chlorine is formed after the first electron transfer in all three pathways. The only difference is the second step in which either a second adsorbed chlorine is formed on the surface or chloronium species are formed. The Volmer–Tafel model involves the adsorption of two chloride ions (Volmer step) followed by the Tafel step, which is the evolution of Cl_2 gas by recombination of two adsorbed chlorine species (Kelly et al., 2002). In the Volmer−Heyrovsky mechanism, the Volmer step is followed by the Heyrovsky step, which is the direct recombination of the adsorbed chlorine species with adjacent chloride ions from the solution with an electron transfer (Exner et al., 2015). In the Krishtalik mechanism, after the Volmer step, a further electron transfer takes place forming chloronium species which in turn recombines with chloride ions from the solution liberating Cl_2 (Krishtalik, 1981).

Density functional theory (DFT) calculations indicate that the Volmer−Heyrovsky pathway is the most likely reaction mechanism on RuO_2 catalyst sites. This mechanism involves a two-step process where the second step requires adsorbed chlorine reacting directly with chloride from the solution to form gaseous chlorine. This step has Gibbs energy changes of 1.23 eV at 0 V and −0.13 eV at 1.36 V, with a small energy loss of 0.13 eV (Exner et al., 2015). In comparison, the Volmer−Tafel mechanism's second step needs higher energy changes of 1.59 eV at 0 V and 0.23 eV at 1.36 V, with a larger energy loss of 0.23 eV (Exner et al., 2015). The Krishtalik mechanism's second step involves forming a chloronium ion, requiring a much higher energy change of 2.21 eV (Exner et al., 2015). Given its lower energy loss, the Volmer−Heyrovsky mechanism is considered the preferred pathway (Exner et al., 2015). Fig. 11.2 shows one such Gibbs energy diagram for CER reaction mechanism over $RuO_2(110)$ (Liu et al., 2023). These diagrams help understand the energy changes during chemical reactions on these catalysts. The researchers in this study focused on two reaction steps (ΔG_1 and ΔG_2) and found that the second step, which involves the formation of molecular Cl_2 from *Cl/*OCl and Cl − , is the most crucial and energy-demanding step (Liu et al., 2023). This step is also known as the potential-determining step (PDS) (Liu et al., 2023).

11.2.2 Selectivity between OER and CER

Insight into the selectivity between OER and CER is an important phenomenon to consider in the development of electrocatalysts for CER. The competition between CER and OER which results in sharing of active sites of the catalyst is the main problem in the Chlor-alkali industry (Baptista et al., 2015). In the Chlor-alkali process, the current efficiency is the primary measure of selectivity for CER (Karlsson & Cornell, 2016). The parasitic OER reaction reduces the catalyst activity, electrode

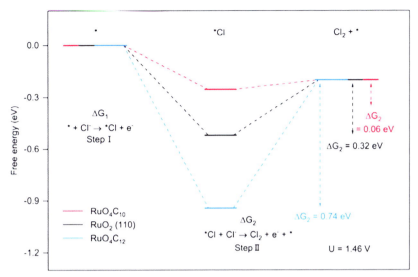

Figure 11.2 Gibbs free-energy diagram for CER over RuO_4C_{10}, RuO_4C_{12}, and RuO_2 (110) as calculated by DFT calculations. Taken from Liu, Y., Li, C., Tan, C., Pei, Z., Yang, T., Zhang, S., Huang, Q., Wang, Y., Zhou, Z., Liao, X., Dong, J., Tan, H., Yan, W., Yin, H., Liu, Z.-Q., Huang, J. & Zhao, S. (2023). Electrosynthesis of chlorine from seawater-like solution through single-atom catalysts. Nature Communications, 14(1), 2475.

stability, and overall process efficiency. It also adds a costly purification step for Cl_2 purification. CER is a preferred reaction kinetically as it requires two-electron transfer versus four-electron transfer for OER. As already discussed, OER is favored thermodynamically as it requires less potential of 1.23 V versus RHE compared to 1.36 V versus SHE of CER. It has been found that the equilibrium potential of CER is mainly pH-dependent. In near-neutral and higher pH, the selectivity of the reaction is shifted toward OER, and at lower pH (acidic), CER is favored compared to OER, which is the reason why the commercial Chlor-alkali process is carried out in acidic conditions (Karlsson & Cornell, 2016). This phenomenon can be explained by Pourbaix diagram shown in Fig. 11.3, which shows different reaction pathways for RuO_2(110) in equilibrium with Cl^-, H^+ and H_2O at 298 K and $a(Cl^-) = 1$ as a function of anolyte pH having NaCl (Exner et al., 2014). The concentration of chlorine gas starts increasing as the pH falls below 3 and keeps increasing as the pH approaches 1 (Exner et al., 2014). It has been found that this stability window changes with process conditions, electrode material (catalyst), and composition of the electrolyte. In the design of an electrocatalyst for

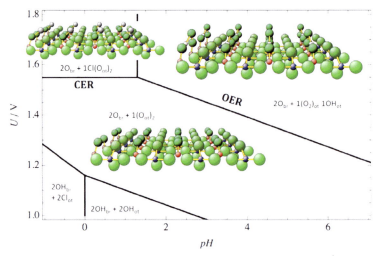

Figure 11.3 Pourbaix diagram for $RuO_2(110)$ in equilibrium with Cl^-, H^+ and H_2O at 298 K and $a(Cl^-) = 1$. *Taken from Exner, K.S., Anton, J., Jacob, T., & Over, H. (2014). Chlorine evolution reaction on RuO2 (110): Ab initio atomistic thermodynamics study-Pourbaix diagrams.* Electrochimica Acta, 120, 460–466.

better selectivity of CER over OER, a linear scaling relationship is used on Gibbs free energy. The scaling relationship occurs when there is a correlation between the atomic or molecular structure of a material and its physical or chemical properties. This would help improve the selectivity because of the greater difference gradient in the Gibbs free energy (Wang et al., 2021a). It has been found that the activity of the Chlorine Evolution Reaction (CER) increases linearly with the concentration of chloride ions ($[Cl^-]$) at all potentials as shown in Fig. 11.4 indicating a reaction order of one (Vos & Koper, 2018). This suggests that OER and CER occur independently and do not share the same active sites on the catalyst, even though previous literature has suggested a relationship between their activities (Vos & Koper, 2018).

11.2.3 Factors affecting the selectivity between OER and CER

In the past many decades, various attempts have been made to understand the factors that influence the complex selectivity between OER and CER reactions. Various factors such as process conditions, contaminants in the electrolyte, and the composition of the anode play a crucial role in the selectivity between OER and CER. These factors and their effect on the selectivity between OER and CER are summarized in Table 11.1.

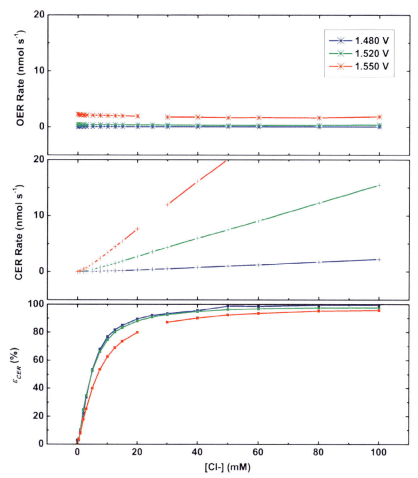

Figure 11.4 Plots of OER and CER reaction rates as function of [Cl$^-$], for three different disk potentials. *Taken from Vos, J. & Koper, M. (2018). Measurement of competition between oxygen evolution and chlorine evolution using rotating ring-disk electrode voltammetry.* Journal of Electroanalytical Chemistry, 819, 260–268.

We have already discussed the role of pH in discussing the reaction mechanism of CER. Among the factors that affect the selectivity between OER and CER, pH plays the most important role for OER as its equilibrium reaction depends on the pH of the solution. OER becomes unfavorable with decreasing pH while keeping the current density unchanged (Bergner, 1990). In the commercial Chlor-alkali industry, it is a common practice to add HCl to eliminate and suppress the OER completely

Table 11.1 Factors affecting the selectivity between OER and CER in concentrated NaCl solutions.

Factor	Effect on selectivity between OER and CER	References
pH	Lower pH favors CER	Bergner (1990)
The concentration of Cl^- in electrolyte	Higher Cl^- increase rate of CER	Jirkovský et al. (2006)
Current density	High current density favors CER	Hardee and Mitchell (1989)
System temperature	Increase in temperature increase the rate of OER in chlorate conditions	Hardee and Mitchell (1989)
Addition of fluoride in electrolyte	Decrease in rate of oxygen evolution reaction	Fukuda et al. (1979)
Addition of phosphate, sulfate, and nitrite in electrolyte	Increase rate of OER	Buné et al. (1985); Jaksić et al. (1972)

(Bergner, 1990). Apart from pH, the concentration of chloride ions also plays an important role. It has been found that an increase in the concentration of Cl^- increases the current efficiency of CER and decreases the selectivity for OER (Jirkovský et al., 2006). The shift toward CER is mainly because of the increased mass transfer for chloride ions on the anodic surface and the suppression of the OER on the anode surface (Jirkovský et al., 2006).

The effect of current density on the selectivity between OER and CER is tangled with the pH of the electrolyte and the electrode's surface area; however, it has been found that higher current density favors the CER reaction and decreases the selectivity for OER (Hardee & Mitchell, 1989). Deceasing selectivity for OER by increasing current density implies that the anodic reactions that generate oxygen has higher Tafel slopes compared with CER. The effect of system temperature has been studied but it is not yet conclusively clear how it affects the selectivity between OER and CER. Temperature affects both the thermodynamics and kinetics of both reactions; therefore, its effect is complicated and poorly understood. However, it has been reported that an increase in temperature increases the rate of OER at chlorate conditions (Hardee & Mitchell, 1989). An increase in temperature causes an increased rate of diffusion of hypochlorite toward the anodic surface.

The electrolyte composition and the effects of additives could be an interesting area to study and control the selectivity between OER and CER; however, very few studies have been conducted in this domain. The addition of fluoride has been found to decrease the rate of OER (Fukuda et al., 1979). Fukuda et al. (1979) investigated the effect of NH_4F addition on the DSA-type anodes and concluded that an increase in the addition of NH_4F caused a decrease in oxygen evolution. Phosphates have been found to increase the rate of OER and phosphates have been found to decrease the efficiency of CER, and thus both phosphates and sulfate improve the selectivity toward OER (Buné et al., 1985). It has been found that both phosphate and sulfate adsorb firmly on the anode surface suppressing the chloride ions (Buné et al., 1985). In contrast, Jaksić et al. (1972) found no significant effect of the addition of phosphates and sulfates on OER; however, they concluded that the addition of nitrates increases the rate of OER. It can be argued that the effect of these species mainly depends on their adsorption on the anodic surface and the concentration of chloride ions. At higher chloride ion concentrations, the increased adsorption of chloride on the electrode surface would overcome the mass transfer resistance. It is known that many metals and metal oxides have catalytic effects on both CER and OER; however, their effect on the selectivity between OER and CER has not been fully explored and more research is required to understand this phenomenon.

The composition of the anode and its effect on the selectivity between OER and CER have been thoroughly investigated. Electrode fabrication and applying a catalyst as the coating is a process where the active catalyst is grown either physically or chemically on the substrate surface. The electrochemical active surface area plays the primary role directly related to the current density in the selectivity between OER and CER. An increase in the electrochemical active surface area is often caused by the doping effect on the anode (Arikawa et al., 1998). It has been reported that the doping of RuO_2 on the anode surface increases the selectivity for OER due to an increase in the electrochemical active surface area (Arikawa et al., 1998). It has already been discussed that RuO_2 is the most active catalyst reported for CER. The effect of doping of various metals and mixed electrodes on the selectivity between OER and CER is complex and it is not yet understood at the atomic level because the developed catalyst is almost equally active for both OER and CER (Karlsson & Cornell, 2016).

11.3 Electrocatalysts for CER

In the last many decades, various materials have been investigated as electrocatalyst to reduce the energy consumption of the Chlor-alkali process. In the development of electrocatalysts, two approaches are mostly adopted; one is nanoelectrocatalysts with active sites and the other is single atom active site electrocatalyst in which the single atom plays an anchoring role as active sites (Ou et al., 2021). In both cases, nanomaterials and in-general nanotechnology play a vital role in the development of electrocatalysts for CER. MMOs have been extensively tried for CER; however, such catalysts are also active for Oxygen Evolution reaction (OER) and thus causing competition between the two anodic reactions (Exner, 2019). As already discussed, the competition and selectivity between OER and CER is the main problem while designing a catalyst for CER. The equilibrium potential of CER is 1.36 V versus 1.23 V for OER under standard conditions and room temperature (Hansen et al., 2010). Thermodynamically OER is more favorable than CER. However, kinetically CER is favored over OER because CER requires two electrons transfer versus four electrons for OER. The competition between OER and CER not only leads to lower selectivity for CER but in such a harsh environment, noble metals forms thermodynamically stable species of Ru with oxides/chlorides which cause catalyst poisoning (Chen et al., 2013). A small quantity of oxygen produced during the process adds to a costly purification step for chlorine gas, which also increases the overall cost of the production.

Therefore, various theoretical and experimental studies have been performed aimed at finding new cheaper alternative electrocatalysts by understanding the reaction pathway of CER and OER. Thermodynamic studies have suggested that engineering the electronic and geometric structure of noble metal electrocatalysts on the atomic scale is crucial for CER selectivity and stability of catalysts (Wang et al., 2021a). Therefore, a variety of other metals have been doped with RuO_2 and they have shown improved performance for CER (Chen et al., 2011a; Wang et al., 2012). Apart from the expensive noble metals, other earth-abundant MMOs of cobalt and antimony have shown active sites for CER; however, most of these oxides show poor corrosion resistance in acidic conditions of CER (Moreno-Hernandez et al., 2019; Zhu et al., 2018a). Alternative to metal-based catalysts, carbon-based nanomaterials have recently shown great potential for high CER activity owing to their high surface area and stability (Zhao et al., 2019).

CER is an intensive energy reaction with an overpotential of 1.36 V versus SHE in a highly acidic condition. Therefore, it is important to design durable electrocatalysts in addition to being efficient and selective for CER. The Chlor-alkali process is well established and the currently used DSA electrode has high performance and selectivity for CER. Currently used DSA also heavily relies on the Nobel metals, which is not sustainable in the long term. Therefore, it is important to explore and find alternative cheap electrocatalysts with high selectivity, efficiency, and durability for CER. In the proceeding section, conventional metal-based electrocatalysts are summarized followed by metal-free electrocatalysts. The results of the electrocatalyst are summarized in Table 11.2.

11.3.1 Noble metal-based nanoelectrocatalysts for CER

Conventionally, precious metals have been used for CER despite their high price mainly because of their high performance and selectivity. Up till now, the DSAs invented by Beer in the 1960s are still the most efficient CER anodes for the Chlor-alkali industry (Beer, 1980). These DSAs, made of TiO_2/RuO_2 and/or IrO_2 composites, have demonstrated exceptional stability, excellent catalytic activity, and selectivity toward CER with high faradic efficiency under high concentration NaCl electrolyte ($> 26\%$) and low pH conditions (pH < 2) (Trasatti, 2000). The only drawback of using a noble metals-based electrocatalyst for CER is the cost factor and sustainability in the long run. Both Ru and Ir work well in acidic conditions; however, the dissolution rate of Ru is much higher than that of Ir and thus limiting its utilization (Chen et al., 2013). It has been reported that the dissolution rate of IrO_2-based electrodes is 20 times less than that of RuO_2 based (Cherevko et al., 2016). However, a limiting factor of IrO_2-based electrodes is that they are less active and selective for CER than RuO_2-based electrodes (Yi et al., 2007). Therefore, the mixing of RuO_2 and IrO_2 together as MMO electrodes has been found to be more stable. In the MMO electrode based on RuO_2/IrO_2, it has been found that IrO_2 helps to limit the oxidation of Ru^{4+} and thus protects the most active catalyst in the harsh acidic environment (Hoseinieh et al., 2010).

In a study by Huang et al. (2020), prepared RuO_2 nanoparticles by a simple hydrothermal growth method followed by electrodeposition had been used to create a RuO_2-TiO_2 electrode with a nanostructured shape to improve the CER, reducing energy consumption and environmental

Table 11.2 Selected electrocatalysts reported for chlorine evolution reaction.

Electrocatalyst	Catalyst category	Electrolyte, pH	Overpotential at 10 mA/cm^2	Current efficiency (%)	References
RuO_2/TNA	RuO_2 based	0.1 M NaCl, pH = 7	40 mV	96	Kim et al. (2018)
RuO_2−TiO_2	RuO_2 based	5 M NaCl, pH = 3	44 mV	NA	Xiong et al. (2016)
RuO_2 NPs@TiO_2 NBs	RuO_2 based	Saturated NaCl, pH = 3.1	70 mV	90	Huang et al. (2020)
RuO_2@TiO_2	RuO_2 based	Saturated NaCl, pH = 2	73 mV	90	Jiang et al. (2017)
RuO_2/lignin carbon electrode	RuO_2 based	Saturated NaCl	64 mV	NA	Chi et al. (2021)
15 wt % RuO_2 on TiO_2	RuO_2 based	4 M NaCl, pH = 3	220 mV	NA	Menzel et al. (2013)
Commercial DSA (Siontech)	Ru−Ir MMO	0.1 M $HClO_4$ + 1 M NaCl, pH = 0.9	105 mV	96	Lim et al. (2020)
Ti−Ru−Ir MMO	Ru−Ir MMO	4 M NaCl, pH = 2	125 mV	97	Zeradjanin et al. (2014)
15 wt % (Ru−Ir) on TiO_2	Ru−Ir MMO	4 M NaCl, pH = 3	140 mV	NA	Menzel et al. (2013)
$Ru_{0.3}Sn_{0.7}O_2$/Ti	Ru−Ti−Sn MMO	3.5 M NaCl, pH = 3.5	30 mV	NA	Chen et al. (2012)
Ti/RuO_2−IrO_2−Sb_2O_5−SnO_2	Ru−Ti−Sn−Sb MMO	Seawater	75 mV[28]	86	Wang et al. (2012)
Ti/RuO_2−Sb_2O_5−SnO_2	Ti−Ru−Sb−Sn MMO	0.5 M Na_2SO_4 + seawater	120 mV	90	Chen et al. (2011a)
($Ru_{0.3}Ti_{0.34}Sn_{0.3}Sb_{0.06}$)$O_2$−$TiO_2$	Ru−Ti−Sn−Sb MMO	5 M NaCl, pH = 2	110 mV	NA	Xiong et al. (2013)
IrO_2/TiO_2 nanosheet arrays	IrO_2 based	Saturated NaCl, pH = 2	44 mV	96	Wang, Xue, et al. (2021)

Ti/Ir$_{0.8}$Nd$_{0.2}$O$_x$	IrO$_2$ based	5 M NaCl, pH = 2	60 mV	91	Hu et al. (2021)
15 wt % IrO$_2$ on TiO$_2$	IrO$_2$ based	4 M NaCl, pH = 3	240 mV	NA	Menzel et al. (2013)
Pt$_1$/carbon nanotubes	Pt based	0.1 M HClO$_4$ + 1 M NaCl, pH = 0.9	97.1 mV	97	Lim et al. (2020)
CoSb$_2$O$_x$	Transition-metal based	4 M NaCl, ph = 2	430 mV	97	Moreno-Hernandez et al. (2019)
NiSb$_2$O$_x$	Transition-metal based	4 M NaCl, ph = 2	450 mV	96	Moreno-Hernandez et al. (2019)
MnSb$_2$O$_x$	Transition-metal based	4 M NaCl, ph = 2	470 mV	90	Moreno-Hernandez et al. (2019)
Co$_3$O$_4$/F-doped SnO$_2$	Transition-metal based	Saturated NaCl pH = 3.1	200 mV	90	Zhu et al. (2018a)

impact compared to the conventional thermal decomposition process. Low-magnification SEM images showed that the TiO$_2$ nanoparticles grew perpendicularly on the substrate with a well-aligned and open nanoarray morphology as shown in Fig. 11.5A. The RuO$_2$ nanoparticles were decorated on the belt-like structures. High-magnification SEM images revealed that the belt-like TiO$_2$ had a smooth surface with a belt width of about 400 nm and length of $1-2$ μm. The prepared

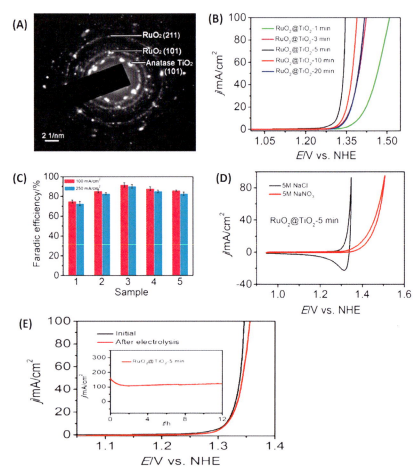

Figure 11.5 (A) SAED patterns, (B) LSV polarization curves, (C) Faradic efficiencies, (D) CV in two different electrolytes, (E) LSV and $j - t$ curve during the long-term electrolysis RuO$_2$ NPs-TiO$_2$. Taken from Huang, J., Hou, M., Wang, J., Teng, X., Niu, Y., Xu, M., & Chen, Z. (2020). RuO$_2$ nanoparticles decorate belt-like anatase TiO$_2$ for highly efficient chlorine evolution. Electrochimica Acta, 339, 135878.

RuO_2-TiO_2 electrode required only a low potential of 1.34 V (overpotential 80 mV) to produce a catalytic current density of 50 mA/cm^2 and had a current efficiency of up to 90% for chlorine evolution (Huang et al., 2020). The selectivity of the RuO_2-TiO_2 electrode toward the CER was compared using two different electrolytes, that is, 5 M NaCl and 5 M $NaNO_3$ solutions. The selectivity was also determined by measuring of Faradaic efficiency for CER through constant current electrolysis using the iodometric titration technique which was around 90%. Furthermore, for the long-term stability for CER, RuO_2-TiO_2 catalytic activity was tested through electrolysis at a constant potential of 1.35 V versus NHE in a saturated NaCl solution (pH = 3.1). Fig. 11.5E demonstrates that the RuO_2-TiO_2 catalytic activity remains almost constant throughout the 12 h electrolysis, indicating negligible degradation (Huang et al., 2020).

Heo et al. (2020) investigated the use of three-dimensional supporting electrodes made of TiO_2 nanotube arrays coated with RuO_2 through pulsed electrodeposition for the purpose of chlorine and hydrogen evolution. FESEM and XRD confirmed that despite undergoing annealing processes in both air and Ar atmospheres, the morphology of the nanotubes hardly undergoes any changes indicating that the as-anodized nanotubes transform from an amorphous state to a single anatase phase (Heo et al., 2020). At 1.090 and 1.125 V versus SCE, the prepared electrode from $RuO_2/b-TiO_2$ demonstrated a current density of 10 and 100 mA/cm^2, respectively in 5 M NaCl electrolyte. DPD method was followed to determine the faradaic efficiency of $RuO_2/b-TiO_2$ nanotubes for CER in which a constant current density of 5 mA/cm^2 was applied during electrolysis in 5.0 M NaCl. Based on this analysis, the faradaic efficiency of the CER was found to be 81.6% (Heo et al., 2020). In another similar study, antimony (Sb) was doped into RuO_2 and TiO_2 nanotube arrays (Cao et al., 2013). The addition of Sb^{3+} to the electrolyte improved the electrodeposition of ruthenium and the thermal oxidation process, resulting in smaller RuO_2 nanoparticles. The Sb-doped RuO_2/TiO_2 electrode showed high electrocatalytic activity for chlorine evolution and reached 170 mA/cm^2 at 1.2 V (Cao et al., 2013). Furthermore, the doping of Sb^{3+} reduced the charge transfer resistance compared to RuO_2/TiO_2 electrode (Cao et al., 2013).

Lim et al. (2021) doped Niobium on the RuO_2-TiO_2 by hydrothermal method in which the prepared catalyst $RuO_2/Nb:TiO_2$-A200, containing 1% of RuO_2 outperformed a DSA with an overpotential of

22 mV and a high Faradaic Efficiency (FE) of 97.3% in 0.6 M NaCl, a seawater-mimicking electrolyte. The morphology results confirmed that ~ 1.7 nm of rutile RuO_2 crystals were effectively deposited and dispersed onto TiO_2 support with sizes of $20-30$ nm. The low-temperature annealing process at $200°C$ created a superior active site covered by a single layer of TiO_2. This ultrathin TiO_2 protection layer greatly improved the electrochemical durability of RuO_2 and prevented degradation in catalytic activity. In contrast, without Nb doping, Ti could not diffuse into RuO_2 nanoparticles at low temperature, resulting in prominent interfaces between RuO_2 and TiO_2. The process of Nb doping is a distinctive method used to facilitate the atomic diffusion of Ti in order to create core-shell structure particles between RuO_2 and TiO_2. This core-shell structure is crucial as it serves as the active site for the generation of active chlorine. By introducing Nb, the diffusion of Ti atoms is enhanced, which allows for the formation of a more uniform distribution of RuO_2 on the TiO_2 support. This uniform distribution, in turn, creates a shell-like structure of RuO_2 on the surface of TiO_2, resulting in a core-shell structure particle. This unique synthesis method of Nb doping thus plays a key role in the development of efficient catalysts for the generation of active chlorine (Lim et al., 2021).

A study by Zeradjanin et al. (2014) investigated the selectivity between CER and OER by differential electrochemical mass spectroscopy on a Ti—Ru—Ir-based dimensionally stable anode (DSA). The working electrode had a diameter of 15 mm, with an exposed diameter of 5 mm during analysis, and was immersed in 3.5 M NaCl electrolyte with a pH of 3. Selectivity was found to be 97% for Ti—Ru—Ir catalyst for CER using scanning electrochemical microscopy (SECM) (Zeradjanin et al. 2014). It was concluded that the enhanced activity and selectivity of the Ti—Ru—Ir-based electrode were due to the stable nanostructure and effective dispersion of the active sites on the substrate. The interaction between the oxide surface and water is important as it affects both the corrosion and catalytic activity of the electrodes. The gas current efficiency remained constant at $80\%-90\%$ for all Cl^- concentrations. The remaining faradaic current was used for oxidizing the catalyst surface, which was revealed by CV measurements to be an important part of the formation of active surface sites, necessary for sustained efficient CER (Zeradjanin et al., 2014).

Although the synergistic effect of Ir and Ru significantly improves the activity and selectivity toward CER, however, the high cost of precious metals like Ir greatly increases the cost of the electrode. For this reason, the focus of recent research has been focused on either entirely replacing the precious metals or mixing them with other cheaper MMOs, or utilizing carbon-based substrates to lower the cost. The reduction in the size of the MMO to the atomic level has recently drawn much attention to increasing the active site utilization and electrochemical active surface area (Moreno-Hernandez et al., 2019).

Recently, Chi et al. (2021) prepared Lignin-based monolithic carbon on which RuO_2 nanospheres were incorporated and the prepared electrode provided high electrochemical active surface area, high activity for CER, and high stability with a very low resistance of $R_S = 0.73\ \Omega$. Fig. 11.6 shows the schematic diagram of the preparation of the electrode. Acetylene black was ground together with dried lignin in different ratios and then carbonization in a tubular furnace followed by Ru nanosphere deposition in a three-electrode electrochemical cell (Chi et al., 2021). The prepared electrode RuO_2/CE- 10% showed a current density of 129 mA/Cm^2 at 1.2 V, which was higher than the benchmark $Ru_{0.3}Ti_{0.7}$/Ti (Chi et al., 2021). The results of this study are crucial as they can open a gateway for the carbon-based electrodes to officially replace the metal-based substrates for CER application.

Recently, the rapid development of single-atom catalysts with isolated active metal sites has shown great potential for reducing precious metal usage while achieving high CER performance in metal-based anodes. Lim et al. fabricated Pt-N_4 sites (Pt_1/CNT) at carbon nanotubes by calcination of a mixture of a Pt precursor (Pt(II) meso-tetraphenylporphine, PtTPP) and acid-treated CNT at 700°C (Lim et al., 2020). The high-angle annular dark-field scanning transmission electron microscopy (HAADF-STEM) and extended X-ray absorption fine structure (EXAFS) analysis confirmed the presence of uniformly dispersed Pt atoms (2.7 wt%). The cyclic voltametric curve of the Pt_1/CNT in 1.0 M NaCl electrolyte exhibited supervisor electrocatalytic activity toward CER, with an overpotential of 50 mV to deliver 10 mA/cm^2 (Fig. 11.7). The CER selectivity on the Pt_1/CNT catalyst was measured as 96.6% with iodometric titration. Chronogamperometry of the Pt_1/CNT has indicated that the 72% of the current value was achieved after 12 hours. Theoretical calculation results indicated that the chlorine adsorption free energy (ΔG) of PtN_4C_{12} site is the lowest (0.09 V) at zero overpotential compared to other active sites; thus, PtN_4C_{12} is the most plausible active site for CER. Moreover, the positive chemical potential difference of Pt atom in PtN_4C_{12} and that on

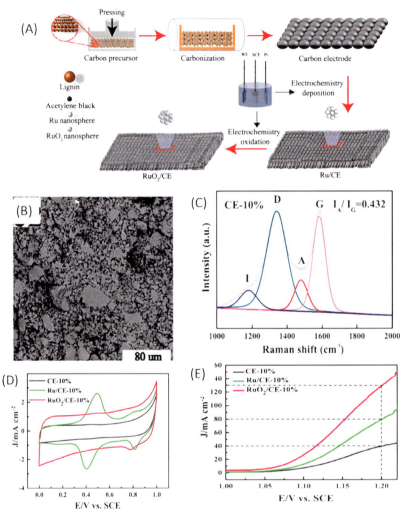

Figure 11.6 Schematic diagram of (A) the RuO$_2$ with lignin-based carbon electrode as substrate, (B) SEM images of carbon electrode (CE)-10%, (C) Raman spectra of CE-10%, (D) CV curves of CE-10%, (E) linear sweep voltammetry of the electrodes at a scan rate of 1 mV/s. *Reproduced from Chi, M., Luo, B., Zhang, Q., Jiang, H., Chen, C., Wang, S., Min, D. (2021). Lignin-based monolithic carbon electrode decorating with RuO$_2$ nanospheres for high-performance chlorine evolution reaction.* Industrial Crops and Products, *159, 113088.*

Nanomaterials for electrochemical chlorine evolution reaction 379

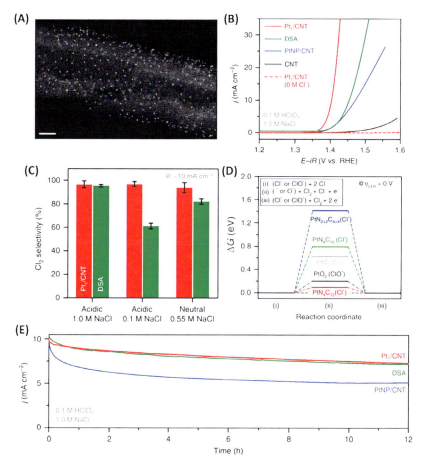

Figure 11.7 (A) HAADF-STEM image of Pt₁/CNT catalyst. Scale bar: 3 nm. (B) CER polarization curves of Pt₁/CNT, PtNP/CNT, DSA, and CNT catalysts obtained in 0.1 M HClO₄ + 1.0 M NaCl at an electrode rotation speed of 1600 rpm and a scan rate of 10 mV/s. (C) CER selectivity of Pt₁/CNT and DSA catalysts measured by iodometric titration under different electrolyte conditions. (D) Free energy diagrams for CER over Pt–N₄ clusters and PtO₂ (110) surface at zero overpotential ($\eta_{CER} = 0$ V). (E) Chronoamperograms of Pt₁/CNT and PtNP/CNT catalysts deposited on a carbon paper (1 cm × 1 cm) and DSA catalyst (1 cm × 1 cm) measured in 0.1 M HClO₄ + 1.0 M NaCl for 12 h with a stirring speed of 300 rpm. *Reproduced from Lim, T., Jung, G. Y., Kim, J. H., Park, S. O., Park, J., Kim, Y. T., Kang, S. J., Jeong, H. Y., Kwak, S. K., & Joo, S. H. (2020). Atomically dispersed Pt–N4 sites as efficient and selective electrocatalysts for the chlorine evolution reaction. Nature Communications, 11(1), 1–11.*

Pt (111) surface implies that dissolution of Pt on PtN_4C_{12} is more energy demanding compared to that on Pt (111), strongly confirming the superior stability of PtN_4C_{12}. Furthermore, the free energy diagrams for OER on PtN_4C_{12} were calculated. The thermodynamic overpotentials for formation of OOH^* at overpotential of 0.13 V is high, certifying a poor activity for OER.

In another study by Liu et al. (2023), ruthenium-based single-atom catalyst with active oxygen was fabricated that achieved a mere ~ 30 mV overpotential for a current density of 10 mA/cm^2 in acidic 1 M NaCl solutions. Remarkably, this catalyst demonstrates over 1000 hours of stable, high-selectivity Cl_2 production at 1000 mA/cm^2. Operando studies and computational analyses show that chloride ions directly adsorb onto the Ru single atoms, reducing the energy barrier and enhancing selectivity compared to RuO_2 electrodes. The remarkable stability exhibited by this chlorine evolution catalyst provides novel insights and opens up opportunities for cost-effective single-atom catalysts to replace the expensive commercial DSA. In another recent study on single atom catalysts, Wang et al. (2024) engineered Ir_1O_4-based single atoms with TiO_2 and the prepared catalyst Ir_1O_4 showed an overpotential of only 71.8 mV with a Tafel slope of 53 mV/dec significantly better than DSA, which had an overpotential of 95.8 mV and tafel slope of 60 mV/dec. Additionally, the Ir_1O_4 anode maintained excellent durability for 200 hours, far surpassing the 2-hour lifespan of Ir_1O_6. Mechanistic insights revealed that the active CER center shifted from the unsaturated Ir in Ir_1O_4 to top-coordinated O in Ir_1O_6, influencing Cl species adsorption energy and activity. The amorphous structure of Ir_1O_4, combined with restricted water dissociation, effectively prevented oxygen permeation through the Ti substrate, ensuring long-term stability (Wang et al., 2024). This research highlights the pivotal role of single-atom coordination structures in catalyst reactivity and presents a straightforward method to create highly active single-atom CER anodes through titanium oxide amorphization.

Mixed metal oxides (MMO) have been investigated for their catalytic properties for the last many decades owing to their unique properties like high exposed surface area, porosity, and their n- or p-semiconductive character (Haralambous et al., 1991). The performance of RuO_2 has been exceptional in wide spectral of operation for CER and many efforts have been made to further enhance its performance. The lack of interest in cheaper MMO has been because of sluggish kinetics for CER, which results in higher reaction potential and lower conductivity. The low conductivity of the bulk MMO is because of its limited accessibility of electron transfer. Therefore, earth-abundant MMOs have found limited success when it comes to the harsh environment of CER. However, there are many ways in which the structure of the MMO can be modified. Cheaper MMO has the advantage of their structure being controlled and hence potentially they can replace precious metals (Chen et al., 2011b). Recently, MMOs have been incorporated with active RuO_2 with an aim to replace expensive metal-based substrates.

Several attempts have been made to incorporate the MMO on established catalysts like RuO_2 with the aim of increasing performance for CER and reducing the cost. The first patent by Beer in 1962 had RuO_2 and TiO_2 at a molar percentage of 50:50 (Beer, 1962). RuO_2 provided the activity for CER while TiO_2 provided the much-needed stability as a conducting substrate increasing the electrode life. In later development, the content of RuO_2 was reduced to 30% and the new ratio provided the best selectivity and stability for CER (Arikawa et al., 1998).

In preparation of electrocatalyst from MMO from transition metals, sol—gel is the most common method followed because of its simplicity and cost-effectiveness (Danial et al., 2015). Furthermore, in the sol—gel method, molecular level network homogeneity of MMO can be easily controlled by controlling the hydrolysis and condensation reaction (Chen et al., 2013). In this way, the nanostructure fabricated has better stability with better dispersion of the active sites. It has been reported that crack-free electrodes fabricated by the sol—gel method have good activity for CER (Chen et al., 2012). A $Ru_{0.3}Sn_{0.7}O_2/Ti$ electrode prepared by the sol—gel method had a low onset potential of $30\,mV$ $10\,mA/cm^2$ in $3.5\,M$ NaCl solution and its activity was also better than the electrode prepared by thermal decomposition (Chen et al., 2012). The improved performance of the prepared $Ru_{0.3}Sn_{0.7}O_2/Ti$ electrode was due to the stable nanoporosity structure (Chen et al., 2012).

Modifying the electronic structure of the active catalyst species such as heteroatom doping on RuO_2 is among the common methods to increase the activity, stability, and selectivity of CER. By Hume Rothery rule, it is possible to combine other metal ions such as Ti^{4+}, Sb^{5+}, and Sn^{4+} with RuO_2 forming a stable rutile solid solution (Chen et al., 2013). The resulted in solid solution holds a stable well-defined nanostructure and even distribution of the active sites. The addition of transitional metal oxide onto the noble metal may induce a synergistic effect resulting in an increase in the activity and stability of the electrode. Doping of TiO_2

significantly stabilizes active RuO_2 on the DSA electrode in the Chlor-alkali process (Wang et al., 2021a).

Tin(IV) oxide (SnO_2) has been tested as an alternative to TiO_2 as it forms a stable crystalline structure with RuO_2. It has been reported that in acidic conditions, an electrode made of RuO_2-SnO_2/Ti showed a longer lifespan than the RuO_2/Ti electrode (Zhang et al., 2019). An electrode made of $Ti/RuO_2-Sb_2O_5-SnO_2$ by pyrolysis of mixed metal salts has shown good activity for CER from seawater electrolysis as shown in Table 11.2. The catalyst $Ti/RuO_2-Sb_2O_5-SnO_2$ showed stability because in theory Sb^{5+} metal ions are well incorporated with Sn^{4+} metal ions owing to their similar ionic radius (Chen et al., 2011a). Similarly, another catalyst ($Ti/RuO_2-IrO_2-Sb_2O_5-SnO_2$) in which Sb^{5+} and Sn^{4+} were combined with RuO_2 and IrO_2 with the same pyrolysis method showed a low onset potential of 75 mV at 10 mA/cm^2 (Wang et al., 2012). The prepared catalyst $RuO_2-Sb_2O_5-SnO_2$ had stable nanopores, which existed in the form of a solution that played the role of active sites for the reaction (Chen et al., 2011a). The results of incorporating Sn and Sb onto RuO_2 and IrO_2 indicate that these multi-component MMOs have great prospects for future chlorine production from seawater electrolysis.

In another study, ($Ru_{0.3}Ti_{0.34}Sn_{0.3}Sb_{0.06}$)$O_2-TiO_2$ nanotube anode was made using anodization, deposition, and annealing (Xiong et al., 2013). The findings showed that well-ordered nanotubes with a large surface area can be infused with active metal oxides. The catalyst effectively attaches to the nanotubes and improves the electrode's electrochemical stability, displaying a high overpotential for oxygen evolution reaction. As a result, the ($Ru_{0.3}Ti_{0.34}Sn_{0.3}Sb_{0.06}$)$O_2-TiO_2$ nanotube anode has a larger potential difference for oxygen and chlorine evolution, leading to improved selectivity toward chlorine evolution compared to traditional dimensionally stable anodes. This improved performance is due to the surface properties and shape of the coated catalyst, as well as the added tin and antimony species affecting electrochemical selectivity.

Apart from SnO_2 and SbO_2, other transition metals like Co, Ti, Mn, Zn, Fe, Cu, and Mg can be successfully doped onto RuO_2 for CER activity. However, their addition could not significantly improve the activity and selectivity for CER. Doping Zn onto the RuO_2 by facile freeze-drying technique has been found to decrease the selectivity for CER (Abbott et al., 2014). Mg has been doped on RuO_2 by spray-freezing freeze-drying technique and has been found to show some limited improved activity for CER; however, Mg-doped catalyst activity was lower than the nondoped RuO_2 (Abbott et al., 2014). Theoretical calculations have concluded that in doping of metals, metals with higher d-electrons than Ru increases the overpotential for OER by lowering the binding strength of the OER intermediates (Saha et al., 2018). Simultaneous OER and CER on Ni dopped Ru ($Ru_{1-x}Ni_xO_{2-y}$) by sol-get method has been tested and it was concluded that the overall activity of Ni-doped catalyst increases for CER with Ni content of 10% (Macounová et al., 2008). Further addition of Ni beyond 10% favors the OER instead of CER. The active sites with Ni-atom showed improved activity for the CER due to its stable nanostructure (Macounová et al., 2008).

11.3.2 Noble metal free nanomaterials for CER

Many attempts have been made to developed noble metal free electrocatalyst for CER. A recent attempt by Gupta et al. (2021) adopted an environmentally friendly method for synthesizing a nonnoble metal/metal oxide (Cu-Fe_2O_3)-doped nitrogen-rich carbon matrix. The Cu-Fe_2O_3 catalyst was created by pyrolyzing a gel mixture of metal salts, nitrogen, and carbon source at high temperature under an inert Ar atmosphere. The gel was subjected to pyrolysis at 800°C in Ar atmosphere, which allowed for the in situ incorporation of Cu in its metallic form, Fe_2O_3, and nitrogen into the carbon composite. The use of a single gel precursor for in situ incorporation of Cu, Fe_2O_3, and nitrogen into the carbon matrix simplified the synthesis process and produced bifunctional activity for HCl electrolysis, where it assists chlorine evolution at the anode and oxygen reduction at the cathode. The combination of Cu and Fe_2O_3 in the N-doped carbon matrix leads to enhanced activity and stability. The performance of the catalyst was optimized by varying the metal content in the precursor (2, 5, and 10 wt.%), resulting in different compositions and flake-like morphologies. The best-performing catalyst had a 5% metal content of Cu-Fe_2O_3, with a higher oxidative current density of 92.1 mA/cm^2 at 1.7 V versus RHE. The synergy between Cu and Fe_2O_3 and the strong interactions between Fe_2O_3 and nitrogen in the carbon matrix contribute to the enhanced activity of the Cu-Fe_2O_3 (5%)/NC catalyst (Gupta et al., 2021).

In another study by Alavijeh et al. (2021), a cost-effective and precious metal-free electrocatalyst ($Ti_{0.35}V_{0.35}Sn_{0.25}Sb_{0.05}$ oxide) was prepared using dip-coating thermal decomposition and the prepared catalyst showed an onset potential of 1.25 V versus RHE with a current efficiency of 88% for CER. The better performance of the catalyst was attributed to the successful incorporation of vanadium pentoxide on the Ti substrate whose presence lowered the band gap of the catalyst as also confirmed by DFT calculations (Alavijeh et al., 2021). The prepared catalyst $Ti_{0.35}V_{0.35}Sn_{0.25}Sb_{0.05}$-oxide showed a flower-like stable nanostructure, which is due to the growth of Sn and V in the coating materials (Alavijeh et al., 2021).

One of the main challenges for precious metal-free electrocatalysts is durability as MMO by itself does not survive in the acidic environment of CER. In a study by Moreno-Hernandez et al. (2019), $CoSb_2O_x$ was found to have good stability and activity for CER. The prepared catalyst $CoSb_2O_x$ showed a potential of 1.804 V versus NHE at 100 mA/cm^2 after 250 hours of operation in a 4 M NaCl solution with a pH of 2.0 (Moreno-Hernandez et al., 2019).

11.3.3 Metal-free electrode for CER

Metal-free electrocatalysts are challenging to fabricate due to their limited stability and low activity compared to metal-based catalysts. Additionally, finding a suitable nonmetal material that can match the performance of metal catalysts is challenging as well. The development of metal-free electrocatalysts requires a deep understanding of the underlying mechanism of electrocatalytic reactions, as well as the design and synthesis of novel nonmetal materials with tailored structures and properties. This becomes even more challenging when it comes to CER as the reaction mechanism has not been fully understood. The current research on metal-free electrocatalysts is focused on improving their performance by enhancing their stability, selectivity, and activity, which are essential properties for practical applications.

There have been many efforts in developing metal-free electrocatalyst for CER. In a research study, differential electrochemical mass spectrometry (DEMS) was used to detect chlorine volatile products produced by the electrolysis of highly saline water (0.6 M NaCl) at a boron-doped diamond electrode under flow-through conditions at various pH levels (Mostafa et al., 2018). The results indicated that oxygen evolution becomes more competitive with increasing pH, as the onset potential shifts to more negative values, becoming more competitive with chloride oxidation at pH 2.0,

7.0, and 12.0 (2.3 V, 1.5 V, and 1 V vs Ag/AgCl, respectively). No volatile Cl2 species were observed at pH 12.0 (Mostafa et al., 2018). These results complement previous studies on the evolution of soluble species and will contribute to a better understanding of the mechanisms involved in the electro chlorination process as an emerging disinfection technology.

Carbon can play a crucial role in the development of noble metal free electrocatalyst for CER. Carbon-based materials have recently shown great potential and emerging as a new class of electrocatalysts for water splitting, OER, and CER applications (Khatun et al., 2021; Kordek et al., 2019). Carbon has many advantages such as low price, stability, high conductivity, and high surface area (Wang et al., 2021b). The main advantage of carbon is that it can be easily converted into a nanostructure electrocatalyst with better stability and a variety of active functional groups can actively be introduced on its surface. Carbon nanomaterials like carbon nanotubes, carbon nanofibers, and carbon nanohorns are very stable materials covalently bonded by carbon atoms. Carbon-based materials have the potential to replace Ti-based metal substrates. There is an increasing interest in the research community on carbon-based electrocatalysts, especially for CER application from seawater. It has been reported that Ti-based substrates with active MMO catalysts can undergo corrosion in the harsh environment of CER (Trasatti, 2000). Therefore, carbon-based materials are considered widely suitable for replacing metal substrates. Furthermore, the carbon surface provides support to a variety of different active species and functional groups due to its unique physicochemical properties which can be ideal for CER.

11.4 Conclusion and future perspectives

The Chlor-alkali process is well established, and the currently used DSA electrode has high performance and selectivity for CER. However, there are several limitations of this process that limit its full potential. Apart from its high overpotential, DSA exhibits low selectivity in low Cl concentration and near-neutral pH meaning it is not effective in seawater electrolysis and is limited to concentrated brine solution from freshwater only. Seawater is one of the most abundant, cheapest, and readily accessible resources on earth, but yet to be directly used for chemicals productions. Water is vital to all living organisms and the existence of human beings on earth. Water comprises about 71% of our planet earth; however, only about 2.5% is fresh drinkable water and the rest is saline seawater. Electrocatalysis can play a vital role in using water potential for our future energy and sustainable chemical production needs. The potential of developing a DSA-like electrode for seawater electrolysis could be huge using renewable electricity from solar, wind, and tidal sources.

Renewable and sustainable energy is crucial for the suitability of life on earth as the demand for energy increases with rapid growth in the world economy and world population. It is evident that a single source of renewable energy would be insufficient to replace the current fossil-fuel-based economy. A comprehensive model where different sources of renewable energies are interconnected together for a greater outcome is the practical scenario. Chlor-alkali process can be replaced by a more environmentally friendly process such as the electrolysis of seawater where renewable electricity is used to drive the electrochemical cell producing Cl_2, NaOH and green hydrogen. Green hydrogen can revolutionize the importance of Chlor-alkali process. Such a model is vital as it can produce green hydrogen while utilizing seawater unlike Chlor-alkali, which depends on freshwater.

Developing enabling technologies to collectively and coherently utilize seawater's bounteous resources is vital for a green and sustainable future for the world. A futuristic concept, as illustrated in Fig. 11.8, the direct seawater electrolysis powered by renewable electricity generated from solar and wind is an ideal model to collectively and coherently utilize

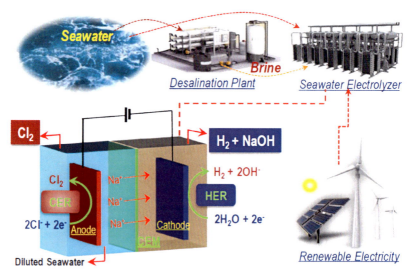

Figure 11.8 Illustrative concept of the seawater electrolysis technology to produce H_2, Cl_2, and NaOH using renewable electricity.

these bounteous resources for the production of high-value clean fuel (H_2) and commodity chemicals (Cl_2 and NaOH). There are nearly 20,000 operational desalination plants in the world to supply 99.8 million m^3 of fresh water every day. On the one hand, desalination mitigates the freshwater shortage; however, on the other hand, it generates a huge waste stream (brines, the concentrated seawater containing 5%—8% NaCl) that imposes a serious impact on the ocean environment. To this end, coupling seawater electrolysis with desalination plants can efficiently treat brines while simultaneously producing H_2, Cl_2, and NaOH (Fig. 11.8). As such, the development of seawater electrolysis will bring electrocatalysis into a new paradigm to collectively address energy and environmental sustainability issues as well as green chemical productions, however, highly challenging due mainly to the lack of electrocatalysts capable of efficiently catalyzing CER.

The Chlor-alkali industry is an established technology and the ongoing research and development aim to develop precious metal-free electrocatalysts for CER with similar performance as that of commercial DSA. Furthermore, there is an increasing interest in developing electrocatalysts that will work for seawater electrolysis without the need for desalination, which would ultimately be more beneficial. In this regard, good progress has been made in the last few decades, and researchers are trying to develop an electrocatalyst that can work in commercial seawater electrocatalysis producing hydrogen, chlorine, and caustic soda.

The development of a future electrocatalyst for CER would depend mostly on its design parameters that could offer high conductivity, high surface area, high resistance in a harsh acidic environment, and high selectivity for CER in comparison with OER. The elements employed for the catalyst fabrication have been largely limited to transition metals, such as Ru, Ir, Pt and Ti. It is expected that the incorporation of s block, p block and f block elements to the catalyst that are rich in chemistry will allow fine tune of the electronic and structural properties and help steering the catalytic performance. Moreover, instead of the traditionally used sol—gel approach for preparing mixed metal oxide, the utilization of coordinated compounds such as metal—organic frameworks (MOFs) with metal active sites as well as chemistry tuneable ligands either as the catalytic active material or its precursor should certainly be considered as a power tool toward novel, efficient catalyst.

To date, tremendous efforts have been devoted to reducing the cost of precious metal-based materials by blending with other transition metal-based dopants or incorporating metal oxides. To further improve the cost-effectiveness of Chlor-alkali industry, the minimized use of noble metal is essential. A recent study on single atom Pt catalyst on carbon nanotube that show promising CER activity could be a gateway for fabricating low-cost CER electrocatalysts (Lim et al., 2020). Also, encouraging activities of the single atom catalysts toward CER have been predicted with theoretical calculation (Liu et al., 2022a, 2022b). Without doubt, more research is needed to further validate the long-term use before their effective commercialization and their application in seawater electrocatalysis. However, a highly selective carbon-based CER electrocatalyst with lower shelf life than commercial DSA may still be cost-effective because of the lower price of carbon. Also, carbon-based material offers much flexibility in terms of its surface modification and addition of highly active functional groups. Further research on the addition of highly active functional groups or single atoms on carbon or other low-cost substrates, for example, metal oxide, is expected for novel CER catalyst design.

Another obstacle in designing of CER electrocatalyst is the poorly understood CER mechanism at the heterogeneous catalytic sites even on the binary or ternary metal oxide materials. Further DFT studies are required to investigate the CER and the competing OER mechanism on a variety of active sites to better understand the design parameters for CER electrocatalyst (Liu et al., 2022a, 2022b). More importantly, more experimental evidence is urgently needed to identify the true active species under operando CER conditions. Thus, in situ spectroscopic characterization experimental techniques is paramount to elucidate the electronic and structural properties and better model and understand the CER mechanism (Ha et al., 2019).

Beside the search for highly active CER site on the nanomaterials, the electrode surface engineering is a nontrivial approach. For example, to create superaerophobic electrode surface by constructing three-dimensional porous nanostructure is conducive to enhanced mass/electron transport and also to reduce the overpotential for chlorine bubble generation, achieving ultrahigh current density (Jiang et al., 2017). Moreover, the introduction of protective top layer (e.g., CeO_x or TiO_2 layer) onto the nanomaterials will assist blocking undesired redox species (e.g., Cl^-) away from catalyst surface, endowing an anticorrosion surface (Obata & Takanabe, 2018). Also, a coating of atomically thin metal oxide layer (e.g., TiO_2) deposited onto the active nanomaterials will tune the surface charge density, and thus the catalytic activity, which could be very promising approach toward reduced operational expenses (Hu et al., 2019; Zhu, 2018b).

The potential for industrial commercialization of seawater electrocatalysis by renewable energy sources could be huge. The current Chloralkali industry relies on saturated NaCl solution made from freshwater which is unsustainable keeping in view our limited freshwater resources. Vigorous research should be carried out in adaptation of the current Chlor-alkali industry on seawater electrocatalysis by economically using renewable electricity sources to fully utilize the potential of renewable electricity. To achieve this, new reactor designs are needed taking inspiration from the currently used membrane-based reactor to deal with the additional reactants and inconsistency of seawater.

References

Abbott, D. F., Petrykin, V., Okube, M., Bastl, Z., Mukerjee, S., & Krtil, P. (2014). Selective chlorine evolution catalysts based on Mg-doped nanoparticulate ruthenium dioxide. *Journal of the Electrochemical Society*, *162*(1), H23.

Alavijeh, M. M., Habibzadeh, S., Roohi, K., Keivanimehr, F., Naji, L., & Ganjali, M. R. (2021). A selective and efficient precious metal-free electrocatalyst for chlorine evolution reaction: An experimental and computational study. *Chemical Engineering Journal*, *421*, 127785.

Arikawa, T., Murakami, Y., & Takasu, Y. (1998). Simultaneous determination of chlorine and oxygen evolving at RuO_2/Ti and RuO_2-TiO_2/Ti anodes by differential electrochemical mass spectroscopy. *Journal of Applied Electrochemistry*, *28*(5), 511–516.

Baptista, F. R., Belhout, S. A., Giordani, S., & Quinn, S. J. (2015). Recent developments in carbon nanomaterial sensors. *Chemical Society Reviews*, *44*(13), 4433–4453.

Beer, H. B. (1962). Method of chemically plating base layers with precious metals of the platinum group. U.S. PATENTS.

Beer, H. B. (1980). The invention and industrial development of metal anodes. *Journal of the Electrochemical Society*, *127*(8), 303C.

Bergner, D. (1990). Reduction of by-product formation in alkali chloride membrane electrolysis. *Journal of Applied Electrochemistry*, *20*(5), 716–722.

Brinkmann, T., Giner Santonja, G., Schorcht, F., Roudier, S., & Sancho, L. D. (2014). *Best available techniques (BAT) reference document for the production of chlor-alkali. Industrial Emissions Directive 2010/75/EU (Integrated Pollution Prevention and Control)*. Publications Office of the European Union.

Buné, N. Y., Filatov, V., Losev, V., & Portnova, M. Y. (1985). Effect of foreign anions on the kinetics of chlorine and oxygen evolution on ruthenium-titanium oxide anodes under the conditions of chlorine electrolysis. *Sov. Electrochem.(Engl. Transl.); (United States)*, *20*(10), 5582768.

Cao, H., Lu, D., Lin, J., Ye, Q., Wu, J., & Zheng, G. (2013). Novel Sb-doped ruthenium oxide electrode with ordered nanotube structure and its electrocatalytic activity toward chlorine evolution. *Electrochimica Acta*, *91*, 234–239.

Chen, R., Trieu, V., Schley, B., Harald, N., Kintrup, J., Bulan, A., Weber, R., & Hempelmann, R. (2013). Anodic electrocatalytic coatings for electrolytic chlorine production: a review. *Zeitschrift für Physikalische Chemie*, *227*(5), 651–666.

Chen, R., Trieu, V., Zeradjanin, A. R., Natter, H., Teschner, D., Kintrup, J., Bulan, A., Schuhmannb, W., & Hempelmann, R. (2012). Microstructural impact of anodic coatings on the electrochemical chlorine evolution reaction. *Physical Chemistry Chemical Physics*, 14(20), 7392−7399.

Chen, S., Zheng, Y., Wang, S., & Chen, X. (2011a). Ti/RuO2−Sb2O5−SnO2 electrodes for chlorine evolution from seawater. *Chemical Engineering Journal*, 172(1), 47−51.

Chen, Z., Higgins, D., Yu, A., Zhang, L., & Zhang, J. (2011b). A review on non-precious metal electrocatalysts for PEM fuel cells. *Energy & Environmental Science*, 4(9), 3167−3192.

Cherevko, S., Geiger, S., Kasian, O., Kulyk, N., Grote, J.-P., Savan, A., Shrestha, B. R., Merzlikin, S., Breitbach, B., Ludwig, A., & Mayrhofer, K. J. J. (2016). Oxygen and hydrogen evolution reactions on Ru, RuO_2, Ir, and IrO_2 thin film electrodes in acidic and alkaline electrolytes: A comparative study on activity and stability. *Catalysis Today*, 262, 170−180.

Chi, M., Luo, B., Zhang, Q., Jiang, H., Chen, C., Wang, S., & Min, D. (2021). Lignin-based monolithic carbon electrode decorating with RuO_2 nanospheres for high-performance chlorine evolution reaction. *Industrial Crops and Products*, 159, 113088.

Crook, J., & Mousavi, A. (2016). The chlor-alkali process: A review of history and pollution. *Environmental Forensics*, 17(3), 211−217.

Danial, A. S., Saleh, M. M., Salih, S. A., & Awad, M. I. (2015). On the synthesis of nickel oxide nanoparticles by sol−gel technique and its electrocatalytic oxidation of glucose. *Journal of Power Sources*, 293, 101−108.

Exner, K. S. (2019). Controlling stability and selectivity in the competing chlorine and oxygen evolution reaction over transition metal oxide electrodes. *ChemElectroChem, 6* (13), 3401−3409.

Exner, K. S., Anton, J., Jacob, T., & Over, H. (2014). Chlorine evolution reaction on RuO_2 (110): Ab initio atomistic thermodynamics study-Pourbaix diagrams. *Electrochimica Acta*, 120, 460−466.

Exner, K. S., Anton, J., Jacob, T., & Over, H. (2015). Microscopic insights into the chlorine evolution reaction on RuO_2 (110): A mechanistic ab initio atomistic thermodynamics study. *Electrocatalysis*, 6(2), 163−172.

Fukuda, K.-I., Iwakura, C., & Tamura, H. (1979). Effect of the addition of NH_4F on anodic behaviors of DSA-type electrodes in H_2SO_4-$(NH_4)_2SO_4$ solutions. *Electrochimica Acta*, 24(4), 367−371.

Gupta, D., Kafle, A., Chaturvedi, A., & Nagaiah, T. C. (2021). Recovery of high purity chlorine by Cu-doped Fe_2O_3 in nitrogen containing carbon matrix: A bifunctional electrocatalyst for HCl electrolysis. *ChemElectroChem*, 8(15), 2858−2866.

Ha, H., Jin, K., Park, S., Lee, K.-G., Cho, K. H., Seo, H., Ahn, H.-Y., Lee, Y. H., & Nam, K. T. (2019). Highly selective active chlorine generation electrocatalyzed by Co_3O_4 nanoparticles: mechanistic investigation through in situ electrokinetic and spectroscopic analyses. *The Journal of Physical Chemistry Letters*, 10(6), 1226−1233.

Hansen, H. A., Man, I. C., Studt, F., Abild-Pedersen, F., Bligaard, T., & Rossmeisl, J. (2010). Electrochemical chlorine evolution at rutile oxide (110) surfaces. *Physical Chemistry Chemical Physics*, 12(1), 283−290.

Haralambous, K., Loizos, Z., & Spyrellis, N. (1991). Catalytic properties of some mixed transition-metal oxides. *Materials letters*, 11(3−4), 133−141.

Hardee, K., & Mitchell, L. (1989). The influence of electrolyte parameters on the percent oxygen evolved from a chlorate cell. *Journal of the Electrochemical Society*, 136(11), 3314.

Heo, S. E., Lim, H. W., Cho, D. K., Park, I. J., Kim, H., Lee, C. W., Ahn, S. H., & Kim, J. Y. (2020). Anomalous potential dependence of conducting property in black titania nanotube arrays for electrocatalytic chlorine evolution. *Journal of Catalysis, 381*, 462−467.

Hoseinieh, S., Ashrafizadeh, F., & Maddahi, M. (2010). A comparative investigation of the corrosion behavior of $RuO_2-IrO_2-TiO_2$ coated titanium anodes in chloride solutions. *Journal of the Electrochemical Society*, 157(4), E50.

Hu, C., Zhang, L., & Gong, J. (2019). Recent progress made in the mechanism comprehension and design of electrocatalysts for alkaline water splitting. *Energy & Environmental Science*, 12(9), 2620−2645.

Hu, Jiajun, Xu, Haoran, Feng, Xiangdong, Lei, Lecheng, He, Yi, & Zhang, Xingwang (2021). Neodymium-doped IrO_2 electrocatalysts supported on titanium plates for enhanced chlorine evolution reaction performance. *ChemElectroChem*, 8(6), 1204−1210.

Huang, J., Hou, M., Wang, J., Teng, X., Niu, Y., Xu, M., & Chen, Z. (2020). RuO_2 nanoparticles decorate belt-like anatase TiO_2 for highly efficient chlorine evolution. *Electrochimica Acta*, 339, 135878.

Jaksić, M., Despic, A. R., Nikolic, B. Z., & Maksic, S. M. (1972). Effect of some anions on the chlorate cell process. *Croatica Chemica Acta*, 44(1), 61−66.

Jiang, M., Wang, H., Li, Y., Zhang, H., Zhang, G., Lu, Z., Sun, X., & Jiang, L. (2017). Superaerophobic RuO_2-based nanostructured electrode for high-performance chlorine evolution reaction. *Small*, 13(4), 1602240.

Jirkovský, J., Hoffmannová, H., Klementová, M., & Krtil, P. (2006). Particle size dependence of the electrocatalytic activity of nanocrystalline RuO_2 electrodes. *Journal of the Electrochemical Society*, 153(6), E111.

Karlsson, R. K., & Cornell, A. (2016). Selectivity between oxygen and chlorine evolution in the chlor-alkali and chlorate processes. *Chemical Reviews*, 116(5), 2982−3028.

Kelly, R. G., Scully, J. R., Shoesmith, D., & Buchheit, R. G. (2002). *Electrochemical techniques in corrosion science and engineering*. CRC Press.

Khatun, S., Hirani, H., & Roy, P. (2021). Seawater electrocatalysis: activity and selectivity. *Journal of Materials Chemistry A*, 9(1), 74−86.

Kim, J., Kim, C., Kim, S., & Yoon, J. (2018). RuO_2 coated blue TiO_2 nanotube array (blue $TNA-RuO_2$) as an effective anode material in electrochemical chlorine generation. *Journal of Industrial and Engineering Chemistry*, 66, 478−483.

Kim, S. K., Shin, D.-M., & Rhim, J. W. (2021). Designing a high-efficiency hypochlorite ion generation system by combining cation exchange membrane aided electrolysis with chlorine gas recovery stream. *Journal of Membrane Science*, 630, 119318.

Kordek, K., Jiang, L., Fan, K., Zhu, Z., Xu, L., Al-Mamun, M., Dou, Y., Chen, S., Liu, P., Yin, H., Rutkowski, P., & Zhao, H. (2019). Two-step activated carbon cloth with oxygen-rich functional groups as a high-performance additive-free air electrode for flexible zinc−air batteries. *Advanced Energy Materials*, 9(4), 1802936.

Krishtalik, L. (1981). Kinetics and mechanism of anodic chlorine and oxygen evolution reactions on transition metal oxide electrodes. *Electrochimica Acta*, 26(3), 329−337.

Lakshmanan, S., & Murugesan, T. (2014). The chlor-alkali process: Work in progress. *Clean Technologies and Environmental Policy*, 16(2), 225−234.

Lim, H. W., Cho, D. K., Park, J. H., Ji, S. G., Ahn, Y. J., Kim, J. Y., & Lee, C. W. (2021). Rational design of dimensionally stable anodes for active chlorine generation. *ACS Catalysis*, 11(20), 12423−12432.

Lim, T., Jung, G. Y., Kim, J. H., Park, S. O, Park, J., Kim, Y.-T., Kang, S. J., Jeong, H. Y., & Kwak, S. K. (2020). Atomically dispersed $Pt-N_4$ sites as efficient and selective electrocatalysts for the chlorine evolution reaction. *Nature communications*, 11(1), 1−11.

Liu, J., Hinsch, J. J., Yin, H., Liu, P., Zhao, H., & Wang, Y. (2022a). TMN4 complex embedded graphene as efficient and selective electrocatalysts for chlorine evolution reactions. *Journal of Electroanalytical Chemistry*, 907, 116071.

Liu, J., Hinsch, J. J., Yin, H., Liu, P., Zhao, H., & Wang, Y. (2022b). Low-dimensional metal−organic frameworks with high activity and selectivity toward electrocatalytic chlorine evolution reactions. *The Journal of Physical Chemistry C*, 126(16), 7066−7075.

Liu, Y., Li, C., Tan, C., Pei, Z., Yang, T., Zhang, S., Huang, Q., Wang, Y., Zhou, Z., Liao, X., Dong, J., Tan, H., Yan, W., Yin, H., Liu, Z.-Q., Huang, J., & Zhao, S. (2023). Electrosynthesis of chlorine from seawater-like solution through single-atom catalysts. *Nature Communications, 14*(1), 2475.

Macounová, K., Makarova, M., Jirkovský, J., Franc, J., & Krtil, P. (2008). Parallel oxygen and chlorine evolution on $Ru_{1-x}NixO_{2-y}$ nanostructured electrodes. *Electrochimica Acta, 53*(21), 6126−6134.

Menzel, N., Ortel, E., Mette, K., Kraehnert, R., & Strasser, P. (2013). Dimensionally stable $Ru/Ir/TiO_2$-anodes with tailored mesoporosity for efficient electrochemical chlorine evolution. *ACS Catalysis, 3*(6), 1324−1333.

Moreno-Hernandez, I. A., Brunschwig, B. S., & Lewis, N. S. (2019). Crystalline nickel, cobalt, and manganese antimonates as electrocatalysts for the chlorine evolution reaction. *Energy & Environmental Science, 12*(4), 1241−1248.

Mostafa, E., Reinsberg, P., Garcia-Segura, S., & Baltruschat, H. (2018). Chlorine species evolution during electrochlorination on boron-doped diamond anodes: In-situ electrogeneration of Cl_2, Cl_2O and ClO_2. *Electrochimica Acta, 281*, 831−840.

Obata, K., & Takanabe, K. (2018). A permselective CeO_x coating to improve the stability of oxygen evolution electrocatalysts. *Angewandte Chemie, 130*(6), 1632−1636.

Ou, H., Wang, D., & Li, Y. (2021). How to select effective electrocatalysts: Nano or single atom? *Nano Select, 2*(3), 492−511.

Saha, S., Kishor, K., & Pala, R. G. (2018). Modulating selectivity in CER and OER through doped RuO_2. *ECS Transactions, 85*(12), 201.

Sohrabnejad-Eskan, I., Goryachev, A., Exner, K. S., Kibler, L. A., Hensen, E. J. M., Hofmann, J. P., & Over, H. (2017). Temperature-dependent kinetic studies of the chlorine evolution reaction over RuO_2 (110) model electrodes. *ACS Catalysis, 7*(4), 2403−2411.

Trasatti, S. (2000). Electrocatalysis: understanding the success of DSA®. *Electrochimica Acta, 45*(15−16), 2377−2385.

Vos, J., & Koper, M. (2018). Measurement of competition between oxygen evolution and chlorine evolution using rotating ring-disk electrode voltammetry. *Journal of Electroanalytical Chemistry, 819*, 260−268.

Wang, Y., Liu, Y., Wiley, D., Zhao, S., & Tang, Z. (2021a). Recent advances in electrocatalytic chloride oxidation for chlorine gas production. *Journal of Materials Chemistry A*.

Wang, C., Shang, H., Jin, L., Xu, H., & Du, Y. (2021b). Advances in hydrogen production from electrocatalytic seawater splitting. *Nanoscale, 13*(17), 7897−7912.

Wang, J., Zhao, L., Zou, Y., Dai, J., Zheng, Q., Zou, X., Hu, L., Hou, W., Wang, R., Wang, K., Shi, Y., Zhan, G., Yao, Y., & Zhang, L. (2024). Engineering the coordination environment of Ir single atoms with surface titanium oxide amorphization for superior chlorine evolution reaction. *Journal of the American Chemical Society, 146*(16), 11152−11163.

Wang, S., Zheng, Y., Wang, S., & Chen, X. (2012). Ti/RuO_2-IrO_2-SnO_2-Sb_2O_5 anodes for Cl_2 evolution from seawater. *Electrochemistry, 80*(7), 507−511.

Wang, Y., Xue, Y., & Zhang, C. (2021). Rational surface and interfacial engineering of IrO_2/TiO_2 nanosheet arrays toward high-performance chlorine evolution electrocatalysis and practical environmental remediation. *Small, 17*(17), 2006587.

Xiong, K., Peng, L., Wang, Y., Liu, L., Deng, Z., Li, L., & Wei, Z. (2016). In situ growth of RuO_2−TiO_2 catalyst with flower-like morphologies on the Ti substrate as a binder-free integrated anode for chlorine evolution. *Journal of Applied Electrochemistry, 46*(8), 841−849.

Xiong, K., Deng, Z., Li, L., Chen, S., Xia, M., Zhang, L., Qi, X., Ding, W., Tan, S., & Wei, Z. (2013). Sn and Sb co-doped RuTi oxides supported on TiO_2 nanotubes anode for selectivity toward electrocatalytic chlorine evolution. *Journal of Applied Electrochemistry, 43*(8), 847−854.

Yi, Z., Kangning, C., Wei, W., Wang, J., & Lee, S. (2007). Effect of IrO_2 loading on $RuO_2-IrO_2-TiO_2$ anodes: A study of microstructure and working life for the chlorine evolution reaction. *Ceramics International, 33*(6), 1087–1091.

Zeradjanin, A. R., Menzel, N., Schuhmann, W., & Strasser, P. (2014). On the faradaic selectivity and the role of surface inhomogeneity during the chlorine evolution reaction on ternary Ti–Ru–Ir mixed metal oxide electrocatalysts. *Physical Chemistry Chemical Physics, 16*(27), 13741–13747.

Zhang, Y., He, P., Jia, L., Zhang, T., Liu, H., Wang, S., Li, C., Dong, F., & Zhou, S. (2019). Dimensionally stable Ti/SnO_2-RuO_2 composite electrode based highly efficient electrocatalytic degradation of industrial gallic acid effluent. *Chemosphere, 224*, 707–715.

Zhao, S., Wang, D.-W., Amal, R., & Dai, L. (2019). Carbon-based metal-free catalysts for key reactions involved in energy conversion and storage. *Advanced Materials, 31*(9), 1801526.

Zhu, X., Wang, P., Wang, Z., Liu, Y., Zheng, Z., Zhang, Q., Zhang, X., Dai, Y., Whangbo, M.-H., & Huang, B. (2018a). Co_3O_4 nanobelt arrays assembled with ultrathin nanosheets as highly efficient and stable electrocatalysts for the chlorine evolution reaction. *Journal of Materials Chemistry A, 6*(26), 12718–12723.

Zhu, Z., Yin, H., He, C. T., Al-Mamun, M., Liu, P., Jiang, L., Zhao, Y., Wang, Y., Yang, H. G., & Tang, Z. (2018b). Ultrathin transition metal dichalcogenide/3d metal hydroxide hybridized nanosheets to enhance hydrogen evolution activity. *Advanced Materials, 30*(28), 1801171.

CHAPTER 12

Photothermal properties of metallic nanostructures for biomedical application

Dorothy Bardhan[1] and Sujit Kumar Ghosh[2]

[1]Department of Chemistry, Assam University, Silchar, Assam, India
[2]Physical Chemistry Section, Department of Chemistry, Jadavpur University, Kolkata, West Bengal, India

12.1 Introduction

The brilliant colors of the dispersions of metallic colloids have been fascinated since antiquity, long before our understanding of light-matter interaction. The ability of noble metal colloids to manipulate light at the nanoscale has pioneered an emerging research area called plasmonics (Ozbay, 2006). The physical origin of the light absorption by metal nanoparticles is the coherent oscillation of the conduction band electrons, coined as, localized surface plasmon resonance (LSPR). The resonance frequency of this LSPR is strongly dependent upon the size, shape, interparticle interactions, dielectric properties, and local environment of the nanostructures (Kreibig & Vollmer, 1995). The unprecedented ability of metallic nanostructures to concentrate light into deep-subwavelength volumes has propelled their use in a vast array of nanophotonics technologies and research endeavors (Schuller et al., 2010). Due to their larger optical cross-sections, compared with those of organic dyes typically used for bioimaging and sensing, plasmonic nanostructures have been extensively utilized for their light scattering as nanoantenna or contrast agents for surface-enhanced Raman scattering, metal-enhanced fluorescence, and optical imaging, such as dark-field and computed tomography. Therefore, the plasmonic nanostructures have received great attention because of their novel properties of enhanced near electromagnetic fields and wavelength tunable light absorption and scattering (van de Hulst, 1981). Apart from radiative scattering of light, absorption of light by nanostructures can also be nonradiatively relaxed and results in significant heat energy or photoluminescence. It has been realized that metal nanostructures can

Electrochemistry and Photo-Electrochemistry of Nanomaterials
DOI: https://doi.org/10.1016/B978-0-443-18600-4.00013-2
© 2025 Elsevier Inc. All rights are reserved, including those for text and data mining, AI training, and similar technologies.

play the role of efficient nanosources of heat, which opens up a new emerging set of applications in nanotechnology, merging the fields of thermodynamics and optics and giving rise to a new promising field that could be designated as thermoplasmonics (Jauffred et al., 2019). The heat generation from these nanoheaters involves not only through absorption of incident photons, but also the conversion of photon energy into heat energy (Gellé et al., 2020; Govorov & Richardson, 2007). The heating effect is, especially, strong for metallic nanostructures since these particles have many mobile electrons and becomes strongly enhanced when the frequency of incident radiation matches the collective plasmon resonance of the nanostructures. A schematic presentation showing the dynamics of an excited spherical plasmonic nanoparticle (Gellé et al., 2020) is shown in Fig. 12.1. The absorption of light by plasmonic nanostructures and their associated temperature increase are exquisitely sensitive to the shape and composition of the structure and to the wavelength of incident light (Gellé et al., 2020). In particular, the conversion of light to thermal energy, known as the photothermal effect can be extensively utilized for plasmonic photothermal therapy of malignant cells that are highly promising for tomorrow's cancer treatment (Lal et al., 2008; Skrabalak et al., 2008). Moreover, since the discrete size and unique shape of nanostructures are directly correlated with their plasmonic properties, particularly with localized surface plasmon resonance (LSPR), and the converted heat energy is highly localized near the nanostructures, the plasmonic photothermal effect can be used as an efficient heat source for controllable and uniform thermal release at a specific excitation wavelength.

Due to the increased electromagnetic near fields produced on plasmonic nanostructures, which have led to a variety of specialized applications

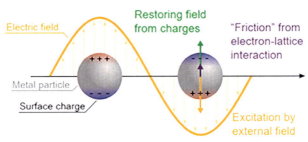

Figure 12.1 Schematic presentation showing the dynamics of an excited spherical plasmonic nanoparticle. *Reprinted with permission from Gellé, A., Jin, T., de la Garza, L., Price, G. D., Besteiro, L. V., & Moores, A. (2020). Applications of plasmon-enhanced nanocatalysis to organic transformations. Chemical Reviews, 120, 986–1041.*

in contemporary cutting-edge research, plasmonic nanostructures have drawn a lot of interest (Jauffred et al., 2019; Kreibig & Vollmer, 1995; Schuller et al., 2010). On the other hand, the epicenter of the development of the field known as thermoplasmonics, plasmonic absorption of electromagnetic radiation can result in significant local heating in metallic nanostructures. When illuminated, the nanostructures take on the role of miniature heaters that can precisely target localized thermal energy with great spatiotemporal precision (Govorov & Richardson, 2007; Skrabalak et al., 2008). The issue is that while heat is not propagative in nature, unlike light, which has propagative characteristics and can be readily detected, investigating temperature has frequently proven difficult. Inserting luminous nanoscale thermometers for real-time applications into the tiny volume of the nanostructures is a feasible solution. In order to create and establish procedures for the detection of photothermal heat generation in the nanostructures, it has been essential to incorporate both theoretical and experimental approaches (Baffou et al., 2020; Bohren & Huffman, 1983; Qiu & Wei, 2014; Zhan et al., 2019). Thus, the goal of the present chapter is to unravel the plasmonic photothermal application in the light of the representative nanostructures that can serve as the source of small-scale control for the formation of photothermal heat for applications that are likely.

Recent advances in the synthetic strategies, characterization techniques, fundamental understanding of the relevant theories and simulation methodologies of plasmonic nanostructures have enabled the researchers to study their optical properties in varieties of microenvironmental conditions. On the other hand, plasmonic absorption of light can lead to significant local heating in metallic nanostructures that has been defined as the subfield of thermoplasmonics. Several efforts have been put forward to separate thermal effects from other plasmonic processes, such as enhancement of electromagnetic field (Bohren & Huffman, 1983) or hot carrier generation (Baffou et al., 2020; Qiu & Wei, 2014; Zhan et al., 2019). However, since heat possesses a nonpropagative nature, dissimilar to light, measuring temperature at the nanoscale dimension has been difficult from fundamental perspectives. The successful synthesis and characterization of high quality and biocompatible plasmonic colloidal nanoparticles has fostered numerous and expanding applications, and has been leveraged in diverse applications. Despite important applications of plasmonic heating in thermal therapy, biosensors, thermal imaging, drug delivery, diagnostics and several other biomedical describing many a current challenges to clinical translation of the technology (Hu et al., 2018; Kennedy et al., 2011; Kim et al., 2019; Wang et al., 2020). Moreover, plasmonic photothermal nanostructures can be exploited in several nonbiomedical contexts, such as subwavelength optical energy transfer waveguides, optical data storage, optical tweezing, and thermoionic conversion (Gu et al., 2016). Although there have been both theoretical approaches to calculate the photothermal heat generation and to exploit plasmonic nanostructures without a priori rationalization of the ability to act as nanoscale heat sources, this, in turns, leads to difficulties in predicting the therapeutic endeavors and comparing the heat generating capability for new plasmonic nanostructures in biomedical applications. Therefore, at the present state-of-the-art research, it is necessary to integrate the experimental and theoretical approaches to quantify the heat generation in nanostructures and to exploit their plausible applications.

In this chapter, we aim to provide an intuitive understanding of the photothermal properties of metallic nanostructures in the context of literature published in the arena of research. Much effort has been put forward to synthesize metallic nanostructures for optimized interaction with the incident light and therefore, to act as controlled nanosources of heat. The experimentally observed plasmonic and thermal properties can be correlated with theoretical perspectives. The knowledge of an accurate estimation of the temperature generated at the nanostructures due to laser heating could offer a basis to design optimized plasmonic nanoheaters toward their possible thermoplasmonic applications.

12.2 Nonradiative properties

Under the influence of external electromagnetic radiation, the conducting free electrons inside metal nanostructures can move; however, this motion is dampened by inelastic collisions of the electrons and the restoring force arising due to the accumulation of surface charges. When the frequency of incident light and the collective oscillation of the conduction electrons are in phase, an intense and broad absorption band appears along the UV-vis-NIR region of the electromagnetic spectrum, often coined as localized surface plasmon resonance (LSPR) band. Under resonance conditions, the enhancement of electric field on the surface of plasmonic nanostructures becomes maximum, and these optoelectronic properties have given birth to outstanding applications in surface enhanced spectroscopies (Holze & Schlücker, 2015). Apart from the enhancement of electric field, other relevant attributes from the plasmonic characteristics evolves from the different relaxation processes occurring within the plasmonic nanostructures. Upon excitation, the energy accumulated in surface plasmons may decay through either radiative (as reemitted photons) or nonradiative (electron—hole pair excitations and electron—electron collisions) pathways (Hartland, 2006; Hugall & Baumberg, 2015; Link & El-Sayed, 2000). Noble metal nanostructures possess absorption cross-sections up to several times larger than their geometrical size and the absorbed electromagnetic radiation can be relaxed through nonradiative decay processes. The nonradiative dissipation may produce local heating while the kinetic energy of the electrons becomes transmitted to the phonons in the metallic lattice. Besides, the plasmon energy can, eventually, generate energetic hot charge carriers that arise due to dephasing of electron oscillations with the surface of the metallic nanostructures (Sánchez & Berdakin, 2022). This contribution to surface scattering, also called "Landau damping," is a solely quantum mechanical process where the plasmon quantum energy can be transferred on the timescale of femtoseconds (1–100 fs) corresponding to a single electron—hole pair excitation (Brongersma et al., 2015; Qin & Bischof, 2012). This phenomenon arises as the law of conservation of linear momentum is not obeyed by the electrons on and near the surface and as a consequence, the electrons can absorb photon quanta with particular energy (Besteiro et al., 2017). A schematic illustration of the time scales of transient events of gold nanoparticles upon activation with pulsed laser irradiation (Qin & Bischof, 2012) is presented in Fig. 12.2. Thus, the conversion of plasmon to thermal energy arising from

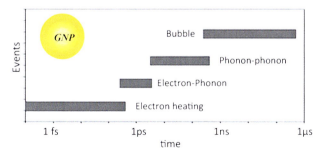

Figure 12.2 Schematic illustration of the time scales of transient events of gold nanoparticles upon activation with pulsed laser irradiation. *Reprinted with permission from Qin, Z., & Bischof, J. C. (2012). Thermophysical and biological responses of gold nanoparticle laser heating.* Chemical Society Reviews, 41, 1191–1217.

nonradiative decay channels of metallic nanostructures can be exploited in plausible photothermal applications.

12.3 Theory of plasmonic heating

Metallic nanostructures exhibit strong scattering and absorption of electromagnetic radiation in visible-NIR regime because of their localized plasmon resonances. The absorbed electromagnetic energy is, subsequently, transformed into thermal energy. Upon pulsed laser irradiation, the transient thermal power engendered inside the nanostructures induces several possible fates, such as ultrafast heating, thermal expansion, photothermal ablation, surface melting, and reshaping into varieties of architectures. The theory of heat transfer upon ultrafast laser heating was proposed in the past century by Kaganov et al. (1957) that considered the relaxation between the electrons and the crystalline lattice (phonons) in a metal describable by equilibrium Fermi and Bose functions as a function of temperature, respectively. The heat transfer coefficient of the electrons was calculated that can be related to "Cerenkov" radiation of sound waves. In these calculations, geometric constraint was imposed on the surface phonons that the energy transfer rate from degenerate hot electrons is order of magnitude less in comparison to that of the bulk. Anisimov et al. (1974) determined the emission current and the charge released from a metal surface upon exposer to a picosecond pulsed laser by considering the competition between the photoelectric and thermionic emissions. It has been shown that the intensities beyond the critical value of the emission current are entirely due to thermionic emission while the emission pulse is not delayed relative to the laser pulse. This is due to the small specific heat of the degenerate electron gas, which is practically thermally insulated from the lattice during the duration of the ultrashort laser pulse. It has been also elucidated that the electron-lattice relaxation kinetics can be investigated by measuring the thermionic emission produced by ultrashort laser pulses. Qiu and Tien (1992) studied the microscopic radiation—metal interactions under short-pulsed laser irradiation and their consequence on the thermal response of the materials. The radiation—metal interactions were treated as two-step processes: first, the absorption of photon energy by the electrons in a metal and second, the subsequent heating of the metallic lattice through electron—phonon collisions.

When a metallic nanoparticle is illuminated with electromagnetic radiation, part of the incident light becomes scattered in the surrounding medium, while the other part is absorbed that gets dissipated into heat. The efficiency of the elastic scattering process and the absorption cross sections can be described by σ_{sca} and σ_{abs}, respectively. Therefore, the extinction cross-section, σ_{ext} can be written as the sum of these two processes as

$$\sigma_{ext} = \sigma_{sca} + \sigma_{abs}. \tag{12.1}$$

The balance between scattering and absorption processes is governed by the geometry of the nanostructures. Upon illumination with a plane wave, the general expression describing the absorption cross-section, σ_{abs}, for a nanoparticle can be described as

$$\sigma_{abs} = \frac{k}{\varepsilon_o |E_o|^2} \int Im(\varepsilon) |E(\boldsymbol{r})|^2 dr, \tag{12.2}$$

where $k = \frac{2\pi n_m}{\lambda_o} = \frac{n_m \omega}{c}$ is the wavevector, ε_m the dielectric constant $(\varepsilon_m = n_m^2)$ of the surrounding medium, ε the dielectric constant of the material under consideration, E_o the incident electric field, $E(\boldsymbol{r})$ the electric field experienced by the particle, λ_o the wavelength of incident radiation, ω the angular frequency of the time-varying electric field and c the velocity of light under vacuum. Calculation of the integral over the nanoparticle volume offers the expression for the power of heat generation, Q inside the nanostructure which is directly proportional to σ_{abs} as

$$Q = \sigma_{abs} I = \sigma_{abs} \frac{n_m c \varepsilon_0}{2} |E_o|^2, \tag{12.3}$$

where $I = \frac{n_m c \varepsilon_0}{2} |E_o|^2$ is the intensity of the incident light, ε_0 being the permittivity of the free space. The absorption coefficient can be calculated with the knowledge of electric field amplitude inside the nanostructures. Putting the value of I and rearrangement gives

$$Q = \frac{n_m^2 \omega}{2} Im(\varepsilon) \int |E(\boldsymbol{r})|^2 dr = \int q(\boldsymbol{r}) dr, \tag{12.4}$$

where $q(\boldsymbol{r}) = \frac{n_m^2 \omega}{2} Im(\varepsilon_\omega) |E(\boldsymbol{r})|^2$ is the volumetric heating power density around the nanostructures and can be estimated employing discrete dipole approximation technique. The power of heat generation, Q inside the nanostructure is linearly related to the absorption efficiency, σ_{abs}, as

$$\sigma_{abs} = \frac{Q}{VNI} = \frac{Q}{NIAd} = \frac{Q}{NP_{avg}d}, \tag{12.5}$$

where N is the number density (i.e., particles per unit volume) of the metallic nanostructures, V the volume of the dispersion containing the colloidal particles, P_{avg} the average power of the irradiated light, and d the path length of the dispersion. This equation, thus, act as the bridge between plasmonic and thermal properties of the nanostructures.

The temperature distribution, $T(r)$ around the nanostructures can be calculated with the help of Poisson equation as

$$\kappa \nabla^2 T(r) = - q(r), \tag{12.6}$$

where κ is the thermal conductivity of the material under consideration and ∇T the temperature gradient. The spatial variation of temperature, δT, throughout the nanostructure can be estimated as

$$\delta T \sim \frac{l^2 Q}{\kappa v_o}. \tag{12.7}$$

However, the Poisson equation can be solved analytically with spherical symmetry. Outside the nanoparticle, it can simply be represented in the form

$$\dot{T}(r) = \dot{T}_o \frac{r_e}{r}, \tag{12.8}$$

where $\dot{T}_o$ is the increase in temperature that can be considered as uniform throughout the nanoparticle, $\dot{T}(r)$ the temperature increase above the room temperature, r the distance between the center of the nanoparticles and r_e the radius of the spherical nanoparticle. Over the spherical boundary, the power of heat generation, Q, can be considered to be equal to the integral of energy current density, $J_{th} = - \kappa_o \nabla T$ and under this condition, the temperature increase inside the nanoparticle, $\dot{T}_o$ can be estimated as

$$\dot{T}_o = \frac{1}{4\pi} \frac{Q}{\kappa_o r_e}, \tag{12.9}$$

where κ_o the thermal conductivity of the vacuum. It has been found that for the gold sphere, the temperature increase is $\dot{T}_o = 13°C$, which is lower that the temperature required for photothermal applications. Since anisotropic nanostructures, due to their sharp corners and edges, are favorable for heat generation and could act as more efficient nanosources of heat, they may be suitable for plasmonic photothermal therapy (Jain et al., 2008).

12.4 Laser heating of nanostructures

Since the past decade, a number of approaches have been adopted to study the effect of laser heating on the different morphologies of metallic nanostructures. For example, Kamat group (Fujiwara et al., 1999) studied the morphological changes of the gold nanoparticles aggregates induced by thionicotinamide upon visible laser pulse excitation. Link et al. (2000) showed that various degrees of shape change in a colloidal dispersion of gold nanorods induced upon excitation with an intense femtosecond and nanosecond laser pulses. The final transformation from the nanorods depends both on the energy and width of the pulsed laser. The higher laser energy leads to fragmentation of the nanorods, partial melting into different odd shapes at an intermediate energy and finally, transformed into spherical nanoparticles at lower laser fluences. The complete melting of the nanorods takes place above the threshold and is about two orders of magnitude higher for nanosecond in comparison with femtosecond laser pulses. Hu and Hartland (2002) investigated the relaxation time of heat dissipation in aqueous dispersion of gold nanoparticles with sizes ranging from 4 to 50 nm to their surrounding with femtosecond-pulsed laser heating. It has been elucidated that the energy relaxation does not follow an exponential manner. The relaxation time is linearly related to the square of the radius but independent of initial temperature of the nanoparticles. The comparable time scale of energy dissipation to that of the electron—phonon coupling indicates substantial amount of energy losses before reaching the thermal equilibrium of electrons and phonons within the nanoparticles. Plech et al. (2004) studied the lattice dynamics of gold nanoparticles in aqueous medium upon excitation with an intense femtosecond-pulsed laser. At low laser fluence, the lattice cooling due to initial heating occurs on the nanosecond time scale; the decay process can be depicted by heat transfer equation that includes the bulk conductivity of water and thermal resistance at the nanoparticle-medium interface. Kotaidis and Plech (2005) described that gold nanoparticles upon ultrafast excitation caused the particle lattice including the surrounding water shell to be thermally activated leading to evolution of nanoscale vapor bubbles around the particles. The bubbles become expanded and collapsed within the formation of first nanoseconds and confirmed the presence of cavitation at the nano- and subnanosecond time scale that can be described in the realm of continuum thermodynamics. Peng et al. (2005) have elucidated the effect of solvent parameters and surface chemistry on the photofragmentation of the toluenic dispersion of gold nanoparticles upon visible pulse laser excitation. Huang, Qian, et al. (2006) have presented that the femtosecond laser pulses in resonance with surface plasmon oscillation used for excitation lead to the equalization of the rates of absorption and heating rather than melting of the nanoparticles. The experiments were performed with assembled gold nanoprisms deposited on a quartz substrate; it was observed that the nanostructures flew away while the particle shape remains preserved but size decreases through atomic sublimation. It was also noted that the displacement from the original position was smaller than the space occupied by the nanostructures prior to the exposure of the laser pulses. Huang, Qian, et al. (2007) have studied the relaxation processes of electron−phonon scattering in plasmonic nanoparticles and found that copper exhibit some size dependence while gold and silver do not. Bulk phonons are responsible for the hot electron relaxation and are dependent on bulk phase structure and polycrystallinity. It was found that the relaxation process is slowed down when prismatic gold nanoparticles are annealed and finally, transformed to single crystalline nanospheres. Plech et al. (2007) have employed synchronous optical sampling technique with coupled femtosecond oscillators and observed thermal phase transition for gold nanoparticles supported on the surface. It was resolved that the Lamb vibrational modes could be identified with variation in annealing temperature. An abrupt phase change at a particular temperature indicates the onset of surface melting. Jones et al. (2018) have adopted anti-Stokes thermometry to investigate the interplay between optical and thermal effects to govern the diffusion of nanocolloids near bowtie nanoantennas. It was found that the motion of the colloidal particles within the plasmonic trap is insensitive to the resonance properties of the nanoantenna. It was also noted that the diffusion of the particles is inhibited by thermophoretic forces and thermal depletion layers are formed extending several microns around the particles. Liu et al. (2006) have developed an optofluidic method for optical to hydrodynamic energy conversion using suspension of photothermal nanoparticles at the liquid-air interface. Chen et al. (2012) have studied transient temperature behavior upon excitation with nanosecond pulsed laser and constructed a heat transfer model to resolve the spatial and temporal variation of temperature in aqueous dispersion of gold nanostructures. Setoura et al. (2013) have studied the laser induced heating of metallic nanostructures on different supported media by using single particle spectroscopy to monitor the temperature rise in nanostructures and found that

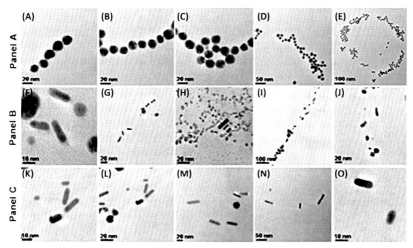

Figure 12.3 Photothermal reshaping of one-dimensional plasmonic polymers: transmission electron micrographs of (A–E) one-dimensional aggregation of gold nanospheres, (F–J) corresponding intermediate molten state, and (K–O) subsequent transformation to rod-like nanostructures. The one-dimensional plasmonic polymers comprise monomeric gold nanospheres of sizes 8, 13, 16, 20, and 32 nm, respectively. Reprinted with permission from Bardhan, D., Chatterjee, H., Sen, D., Sengupta, M., & Ghosh, S. K. (2022). Photothermal reshaping of one-dimensional plasmonic polymers: From colloidal dispersion to living cells. ACS Omega, 7, 11501–11509.

temperature increase was dependent on both the surrounding medium and supporting substrate. Zhang et al. (2013) have investigated the optically heated gold nanoparticle pairs to determine the distance dependence on particle temperature and formulated plasmon ruler equation based on nanoscale distance changes. Recently, Bardhan et al. (2022) have shown that the photothermal reshaping of the plasmonic polymers obtained through interdigitation of size-selective constituent monomers leads to the transformation to quasi-rod-like nanostructures with different aspect ratios. Transmission electron micrographs upon photothermal restructuring of one-dimensional plasmonic polymers comprised of monomeric gold nanoparticles of variable sizes (Bardhan et al., 2022) are exhibited in Fig. 12.3.

12.5 Biological tissue transparency window

The photothermal effect of metallic nanostructures increases the temperature in the surrounding medium and has been recognized as hyperthermia for the ablation of tumors. Various heat sources, such as ultrasound, microwaves and laser light can be used to augment the photothermal destruction of malignant cells. The challenge in the implementation of lethal hyperthermia is that generation of high temperatures may cause cell death in the nearby healthy tissues surrounding the exposed region. Since, metallic nanostructures are very inefficient fluorophores, high absorption-to-scattering ratio of plasmonic metal nanostructures renders noninvasive photon excitation modalities in the near-infrared "theranostic window" (700–1300 nm) (Weissleder, 2001). The optical characteristics of tissue components in the regime of visible to nearinfrared (Qin & Bischof, 2012) are displayed in Fig. 12.4. However, laser-induced photothermal treatment has often been preferred because of its capability to release particular amount of energy unswervingly to cancerous tissues. Near-infrared (NIR) radiation has been considered as the ideal electromagnetic source due to its ability to transmit deeply in biological tissues (Rotomskis, 2008). Based on these conceptive, thermographic camera (also called a thermal imaging camera) has been devised that can form an image utilizing infrared radiation operating in the wavelength regime up to 14,000 nm. The thermal imaging camera is a noncontact device that is able to detect the heat released by the object under consideration (Liu et al., 2019).

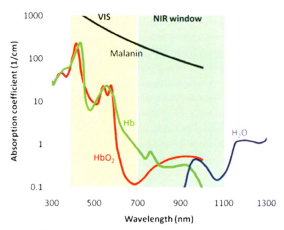

Figure 12.4 Optical characteristics of tissue components in the regime of visible to near infrared based on the data provided by Scott Prahl at the Oregon Medical Laser Center. *Reprinted with permission from Qin, Z., & Bischof, J. C. (2012). Thermophysical and biological responses of gold nanoparticle laser heating.* Chemical Society Reviews, 41, 1191–1217.

12.6 Photothermal nanostructures

The judicious selection of suitable materials has been important to implement their application as photothermal conversion agent. Amongst the various materials, noble metal nanostructures, including, copper (Wang et al., 2021), silver (Aiello et al., 2021; Bian et al., 2018), gold (Daniel & Astruc, 2004), palladium (Wu, Deng, et al., 2019), platinum (Song et al., 2018) have often been employed as the efficient photosensitizers and excellent biocompatibility in the physiological environment. The thermal stress induced by the nanostructures possesses numerous consequences at the cellular and tissue levels in the physiological environments. The nanostructures of gold possess the ability to convert NIR light into heat on a picosecond timescale that results in an increase of temperature in the surrounding dielectric environment, and thus, can render the decease of malignant cells. Over the past decade, numerous ideal (spherical) and nonideal (nonspherical) gold nanostructures, such as nanospheres (Ali et al., 2019; Zhang et al., 2012), nanorods (Alkilany et al., 2012; Huang et al., 2009; Huang, El-Sayed, et al., 2006; Li et al., 2018; Wang et al., 2014), nanoprisms (Ma et al., 2015; Moros et al., 2020; Perez-Hernandez et al., 2015), nanopyramids (Hasan et al., 2009), nanobipyramids (Lee et al., 2017), nanostars (Chatterjee et al., 2018; Wang et al., 2014; Wang, Huang, et al., 2013; Yuan et al., 2012), nanoshells (Loo et al., 2005; Melancon et al., 2008), nanocages (Au et al., 2008; Chen et al., 2007; Yavuz et al., 2009), nanostoves (Stehr et al., 2008), nanovalves (Croissant & Zink, 2012), and also nanoporous metals (Koya et al., 2021) have been adopted as the photosensitizers to transduce light into heat. The light-induced heating of a gold nanosphere in an aqueous dispersion and consequent changes in the optothermal properties (Chen et al., 2012) are presented in Fig. 12.5. Although near-infrared electromagnetic radiation possesses the capability and compatibility of stronger tissue penetration compared to the visible light, the mere absorption of light by spherical gold nanoparticles augments low photothermal conversion efficiency. The interparticle coupling effect of the small metallic particulates can also shift the absorption of electromagnetic radiation to the near-infrared and render the particles as efficient nanosources of heat (Ghosh & Pal, 2007). In addition, various kinds of other inorganic and organic nanomaterials, including Cu_2O (Zhang et al., 2021), CuX (X = S, Se, Te) (Gao et al., 2018; Ren et al., 2022; Tian et al., 2011; Yan et al., 2017), Cu_3P (Qi et al., 2021), Fe_3O_4 (Shi et al., 2009; Zhu et al., 2014), black phosphorus

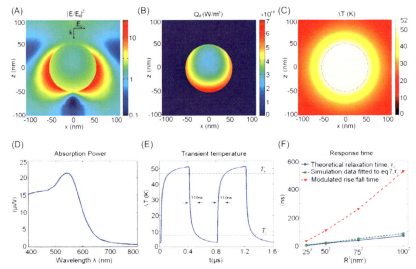

Figure 12.5 Photothermal effect of a gold nanosphere (∼100 nm radius) in aqueous dispersion: (A) the intensity of the electric field normalized to the incident field; (B) volume density of the heat power (Q_d); (C) steady-state increase of temperature upon 530 nm light excitation; (D) absorption power spectrum (the intensity of incident light at 1 mW/μm^2) and (E) transient temperature due to excitation with on−off modulated light and (F) comparison between the theoretical (τ_e) and simulated (τ_s) time and modulation of rise/fall time as a function of (radius)^2. *Reprinted with permission from Chen, X., Chen, Y., Yan, M., & Qiu, M. (2012). Nanosecond photothermal effects in plasmonic nanostructures. ACS Nano, 6, 2550−2557.*

(Chen et al., 2017), and polymeric (Cheng et al., 2012) and carbon-based nanostructures, such as single-walled carbon nanotubes (Markovic et al., 2011; Neves et al., 2013; Wang, Shi, et al., 2013), graphene oxide (Li et al., 2012; Sheng et al., 2013; Yang et al., 2012), and fullerene-carbon nanotube composites (Shen et al., 2010) have been employed for the photothermal destruction of cancerous cells.

12.7 Applications

Photothermal effects in plasmonic metal nanostructures have attracted significant attention owing to their ability to provide specific and localized heat energy. Although, for certain specific applications, nonradiative losses are considered as limitations, these processes exhibit great potentials toward a diverse range of niche scientific and technological applications. The unique ability of the metallic nanostructures to selectively light-to-heat conversion in a spatiotemporal fashion is the epicenter plasmonic photothermal therapy of the cancerous cells (Hirsch et al., 2003). The primary basis to such therapy is that the nanostructures can absorb incident light and subsequently, convert into heat that can be released to the specific tissues (Ali et al., 2019; Huang et al., 2008; Jaque et al., 2014). The application of gold nanocages as the photothermal transducers toward the treatment of cancerous cells (Chen et al., 2010) is shown in Fig. 12.6. Thus, utilization of plasmonic photothermal nanostructures has also been extended in a variety of biomedical, such as biosensing (Brus, 2008; Palermo & Strangi, 2020; Zhou et al., 2020), bioimaging (Berciaud et al., 2006), drug delivery and release (Croissant & Zink, 2012; Yavuz et al., 2009), as well as nonbiomedical realms, such as solar steam generation (Neumann et al., 2017; Neumann, Feronti, et al., 2013; Neumann, Urban, et al., 2013), photocatalysis (Adleman et al., 2009; Gargiulo et al., 2019; Gelle' & Moores, 2019; Zhou et al., 2018), environmental remediation (MacPhee et al., 2021), interphase chemical separation (Boyd et al., 2008), nanofluidics (Liu et al., 2005), and even synthesis (Gargiulo et al., 2017), fabrication (Kuznetsov et al., 2011; Sun et al., 2005), or deposition on the substrates (Boyd et al., 2006) of the nanostructures. Since the recent past, the outbreak of the coronavirus disease (COVID-19) has posed a threat to public health all over the globe and reliable laboratory diagnosis has been indispensable to get rid of such devastating interventions. The thermoplasmonic heat generated on two-dimensional gold nanoislands illuminated at the frequency corresponding to the localized surface plasmon resonance enables the performance of dual-functional LSPR biosensor to exhibit high sensitivity toward the SARS-CoV-2 sequences and allows precise discrimination of two similar gene sequences in a multigene mixture (Qiu et al., 2020) as shown in Fig. 12.7.

The plasmonic heat energy remains highly localized near the nanostructures and therefore, can be utilized as the stable heat source for uniform and controllable thermal processing (Xing et al., 2015, 2016). The biomedical applications of plasmonic photothermal nanostructures have also been extended to photothermal microscopy (Adhikari et al., 2020; Selmke et al., 2012), anti-Stokes imaging (Cai et al., 2019), anti-Stokes thermometry (Tran et al., 2019), diagnosis of diseases (Huang, El-Sayed, et al., 2007), and combinational theranostics (Xu & Pu, 2021). The photothermal properties of plasmonic nanostructures can also be realized for the design of nanoscale thermometry (Carattino et al., 2018; Xie & Cahill, 2016), thermoionic conversion (Wu, Hogan, et al., 2019), optical

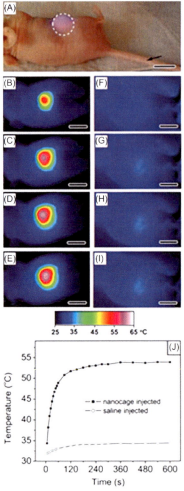

Figure 12.6 Photothermal treatment of malignant cells: (A) photograph of a mouse with tumor; thermal images of (B−E) nanocage- and (F−I) saline-injected tumor-bearing mice upon irradiation with a diode laser at different time intervals; and (J) plot of average temperature within the tumor as a function of irradiation time. All the scale bars are 1 cm. *Reprinted with permission from Chen, J., Glaus, C., Laforest, R., Zhang, Q., Yang, M., Gidding, M., Welch, M. J., & Xia, Y. (2010). Gold nanocages as photothermal transducers for cancer treatment.* Small (Weinheim an der Bergstrasse, Germany), 6, 811−817.

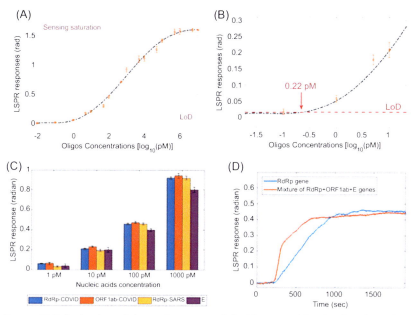

Figure 12.7 Evaluation of the performance of dual-functional localized surface plasmon resonance (LSPR) biosensors: (A) plot of LSPR responses as a function of RNA-dependent RNA polymerase (RdRp) sequence -COVID oligos concentrations; (B) zoom-in view in the low concentration regime for the determination of limit of detection; (C) plot of LSPR responses as a function of concentrations of different viral oligos; and (D) plot of LSPR responses at different time interval for the comparison of single analyte (RdRp-COVID) with the mixture of multiple sequences measured using the plasmonic photothermal effect of LSPR biosensors. The error bars in the panels (A) and (B) refer to the standard deviations of the LSPR responses after attaining the steady state following the flushing of buffer solution. *Reprinted with permission from Qiu, G., Gai, Z., Tao, Y., Schmitt, J., Kullak-Ublick, G. A., & Wang, J. (2020). Dual-functional plasmonic photothermal biosensors for highly accurate severe acute respiratory syndrome coronavirus 2 detection. ACS Nano, 14, 5268 – 5277.*

data storage (Zijlstra et al., 2009), optical tweezing intertwined by optical and thermal effects (Cuche et al., 2013; Zemánek et al., 2019), and photothermal chirality (characterized by differential circular extinction and scattering) (Miandashti et al., 2020). Fig. 12.8 corresponds to the detection limits in photothermal microscopy through systematic optimization of signal-to-noise ratios with single gold nanoparticles immobilized on a glass slide immersed in glycerol (Gaiduk et al., 2010).

The functionality of optical manipulation can be improved by inducing synergistic integrations of plasmonic heating with other fields, such as

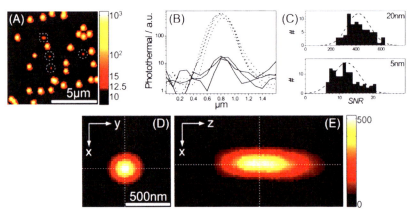

Figure 12.8 Photothermal microscopy image of gold nanoparticles: (A) photothermal signal color-coded on a logarithmic scale of a mixture of 5 nm and 20 nm gold nanoparticles immobilized on a glass slide immersed in glycerol; (B) several cross-sections of along the horizontal plane corresponding to the slow scan axis; (C) histograms of the photothermal signal-to-noise ratios; (D) photothermal image of a single ~20 nm gold sphere; and (E) photothermal image of a single ~20 nm gold sphere taken along the perpendicular to the glass surface. *Reprinted with permission from Gaiduk, A., Ruijgrok, P. V., Yorulmaz, M., & Orrit, M. (2010). Detection limits in photothermal microscopy. Chemical Science, 1, 343–350.*

chemical field (Sanchez et al., 2011), electric field (Ndukaife et al., 2016), fluid dynamics (Duhr & Braun, 2006), and acoustic field (Shin et al., 2017). Toward an optimal exploitation of the plasmonic photothermal effect in catalytic reactions, the photocatalysts must fulfill several prerequisites, such as intense light absorption over the range of solar spectrum, highly efficient charge carrier generation (Boerigter et al., 2016) and high efficiency for heat generation and/or transfer (Lalisse et al., 2015). To lower the parasitic energy demand, CO_2 capture technologies has been considered as an alternative energy as the regenerative, cheaper, and renewable energy sources (Luo et al., 2021). The reaction between CO_2 and H_2O (actually the reaction between CO_2 and H_2 obtained from H_2O deriving green energy sources) can be considered as one of the most promising alternative for the possible production of desired chemicals or synthetic fuels to achieve CO_2 neutrality for the transportation sector and in the chemical industries. Photothermal nanostructures have also been applied in catalysis, including, CO_2 conversion through artificial photosynthesis, reverse water gas shift reaction, methanol synthesis and methane production, Fischer–Tropsch synthesis, activation of methane, synthesis/decomposition of NH$_3$, hydrogen evolution, Suzuki coupling reactions, and other physicochemical processes (Ghoussoub et al., 2019; Nguyen et al., 2014; Wang, Li, et al., 2013).

To meet the increasing clean water demand of both the rural and urban communities in the developing and developed countries, photothermal membrane provides highly efficient, scalable, and cost-effective technologies toward the clean water production (Jun et al., 2019). The presence of trace amounts of micropollutants, such as pharmaceuticals, pesticides, and perfluorooctanoic acid, has posed significant risks to human health and photothermal membrane can be utilized to develop the water purification technologies (Irshad et al., 2021). Nanotechnology has proved significant benefits in soil remediation through the removal of contaminants and the upgradation of quality and fertility of the soil (Brandl et al., 2015). The photothermal properties of the metal nanostructures have also been extended to other environmental applications, such as solar heat production (Deng et al., 2017; Lin et al., 2019), bacterial remediation (Merkl et al., 2021), and the degradation of volatile organic compounds (Cai et al., 2018). The photothermocatalytic activities of Pt/γ-Al$_2$O$_3$ for the degradation of the volatile organic compound (toluene) under the full solar spectrum irradiation (Cai et al., 2018) are presented in Fig. 12.9. Thus, the photothermal properties of plasmonic metal nanostructures can be utilized to serve many a challenge of the sustainable development of the global society.

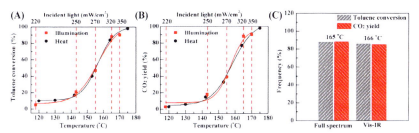

Figure 12.9 Photothermocatalytic activities of Pt/γ-Al$_2$O$_3$ for the degradation of toluene under the full solar spectrum irradiation: (A) toluene conversion and (B) CO$_2$ yield at different (red) light intensities (220, 250, 270, 320 and 350 mW/cm^2) and heating temperatures (black); and (C) catalytic activities under the full range of solar spectrum (320 mW/cm^2) and visible-infrared (390 mW/cm^2) irradiation of the simulated sunlight. *Reprinted with permission from Cai, S. -C., Li, J. -J., Yu, E. -Q., Chen, X., Chen, J., & Jia, H. -P. (2018). Strong photothermal effect of plasmonic Pt nanoparticles for efficient degradation of volatile organic compounds under solar light irradiation. ACS Applied Nano Materials, 1, 6368−6377.*

12.8 Challenges that lie ahead

While biology talks about the laws of life, physics deals with the laws of nature. It is difficult to predict therapeutic endeavors and compare the heat generating capability for new plasmonic nanostructures in biomedical applications, despite the fact that both physicists and biologists have used theoretical approaches to calculate the photothermal heat generation and to exploit plasmonic nanostructures. The present context provides an intuitive knowledge of the photothermal characteristics of a group of nanostructures that might provide a landscape for designing optimal plasmonic nanoheaters toward potential applications in order to bridge the gap between the two modalities. A reliable description of the plasmonic photothermal response and the subsequent heat dissipation in the surrounding environment has been challenging from both the theoretical and experimental perspectives. The study of anti-Stokes thermometry of individual spherical gold nanoparticles has enabled the design of nanothermometers through in situ photothermal characterization (Barella et al., 2021) as has been presented in Fig. 12.10.

Due to their capacity for effective light-to-heat conversion, representative plasmonic nanostructures have been carefully chosen as the suitable materials for optothermal manipulation. To manage the photothermal

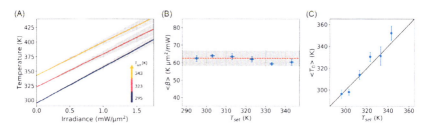

Figure 12.10 Gold nanoparticles as nanothermometers: (A) plot of temperature as a function of irradiance; (B) plot of extracted photothermal coefficient; ⟨β⟩ as a function of set temperature, T_{set}; and (C) plot of initial temperature, T_0 as a function of set temperature, T_{set} corresponding to a particular photothermal coefficient for an 80 nm gold nanoparticle on a glass substrate, immersed in water. The red dashed line in the panel (B) indicates photothermal coefficient, ⟨β⟩ at room temperature (62.5 K μm^2/mW) while the gray band represents ±1 standard deviations (4.6 K μm^2/mW). The error bars in the panels (B) and (C) represent the standard deviations to the corresponding mean values. *Reprinted with permission from Barella, M., Violi, I. L., Gargiulo, J., Martinez, L. P., Goschin, F., Guglielmotti, V., Pallarola, D., Schlücker, S., Pilo-Pais, M., Acuna, G. P., Maier, S. A., Cortes, E., & Stefani, F. D. (2021). In situ photothermal response of single gold nanoparticles through hyperspectral imaging anti-Stokes thermometry. ACS Nano, 15, 2458–2467.*

manipulation that is directly connected to one another, recent advancements in the synthesis of plasmonic nanostructures into different predefined designs have been used to manipulate the localized surface plasmon resonances. Because spherical nanoparticles are isotropic by nature, a significant shielding effect prevents heat from being generated on the outer surface of the particles, which is exposed to incoming electromagnetic radiation. The inner portion of the nanostructure is closer to the outside portion in the case of elongated nanorods, and the entire volume is affected by the photothermal processes. The nanostructures are increasingly appropriate as nanoheaters for plasmonic photothermal applications as the anisotropy rises. Another option for adjusting plasmonic response and therefore achieving the necessary heat-mediated optical manipulations for the intended applications is to induce electromagnetic interaction among the nanostructures. As a result, variations in particle size, shape, and interparticle interaction may make it easier for optical and thermal fields to influence the photothermal properties of nanostructures.

The creation of nanoscale thermometers based on the utilization of particle size in one-dimensional plasmonic polymers inside live cells for photothermal ablation of malignant cells can result from manipulating the radiative and nonradiative characteristics of noble metal nanostructures. The dependency of gold nanorod dimers on their orientation angles makes it possible to manipulate the fringing field, which has been used to manipulate the apoptosis of malignant cells. It has been shown that when nanostructures are aggregated randomly, altering the particle size inside the aggregates may tune the plasmon resonance and provide real-time temperature control for photothermal applications. It is conceivable that there may be fascinating opportunities for measuring temperature rise caused by laser heating using a wide range of other anisotropic nanostructures, such as rice, triangle, prism, tetrapod, cubic, and so on, that could provide a framework for the design of miniature plasmonic nanoheaters for photothermal ablation of malignant cells. Exciting possibilities exist for controlling temperature by using various nanostructure forms when laser heating is used, and there may be potential to take advantage of the use of the nanostructures in diverse optothermal applications. It is envisaged that a number of nanostructures will soon be used for sophisticated studies, allowing for the simultaneous achievement of photothermal heating, nanoparticle tracking, temperature sensing, and biological parameter recording. The linked plasmons convey various coherence effects when the nanostructures are aggregated, and their impact on laser heating is a subject of extensive research. Furthermore, the development of one-dimensional nanostructures with a variety of distinctive characteristics can direct the development of organized structures at various macroscopic length scales. In order to develop the blueprint for bioheat transfer inside physiological systems, plasmonic polymers made of a specific number of metallic nanostructures can be used to enhance nanoscale thermometers that can measure the temperature of various nano- and microenvironments, even when they are subjected to physiological conditions. In order to meet the challenges that lie ahead for the cutting-edge theranostics applications, photothermal motors can be produced and then delivered to the apoptosis of cancer cells. It is evident that there is sufficient room at the bottom for efficient heat generation and thermal management, which might be improved to create new opportunities for a range of biological applications. Our forecasts show that one of the major study areas in the domains of nanoscience and nanotechnology will soon be the development of multifunctional nanostructures capable of heating, following, moving, and sensing.

12.9 Conclusions and outlook

In conclusion, we have presented a critical and comprehensive review on the photothermal aspects of plasmonic metal nanostructures and their plausible utilizations in a diverse range of niche applications. The theory of plasmonic heating has been elucidated from the fundamental perspectives of light-matter interaction upon laser irradiation. Upon pulsed laser irradiation, the metallic nanostructures experience several possible fates, including, reshaping into varieties of geometries. Several isotropic and anisotropic nanostructures of various dimensions and geometries and their aggregates have been adopted as the adequate photosensitizers for efficient light-to-heat conversion. Moreover, a wide gamut of inorganic and organic (polymeric and carbon-based) semiconductor nanostructures has also been employed for the reduction of irradiation energy. Although there have been numerous theoretical and experimental investigation to elucidate the consequences of plasmonic metal nanostructures upon laser irradiation, however, many a question still remain. The challenge is has often been to mitigate photothermal nanostructures that can absorb electromagnetic radiation in the "biological tissue transparency window" that can act as the epicenter of plasmonic photothermal therapy of the destruction of cancerous cells. The photothermal properties of plasmonic metal nanostructures could also be utilized in diverse range of biomedical and nonbiomedical applications which all require particular and localized thermal energy. Thus, the present chapter provides a critical correlation between the plasmonic and thermal properties of the metallic nanostructures based on the literature to design a landscape toward their plausible applications.

Thermoplasmonics has been demonstrated to be one of the most significant subdisciplines in the field of nanoscience and nanotechnology. Surface improved spectroscopy to energy harvesting is only a few of the specialized uses for the photothermal nanostructures. Materials science, phononics, nanofluidics, drug transport, and plasmonic photothermal treatment of cancer cells are only a few possible applications of the thermoplasmonics idea. The foundation of suitable analytical and numerical simulation techniques, the condensation between the experimental observables with the calculated parameters, and prudent advancements in the synthesis of nanostructures and the design and fabrication of hierarchical assemblies could open the doors for their potential application to meet future challenges.

References

Adhikari, S., Spaeth, P., Kar, A., Baaske, M. D., Khatua, S., & Orrit, M. (2020). Photothermal microscopy: Imaging the optical absorption of single nanoparticles and single molecules. *ACS Nano*, *14*, 16414−16445.

Adleman, J. R., Boyd, D. A., Goodwin, D. G., & Psaltis, D. (2009). Heterogenous catalysis mediated by plasmon heating. *Nano Letters*, *9*, 4417−4423.

Aiello, M. B. R., Azcárate, J. C., Zelaya, E., Gara, P. D., Bosio, G. N., Gensch, T., & Mártire, D. O. (2021). Photothermal therapy with silver nanoplates in HeLa cells studied by in situ fluorescence microscopy. *Biomaterials Science*, *9*, 2608−2619.

Ali, M. R. K., Wu, Y., & El-Sayed, M. A. (2019). Gold-nanoparticle-assisted plasmonic photothermal therapy advances toward clinical application. *The Journal of Physical Chemistry C*, *123*, 15375−15393.

Alkilany, A. M., Thompson, L. B., Boulos, S. P., Sisco, P. N., & Murphy, C. J. (2012). Gold nanorods: Their potential for photothermal therapeutics and drug delivery, tempered by the complexity of their biological interactions. *Advanced Drug Delivery Reviews*, *64*, 190−199.

Anisimov, S. I., Kapeliovich, B. L., & Perel'Man, T. L. (1974). Electron emission from metal surfaces exposed to ultra-short laser pulses. *Soviet Physics—JETP*, *39*, 375−377.

Au, L., Zheng, D., Zhou, F., Li, Z.-Y., Li, X., & Xia, Y. (2008). A quantitative study on the photothermal effect of immuno gold nanocages targeted to breast cancer cells. *ACS Nano*, *2*, 1645−1652.

Baffou, G., Bordacchini, I., Baldi, A., & Quidant, R. (2020). Simple experimental procedures to distinguish photothermal from hot-carrier processes in plasmonics. *Light: Science & Applications*, *9*, 1−16, 108.

Bardhan, D., Chatterjee, H., Sen, D., Sengupta, M., & Ghosh, S. K. (2022). Photothermal reshaping of one-dimensional plasmonic polymers: From colloidal dispersion to living cells. *ACS Omega, 7*, 11501−11509.

Barella, M., Violi, I. L., Gargiulo, J., Martinez, L. P., Goschin, F., Guglielmotti, V., Pallarola, D., Schlücker, S., Pilo-Pais, M., Acuna, G. P., Maier, S. A., Cortes, E., & Stefani, F. D. (2021). In Situ photothermal response of single gold nanoparticles through hyperspectral imaging anti-Stokes thermometry. *ACS Nano, 15*, 2458−2467.

Berciaud, S., Lasne, D., Blab, G. A., Cognet, L., & Lounis, B. (2006). Photothermal heterodyne imaging of individual metallic nanoparticles: Theory versus experiment. *Physical Review B, 73*, 1−8, 045424.

Besteiro, L. V., Kong, X.-T., Wang, Z., Hartland, G. V., & Govorov, A. O. (2017). Understanding hot-electron generation and plasmon relaxation in metal nanocrystals: Quantum and classical mechanisms. *ACS Photonics, 4*, 2759−2781.

Bian, K., Zhang, X., Liu, K., Yin, T., Liu, H., Niu, K., Cao, W., & Gao, D. (2018). peptide-directed hierarchical mineralized silver nanocages for anti-tumor photothermal therapy. *ACS Sustainable Chemistry & Engineering, 6*, 7574−7588.

Boerigter, C., Campana, R., Morabito, M., & Linic, S. (2016). Evidence and implications of direct charge excitation as the dominant mechanism in plasmon-mediated photocatalysis. *Nature Communications, 7*, 1−9, 10545.

Bohren, C. F., & Huffman, D. R. (1983). *Absorption and scattering of light by small particles.* New York: Wiely Interscience.

Boyd, D. A., Adleman, J. R., Goodwin, D. G., & Psaltis, D. (2008). Chemical separations by bubble-assisted interphase mass-transfer. *Analytical Chemistry, 80*, 2452−2456.

Boyd, D. A., Greengard, L., Brongersma, M., El-Naggar, M. Y., & Goodwin, D. G. (2006). Plasmon-assisted chemical vapor deposition. *Nano Letters, 6*, 2592−2597.

Brandl, F., Bertrand, N., Martins, E., & Langer, L. R. (2015). Nanoparticles with photoinduced precipitation for the extraction of pollutants from water and soil. *Nature Communications, 6*, 1−10, 7765.

Brongersma, M. L., Halas, N. J., & Nordlander, P. (2015). Plasmon-induced hot carrier science and technology. *Nature Nanotechnology, 10*, 25−34.

Brus, L. (2008). Noble metal nanocrystals: Plasmon electron transfer photochemistry and single-molecule Raman spectroscopy. *Accounts of Chemical Research, 41*, 1742−1749.

Cai, S.-C., Li, J.-J., Yu, E.-Q., Chen, X., Chen, J., & Jia, H.-P. (2018). Strong photothermal effect of plasmonic Pt nanoparticles for efficient degradation of volatile organic compounds under solar light irradiation. *ACS Applied Nano Materials, 1*, 6368−6377.

Cai, Y. Y., Sung, E., Zhang, R., Tauzin, L. J., Liu, J. G., Ostovar, B., Zhang, Y., Chang, W. S., Nordlander, P., & Link, S. (2019). Anti-Stokes emission from hot carriers in gold nanorods. *Nano Letters, 19*, 1067−1073.

Carattino, A., Caldarola, M., & Orrit, M. (2018). Gold nanoparticles as absolute nanothermometers. *Nano Letters, 18*, 874−880.

Chatterjee, H., Rahman, D. S., Sengupta, M., & Ghosh, S. K. (2018). Gold nanostars in plasmonic photothermal therapy: The role of tip heads in the thermoplasmonic landscape. *The Journal of Physical Chemistry C, 122*, 13082−13094.

Chen, J., Glaus, C., Laforest, R., Zhang, Q., Yang, M., Gidding, M., Welch, M. J., & Xia, Y. (2010). Gold nanocages as photothermal transducers for cancer treatment. *Small (Weinheim an der Bergstrasse, Germany), 6*, 811−817.

Chen, J., Wang, D., Xi, J., Au, L., Siekkinen, A., Warsen, A., Li, Z.-Y., Zhang, H., Xia, Y., & Li, X. (2007). Immuno gold nanocages with tailored optical properties for targeted photothermal destruction of cancer cells. *Nano Letters, 7*, 1318−1322.

Chen, W. S., Ouyang, J., Liu, H., Chen, M., Zeng, K., Sheng, J. P., Liu, Z. J., Han, Y. J., Wang, L. Q., Li, J., Deng, L., Liu, Y. −N., & Guo, S. (2017). Black phosphorus nanosheet based drug delivery system for synergistic photodynamic/photothermal/chemotherapy of cancer. *Advanced Materials*, *29*, 1−7, 1603864.

Chen, X., Chen, Y., Yan, M., & Qiu, M. (2012). Nanosecond photothermal effects in plasmonic nanostructures. *ACS Nano*, *6*, 2550−2557.

Cheng, L., Yang, K., Chen, Q., & Liu, Z. (2012). Organic stealth nanoparticles for highly effective in vivo near-infrared photothermal therapy of cancer. *ACS Nano*, *6*, 5605−5613.

Croissant, J., & Zink, J. I. (2012). Nanovalve-controlled cargo release activated by plasmonic heating. *Journal of the American Chemical Society*, *134*, 7628−7631.

Cuche, A., Canaguier-Durand, A., Devaux, E., Hutchison, J. A., Genet, C., & Ebbesen, T. W. (2013). Sorting nanoparticles with intertwined plasmonic and thermo-hydrodynamical forces. *Nano Letters*, *13*, 4230−4235.

Daniel, M. C., & Astruc, D. (2004). Gold nanoparticles: Assembly, supramolecular chemistry, quantum-size-related properties, and applications toward biology, catalysis, and nanotechnology. *Chemical Reviews*, *104*, 293−346.

Deng, Z., Zhou, J., Miao, L., Liu, C., Peng, Y., Sun, L., & Tanemura, S. (2017). The emergence of solar thermal utilization: Solar-driven steam generation. *Journal of Materials Chemistry A*, *5*, 7691−7709.

Duhr, S., & Braun, D. (2006). Thermophoretic depletion follows Boltzmann distribution. *Physical Review Letters*, *97*, 1−4, 038103.

Fujiwara, H., Yanagida, S., & Kamat, P. V. (1999). Visible laser induced fusion and fragmentation of thionicotinamide-capped gold nanoparticles. *The Journal of Physical Chemistry. B*, *103*, 2589−2591.

Gaiduk, A., Ruijgrok, P. V., Yorulmaz, M., & Orrit, M. (2010). Detection limits in photothermal microscopy. *Chemical Science*, *1*, 343−350.

Gao, W., Sun, Y., Cai, M., Zhao, Y., Cao, W., Liu, Z., Cui, G., & Tang, B. (2018). Copper sulfide nanoparticles as a photothermal switch for TRPV1 signaling to attenuate atherosclerosis. *Nature Communications*, *9*, 1−10, 231.

Gargiulo, J., Berté, R., Li, Y., Maier, S. A., & Cortés, E. (2019). From optical to chemical hot spots in plasmonics. *Accounts of Chemical Research*, *52*, 2525−2535.

Gargiulo, J., Brick, T., Violi, I. L., Herrera, F. C., Shibanuma, T., Albella, P., Requejo, F. G., Cortés, E., Maier, S. A., & Stefani, F. D. (2017). Understanding and reducing photothermal forces for the fabrication of Au nanoparticle dimers by optical printing. *Nano Letters*, *17*, 5747−5755.

Gelle, A., Jin, T., de la Garza, L., Price, G. D., Besteiro, L. V., & Moores, A. (2020). Applications of plasmon-enhanced nanocatalysis to organic transformations. *Chemical Reviews*, *120*, 986−1041.

Gelle, A., & Moores, A. (2019). Plasmonic nanoparticles: Photocatalysts with a bright future. *Current Opinion in Green and Sustainable Chemistry*, *15*, 60−66.

Ghosh, S. K., & Pal, T. (2007). Interparticle coupling effect on the surface plasmon resonance of gold nanoparticles: From theory to applications. *Chemical Reviews*, *107*, 4797−4862.

Ghoussoub, M., Xia, M., Duchesne, P. N., Segal, D., & Ozin, G. (2019). Principles of photothermal gas-phase heterogeneous CO2 catalysis. *Energy & Environmental Science*, *12*, 1122−1142.

Govorov, A. O., & Richardson, H. H. (2007). Generating heat with metal nanoparticles. *Nano Today*, *2*, 30−38.

Gu, M., Zhang, Q., & Lamon, S. (2016). Nanomaterials for optical data storage. *Nature Reviews Materials*, *1*, 1−14, 16070.

Hartland, G. V. (2006). Coherent excitation of vibrational modes in metallic nanoparticles. *Annual Review of Physical Chemistry*, *57*, 403−430.

Hasan, W., Stender, C. L., Lee, M. H., Nehl, C. L., & Lee, J. (2009). Tailoring the structure of nanopyramids for optimal heat generation. *Nano Letters*, 9, 1555−1558.

Hirsch, L. R., Stafford, R. J., Bankson, J. A., Sershen, S. R., Rivera, B., Price, R. E., Hazle, J. D., Halas, N. J., & West, J. L. (2003). Nanoshell-mediated near-infrared thermal therapy of tumors under magnetic resonance guidance. *Proceedings of the National Academy of Sciences of the United States of America*, 100, 13549−13554.

Holze, R., & Schlücker, S. (2015). Surface-enhanced spectroscopies. *Physical Chemistry Chemical Physics: PCCP*, 17, 21045-21045.

Hu, J. −J., Cheng, Y. −J., & Zhang, X. −Z. (2018). Recent advances in nanomaterials for enhanced photothermal therapy of tumors. *Nanoscale*, 10, 22657−22672.

Hu, M., & Hartland, G. V. (2002). Heat dissipation for Au particles in aqueous solution: Relaxation time versus size. *The Journal of Physical Chemistry. B*, 106, 7029−7033.

Huang, W., Qian, W., & El-Sayed, M. A. (2006). Gold nanoparticles propulsion from surface fueled by absorption of femtosecond laser pulse at their surface plasmon resonance. *Journal of the American Chemical Society*, 128, 13330−13331.

Huang, W., Qian, W., El-Sayed, M. A., Ding, Y., & Wang, Z. L. (2007). Effect of the lattice crystallinity on the electron − phonon relaxation rates in gold nanoparticles. *The Journal of Physical Chemistry C*, 111, 10751−10757.

Huang, X., El-Sayed, I. H., Qian, W., & El-Sayed, M. A. (2006). Cancer cell imaging and photothermal therapy in the near-infrared region by using gold nanorods. *Journal of the American Chemical Society*, 128, 2115−2120.

Huang, X., El-Sayed, I. H., Qian, W., & El-Sayed, M. A. (2007). Cancer cells assemble and align gold nanorods conjugated to antibodies to produce highly enhanced, sharp, and polarized surface raman spectra: A potential cancer diagnostic marker. *Nano Letters*, 7, 1591−1597.

Huang, X., Jain, P. K., El-Sayed, I. H., & El-Sayed, M. A. (2008). Plasmonic photothermal therapy (PPTT) using gold nanoparticles. *Lasers in Medical Science*, 23, 217−228.

Huang, X., Neretina, S., & El-Sayed, M. A. (2009). Gold nanorods: From synthesis and properties to biological and biomedical applications. *Advanced Materials*, 21, 4880−4910.

Hugall, J. T., & Baumberg, J. J. (2015). Demonstrating photoluminescence from Au 1s electronic inelastic light scattering of a plasmonic metal: The origin of SERS backgrounds. *Nano Letters*, 15, 2600−2604.

van de Hulst, H. C. (1981). *Light scattering by small particles*. New York: Dover.

Irshad, M. S., Arshad, N., & Wang, X. (2021). Nanoenabled photothermal materials for clean water production. *Global Challenges*, 5, 1−21, 2000055.

Jain, P. K., Huang, X., El-Sayed, I. H., & El-Sayed, M. A. (2008). Noble metals on the nanoscale: Optical and photothermal properties and some applications in imaging, sensing, biology, and medicine. *Accounts of Chemical Research*, 41, 1578−1586.

Jaque, D., Martinez Maestro, L., del Rosal, B., Haro-Gonzalez, P., Benayas, A., Plaza, J. L., Martin Rodriguez, E., & Garcia Sole, J. (2014). Nanoparticles for photothermal therapies. *Nanoscale*, 6, 9494−9530.

Jauffred, L., Samadi, A., Klingberg, H., Bendix, P. M., & Oddershede, L. B. (2019). Plasmonic heating of nanostructures. *Chemical Reviews*, 119, 8087−8130.

Jones, S., Andrén, D., Karpinski, P., & Käll, M. (2018). Photothermal heating of plasmonic nanoantennas: Influence on trapped particle dynamics and colloid distribution. *ACS Photonics*, 5, 2878−2887.

Jun, Y. −S., Wu, X., Ghim, D., Jiang, Q., Cao, S., & Singamaneni, S. (2019). Photothermal membrane water treatment for two worlds. *Accounts of Chemical Research*, 52, 1215−1225.

Kaganov, M. I., Lifshitz, I. M., & Tanatarov, L. V. (1957). Relaxation between electrons and crystalline lattices. *Soviet Physics—JETP*, 4, 173−178.

Kennedy, L. C., Bickford, L. R., Lewinski, N. A., Coughlin, A. J., Hu, Y., Day, F. S., West, J. L., & Drezek, R. A. (2011). A new era for cancer treatment: Gold-nanoparticle-mediated thermal therapies. *Small (Weinheim an der Bergstrasse, Germany)*, 7, 169–183.

Kim, M., Lee, J. —H., & Nam, J. —M. (2019). Plasmonic photothermal nanoparticles for biomedical applications. *Advancement of Science*, 6, 1–23, 1900471.

Kotaidis, V., & Plech, A. (2005). Cavitation dynamics on the nanoscale. *Applied Physics Letters*, 87, 1–3, 213102.

Koya, A. N., Zhu, X., Ohannesian, N., Yanik, A. A., Alabastri, A., Zaccaria, R. P., Krahne, R., Shih, W. —C., & Garoli, D. (2021). Nanoporous metals: From plasmonic properties to applications in enhanced spectroscopy and photocatalysis. *ACS Nano*, 15, 6038–6060.

Kreibig, U., & Vollmer, M. (1995). *Optical properties of metal clusters*. Berlin: Springer.

Kuznetsov, A. I., Evlyukhin, A. B., Goncalves, M. R., Reinhardt, C., Koroleva, A., Arnedillo, M. L., Kiyan, R., Marti, O., & Chichkov, B. N. (2011). Laser fabrication of large-scale nanoparticle arrays for sensing applications. *ACS Nano*, 5, 4843–4849.

Lal, S., Clare, S. E., & Halas, N. J. (2008). Nanoshell-enabled photothermal cancer therapy: Impending clinical impact. *Accounts of Chemical Research*, 41, 1842–1851.

Lalisse, A., Tessier, G., Plain, J., & Baffou, G. (2015). Quantifying the efficiency of plasmonic materials for near-field enhancement and photothermal conversion. *The Journal of Physical Chemistry C*, 119, 25518–25528.

Lee, J. H., Cheglakov, Z., Yi, J., Cronin, T. M., Gibson, K. J., Tian, B. Z., & Weizmann, Y. (2017). Plasmonic photothermal gold bipyramid nanoreactors for ultrafast real-time bioassays. *Journal of the American Chemical Society*, 139, 8054–8057.

Li, J. L., Bao, H. C., Hou, X. L., Sun, L., Wang, X. G., & Gu, M. (2012). Graphene oxide nanoparticles as a nonbleaching optical probe for two-photon luminescence imaging and cell therapy. *Angewandte Chemie International Edition*, 51, 1830–1834.

Li, X., Zhou, J., Dong, X., Cheng, W.-Y., Duan, H., & Cheung, P. C. K. (2018). In vitro and in vivo photothermal cancer therapeutic effects of gold nanorods modified with mushroom β-glucan. *Journal of Agricultural and Food Chemistry*, 66, 4091–4098.

Lin, Y., Xu, H., Shan, X., Di, Y., Zhao, A., Hu, Y., & Gan, Z. (2019). Solar steam generation based on the photothermal effect: from designs to applications, and beyond. *Journal of Materials Chemistry A*, 7, 19203–19227.

Link, S., Burda, C., Nikoobakht, B., & El-Sayed, M. A. (2000). Laser-induced shape changes of colloidal gold nanorods using femtosecond and nanosecond laser pulses. *The Journal of Physical Chemistry. B*, 104, 6152–6616.

Link, S., & El-Sayed, M. A. (2000). Shape and size dependence of radiative, non-radiative and photothermal properties of gold nanocrystals. *International Reviews in Physical Chemistry*, 19, 409–453.

Liu, G. L., Kim, J., Lu, Y., & Lee, L. P. (2005). Optofluidic control using photothermal nanoparticles. *Nature Materials*, 5, 27–32.

Liu, G. L., Kim, J., Lu, Y., & Lee, L. P. (2006). Optofluidic control using photothermal nanoparticles. *Nature Materials*, 5, 27–32.

Liu, Y., Bhattarai, P., Dai, Z., & Chen, X. (2019). Photothermal therapy and photoacoustic imaging via nanotheranostics in fighting cancer. *Chemical Society Reviews*, 48, 2053–2108.

Loo, C., Lowery, A., Halas, N. J., West, J. L., & Drezek, R. (2005). Immunotargeted nanoshells for integrated cancer imaging and therapy. *Nano Letters*, 5, 709–711.

Luo, S., Ren, X., Lin, H., Song, H., & Ye, J. (2021). Plasmonic photothermal catalysis for solar-to-fuel conversion: Current status and prospects. *Chemical Science*, 12, 5701–5719.

Ma, X., Cheng, Y., Huang, Y., Tian, Y., Wang, S., & Chen, Y. (2015). PEGylated gold nanoprisms for photothermal therapy at low laser power density. *RSC Advances*, 5, 81682–81688.

MacPhee, J., Kinyenye, T., MacLean, B. J., Bertin, E., & Hallett-Tapley, G. L. (2021). Investigating the photothermal disinfecting properties of light-activated silver nanoparticles. *Industrial & Engineering Chemistry Research*, 60, 17390−17398.

Markovic, Z. M., Harhaji, L. M., Todorovic, B. M., Kepic, D. P., Arsikin, K. M., Jovanovic, S. P., Pantovic, A. C., Dramicanin, M. D., & Trajkovic, V. S. (2011). In vitro comparison of the photothermal anticancer activity of graphene nanoparticles and carbon nanotubes. *Biomaterials*, 32, 1121−1129.

Melancon, M. P., Lu, W., Yang, Z., Zhang, R., Cheng, Z., Elliot, A. M., Stafford, J., Olson, T., Zhang, J. Z., & Li, C. (2008). In vitro and in vivo targeting of hollow gold nanoshells directed at epidermal growth factor receptor for photothermal ablation therapy. *Molecular Cancer Therapeutics*, 7, 1730−1739.

Merkl, P., Zhou, S., Zaganiaris, A., Shahata, M., Eleftheraki, A., Thersleff, T., & Sotiriou, G. A. (2021). Plasmonic coupling in silver nanoparticle aggregates and their polymer composite films for near-infrared photothermal biofilm eradication. *ACS Applied Nano Materials*, 4, 5330−5339.

Miandashti, A. R., Khorashad, L. K., Kordesch, M. E., Govorov, A. O., & Richardson, H. H. (2020). Experimental and theoretical observation of photothermal chirality in gold nanoparticle helicoids. *ACS Nano*, 14, 4188−4195.

Moros, M., Lewinska, A., Merola, F., Ferraro, P., Wnuk, M., Tino, A., & Tortiglione, C. (2020). Gold nanorods and nanoprisms mediate different photothermal cell death mechanisms in vitro and in vivo. *ACS Applied Materials & Interfaces*, 12, 13718−13730.

Ndukaife, J. C., Kildishev, A. V., Nnanna, A. G. A., Shalaev, V. M., Wereley, S. T., & Boltasseva, A. (2016). Long-range and rapid transport of individual nano-objects by a hybrid electrothermoplasmonic nanotweezer. *Nature Nanotechnology*, 11, 53−59.

Neumann, O., Feronti, C., Neumann, A. D., Dong, A., Schell, K., Lu, B., Kim, E., Quinn, M., Thompson, S., Grady, N., Nordlander, P., Oden, M., & Halas, N. J. (2013). Compact solar autoclave based on steam generation using broadband light-harvesting nanoparticles. *Proceedings of the National Academy of Sciences of the United States of America*, 110, 11677−11681.

Neumann, O., Neumann, A. D., Tian, S., Thibodeaux, C., Shubhankar, S., Müller, J., Silva, E., Alabastri, A., Bishnoi, S. W., Nordlander, P., & Halas, N. J. (2017). Combining solar steam processing and solar distillation for fully off-grid production of cellulosic bioethanol. *ACS Energy Letters*, 2, 8−13.

Neumann, O., Urban, A. S., Day, J., Lal, S., Nordlander, P., & Halas, N. J. (2013). Solar vapor generation enabled by nanoparticles. *ACS Nano*, 7, 42−49.

Neves, L. F., Krais, J. J., Van Rite, B. D., Ramesh, R., Resasco, D. E., & Harrison, R. G. (2013). Targeting single-walled carbon nanotubes for the treatment of breast cancer using photothermal therapy. *Nanotechnology*, 24, 1−12, 375104.

Nguyen, D. T., Truong, R., Lee, R., Goetz, S. A., & Esser-Kahn, A. P. (2014). Photothermal release of CO2 from capture solutions using nanoparticles. *Energy & Environmental Science*, 7, 2603−2607.

Ozbay, E. (2006). Plasmonics: Merging photonics and electronics at nanoscale dimensions. *Science (New York, N.Y.)*, 311, 189−193.

Palermo, G., & Strangi, G. (2020). Thermoplasmonic-biosensing demonstration based on the photothermal response of metallic nanoparticles. *Journal of Applied Physics*, 128, 1−5, 164302.

Peng, Z., Walther, T., & Kleinermanns, K. (2005). Photofragmentation of phase-transferred gold nanoparticles by intense pulsed laser light. *The Journal of Physical Chemistry B*, 109, 15735−15740.

Perez-Hernandez, M., del Pino, P., Mitchell, S. G., Moros, M., Stepien, G., Pelaz, B., Parak, W. J., Galvez, E. M., Pardo, J., & de la Fuente, J. M. (2015). Dissecting the molecular mechanism of apoptosis during photothermal therapy using gold nanoprisms. *ACS Nano*, *9*, 52−61.

Plech, A., Cerna, R., Kotaidis, V., Hudert, F., Bartels, A., & Dekorsy, T. (2007). A surface phase transition of supported gold nanoparticles. *Nano Letters*, *7*, 1026−1031.

Plech, A., Kotaidis, V., Grésillon, S., Dahmen, C., & von Plessen, G. (2004). Laser-induced heating and melting of gold nanoparticles studied by time-resolved X-ray scattering. *Physical Review B*, *70*, 1−7, 195423.

Qi, F., Chang, Y., Zheng, R., Wu, X., Wu, Y., Li, B., Sun, T., Wang, P., Zhang, H., & Zhang, H. (2021). Copper phosphide nanoparticles used for combined photothermal and photodynamic tumor therapy. *ACS Biomaterials Science & Engineering*, *7*, 2745−2754.

Qin, Z., & Bischof, J. C. (2012). Thermophysical and biological responses of gold nanoparticle laser heating. *Chemical Society Reviews*, *41*, 1191−1217.

Qiu, G., Gai, Z., Tao, Y., Schmitt, J., Kullak-Ublick, G. A., & Wang, J. (2020). Dual-functional plasmonic photothermal biosensors for highly accurate severe acute respiratory syndrome coronavirus 2 detection. *ACS Nano*, *14*, 5268−5277.

Qiu, J., & Wei, W. D. (2014). Surface plasmon-mediated photothermal chemistry. *The Journal of Physical Chemistry C*, *118*, 20735−20749.

Qiu, T., & Tien, C. (1992). Short-pulse laser heating on metals. *International Journal of Heat and Mass Transfer*, *35*, 719−726.

Ren, Y., Yan, B., Wang, P., Yu, Y., Cui, L., Zhou, M., & Wang, Q. (2022). Construction of a rapid photothermal antibacterial silk fabric via QCS-guided in situ deposition of CuS NPs. *ACS Sustainable Chemistry & Engineering*, *10*, 2192−2203.

Rotomskis, R. (2008). Optical biopsy of cancer: Nanotechnological aspects. *Tumori*, *94*, 200−205.

Sanchez, S., Ananth, A. N., Fomin, V. M., Viehrig, M., & Schmidt, O. G. (2011). Microbots swimming in the flowing streams of microfluidic channels. *Journal of the American Chemical Society*, *133*, 14860−14863.

Schuller, J., Barnard, E. S., Cai, W., Jun, Y. C., White, J. S., & Brongersma, M. I. (2010). Plasmonics for extreme light concentration and manipulation. *Nature Materials*, *9*, 193−204.

Selmke, M., Braun, M., & Cichos, F. (2012). Photothermal single-particle microscopy: Detection of a nanolens. *ACS Nano*, *6*, 2741−2749.

Setoura, K., Okada, Y., Werner, D., & Hashimoto, S. (2013). Observation of nanoscale cooling effects by substrates and the surrounding media for single gold nanoparticles under CW-laser illumination. *ACS Nano*, *7*, 7874−7885.

Shen, Y., Skirtach, A. G., Seki, T., Yagai, S., Li, H., Möhwald, H., & Nakanishi, T. (2010). Assembly of fullerene-carbon nanotubes: Temperature indicator for photothermal conversion. *Journal of the American Chemical Society*, *132*, 8566−8568.

Sheng, Z., Song, L., Zheng, J., Hu, D., He, M., Zheng, M., Gao, G., Gong, P., Zhang, P., & Ma, Y. (2013). Protein-assisted fabrication of nano-reduced graphene oxide for combined in vivo photoacoustic imaging and photothermal therapy. *Biomaterials*, *34*, 5236−5243.

Shi, D. L., Cho, H. S., Chen, Y., Xu, H., Gu, H. C., Lian, J., Wang, W., Liu, G. K., Huth, C., Wang, L., Ewing, R. C., Budko, S., Pauletti, G. M., & Dong, Z. (2009). Fluorescent polystyrene−Fe3O4 composite nanospheres for in vivo imaging and hyperthermia. *Advanced Materials*, *21*, 2170−2173.

Shin, J. H., Seo, J., Hong, J., & Chung, S. K. (2017). Hybrid optothermal and acoustic manipulations of microbubbles for precise and on-demand handling of micro-objects. *Sensors and Actuators B*, *246*, 415−420.

Skrabalak, S. E., Chen, J., Sun, Y., Lu, X., Au, L., Cobley, C. M., & Xia, Y. (2008). Gold nanocages: Synthesis, properties, and applications. *Accounts of Chemical Research*, *41*, 1587−1595.

Song, H., Meng, X., Dao, T. D., Zhou, W., Liu, H., Shi, L., Zhang, H., Nagao, T., Kako, T., & Ye, J. (2018). Light-enhanced carbon dioxide activation and conversion by effective plasmonic coupling effect of Pt and Au nanoparticles. *ACS Applied Materials & Interfaces, 10*, 408−416.

Stehr, J., Hrelescu, C., Sperling, R. A., Raschke, G., Wunderlich, M., Nichtl, A., Heindl, D., Kürzinger, K., Parak, W. J., Klar, T. A., & Feldmann, J. (2008). Gold nanostoves for microsecond DNA melting analysis. *Nano Letters, 8*, 619−623.

Sun, F., Cai, W., Li, Y., Duan, G., Nichols, W., Liang, C., Koshizaki, N., Fang, Q., & Boyd, I. (2005). Laser morphological manipulation of gold nanoparticles periodically arranged on solid supports. *Applied Physics. B, Lasers and Optics, 81*, 765−768.

Sánchez, C. G., & Berdakin, M. (2022). Plasmon-induced hot carriers: An atomistic perspective of the first tens of femtoseconds. *The Journal of Physical Chemistry C, 126*, 10015−10023.

Tian, Q., Tang, M., Sun, Y., Zou, R., Chen, Z., Zhu, M., Yang, S., Wang, J., Wang, J., & Hu, J. (2011). Hydrophilic flower-like CuS superstructures as an efficient 980 nm laser-driven photothermal agent for ablation of cancer cells. *Advanced Materials, 23*, 3542−3547.

Tran, T. T., Regan, B., Ekimov, E. A., Mu, Z., Zhou, Y., Gao, W. −B., Narang, P., Solntsev, A. S., Toth, M., Aharonovich, I., & Bradac, C. (2019). Anti-Stokes excitation of solid-state quantum emitters for nanoscale thermometry. *Science Advances, 5*, 1−6, eaav9180.

Wang, F., Li, C., Chen, H., Jiang, R., Sun, L. D., Li, Q., Wang, J., Yu, J. C., & Yan, C. H. (2013). Plasmonic harvesting of light energy for Suzuki coupling reactions. *Journal of the American Chemical Society, 135*, 5588−5601.

Wang, L., Shi, J., Zhang, H., Li, H., Gao, Y., Wang, Z., Wang, H., Li, L., Zhang, C., Chen, C., Zhang, Z., & Zhang, Y. (2013). Synergistic anticancer effect of RNAi and photothermal therapy mediated by functionalized single-walled carbon nanotubes. *Biomaterials, 34*, 262−274.

Wang, S., Huang, P., Nie, L., Xing, R., Liu, D., Wang, Z., Lin, J., Chen, S., Niu, G., & Lu, G. (2013). Single continuous wave laser induced photodynamic/plasmonic photothermal therapy using photosensitizer-functionalized gold nanostars. *Advanced Materials, 25*, 3055−3061.

Wang, X., Li, G., Ding, Y., & Sun, S. (2014). Understanding the photothermal effect of gold nanostars and nanorods for biomedical applications. *RSC Advances, 4*, 30375−30383.

Wang, X., Shi, Q., Zha, Z., Zhu, D., Zheng, L., Shi, L., Wei, X., Lian, L., Wu, K., & Cheng, L. (2021). Copper single-atom catalysts with photothermal performance and enhanced nanozyme activity for bacteria-infected wound therapy. *Bioactive Materials, 6*, 4389−4401.

Wang, Y., Du, W., Zhang, T., Zhu, Y., Ni, Y., Wang, C., Raya, F. M. S., Zou, L., Wang, L., & Liang, G. (2020). Self-evaluating photothermal therapeutic nanoparticle. *ACS Nano, 14*, 9585−9593.

Weissleder, R. (2001). A clearer vision for in vivo imaging. *Nature Biotechnology, 19*, 316−317.

Wu, D., Deng, K., Hu, B., Lu, Q., Liu, G., & Hong, X. (2019). Plasmon-assisted photothermal catalysis of low-pressure CO2 hydrogenation to methanol over Pd/ZnO catalyst. *ChemCatChem, 11*, 1598−1601.

Wu, N., Hogan, N., & Sheldon, M. (2019). Hot electron emission in plasmonic thermionic converters. *ACS Energy Letters, 4*, 2508−2513.

Xie, X., & Cahill, D. G. (2016). Thermometry of plasmonic nanostructures by anti-Stokes electronic Raman scattering. *Applied Physics Letters, 109*, 1−5, 183104.

Xing, R. R., Jiao, T. F., Yan, L. Y., Ma, G. H., Liu, L., Dai, L. R., Li, J. B., Möhwald, H., & Yan, X. H. (2015). Colloidal gold—collagen protein core — shell nanoconjugate: One—step biomimetic synthesis, layer-by-layer assembled film, and controlled cell growth. *ACS Applied Materials & Interfaces*, 7, 24733—24740.

Xing, R. R., Liu, K., Jiao, T. F., Zhang, N., Ma, K., Zhang, R. Y., Zou, Q. L., Ma, G. M., & Yan, X. H. (2016). An injectable self-assembling collagen — gold hybrid hydrogel for combinatorial antitumor photothermal/photodynamic therapy. *Advanced Materials*, 28, 3669—3676.

Xu, C., & Pu, K. (2021). Second near-infrared photothermal materials for combinational nanotheranostics. *Chemical Society Reviews*, 50, 1111—1137.

Yan, C., Tian, Q., & Yang, S. (2017). Recent advances in the rational design of copper chalcogenide to enhance the photothermal conversion efficiency for the photothermal ablation of cancer cells. *RSC Advances*, 7, 37887—37897.

Yang, K., Hu, L. L., Ma, X. X., Ye, S. Q., Cheng, L., Shi, X. Z., Li, C. H., Li, Y. G., & Liu, Z. (2012). Multimodal imaging guided photothermal therapy using functionalized graphene nanosheet anchored with magnetic nanoparticles. *Advanced Materials*, 24, 1868—1872.

Yavuz, M. S., Cheng, Y., Chen, J., Cobley, C. M., Zhang, Q., Rycenga, M., Xie, J., Kim, C., Song, K. H., Schwartz, A. G., Wang, L. V., & Xia, Y. (2009). Gold nanocages covered by smart polymers for controlled release with near-infrared light. *Nature Materials*, 8, 935—939.

Yuan, H., Khoury, C. G., Wilson, C. M., Grant, G. A., Bennett, A. J., & Vo-Dinh, T. (2012). In vivo particle tracking and photothermal ablation using plasmon-resonant gold nanostars. *Nanomedicine: Nanotechnology, Biology, and Medicine*, 8, 1355—1365.

Zemánek, P., Volpe, G., Jonáš, A., & Brzobohatý, O. (2019). Perspective on light-induced transport of particles: From optical forces to phoretic motion. *Advances in Optics and Photonics*, 11, 577—678.

Zhan, C., Liu, B. W., Huang, Y. F., Hu, S., Ren, B., Moskovits, M., & Tian, Z. Q. (2019). Disentangling charge carrier from photothermal effects in plasmonic metal nanostructures. *Nature Communications*, 10, 1—8, 2671.

Zhang, S., Zhao, Y., Shi, R., Zhou, C., Waterhouse, G. I. N., Wang, Z., Weng, Y., & Zhang, T. (2021). Sub-3 nm ultrafine Cu2O for visible light driven nitrogen fixation. *Angewandte Chemie International Edition*, 60, 2554—2560.

Zhang, W., Li, Q., & Qiu, M. (2013). A plasmon ruler based on nanoscale photothermal effect. *Optics Express*, 21, 172—181.

Zhang, X. — D., Wu, D., Shen, X., Chen, J., Sun, Y. — M., Liu, P. — X., & Liang, X. — J. (2012). Size-dependent radiosensitization of PEG-coated gold nanoparticles for cancer radiation therapy. *Biomater.*, 33, 6408—6419.

Zhou, L., Swearer, D. F., Zhang, C., Robatjazi, H., Zhao, H., Henderson, L., Dong, L., Christopher, P., Carter, E. A., Nordlander, P., & Halas, N. J. (2018). Quantifying hot carrier and thermal contributions in plasmonic photocatalysis. *Science (New York, N. Y.)*, 362, 69—72.

Zhou, W., Hu, K., Kwee, S., Tang, L., Wang, Z., Xia, J., & Li, X. J. (2020). Gold nanoparticle aggregation-induced quantitative photothermal biosensing using a thermometer: a simple and universal biosensing platform. *Analytical Chemistry*, 92, 2739—2747.

Zhu, C. H., Lu, Y., Chen, J. F., & Yu, S. H. (2014). Photothermal poly(N-isopropylacrylamide)/Fe3O4 nanocomposite hydrogel as a movable position heating source under remote control. *Small (Weinheim an der Bergstrasse, Germany)*, 10, 2796—2800.

Zijlstra, P., Chon, J. W. M., & Gu, M. (2009). Five-dimensional optical recording mediated by surface plasmons in gold nanorods. *Nature*, 459, 410—413.

CHAPTER 13

Nanophoto/electrochemistry for green energy production

Rana Ahmed Aly[1], Abdulwahab Alaamer[1], Tala Ashira[1], Saeed Najib Alkhajeh[1], Abdullah Ali[1], Amani Al-Othman[1] and Muhammad Tawalbeh[2,3]

[1]Department of Chemical and Biological Engineering, American University of Sharjah, Sharjah, United Arab Emirates
[2]Sustainable and Renewable Energy Engineering Department, University of Sharjah, Sharjah, United Arab Emirates
[3]Sustainable Energy & Power Systems Research Centre, RISE, University of Sharjah, Sharjah, United Arab Emirates

13.1 Introduction

Nanoscience and nanotechnology are fast gaining prominence in the scientific world. In 1959, Richard Feynman first introduced the idea of nanoscience while presenting a talk tile "There is still plenty of room at the bottom" at the American Physical Society meeting at Caltech (Patel & Pathak, 2021). In his presentation, Feynman introduced the idea that individual atoms and molecules could be manipulated and may even be developed using precise tools. He explained that when doing so, gravity would have a limited influence, and the surface effect would have a significant impact on the created or manipulated individual atoms of molecules. The idea of nanotechnology was first presented and defined by Norio Taniguchi in 1974, who explained that it involved nanoscale reduction, fabrication, and manipulation of materials (Mehnath et al., 2021). In the 1980s, the scanning tunneling microscope (STM) and the atomic force microscope (AFM) were developed, which played a crucial role in studying atoms and molecules (Bottomley et al., 2005). Scientists were able to understand the nature and form of atoms and molecules and could use the STM and AFM to comprehensively study various features and properties that could help manipulate individual atoms and molecules. The introduction of the tools and ideas prompted further research and development, which have led to the current state of affairs where nanoscience and nanotechnology are playing a critical role in the development of various highly valuable products and materials in society.

Electrochemistry and Photo-Electrochemistry of Nanomaterials
DOI: https://doi.org/10.1016/B978-0-443-18600-4.00014-4
© 2025 Elsevier Inc. All rights reserved, including those for text and data mining, AI training, and similar technologies.

Nanomaterials and conventional nanomaterials have been borne out of the paradigm-shifting research activities that have been undertaken ever since the ideas of nanoscience and nanotechnology were presented by Feynman and Taniguchi. Nanomaterials have been defined as any material with at least one of its dimensions measuring less than 100 nm on the nanoscale (Kolahalam et al., 2019). Conventional nanomaterials are created following the completion of manufacturing processes that can either be in top-down or bottom-up formats (Baig et al., 2021). In the top-down manufacturing approach of conventional nanomaterials, larger particles are ground or milled down smaller particulates to the nanomaterial size. A conventional nanomaterial made through the top-down manufacturing process is graphene, which has multiple applications in various fields (Nauman Javed et al., 2022). The bottom-up manufacturing process of conventional nanomaterials focuses on synthesizing atomic or molecular species through chemical reactions to generate materials that can be identified as nanomaterials. The chemical reactions enable the precursor atoms or molecules to grow in size to such an extent that at least one of their dimensions will measure between 1 and 100 nm on the nanoscale, which will qualify them to be nanomaterials. The chemical reactions adopted in the bottom-up approach to creating nanomaterials enable the precursor atoms or molecules to grow in size to get to the scale where they can be classified as nanomaterials (Coroş et al., 2019). A conventional nanomaterial manufactured through the bottom-up processes is sol−gel, which has many significant applications in various fields. From the detailed definitions of nanomaterials and conventional nanomaterials, it is now possible to undertake a comprehensive review of their characteristic properties and features.

13.2 Characteristic properties of nanomaterials and conventional nanomaterials

Nanomaterials and conventional nanomaterials have relatable characteristic properties. The only major difference is that nanomaterials occur naturally while conventional nanomaterials are engineered and manufactured using chemical processes and procedures. As already indicated, conventional nanomaterials can be engineered through top-down or bottom-up processes (Baig et al., 2021). For instance, when the top-down process is applied, larger materials are milled and grounded to generate nanomaterials, while when the bottom-up approach is used, atomic or molecular species are exposed to chemical conditions and processes that lead to the creation of nanomaterials. From such a context, once materials that can be defined as nanomaterials have been created, they adopt characteristic properties that fit the definition of naturally occurring nanomaterials. Such characteristic properties include the following:

13.2.1 Number of dimensions

A major characteristic property of nanomaterials is the number of dimensions. Different types of nanomaterials have been categorized based on the number of dimensions (Cai et al., 2019). For instance, zero-dimensional nanostructures have been identified, and they are usually in the form of spheres. Further, one-dimensional nanostructures have been identified, and they are mostly applied in film, coatings, and multilayer applications. Two-dimensional nanostructures have been identified, and they are mainly applied in tubes, fibers, wires, and platelets. Besides, three-dimensional nanostructures have been identified, mainly in the form of particles, quantum dots and hollow spheres. Understanding the number of dimensions on nanostructures plays a critical role in enhancing the comprehension of the characteristic properties.

13.2.2 Phase composition

The characteristic properties of nanomaterials can also be understood based on their phase composition (Chen et al., 2020). In most instances, single-phase solids are nanomaterials with features that lead to their identification as crystalline and amorphous particles and layers. Multiphase solids are nanomaterials that possess features and attributes that lead to them being classified as matrix composites and coated particles. On the other hand, multiphase systems are nanomaterials that possess features and attributes that lead them to be classified as colloids, aerogels, and ferrofluids. Therefore, understanding the phase composition of nanomaterials makes it possible to understand their characteristic properties.

13.2.3 Manufacturing process

The manufacturing processes, which generate nanomaterials, play a crucial role in understanding their characteristic properties. Both naturally occurring and conventional nanomaterials undergo spontaneous or intentional chemical processes, which lead to their generation. From such a context, nanomaterials can be classified as those produced from gas-phase reaction processes, liquid phase reaction and mechanical procedures. Examples of gas phase reaction manufacturing processes include flame synthesis, condensation, or chemical vapor deposition (CVD) (Swihart, 2003). Liquid phase reaction examples include sol—gel, chemical etching, laser ablation, and hydrothermal processing (Karatutlu et al., 2018). Finally, ball milling is a common mechanical procedure to produce nanomaterials (Prasad Yadav et al., 2012). Understanding the manufacturing process of nanomaterials is crucial to understanding the specific characteristic properties they hold. Some of the adjustable properties that all nanomaterials possess (Sulaiman & Santuraki, 2018) include:

- Catalytic: Higher surface to volume ratio achieved through enhancement of the catalytic efficiency.
- Electrical: Electric resistance can be lowered in metals and conductivity can be increased in ceramics and nanocomposites.
- Magnetic: Nanomaterials can demonstrate superparamagnetic behavior due to their increased magnetic coercivity.
- Mechanical: Ductility and superplasticity can be enhanced for ceramics while for the metal alloys the toughness and the hardness can be increased.
- Optical: Optical absorption and fluorescence properties can undergo a spectral shift and the quantum efficiency of semiconductor crystals can be escalated.
- Sterical: Selectivity can be enhanced; hollow spheres can be created that enable drug transportation and controlled release.
- Biological: Permeability across biological membranes can be increased, along with biocompatibility.

Based on the detailed review of the characteristic properties of nanomaterials covered in the sections above, and according to Luther, the following are the core features and attributes, which can summarize all forms of nanomaterials regardless of the nature of occurrence or synthesis (Luther, 2004):

- Origin: they can occur naturally, released unintentionally, or be manufactured through chemical processes.
- Dispersion: they can be dispersed in gases such as aerosols, liquids such as gels and ferrofluids, and solids such as matrix materials.
- Shape/structure: nanomaterials can adopt the shape or structure of needles, spheres, platelets, or tubes.
- Surface modification: the surface of nanomaterials can be untreated, coated or possess cores or shell particles.

- Aggregation state: nanomaterials can be single particles, aggregates, or agglomerates.
- Chemical composition: nanomaterials can be metals or metal oxides, polymers or carbon compounds, semiconductors, biomolecules, or other compounds.

The detailed review of the characteristic properties of nanomaterials covered in the current section demonstrates how unique they are compared to other substances that occur naturally or can be synthesized through chemical or mechanical means. Therefore, it is important to consider the provided guidelines when identifying and classifying nanomaterials for use and application in photoelectrochemistry or any other field and technology.

13.3 Electrochemistry and photoelectrochemistry of nanomaterials technologies

13.3.1 Electrochemistry of nanomaterials

Nanotechnology studies relating to nanomaterials have demonstrated that they possess unique physical, chemical, and catalytic properties that lead them to be used in various application areas. Nanomaterial electrochemistry is now emerging to be an important field of study as experts seek to expand and enhance their application (Curulli, 2020; Serrà & García-Torres, 2021). In the current section, the focus is on reviewing the various features, attributes, and properties of nanomaterials, which influence their electrochemistry and render them applicable in various activities. The section will identify the different categories of nanomaterials and review their features and properties, which inform their application in different operations. It is expected that with the understanding of the different properties of nanomaterials, it will be possible to understand their applications while enhancing knowledge that can improve the processes and outcomes of the experiments being undertaken on the electrochemical and physical properties of nanomaterials.

13.3.1.1 Categories of nanomaterials

In studying the electrochemical properties of nanomaterials, it is critical to consider their various categories, including inorganic-based nanomaterials, carbon-based nanomaterials, organic-based nanomaterials, and composite-based nanomaterials (Jeevanandam et al., 2018). The inorganic-based nanomaterials include metal and metal oxide nanomaterials, which are

432 Electrochemistry and Photo-Electrochemistry of Nanomaterials

applied in various electrochemical operations. Examples of metal-based inorganic materials include silver (Ag), alumina, copper (Cu), iron (Fe), zinc (Zn), while the metal oxide-based inorganic materials include zinc oxide (ZnO), titanium dioxide (TiO_2), iron oxide (Fe_2O_3), aluminum oxide (Al_2O_3), silica (SiO_2), and others that acquire the features of nanomaterials during their metallic oxidation (Saleh, 2020). Graphene, fullerene, carbon nanotubes, nanodiamonds, graphene oxide, and carbon-based quantum dots are common examples of carbon-based nanomaterials (Patel et al., 2019). Dendrimers, liposomes, and micelles are common examples of organic-based nanomaterials, which are formed from organic materials that exclude carbon in their structure and composition (Jeevanandam et al., 2018). Composite nanomaterials are combinations of metal-based, carbon-based, organic-based or metal-oxide-based nanomaterials (Majhi & Yadav, 2021). Such a detailed review of the different categories, types, and examples of nanomaterials provides an overview of the nature of chemical and electrochemical properties that they can be expected to have in their form and existence.

13.3.1.2 Dimensional structures of nanomaterials

Nanomaterials' dimensional structures also provide a basis for studying and understanding their physical and electrochemical properties. Nanomaterials have been identified and classified based on the number of dimensions that include zero-dimensional (0-D), one-dimensional (1-D), two-dimensional (2-D), and three-dimensional (3-D) nanomaterials. The 0-D nanostructured materials include solid, hollow, and core-shell nanoparticles, quantum dots, nanoonions, and core-shell-like carbon nanospheres (Saleh, 2020). The 1-D nanostructured materials include homostructures such as carbon nanotubes, carbon nanofibers, titania nanorodes, nanowires, and heterostructures such as Si-SIC core-shell. The 2-D nanostructured materials include homostructures such as graphene, metal carbides, metal nitrides, layered solids like MoS_2, $CaGe_2$, and $CaSi_2$, and flower-like MoS_2 nanosheets. Nanosheets have a thickness of only a few atomic layers and several advantages, such as a large specific surface area and high flexibility. These advantages are very important in electrochemical systems, sensors, and flexible electronics (Tawalbeh et al., 2023). The 3-D nanostructured materials have been identified as graphene aerogel, crumpled graphene balls, nanoflowers such as NiO nanoflowers, Co_3O_4 urchin-like nanostructures, and VA-CNT-graphene hybrid architecture. From such a context, understanding the nature of dimensions of nanostructured materials can enhance the comprehension of their physical and electrochemical properties. The properties and attributes of 0-D, 1-D, 2-D, and 3-D nanostructured materials are different, which affects their electrochemical properties. For instance, 0-D nanostructured materials have been identified as amorphous or crystalline, metallic, ceramic, or polymeric. The 0-D nanostructured materials have also been observed to be composed of single or multichemical elements, which affects their electrochemical properties. The 0-D nanostructured materials exhibit different shapes and forms depending on their environment, and they exist individually or are incorporated to form a matrix. The 1-D nanostructured materials are defined as materials with two dimensions (x and y) at the nanoscale, with another dimension beyond the nanoscale (> 100 nm). The core features that define the 1-D nanostructured materials include amorphous or crystalline and single crystalline or polycrystalline. The 1-D nanostructured materials have been chemically pure or impure depending on the state in which they are identified. Besides, the 1-D nanostructured materials can either be standalone or be embedded within some form of medium. The 1-D nanomaterials can also be metallic, ceramic, or polymeric, which gives them the different forms and shapes they take based on their environment, affecting their electrochemical properties.

The features and attributes of 2-D nanostructured materials have also been studied comprehensively. Such materials are defined as having one dimension at the nanoscale and two dimensions that are not confined to the nanoscale, identified as (> 100 nm). The 2-D nanostructured materials often adopt a platelet-like structure, and they possess the following attributes. Mainly, they are either amorphous or crystalline and are made up of different chemical compositions. In most instances, they are used either as single-layer or multilayer structures, considering their platelet-like form. Usually, the 2-D nanostructured materials are deposited on a substrate and integrated into the surrounding matrix materials. The 2-D nanostructured materials can be metallic, ceramic, or polymeric, which influences the various electrochemical attributes that they demonstrate in their applications. Examples of nanomaterials as classified under this category include nanofilms, fullerene films, nanowalls, nanostraw, nanolayers, and layered films amongst many (Krishnan et al., 2014).

The features of the 3-D structured nanomaterials have also been reviewed quite extensively. Mainly, the 3-D structured nanomaterials have three arbitrary dimensions that above the nanoscale (> 100 nm),

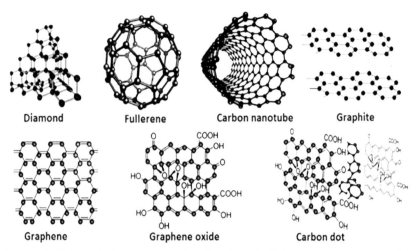

Figure 13.1 Chemical structures of carbon-based allotropes and nanomaterials (Sudha et al., 2018).

but they hold a nanocrystalline structure, which involves the features of the presence of the nanoscale. They usually occur as arrangements of nanosized crystals that possess different orientations. The 3-D nanoscale structures include the skeletons of fibers and nanotubes, fullerites, honeycombs, and layers of skeleton buildings and forms. Layer-fiber skeletons, layer fiber composites, particles in a matrix, and membrane and powder fiber nanoparticles are all examples of 3-D structured nanoparticles (Pokropivny & Skorokhod, 2007). Understanding the different attributes of the 3-D nanostructured materials enhances the ability to understand their physical and electrochemical properties, which influence their application in different operations and activities. Fig. 13.1 illustrates the chemical structures of selected allotropes and nanomaterials of carbon (Sudha et al., 2018).

13.3.1.3 Applications of nanomaterials based on the properties

Nanomaterials have unique properties that facilitate their effectiveness in the different applications in which they are adopted in electrochemical operations. For instance, nanomaterials have unique physical, optical, catalytic, magnetic, and sensory properties and attributes, differentiating them from their bulk analogs. They are also unique based on the dependence of their properties on their degrees of dispersion, which is referred to as the size effect. The size effect demonstrated by nanomaterials is attributed to the surface excess energy, which results from uncompensated bonds between the surface and subsurface atoms that have experience proportion increases with significantly smaller particle sizes. The smaller particle sizes of the nanomaterials cause a growth in the degree of defectiveness of their crystalline lattice. All such properties affect their quantum effects, which influence the optical properties they demonstrate. Such features make the nanomaterials significantly different from the common materials and substances hence their applications in various electrochemical operations.

The energy state of nanomaterials is an important attribute to study in reviewing their properties. It has been established that the energy state of nanoparticles is influenced by their size and the Gibbs free energy, which can be represented as $\Delta G°$. The energy state of the nanomaterials affects their kinetics and thermodynamics, which causes their electrooxidation that affects the processes which occur on their surfaces. A physical-mathematical model has been proposed in which the effect of the energy state of nanomaterials can be measured, and its impact on the surface and subsurface reactions determined quite accurately. In the model, the electrochemical reactions of the dissolved substances, which diffuse through the nanomaterial surfaces, can be predicted reliably. The electrochemical processes occurring internally in nanomaterials after diffusion consider electron transfer, account for possible chemical and catalytic reactions, consider the passivation of the electrode surface based on the products of the electrode reactions, and the nanoeffect manifestation. From such a context, it is possible to effectively measure and ascertain the electrochemical properties and features of nanomaterials, which could influence the results in their applications.

Nanomaterials are often used as transducers in ion-selective electrodes (Mohamed et al., 2021). Such an application is possible due to their property relating to their high surface area and conductivity. The number of immobilized selective ionophores that arises from the inclusion of nanomaterials in sensitive membrane ion-selective electrodes makes the nanomaterials highly effective in their application as transducers in ion-selective electrodes. Nanomaterials provide such a unique application, enabling them to be used in areas where common substances and chemical elements can be used. The unique reactivity and products from electrochemical reactions of nanomaterials make them highly effective for application and usage as transducers and ion-selective electrodes.

Nanomaterials have been identified to be having a small surface to volume ratio. Such attributes make metal nanoparticles such as gold, silver, and platinum significantly important in the various application in which they are usually deployed (Al-Othman et al., 2020). Such metals have localized surface plasmon resonance characteristics, making them irreplaceable in chemical sensing and biosensing applications. Metal nanoparticles can be used to create nanoporous films that possess unique properties owing to their enhanced corrosion resistance, electrical conductivity and mechanical stability (Ding & Chen, 2009; Tawalbeh et al., 2022). Nanoparticles have also been observed to have a large surface area to the high density of active sites, making them vital catalysts used in photocatalysis and electrocatalysis (Almomani et al., 2023). Owing to such important and unique features of nanoparticles and nanomaterials, their continued application in various fields and activities remains irreplaceable. Continuous absorption bands, emission spectra that are both narrow and intensive, and high photobleaching stability and chemical stability, processability, and surface functionality are characteristics typically associated with semiconductor nanomaterials (Suresh, 2013). Such unique features and attributes lead them to be used in single electron and imaging devices. Fig. 13.2 classifies certain carbon nanomaterials based on the dimension and its potential application (Xie, 2021).

In conclusion, the detailed review of the properties of nanomaterials has demonstrated their unique features when viewed from an electrochemical perspective. The 0-D, 1-D, 2-D, and 3-D nanostructured materials have unique features and attributes, which enable the nanomaterials to be applied in various activities. Further research is recommended to enhance the levels of knowledge of nanomaterials considering that vast details are yet to be understood regarding the electrochemical and physical properties of nanomaterials, especially when they are exposed to various conditions or when they are manipulated in various ways. Therefore, advances in nanotechnology need to be deployed to help experts understand the various issues relating to the properties of nanomaterials, which can improve their application in various activities and devices, as already described in the current chapter.

13.3.2 Photoelectrochemistry of nanomaterials

While electrochemistry entails studying electricity and its role in chemical reactions, photoelectrochemistry goes beyond electricity by

Nanophoto/electrochemistry for green energy production 437

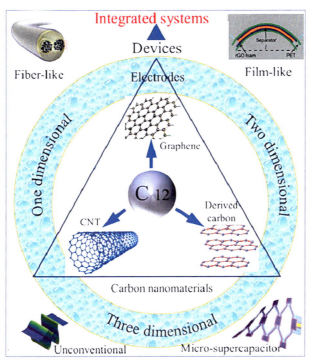

Figure 13.2 Application of different types of nanomaterials (Xie, 2021).

including light in electrochemical systems. According to Barham and König (2020), photoelectrochemistry is a term applied to a hybrid field of chemistry employing techniques which combine photochemical and electrochemical methods for the study of the oxidation-reduction chemistry of the ground or excited states of molecules or ions. It may be simply defined as the chemistry generated from the interaction of electrochemical systems with light.

This multidisciplinary field combines solid-state physics, optics, electrochemistry, and surface science. Although photoelectrochemistry is a relatively new doctrine, it has been intensively studied in previous decades due to its high potential for lasting solutions. For instance, the unsustainability of fossil fuels compels scientists and researchers to find other effective clean energy sources. Some proposals entailing photoelectrochemistry are regenerative solar cells and photoelectrochemical water splitting (Jin, 2018). More importantly, there has been intensive research regarding the photoelectrochemistry of nanomaterials, which are at the core of several emerging technologies, including catalysts, imaging gadgets, memory devices, solar cells, sensors, and other optoelectronic devices.

Nanomaterials are made of single units and usually range between 1 and 100 nm in size. Nanomaterials come in different sizes and shapes and can be classified into different categories, such as fullerenes, metal, and polymeric nanoparticles (Khan et al., 2019). Like other materials, nanomaterials have unique chemical and physical characteristics. They have a high surface area when viewed in correlation with their microscale. Additionally, size is a crucial attribute since it affects the optical properties and, in turn, color of the nanomaterials, thereby, rendering them attractive for applications in bioimaging (Dreaden et al., 2012). Modern technology enables inventors to create hybrid nanoparticles with enhanced nanostructures to provide new physical and chemical properties (Tri et al., 2019). Photoelectrochemistry is one of the methods that technology experts can use to modernize and impose new chemical capabilities on electrical systems.

Unlike other bulk materials, nanowires and nanomaterials have distinct attributes that make them have numerous practical applications. For instance, the radical dimensions of nanomaterials can be fine-tuned to make the diameter smaller for use in various applications, including the diffusion spans of photoexcited carriers and photon wavelengths (Deng et al., 2019). Moreover, nanomaterials have abundant active sites due to their large surface-to-volume ratio, facilitating chemical reactions. This property also makes nanomaterials ideal for yielding biological nanohybrids and interfaces for biological moieties. A multitude of synthetic and fabrication techniques have been developed for the creation of increased fidelity and targeted functionality nanowires courtesy the favorable characteristics possessed by the nanowire morphology (Dasgupta, 2014; Yang, 2012). Overall, nanomaterials and the potential for altering their morphology have set the stage for numerous innovative applications in many fields, including biotechnology, energy, and optoelectronics.

The concept of photoelectrochemistry occurs when an interface between a liquid electrolyte and a semiconductor produces an optoelectronic effect. Fermi level is the maximum energy that an electron obtains while the temperature is at absolute zero. The contact between an electrolyte and a semiconductor produces a difference in Fermi levels, resulting in two main outcomes. First, the setup causes a band bending at the interface (Zhang et al., 2021). Second, the setup creates a built-in electric field. Photoelectrochemistry happens when photon irradiation is introduced, resulting in the generation of additional electron—hole pairs (e$^-$/h$^+$). The high mobility of the carriers and the built-in electric field combine to create a spatial separation of the photo-excited carriers (Deng et al., 2019). While some careers assemble around the electrolyte/semiconductor interface, most of them move through the semiconductor, initiating a redox half-reaction. Notably, this reaction would be nonexistent in the absence of photon irradiation.

The photoelectrochemical cell is one of the devices that explains the concept of photoelectrochemistry. Once the light is introduced into the electrical system and absorbed by the semiconductor electrode, the electrons in the system become excited and gain higher energy levels, as shown in Fig. 13.3 (Sherman et al., 2016). The electron—hole pairs and the electrons themselves are involved in redox reactions at the interface between the electrolyte and the semiconductor (Deng et al., 2019). A redox reaction occurs when there is a transfer of electrons. Two types of reactions happen within the photoelectrochemical cell. First, there is oxidation at the electron—hole pairs, which creates a molecule. Second, a reduction reaction occurs at the counter electrode converting molecules to anions. While this setup emulates the electrochemical cell, there is a slight difference since light is a key factor in the photoelectrochemical cell. Without light, redox reactions would not happen due to the absence of energy.

Photoelectrochemistry of nanomaterials has several benefits due to one-unit morphology that can be reorganized for optimal functionality. First, nanomaterials can be used to trap light for enhanced photon absorption. The interface between semiconductors and electrolytes provides a gradient of refractive index. This feature makes nanowires ideal for use as antireflection layers, which enhances incident light scattering and suppresses

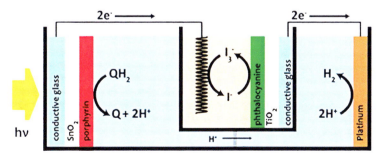

Figure 13.3 The photoelectrochemical cell for hydrogen production (Sherman et al., 2016).

reflection (Huang, 2007; Xi, 2007). This light-trapping effect enables nanomaterials to possess a high absorption coefficient. Moreover, crystalline nanomaterials are well versed in absorbing incident photons due to their unique optical qualities. When used in semiconductors, nanomaterials drive down the cost, particularly silicon semiconductors requiring hundreds of micrometers to make considerable light absorption.

Second, nanomaterials using photoelectrochemistry exhibit enhanced efficacy when it comes to charge separation. The morphology of nanomaterials allows them to orthogonalize the process of charge separation and direction of light absorption. Conventionally, incident photons receive illumination when inclined at an angle to the electrode of a semiconductor. Under this setup, minority carriers move a distance equal to the absorption depth of the photons. On the contrary, many indirect light-absorbing semiconductors have shorter carrier diffusion lengths than the absorption depth of the photons (Deng et al., 2019). This scenario results in lower energy inefficiencies due to the recombination of carriers. Nanomaterials provide a practical solution to this problem by allowing carriers to move for shorter distances to the semiconductor/liquid interface. The outcome of this phenomenon is increased energy and quantum efficiency of photoelectrochemistry processes and higher tolerance of impurities for electrode materials.

Lastly, nanomaterials have higher reaction turnover due to their large surface area. Attaining high efficiency in solar-to-chemical processes requires a high turnover rate at the interface separating the electrolyte and the semiconductor. Each redox half-reaction has its unique reactivity. Nevertheless, nanomaterials have larger surface areas, providing more active sites for light-induced reactions (Khan et al., 2019). This attribute results in energy efficiency by boosting surface-associated recombination and reducing the flux density and overpotentials for half-reactions. Additionally, the relatively larger surface area for nanomaterials unlocks other potentials, such as extra functionality in solar-to-chemical processes and more space for catalytic action. These benefits are crucial in preserving precious metals since catalysts can be added to earth-abundant materials to increase performance.

Although scientists and researchers believe that photoelectrochemistry holds great potential for industrial applications, the existing knowledge in this field remains somewhat elusive. Researchers argue that the model disregards potential heterogeneity on the material surface preventing the model from being completely refined at the nanoscopic scale (Deng et al., 2019).

For instance, researchers are not certain as to what happens when a catalyst is added at the electrolyte–semiconductor interface. It is expected that future research will avail more complete answers and applications for photoelectrochemistry.

13.3.3 Integrated methods toward green approaches

The growing global population has put pressure on natural resources, such as energy and water, increasing energy demand. It was recorded in 2017 that fossil fuels shared about 80% of the global energy demand even though the utilization renewable energy sources has been escalating (Johnsson et al., 2019). This large quantity of fossil fuels continues to serve the global demand. Other large reserves of fossil fuels are unexploited, including petroleum and coal. However, the increasing demand is likely to result in the quick depletion of fossil fuels. It has been estimated that it will only remain approximately 14% of oil proven reserves, 18% of gas proven reserves and 72% of coal proven reserves by 2050 (Martins et al., 2019). This problem is aggravated by the increasing consumer awareness of the negative impacts of fossil fuels on climate change, where more people are prioritizing green energy over fossil fuels. People are now aware that the continued use of fossil fuels results in the depletion of the ozone layer, rise in sea levels, and increased accumulation of greenhouse gases, all of which worsen the problem of climate change. Nevertheless, the challenge is insufficient renewable energy sources to serve the existing demand. Fortunately, several technologies hold great potential in the large-scale production of green energy, including solar, wind, and geothermal energy. Electrochemistry and photoelectrochemistry of nanomaterials is a combination that can be used to generate renewable energy.

Water splitting to generate hydrogen is one of the methods that electrochemistry and photoelectrochemistry of nanomaterials can be harnessed. Over the last decades, single-unit materials have been widely researched for industrial applications, and their use in electrochemical and photoelectrochemical devices can be harnessed in water splitting. Compared to conventional bulk films, virtues of nanostructured photoelectrodes include reduced carrier diffusion length, greater surface area, decreased light reflection loss, and adjustable electronic structure and optical bandgap (Yao et al., 2019). Light-triggered redox reactions occur at electrode/electrolyte interfaces and the relatively larger surface area of nanoelectrodes increases the excitement of electrons.

Hydrogen gas is considered a green and renewable energy source with high-energy efficiency, high gravimetric energy density, superior portability, and overwhelming environmental benefits. Despite these numerous benefits, production of hydrogen in and of itself contributes to the detriment of the environment since a large proportion of it is produced by reforming natural gas (Agyekum et al., 2022). Notably, this method is associated with fossil fuels since it releases environment-damaging carbon byproducts, such as carbon monoxide (CO) and carbon dioxide (CO_2). The generation of hydrogen using photoelectrochemical water splitting emulates the photosynthesis process in plants. Just like in photosynthesis, photoelectrochemical water splitting converts energy from the sun into chemical energy. Fortunately, water and sunlight are abundant natural resources that can be harvested for clean hydrogen production. Water splitting entails separating the hydrogen and oxygen molecules. As a result, oxygen is the only byproduct, and it is useful rather than harmful to the environment.

The advancement of nanomaterial technologies supports photoelectrochemical water splitting due to the superior performance of nanostructured semiconductor materials and nanoparticles. Examples of photoactive semiconductor materials that have exhibited superior photoelectrochemical performance, with dissimilar nanostructures, including 0-D nanoparticles, 1-D nanowires/nanotubes/nanorods, 2-D nanosheets, and 3-D complex structures, are ZnO, Fe_2O_3, Si, TiO_2, WO_3, Cu_2O, and Ta_3N_5 (Yao et al., 2019). Three distinctive characteristics make nanomaterials better alternatives for photoelectrodes. First, single-unit materials have a quantum confinement effect, which increases electron–hole exchange interaction and electron–hole overlap factor, both of which yield better light absorption efficiency and higher oscillator length. Second, nanomaterials have a relatively larger surface area than bulk materials, which means that they have more active sites for surface redox reactions. Lastly, nanomaterials have single layers, meaning that photoexcited carriers move for reduced diffusion lengths. Therefore, nanomaterials are excellent for use in electrochemical and photoelectrochemical devices for hydrogen production using the water-splitting method.

Semiconductor electrochemistry is another method that can help in the usage of sustainable energy using nanomaterials. This concept combines chemistry and physics to explore the photo-driven reactions at the interfaces between nanoelectrodes and electrolytes. The utilization of polycrystalline semiconductors for converting solar energy into electricity and chemical fuels was first discussed by Heller and Gerischer in the mid-1970s (Wang et al., 2019). Regardless of the concepts they use to function, semiconductors are essential systems in all electronic devices. Their widespread production and improved efficiency would help humans to sustainably tap into renewable energy sources, such as solar and wind energy. For instance, semiconductors are vital components in the automobile industry, and their shortage or inefficiency would affect the performance of the final products. Semiconductors made using electrochemistry typically convert solar energy to chemical energy. The absorbers in solar panels contain nanomaterials to collect solar energy and make it available for human use in the form of electricity.

Photoelectrochemistry further enhances the use of nanomaterials in semiconductor production thanks to the single-unit morphology that optimizes functionality. Nanomaterials are ideal for light trapping due to their enhanced photon absorption properties. Nanomaterials are remarkably suitable as an antireflection layer, which suppresses reflection and enhances nondirectional scattering of incident light. This phenomenon is termed as the light trapping effect. It permits nanowires to demonstrate greater equivalent absorption coefficients across a wide-spanning spectrum range (Deng et al., 2019).

These properties make nanomaterials suitable for antireflection layers since they suppress reflection and enhance incident light scattering. Moreover, nanomaterials with crystalline morphology have exceptional optical qualities, which make them ideal for incident photon absorption. Overall, the use of nanomaterials in photo-driven semiconductors significantly reduces the cost, especially the semiconductors that require numerous micrometers to function optimally.

The CO_2 is one of the greenhouse gases and a primary contributor to global warming that continually damages the environment (Tawalbeh et al., 2021). The electrochemical reduction of this gas is a promising solution to global warming. The CO_2 is a kinetically and thermodynamically stable molecule with a linear structure (Yang et al., 2018). This gas is a common product in many chemical and biological processes. When large volumes of CO_2 accumulate in the earth's atmosphere, they cause far-ranging environmental effects, including air pollution, global warming, and increased wildfires (Alami et al., 2020). Electrochemical reduction of CO_2 entails using electrochemical energy to convert the harmful CO_2 to harmless and useful products, such as methane (CH_4), ethylene (C_2H_4), and methanol (CH_3OH). Due to the high thermodynamic and kinetic stability of the CO_2 molecule, a lot of energy is required to break it down. Nevertheless, electrochemistry presents an efficient method of reducing CO_2 to useful products.

Converting CO_2 into feedstocks and fuels is one of the fundamental processes in saving the planet from climate change (Alami et al., 2021). While it is difficult to break down CO_2 molecules using conventional methods, electrochemistry and photoelectrochemistry of nanomaterials present a viable solution. The electrochemical CO_2 reduction reaction (CO_2RR) is the most attractive method due to its mild reaction conditions and capacity for renewable electricity storage (Verma et al., 2019). Low-value C_1 species such as CO and formate ($HCOO^-$) are the common products of the CO_2RR given the slow kinetics of the $C-C$ coupling reaction (Liu, 2022). The electrochemical approach is ranked the simplest in reducing CO_2 into low-carbon chemicals. In this method, renewable energy from the sun and wind drives the reduction process with the help of different types of electrocatalysts. Different electrocatalysts have been widely researched for their high potential in CO_2 reduction. Metallic nanostructures, metal oxides, and metals are among the materials that have shown great electrocatalytic potential. However, the electrochemical method has some flaws due to poor efficiency, low activity, and low yield of products.

Unlike the electrochemical method, the photoelectrochemical reduction of CO_2 has high efficiency. This method emulates the natural photosynthesis process that uses light to convert CO_2 into useful, sustainable energy sources. The photoelectrochemical approach yields several benefits, including environmental friendliness due to the use of green solar energy, the flexibility of products used, and economic feasibility (Almomani, et al., 2019; Kumaravel, et al., 2020). As natural photosynthesis utilizes sunlight to perform endothermic reactions, artificial photosynthesis in CO_2 reduction is driven by light irradiation and electricity. Researchers argue that better semiconductors with superior charge separation and transportation efficiency, excellent light-capturing, and good charge formation ability should be designed given the low photoconversion efficiency of the majority of semiconductors (Yang et al., 2018). Nanomaterials with single layers enhance the redox reactions, resulting in better efficiency. Therefore, photoelectrochemical reduction of CO_2 is one of the potential solutions to climate change since it efficiently converts harmful CO_2 into harmless substances ideal for energy uses.

13.4 Industrial applications, challenges, and limitations

13.4.1 Greener industrial applications

Nanomaterials have unique characteristics that make them unique from bulk materials. They can be fine-tuned and redesigned to change size, shape, and morphology. In other words, these materials can be customized for optimal use in various industrial applications. Photoelectrochemistry has been a subject of intensive study in the previous decades, and researchers have found several befitting applications. The favorable properties of nanomaterials, combined with the ability for photoelectrochemical technology to convert light and solar energy into chemicals, make them ideal for a wide range of domestic and commercial applications, including semiconductors, medical uses, imaging, and catalysts. It is important to note that most photoelectrochemistry applications are under-developed since this field is relatively new.

Semiconductor photochemistry is an interesting field that combines physics and chemistry in the study of photo-driven reactions at electrode/electrolyte interfaces. This field is a subject of interest for materials scientists, theorists, electrochemists, and spectroscopists, all of who strive to study the applied problems and potentials of solar conversion to chemical fuels and electricity (Wang et al., 2019). Semiconductors are vital components due to their widespread use in electronic devices. Their use is even more important when viewed in the context of diminishing fossil fuels, necessitating humanity to find other cleaner sources of energy. Converting solar energy into useful electricity is one of the many uses of semiconductors utilizing photoelectrochemistry. Solar absorbers utilize nanomaterials and photoelectrochemistry to collect solar energy and convert it into useful electricity. Fig. 13.4 illustrates a simple working mechanism of a typical photoelectrochemical cell (Liao et al., 2012).

Photoelectrochemistry can be combined with bioelectrochemistry to form an interdisciplinary research field called biophotoelectrochemistry (BPEC). The BPEC enables the use of light harvesters and biomachineries in energy production, predominantly using solar energy (Ye et al., 2021). For this reason, BPEC can be used in waste minimization and green chemistry efforts. For instance, it holds great potential for use in pollution control, hydrogen production, and CO_2 reduction. Climate change and global warming are undesirable outcomes of increased carbon emissions into the earth's atmosphere. Capturing and utilizing CO_2 is one of the methods to alleviate the unwanted outcomes of global warming.

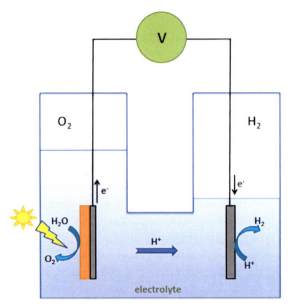

Figure 13.4 Schematic of a photoelectrochemical cell (Liao et al., 2012).

The two-electron reduction of CO_2 to CO (-0.52 V vs. SHE, pH 7.0) and formate (-0.39 V vs. SHE, pH 7.0) can be driven by the two bio photocatalysts CO dehydrogenase (CODH) and formate dehydrogenase (FDH) respectively (Ye et al., 2021), Where SHE is the standard hydrogen electrode. The basic objective is to convert harmful CO_2 into useful or harmless products, such as methanol and methane as illustrated through the reaction shown in Fig. 13.5 (Amao, 2018). This application holds great potential for reversing the negative impacts of global warming.

Hydrogen is viewed as a clean source of energy, and photoelectrochemistry can help in its bulk production through water splitting methods. Photoelectrochemical water splitting represents a green and environmentally friendly method for producing solar hydrogen. Current research is focused toward the development of efficient semiconductor photoelectrodes demonstrating high photocurrent densities (Chen, 2022).

Notably, most redox reactions occur at the surface of electrodes. For this reason, the chemical structure, and physical properties of photoelectrode materials can affect their photoelectrochemical performance. Several surface engineering tactics can be adopted to enhance the performance of nanostructured electrodes such as the deposition of a functional overlayer

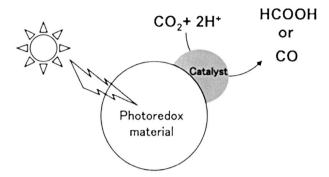

Figure 13.5 Reduction of CO_2 using photoelectrochemistry (Amao, 2018).

of sensitizers, protective and catalytic materials, and plasmonic metallic structures along with alterations to the nanomaterial surface morphology, defect, doping concentrations and crystal facet (Yao et al., 2019). Besides, nanomaterials have relatively larger surface areas, better electronic structures, and shorter carrier diffusion lengths compared to their bulk counterparts. The use of nanomaterials in semiconductor photoelectrochemistry to split water molecules and generate hydrogen gas offers humanity cleaner sources of energy. Fig. 13.6 illustrates the photoelectrochemical cell and lists examples of carbon nanomaterials utilized (Ke, 2021).

13.4.2 Challenges

Despite the numerous promising applications of photoelectrochemistry, several challenges have to be overcome for optimal performance. First, the assembly of photoelectrochemical solar fuels necessitates the use of prototypes, which vary in sophistication and design (Spitler, 2020). An analytical facility is essential to verify the efficiency and performance of these prototypes. Second, the photoelectrochemistry of nanomaterials combines two largely underdeveloped fields; applications are still at the concept stage. It may take time before applications become mainstream.

13.4.3 Limitations

The most significant limitation of photoelectrochemistry of nanomaterials is the inefficiencies of earth-abundant materials and devices. In other

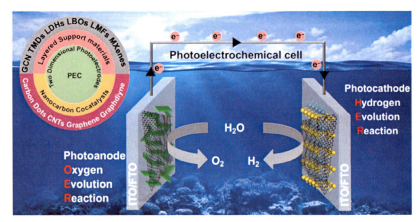

Figure 13.6 Application of nanomaterials in photoelectrochemistry to split water molecules (Ke, 2021).

words, most nanomaterials have suboptimal performance. For instance, some have shunt resistance, series resistance, and poor light absorption. For these materials to attain optimal performance, they need modification to their density and morphology. Therefore, the abundance of inefficient nanomaterials limits the potential of photoelectrochemistry due to the need for additional customization before use.

13.5 Conclusions

In recent years, the research of photoelectrochemistry using nanomaterials has attracted great attention. Nanomaterials have been used in numerous fields and processes where no other substance, chemical element, or compound may be used due to people's extensive knowledge of their properties and traits. The photoelectrochemical characteristics of nanostructured materials are yet unknown. Nanomaterials offer unique features that make them more effective in a variety of electrochemical applications. Solid-state physics, optics, electrochemistry, and surface science are all part of this multidisciplinary field. One of the devices that explain the principle of photoelectrochemistry is the photoelectrochemical cell that can be used in hydrogen production to fulfill some of the objectives toward decarbonization policy.

References

Agyekum, E. B., Nutakor, C., Agwa, A. M., & Kamel, S. (2022). A critical review of renewable hydrogen production methods: Factors affecting their scale-up and its role in future energy generation. *Membranes (Basel)*, *12*(2), 173. Available from https://doi.org/10.3390/membranes12020173.

Alami, A. H., et al. (May 2020). Materials and logistics for carbon dioxide capture, storage and utilization. *The Science of the Total Environment*, *717*, 137221. Available from https://doi.org/10.1016/j.scitotenv.2020.137221.

Alami, A. H., Tawalbeh, M., Alasad, S., Ali, M., Alshamsi, M., & Aljaghoub, H. (2021). Cultivation of nannochloropsis algae for simultaneous biomass applications and carbon dioxide capture. *Energy Sources Part A: Recovery, Utilization, and Environmental Effects*, 1−12. Available from https://doi.org/10.1080/15567036.2021.1933267.

Almomani, F., Al-Rababah, A., Tawalbeh, M., & Al-Othman, A. (2023). "A comprehensive review of hydrogen generation by water splitting using 2D nanomaterials: Photo vs electro-catalysis. *Fuel*, *332*, 125905. Available from https://doi.org/10.1016/j.fuel.2022.125905.

Almomani, F., Bhosale, R., Khraisheh, M., Kumar, A., & Tawalbeh, M. (2019). Photocatalytic conversion of CO_2 and H_2O to useful fuels by nanostructured composite catalysis. *Applied Surface Science*, *483*, 363−372. Available from https://doi.org/10.1016/j.apsusc.2019.03.304.

Al-Othman, A., Al-Nashash, H., Tawalbeh, M., Elhariri, Y., Alami, A. H., & Salameh, T. (2020). Bio-electrodes based on poly(methyl methacrylate) (PMMA) for neural sensing. *2020 IEEE 20th International Conference on Nanotechnology (IEEE-NANO*, 253−256. Available from https://doi.org/10.1109/NANO47656.2020.9183632.

Amao, Y. (2018). Formate dehydrogenase for CO2 utilization and its application. *Journal of CO_2 Utilization*, *26*, 623−641. Available from https://doi.org/10.1016/j.jcou.2018.06.022.

Baig, N., Kammakakam, I., & Falath, W. (2021). Nanomaterials: a review of synthesis methods, properties, recent progress, and challenges. *Materials Advances*, *2*(6), 1821−1871. Available from https://doi.org/10.1039/D0MA00807A.

Barham, J. P., & König, B. (2020). Synthetic Photoelectrochemistry. *Angewandte Chemie International Edition*, *59*(29), 11732−11747. Available from https://doi.org/10.1002/anie.201913767.

Bottomley, L. A., Gadsby, E. D., & Poggi, M. A. (2005). MICROSCOPY TECHNIQUES | atomic force and scanning tunneling microscopy. *Encyclopedia of analytical science*, 143−151. Available from https://doi.org/10.1016/B0-12-369397-7/00386-1, 2nd edn.

Cai, Z., Yao, Q., Chen, X., & Wang, X. (2019). Nanomaterials with different dimensions for electrocatalysis. In X. Wang & X. Chen (Eds.), Novel nanomaterials for biomedical, environmental and energy applications (pp. 435−464). Elsevier. Available from https://doi.org/10.1016/B978-0-12-814497-8.00014-X.

Chen, S., et al. (2022). Recent progress and perspectives on Sb2Se3-based photocathodes for solar hydrogen production via photoelectrochemical water splitting. *Journal of Energy Chemistry*, *67*, 508−523. Available from https://doi.org/10.1016/j.jechem.2021.08.062.

Chen, Y., et al. (May 2020). Phase engineering of nanomaterials. *Nature Reviews Chemistry*, *4*(5), 243−256. Available from https://doi.org/10.1038/s41570-020-0173-4.

Coroş, M., Pogăcean, F., Măgeruşan, L., Socaci, C., & Pruneanu, S. (2019). A brief overview on synthesis and applications of graphene and graphene-based nanomaterials. *Frontiers of Materials Science*, *13*(1), 23−32. Available from https://doi.org/10.1007/s11706-019-0452-5.

Curulli, A. (2020). Nanomaterials in electrochemical sensing area: Applications and challenges in food analysis. *Molecules (Basel, Switzerland)*, *25*(23), 5759. Available from https://doi.org/10.3390/molecules25235759.

Dasgupta, N. P., et al. (2014). 25th Anniversary Article: Semiconductor nanowires—synthesis, characterization, and applications. *Advanced Materials*, *26*(14), 2137−2184. Available from https://doi.org/10.1002/adma.201305929.

Deng, J., Su, Y., Liu, D., Yang, P., Liu, B., & Liu, C. (2019). Nanowire photoelectrochemistry. *Chemical Reviews*, *119*(15), 9221−9259. Available from https://doi.org/10.1021/acs.chemrev.9b00232.

Ding, Y., & Chen, M. (2009). Nanoporous metals for catalytic and optical applications. *MRS Bulletin/Materials Research Society*, *34*(8), 569−576. Available from https://doi.org/10.1557/mrs2009.156.

Dreaden, E. C., Alkilany, A. M., Huang, X., Murphy, C. J., & El-Sayed, M. A. (2012). The golden age: gold nanoparticles for biomedicine. *Chemical Society Reviews*, *41*(7), 2740−2779. Available from https://doi.org/10.1039/C1CS15237H.

Huang, Y.-F., et al. (2007). Improved broadband and quasi-omnidirectional anti-reflection properties with biomimetic silicon nanostructures. *Nature Nanotechnology*, *2*(12), 770−774. Available from https://doi.org/10.1038/nnano.2007.389.

Jeevanandam, J., Barhoum, A., Chan, Y. S., Dufresne, A., & Danquah, M. K. (2018). Review on nanoparticles and nanostructured materials: History, sources, toxicity and regulations. *Beilstein Journal of Nanotechnology*, *9*, 1050−1074. Available from https://doi.org/10.3762/bjnano.9.98.

Jin, S. (2018). What else can photoelectrochemical solar energy conversion do besides water splitting and CO_2 reduction? *ACS Energy Letters*, *3*(10), 2610−2612. Available from https://doi.org/10.1021/acsenergylett.8b01800.

Johnsson, F., Kjärstad, J., & Rootzén, J. (2019). The threat to climate change mitigation posed by the abundance of fossil fuels. *Climate Policy*, *19*(2), 258−274. Available from https://doi.org/10.1080/14693062.2018.1483885.

Karatutlu, A., Barhoum, A., & Sapelkin, A. (2018). *Liquid-phase synthesis of nanoparticles and nanostructured materials. In A. Barhoum & A. S. H. Makhlouf (Eds.),* Emerging applications of nanoparticles and architectural nanostructures: Current prospects and future trends (pp. 1−28). Elsevier Inc. Available from https://doi.org/10.1016/B978-0-323-51254-1.00001-4.

Ke, J., et al. (2021). Nanocarbon-enhanced 2D photoelectrodes: A new paradigm in photoelectrochemical water splitting. *Nano-Micro Letters*, *13*(1), 24. Available from https://doi.org/10.1007/s40820-020-00545-8.

Khan, I., Saeed, K., & Khan, I. (2019). Nanoparticles: Properties, applications and toxicities. *Arabian Journal of Chemistry*, *12*(7), 908−931. Available from https://doi.org/10.1016/j.arabjc.2017.05.011.

Kolahalam, L. A., Kasi Viswanath, I. V., Diwakar, B. S., Govindh, B., Reddy, V., & Murthy, Y. L. N. (2019). Review on nanomaterials: Synthesis and applications. *Materials Today: Proceedings*, *18*, 2182−2190. Available from https://doi.org/10.1016/j.matpr.2019.07.371.

Krishnan, D., Raidongia, K., Shao, J., & Huang, J. (2014). Graphene oxide assisted hydrothermal carbonization of carbon hydrates. *ACS Nano*, *8*(1), 449−457. Available from https://doi.org/10.1021/nn404805p.

Kumaravel, V., Bartlett, J., & Pillai, S. C. (2020). Photoelectrochemical conversion of carbon dioxide (CO_2) into fuels and value-added products. *ACS Energy Letters*, *5*(2), 486−519. Available from https://doi.org/10.1021/acsenergylett.9b02585.

Liao, C. H., Huang, C. W., & Wu, J. C. S. (2012). Hydrogen production from semiconductor-based photocatalysis via water splitting. *Catalysts*, *2*(4), 490−516. Available from https://doi.org/10.3390/catal2040490, MDPI AG.

Liu, W., et al. (2022). Electrochemical CO_2 reduction to ethylene by ultrathin CuO nanoplate arrays. *Nature Communications*, *13*(1), 1877. Available from https://doi.org/10.1038/s41467-022-29428-9.

Luther, W. (2004). *Industrial application of nanomaterials—Chances and risks*. Future Technologies Division of VDI Technologiezentrum GmbH.

Majhi, K. C., & Yadav, M. (2021). Synthesis of inorganic nanomaterials using carbohydrates. In: Inamuddin, R. Boddula, M.I. Ahamed, & A.M. Asiri (Eds.), *Green sustainable process for chemical and environmental engineering and science* (pp. 109—135). Elsevier. Available from https://doi.org/10.1016/B978-0-12-821887-7.00003-3.

Martins, F., Felgueiras, C., Smitkova, M., & Caetano, N. (2019). Analysis of fossil fuel energy consumption and environmental impacts in european countries. *Energies*, *12* (6). Available from https://doi.org/10.3390/en12060964.

Mehnath, S., Das, A. K., Verma, S. K., & Jeyaraj, M. (2021). Biosynthesized/green-synthesized nanomaterials as potential vehicles for delivery of antibiotics/drugs. *Comprehensive Analytical Chemistry*, *94*, 363—432. Available from https://doi.org/10.1016/bs.coac.2020.12.011.

Mohamed, O., Al-Othman, A., Al-Nashash, H., Tawalbeh, M., Almomani, F., & Rezakazemi, M. (2021). Fabrication of titanium dioxide nanomaterial for implantable highly flexible composite bioelectrode for biosensing applications. *Chemosphere*, *273*, 129680. Available from https://doi.org/10.1016/j.chemosphere.2021.129680.

Nauman Javed, R. M., Al-Othman, A., Tawalbeh, M., & Olabi, A. G. (2022). Recent developments in graphene and graphene oxide materials for polymer electrolyte membrane fuel cells applications. *Renewable and Sustainable Energy Reviews*, *168*, 112836. Available from https://doi.org/10.1016/j.rser.2022.112836.

Patel, K. J., & Pathak, V. Y. (2021). *Emerging technologies for nanoparticle manufacturing*. Cham: Springer International Publishing. Available from https://doi.org/10.1007/978-3-030-50703-9.

Patel, K. D., Singh, R. K., & Kim, H.-W. (2019). Carbon-based nanomaterials as an emerging platform for theranostics. *Materials Horizons*, *6*(3), 434—469. Available from https://doi.org/10.1039/C8MH00966J.

Pokropivny, V. V., & Skorokhod, V. V. (2007). Classification of nanostructures by dimensionality and concept of surface forms engineering in nanomaterial science. *Materials Science and Engineering: C*, *27*(5—8), 990—993. Available from https://doi.org/10.1016/j.msec.2006.09.023.

Prasad Yadav, T., Manohar Yadav, R., & Pratap Singh, D. (2012). Mechanical milling: a top down approach for the synthesis of nanomaterials and nanocomposites. *Nanoscience and Nanotechnology*, *2*(3), 22—48. Available from https://doi.org/10.5923/j.nn.20120203.01.

Saleh, A. T. (2020). Nanomaterials: Classification, properties, and environmental toxicities. *Environmental Technology & Innovation*, *20*, 101067. Available from https://doi.org/10.1016/j.eti.2020.101067.

Serrà, A., & García-Torres, J. (2021). Electrochemistry: A basic and powerful tool for micro- and nanomotor fabrication and characterization. *Applied Materials Today*, *22*, 100939. Available from https://doi.org/10.1016/j.apmt.2021.100939.

Sherman, B. D., Bergkamp, J. J., Brown, C. L., Moore, A. L., Gust, D., & Moore, T. A. (2016). A tandem dye-sensitized photoelectrochemical cell for light driven hydrogen production. *Energy & Environmental Science*, *9*(5), 1812—1817. Available from https://doi.org/10.1039/C6EE00258G.

Spitler, M. T., et al. (2020). Practical challenges in the development of photoelectrochemical solar fuels production. *Sustain. Energy Fuels*, *4*(3), 985—995. Available from https://doi.org/10.1039/C9SE00869A.

Sudha, P. N., Sangeetha, K., Vijayalakshmi, K., & Barhoum, A. (2018). *Nanomaterials history, classification, unique properties, production and market. Emerging applications of nanoparticles and*

452 Electrochemistry and Photo-Electrochemistry of Nanomaterials

architecture nanostructures (pp. 341−384). Elsevier. Available from https://doi.org/10.1016/B978-0-323-51254-1.00012-9.

Sulaiman, M. B. & Santuraki, A. H. (2018). *Applications and implications of environmental nanotechnology.* In research trends in environmental science (Chapter 4). AkiNik Publications.

Suresh, S. (2013). Semiconductor nanomaterials, methods and applications: A review. *Nanoscience and Nanotechnology, 3*(3), 62−74. Available from https://doi.org/10.5923/j.nn.20130303.06.

Swihart, M. T. (2003). Vapor-phase synthesis of nanoparticles. *Current Opinion in Colloid & Interface Science, 8*(1), 127−133. Available from https://doi.org/10.1016/S1359-0294(03)00007-4.

Tawalbeh, M., Al-Ismaily, M., Kruczek, B., & Tezel, F. H. (2021). Modeling the transport of CO_2, N_2, and their binary mixtures through highly permeable silicalite-1 membranes using Maxwell − Stefan equations. *Chemosphere, 263,* 127935. Available from https://doi.org/10.1016/j.chemosphere.2020.127935.

Tawalbeh, M., Khan, H. A., & Al-Othman, A. (2023). Insights on the applications of metal oxide nanosheets in energy storage systems. *Journal of Energy Storage, 60,* 106656. Available from https://doi.org/10.1016/j.est.2023.106656.

Tawalbeh, M., Muhammad Nauman Javed, R., Al-Othman, A., & Almomani, F. (2022). The novel advancements of nanomaterials in biofuel cells with a focus on electrodes' applications. *Fuel, 322,* 124237. Available from https://doi.org/10.1016/j.fuel.2022.124237.

Tri, P. N., Rtimi, S., Nguyen, T. A., & Vu, M. T. (2019). Physics, electrochemistry, photochemistry, and photoelectrochemistry of hybrid nanoparticles. In S. Mohapatra, T. A. Nguyen, & P. Nguyen-Tri (Eds.), *(Eds.),* Noble metal-metal oxide hybrid nanoparticles (pp. 95−123). Elsevier. Available from https://doi.org/10.1016/B978-0-12-814134-2.00005-X.

Verma, S., Lu, S., & Kenis, P. J. A. (2019). Co-electrolysis of CO_2 and glycerol as a pathway to carbon chemicals with improved technoeconomics due to low electricity consumption. *Nature Energy, 4*(6), 466−474. Available from https://doi.org/10.1038/s41560-019-0374-6.

Wang, L., Schmid, M., & Sambur, J. B. (2019). Single nanoparticle photoelectrochemistry: What is next. *The Journal of Chemical Physics, 151*(18), 180901. Available from https://doi.org/10.1063/1.5124710.

Xi, J.-Q., et al. (2007). Optical thin-film materials with low refractive index for broadband elimination of Fresnel reflection. *Nature Photonics, 1*(3), 176−179. Available from https://doi.org/10.1038/nphoton.2007.26.

Xie, P., et al. (2021). Advanced carbon nanomaterials for state-of-the-art flexible supercapacitors. *Energy Storage Materials, 36,* 56−76. Available from https://doi.org/10.1016/j.ensm.2020.12.011.

Yang, P. (2012). Semiconductor nanowire building blocks: From flux line pinning to artificial photosynthesis. *MRS Bulletin/Materials Research Society, 37*(9), 806−813. Available from https://doi.org/10.1557/mrs.2012.200.

Yang, Y., Ajmal, S., Zheng, X., & Zhang, L. (2018). Efficient nanomaterials for harvesting clean fuels from electrochemical and photoelectrochemical CO_2 reduction. *Sustainable Energy & Fuels, 2*(3), 510−537. Available from https://doi.org/10.1039/C7SE00371D.

Yao, B., Zhang, J., Fan, X., He, J., & Li, Y. (2019). Surface engineering of nanomaterials for photo-electrochemical water splitting. *Small (Weinheim an der Bergstrasse, Germany), 15*(1), 1803746. Available from https://doi.org/10.1002/smll.201803746.

Ye, J., Hu, A., Ren, G., Chen, M., Zhou, S., & He, Z. (2021). Biophotoelectrochemistry for renewable energy and environmental applications. *iScience, 24*(8), 102828. Available from https://doi.org/10.1016/j.isci.2021.102828.

Zhang, Y., Guo, W., Zhang, Y., & Wei, W. D. (2021). Plasmonic photoelectrochemistry: In view of hot carriers. *Advanced Materials, 33*(46), 2006654. Available from https://doi.org/10.1002/adma.202006654.

Index

Note: Page numbers followed by "*f*" and "*t*" refer to figures and tables, respectively.

A

Absorption processes, 401–402
AC. *See* Activated carbon (AC)
ACC. *See* Activated carbon cloth (ACC)
Activated carbon (AC), 262–263
Activated carbon cloth (ACC), 317–319
Active catalyst species, 381–382
Active chlorine, generation of, 375–376
Adenosine monophosphate (AMP), 165–166
Adsorbed chlorine, 363–364
Advanced nanomaterials, 97
AFM. *See* Atomic force microscopy (AFM)
AgBiS$_2$ quantum dots, 113
Algal toxins
 detection and decontamination, 346–347
 detection methods of, 346–347
Alkaline fuel cells (AFCs), cathode of, 134–135
Alkaline phosphate activity (APA)
 pollutant's detection using, 42*f*
All-inorganic perovskites, 118–119
Allotropes, 433–434
Amorphous carbon, 296
AMP. *See* Adenosine monophosphate (AMP)
Amperometric biosensor, 35–36, 38–39
 for antigen, 46*f*
 electrode, configuration of, 38–39, 38*f*
 used for feedback control loop, 48*f*
 use of, 39
Amperometric enzyme, 35
Amperometric sensors, 36
Anion exchange membranes, 338–339
Anisotropic nanostructures, 402
Anode materials, nanoelectrochemistry in, 222–232
 alloy-type materials and nanoelectrochemistry, 223–230

conversion materials and
 nanoelectrochemistry, 230–232
Anodes
 composition of, 369
 extracellular electron transfer, 163–167
Anodic electrochromic materials, 11–12
Anodic modification materials, 177–191
 biomass-derived anodes, 178–179
 biomimetic anodes, 189–191
 carbon nanotubes-based anodes, 179–180
 commercial anodes, 177–178
 graphene and reduced graphene oxides-based anodes, 181
 heteroatom-doped carbonaceous anodes, 181–184
 metal- and metal compound-derived anodes, 184–186
 MXene-based anodes, 188–189
 polymer-based anodes, 186–188
Anodic nanoelectrochemistry, 170–177
 hierarchical porous structures, 173–176
 surface micro-/nanostructures, 176–177
 surface roughness, 171–172
 surface wettability, 172–173
Antibacterial materials, 195–196
Anti-reflective coating, 79–80
Anti-Stokes imaging, 409–411
Anti-Stokes thermometry, 403–405
Ant-nest-like microscale porous Si (AMPSi), 226–227
APA. *See* Alkaline phosphate activity
Applications, electrochemical biosensing
 cancer detection, 44–46
 environmental contamination detection, 48–49
 food contamination detection, 51
 glucose detection, 46–48
 herbicide contamination detection, 50–51
 pathogen detection, 49–50

454 Index

Aptamers, 346–347
Aquatic systems, algal toxin
  decontamination in, 347
Ascorbic acid, 53
ASCs, 265–266
Atomic force microscopy (AFM), 124, 427

B

Bacteria/bacterial
  adhesion, micro-/nanostructured surface
    effect on, 176f
  biomineralization, 194–195
  mineralization of, 196–197
  remediation, 413
Battery
  cathode, capacity of, 219
  materials, nanostructure of, 211
  performance, 223
  technology, 215–216
  type electrode materials, 252
Biofilms
  formation of, 170–171
  research on, 170
Bioimaging, applications in, 438
Biological materials, carbonization process
  of, 296–297
Biological tissue transparency window,
  405–406
Biomarkers, 44–45
Biomass anode, 179f
Biomass-derived anodes, 178–179
Biomedical application, metallic
    nanostructures for
  applications, 408–413
  biological tissue transparency window,
    405–406
  challenges, 414–416
  description of, 395–398
  laser heating of nanostructures, 403–405
  nonradiative properties, 398–400
  photothermal nanostructures, 407–408
  theory of plasmonic heating, 400–402
Biomedical applications, 414
Biomimetic anodes, 189–191, 190f
Biophotoelectrochemistry (BPEC),
    445–446

Biosensors, 35
  amperometric, 38–39
  characteristics of, 52–53
  conductometric, 41–42
  demonstration of, 35
  description of, 35
  design and development, 52–53
  GOx in, 47–48
  impedimetric, 43
  potentiometric, 39–41
  representation of elements in, 37f
Biotic-abiotic interface, 167–168
Bottom-up methods, 120–121
  chemical vapor deposition (CVD), 121
  electrospinning, 121
  hydrothermal/solvothermal synthesis,
    121
Bulk
  analogs, 434–435
  electrochemistry, 333–334
  materials, 438
  molecules, 64–65
  phonons, 403–405

C

Cadmium sulfide (CdS) quantum dots, 112
Cancer detection, 44–46
  sequence of, 45f
Carbon, 385
  materials, 181–182
    electrochemical performance of, 304
  microflowers, 184
  nanomaterials, 335–336, 446–447
    structures, 336f
  nanostructures, 311–312
  shell, structural integrity of, 226
Carbonaceous materials, 293–297
Carbon-based allotropes, chemical
    structures of, 434f
Carbon-based nanomaterials, 94f, 106–109
  carbon nanotubes (CNTs), 107–108
  fullerenes, 108–109
  graphene, 107
Carbon dots (CDs), 192–194, 193f
  optical properties of, 194
Carbon nanofibers (CNFs), 180, 317

Carbon nanotubes (CNTs), 67, 77–78, 107–108, 144–146, 179–180, 184–186, 301f, 335–336, 339–340
  based anodes, 179–180
  concentration of, 179–180
  modified anodes, 179f, 180
  single-walled, 179–180
Carbon paper (CP) substrate, 261–262
Carcinoembryonic antigen (CEA), 44–45
Castner–Kellner process, 360
Catalyst surface, 376
Catalytic reactions, 411–413
Cathode electrolyte interphase (CEI), 233
Cathode materials
  electrochemical performance of, 218
  energy densities of, 212f
  fundamental chemical and physical properties of, 213
  nanoelectrochemistry in
    challenges and future perspectives i, 220–222
    field of, 221
    fundamental properties of, 212–213
    nanoelectrochemical mechanisms in, 216–219
    nanoengineering for high-performance cathodes, 219–220
    production of, 221
    synthesis of nanostructured, 213–215
    types of advanced, 215–216
  traditional and nanoscale, 213
Cell metabolism, 167–168
CER. See Chlorine evolution reaction (CER)
"Cerenkov" radiation of sound waves, 400
Charge
  collection, 99–101
  discharge cycles, 230–231
  separation and transport, 99
  storage capability, 252
  transfer kinetics, 253
  transfer process, 217, 218f
  transport, 101
Chemical bonding, 291–292
Chemical processes, design of, 336–337

Chemical stability, 220
Chemical vapor deposition (CVD), 121, 429–430
Chlor-alkali process
  description of, 359–369
  electrocatalysts for CER, 370–385
    metal-free electrode for, 384–385
    noble metal-based nanoelectrocatalysts, 371–383
    noble metal free nanomaterials for, 383–384
  energy consumption of, 370
  reaction mechanism of CER, 363–364
  selectivity between OER and CER, 364–366
    factors affecting, 366–369
Chlorella Vulgaris, 41–42
Chlorine evolution reaction (CER), 366–368, 372t
  activity and selectivity, 376–377
  electrocatalysts for, 370–385
    metal-free electrode for, 384–385
    noble metal-based nanoelectrocatalysts, 371–383
    noble metal free nanomaterials for, 383–384
  Gibbs free-energy diagram for, 365f
  mechanism of, 362–364
  metal free electrocatalyst for, 383
  OER and, 364–369
  pathway of, 370
  polarization curves, 379f
  rates, 367f
  selectivity, 368t
Chlorine gas, 361
Chronoamprometric (CA) measurements, 17–18
Climate change, 59–60
CNTs. See Carbon nanotubes (CNTs)
CO dehydrogenase (CODH), 445–446
Coloration efficiency, 4–5
Color storage memory, 5–6
Combined Heat and Power (CHP) distributed generation, 140
Commercial anodes, 177–178
Commercial solar panels, 59–60
Composite nanomaterials, 431–432

456 Index

Composite polymer electrolytes (CPE), 236–237
electrochemical properties of, 236
Comproportionation, 27–28
Conductimetric biosensors, 35–36
Conductive polymers (CPs), 25
conductive polymers, 26t
modified electrodes, 173
Conductometric biosensors, 41–42
Conjugated polymer, 25–26
Consumer electronics, 211
Contact angle measurements, 125
Continuous absorption bands, 436
Continuous glucose monitoring system (CGMS), 47
Conventional catalysts, 347–348
Conventional internal combustion engines, 140
Conventional nanomaterials, 428
Copper indium gallium selenide (CIGS) quantum dots, 113
Copper indium selenide (CIS), 113
Core-shell nanocatalysts, 336–337
Cost-efficient coordination materials, 263–265
Cost reduction, 66, 80, 95
Coulombic efficiency, 291
Counter electrode (CE), 35–36, 96, 102
Crystalline, 432–433
lattice, 400, 434–435
synergistic mechanism of, 10
Curled graphene, 295–296
Current generation, 99–101
CVD. *See* Chemical vapor deposition (CVD)
Cyclic voltammogram, 13*f*
Cyclodextrine (CD), 235
Cytochrome C, 166–167

D

Defect engineering, 278–282
Degradation, 347–348
Density functional theory (DFT) calculations, 364
Derivatives, 116
Detection
cancer, 44–46

environmental contamination, 48–49
food contamination, 51
glucose, 46–48
herbicide contamination, 50–51
pathogen, 49–50
Detoxification, photocatalysts for, 347–349
Differential electrochemical mass spectrometry (DEMS), 384–385
Dimensionally stable anode (DSA), 359, 361
discovery and use of, 361
Dimerization, 27–28
Dimethylacetamide (DMAC), 233
Direct electron transfer (DET), 165
mechanism, 167–168
schematic diagram of, 165
Direct methanol fuel cells (DMFCs), 138–139, 144
anodes, 144
merits of, 139
problem with, 144
Doping and surface functionalization of nanomaterials, 122–123
atomic layer deposition (ALD), 122–123
ion implantation, 122
plasma treatment, 123
Double perovskites, 120
Dual-atom doping, 184
Dual chamber microbial fuel cell, 162*f*
Dye, 102
pollutants, 343
regeneration of, 101
Dye-sensitized solar cells (DSSCs), 63, 89, 96

E

ECDs. *See* Electrochromic devices (ECDs)
Economic viability, 128
EET. *See* Extracellular electron transfer (EET)
Elastic scattering process, 400–401
Electrical characterization
current-voltage (I-V) measurements, 125
electrochemical impedance spectroscopy (EIS), 125
quantum efficiency measurements, 125
Electrical conductivity, limitations in, 231–232

Electrical deposition parameters, 18–19
Electrical double layer capacitors (EDLCs), 253, 255–257, 256f
  carbon-based electrodes used in, 257
  CV curves of, 258f
Electric charge density, 5–6
Electric vehicles (EV), 211
Electroactive components, aggregation of, 276–277
Electrocatalysts, 444
  for CER, 370–385
    metal-free electrode for, 384–385
    noble metal-based nanoelectrocatalysts, 371–383
    noble metal free nanomaterials for, 383–384
Electrochemical biosensing
  advantages, 51–52
  applications
    cancer detection, 44–46
    environmental contamination detection, 48–49
    food contamination detection, 51
    glucose detection, 46–48
    herbicide contamination detection, 50–51
    pathogen detection, 49–50
  biosensors, types of
    amperometric biosensors, 38–39
    conductometric biosensors, 41–42
    impedimetric biosensors, 43
    potentiometric biosensors, 39–41
  description of, 35–37
  limitations, 52–53
Electrochemical biosensors, 36, 45, 49
  groups of, 35–36
Electrochemical deposition method, 19–21
Electrochemical impedance spectroscopy (EIS), 43, 50
Electrochemical oxidation, 8
Electrochemical sensing devices, 345
Electrochemical supercapacitors, 254–263
  cost-efficient coordination materials for, 263–265
  hybrid supercapacitors (HSCs), 262–263
  pseudocapacitors, 258–262
Electrochemistry, definition of, 333–334

Electrochromic devices (ECDs), 3–4, 17–18
  functional properties of, 17–18
  ion transport for, 4
  multicolor transparent properties of, 12–14
  performance, 4–6
  transmittance vs. wavelength for, 18f
Electrochromic materials, 2, 6
  characteristics of, 2–3
Electrochromic metal oxides, color change of, 7t
Electrochromic minerals, preparation of, 7–8
Electrochromic nanomaterials, organic class of, 27–28
Electrochromism, 2
Electrodes, 255–256, 345–346
  fabrication, 369
  materials, 253
  reduction and oxidation reactions at, 342–343
  surface of, 446–447
Electrolytes, 96, 102, 439–440, 442–443
  composition, 369
  ionic conductivity of, 4–5
  wettability of, 239–240
Electron acceptors, 159–160
Electronegativity, 266–268
Electron-hole pairs, 439
Electronic mediators, 166, 168
Electron redistribution, 277
Electron transfer, 166–167
Electrochemical supercapacitors (ESCs), 251, 256
  electrode for, 252
Electrospinning, 121
Electrothermophoresis (ETP), 21
Elemental doping, 265–268
Elongated nanorods, 414–415
Emergency back-up power supply (EPS), 140
Endotoxin
  degradation of, 348–349
Energy conversion efficiency, 100
Energy-dispersive X-ray spectroscopy (EDS), 19–21, 124

458 Index

Energy loss, 1–2
Energy shortage, 159
Energy storage capability of electrodes,
265–282
defect engineering, 278–282
elemental doping, 265–268
heterointerface engineering, 274–278
hybridization with carbons, 272–274
surface functionalization, 268–271
Energy storage system (ESS), 211
Energy-storing appliances, 251
Engineering cathode materials, 214
Engineering porosity, 219
Enhanced light absorption, 95
Environmental applications, nano/
photoelectrochemistry for
algal toxins detection and
decontamination, 346–347
carbon nanomaterials, 335–336
description of, 333–334
heavy metals detection, 344–346
mechanism of photoelectrochemical
reduction of, 342–343
modified nanomaterials, 337–338
nanocatalysts, 336–337
photochemical enzymatic systems in
wastewater, 349–351
photoelectrochemical TiO_2 catalyst on
pollutant degradation, 343–344
reduction of CO_2 to CH_3OH, 341–343
use of photocatalysts for detoxification,
347–349
water splitting, 338–341
basic fundaments of, 339–340
for H_2 production, 341
photoelectrochemical for solar
hydrogen production, 339
Environmental contamination detection,
48–49
Environmental pollution, 159
Environmental sustainability, 333–334
Enzyme immobilization technique, 37
Epitope, 49–50
Equivalent series resistance (ESR),
254–255
Eroatom doping, 182–183
Exciton generation, 99–100

Exoelectrogens, 167–169
and metal compounds, 167–169
Extracellular electron transfer (EET)
between bacteria and electrode surfaces,
168
of electrically active bacteria, 165
indirect pathway for, 170–171
mechanisms, 164f
paths, 195f
pathways in microorganisms, 191–197
internalize/biomineralize
nanomaterials, 192–197
polymer coated microorganisms,
191–192
process, 168–169
schematic illustration of, 195f
Extracellular polymers
nanometers of, 166

F

Fabrication strategy, 275f
Fabrication techniques, 438
Faradaic redox reactions, 268–269
Feedstocks, 444
Feed substrate sodium acetate, 163
Fe metal particles, 230–231
Femtosecond-pulsed laser heating, 403–405
Fermi levels, 438–439
Ferric hexacyanoferrite, 14–15
FESEM images, 22f
Film surface, 10–11
Fimbriae, 168
First-generation nanomaterials, 97
5 lithium–sulfur batteries, 293–308
carbonaceous materials, 293–297
metal-organic frameworks (MOFs),
297–302
MXenes, 302–305
polymeric materials, 305–308
Flagella, 168
Flavin mononucleotide (FMN), 165–166
Fluorine-doped tin dioxide (FTO), 4, 14f
Food contamination detection, 51
Formate dehydrogenase (FDH), 445–446
Fossil fuels
negative impacts of, 441
producing electricity using, 59–60

Fourier transform infrared spectroscopy (FTIR), 125
FPEEK electrospun separators, 242f
Fuel cells' electrodes
applications of, 139–140
description of, 133–134
novel materials for nanoelectrodes
graphene and graphene oxide, 146–148
metal-organic frameworks (MOFs), 148–149
metal oxide nanosheets, 149–151
novel materials for nanoelectrodes in, 144–151
carbon nanotubes (CNTs), 144–146
state-of-the-art nanomaterials for, 140–144
DAFCs, 144
PEMFCs, 142–143
SOFCs, 141
types and applications of, 134–139
alkaline fuel cells, 134–135
direct methanol fuel cells, 138–139
Molten carbonate fuel cells, 136–137
phosphoric acid fuel cells, 135–136
proton exchange membrane fuel cell, 137–138
solid oxide fuel cells, 136
Fuel cells
catalysts, 141
disadvantages of, 136
theoretical capability of, 135
types of, 145f
Fuels, 444
Fullerenes, 108–109
Fullerenes nanoparticles, 438

G

Galvanostatic charge-discharge (GCD), 260f
Galvanostatic charge/discharge curves, 317f
Gibbs energy, 364, 365f
Glass or transparent plastic, 3–4
Global warming, 59–60, 443–444
Glucose, 51
Glucose detection, 46–48
Gold nanocages, application of, 408–409

Gold nanoislands, 408–409
Gold nanoparticles, 403–405
lattice dynamics of, 403–405
as nanothermometers, 414f
photothermal microscopy image of, 412f
transient events of, 398–400, 399f
Gold nanorods, 403–405
Gold nanospheres, 405f
photothermal effect of, 408f
Graphene, 70, 107, 146–148
chemical structures of, 150f
family nanomaterials, 335–336
nanosheets, 146–147
nitrogen-doped, 294f
pyrrole modification of, 294–295
Graphene oxide, 146–148
based anodes, 181
chemical structures of, 150f
Graphite, 223–224, 293–294
anodes, 231–232
Gratzel cells, 89, 96
Gravimetric energy density, 288f
Green energy production
challenges, 447
description of, 427–428
industrial applications, 445–447
integrated methods toward green approaches, 441–444
nanomaterials and conventional nanomaterials, 428–431
electrochemistry of, 431–436
manufacturing process, 429–431
number of dimensions, 429
phase composition, 429
Greenhouse gas emissions, 59–60, 341–342, 443–444

H

Heavy metals detection, 335f, 344–346
Herbicide contamination detection, 50–51
Heteroatom-doped carbonaceous anodes, 181–184
Heteroatomic implanting, 265–266
Heterogeneous physiochemical properties, 350–351
Heterointerface engineering, 274–278
Hetero-nanostructures, 278

460 Index

Hetero-structurization, 272–274
Hierarchical nanostructured materials, 253
Hierarchical porous
anode, 175f
structures, 173–176
High-concentration electrolytes (HCEs), 232–233
disadvantages of, 233
High-energy density batteries, 222, 293
High-performance batteries, 219–220
Hole transport layers (HTL), 70
Hollow carbon nanospheres (HC), 311–312
Horseradish peroxidase (HRP), 45–46
Host materials, requirements for, 291–293
HPCFs, hierarchical porous anode of, 175–176, 175f
HSCs. See hybrid supercapacitors (HSCs)
Hybrid and composite nanostructures, 97–98
Hybridization with carbons, 272–274
Hybrid supercapacitors (HSCs), 253, 255, 256f, 262–263
Hydrocarbon contaminated ditches, 161
Hydroelectric power plant, 287–289
Hydrogen, 446
gas, 442
ions, 40–41
production, 339–340, 439f
proton, 138
Hydrogen Evolution Reaction (HER), 149
Hydrophobicity, 172
Hydropower plants, 287–289
Hydrothermal/solvothermal synthesis, 121
Hydroxide ions, 133–134
Hydroxypropyl cellulose (HPC) membrane, 17–18

I

Impedimetric biosensors, 35–36, 43
between conductive electrodes, 43f
Indirect electron transfer (MET)
between cytochrome c/electron mediator complexes, 168–169
process, 165–166
Indium tin oxide (ITO), 4

Inductively coupled plasma mass spectrometry (ICP-Ms), 124
Inkjet printing, 122
Innovative deposition technique, 21
Inorganic nanoparticle incorporated composite polymer electrolytes, 236–237
Inorganic nanoparticles molecules, 347
InP (indium phosphide) quantum dots, 113
In situ photothermal characterization, 414
Intercalation-based electrodes, 230–231
Interface engineering, 274–276
Interfacial electrochemistry, 333–334
Internalize/biomineralize nanomaterials, 192–197
International Agency for Research on Cancer (IARC), 44–45
Ion conducting layer, 3–4
Ion-exchange membrane, 362–363
Ion-selective electrodes, 435
Ion-sensitive field effect transistors (ISFET), 40–41
Iron(III) oxide, 104–105
Ithium lanthanum tantalum zirconium oxide (LLZTO), 236

K

Krishatalik pathway, 363
Krishtalik mechanism, 363–364

L

Lamb vibrational modes, 403–405
Landau damping, 398–400
Langmuir–Blodgett (LB) techniques, 8–9
Laser heating of nanostructures, 403–405
Laser pulse excitation, 403–405
Layer-by-layer assembly, 122
Layered oxide cathode materials, 215–216
Layer-fiber skeletons, 433–434
Lead-free perovskites, 119
Leucine-arginine (LR), 346–347
Light
absorption, 76–77, 99–100, 440
forms of, 2–3
to-heat conversion, 414–415
triggered redox reactions, 441

Lignin
  based carbon electrode, 378f
  based monolithic carbon, 377
Lithium-air (Li-air) batteries, 216
Lithium aluminum titanium phosphate
    (LATP), 236
Lithium-ion batteries (LIBs), 222,
    237–238, 287–289
  rechargeable, 232–233
  separator, 238–239
  theoretical capacity of, 228–229
  transport, 231–232
Lithium ions, 212–213
  intercalation, active sites for, 219
Lithium lanthanum zirconium oxide
    (LLZO), 236
Lithium-sulfur (Li-S) batteries, 216, 221
Lithography, 121
LiX matrix, 230–231
Localized high-concentration electrolytes
    (LHCEs), 232–233
  nonflammable, 233
Localized surface plasmon resonance
    (LSPR), 395–396
  band, 398–400
  biosensors, dual-functional, 411f
Los Alamos National Lab, 65
Lower dimensional perovskites, 119–120
Low ionic conductivity, 233–234

M

Malignant cells
  photothermal treatment of, 410f
  plasmonic photothermal therapy of,
    395–396
Membrane cells, 362–363
  technology, 360–363
Membrane Electrode Assembly (MEA),
    133–134
Mesoporous iron oxide nanoparticle
    clusters (MIONCs), 231–232
Metal- and metal compound-derived
    anodes, 184–186
Metal-based electrocatalysts, 371
Metal-based inorganic materials,
    431–432
Metal compounds, 167–169

Metal-free electrocatalyst, 383–384
Metal-free electrode, 384–385
Metal Hexacyano, 16
Metal-ion capacitors, 262–263
Metallic colloids, dispersions of, 395–396
Metallic nanoparticle, 400–401
Metallic nanostructures, 397–398, 400,
    403–405, 444
  photothermal effect of, 405–406
  photothermal properties of, 398
Metallophthalocyanines, 26–27
Metal nanoparticles, 436, 438
Metal–organic frameworks (MOFs), 70,
    144, 148–149, 297–302, 312
  compound, 300–301
  electrochemical performance of, 303t
  electrochemical properties of, 302
  materials, 299–300
  nanocomposites, 148–149
  porphyrin-based, 301–302
  structure, 151f
Metal oxide nanomaterials, 102–106
  iron(III) oxide, 104–105
  titanium dioxide, 103
  tungsten trioxide, 105–106
  zinc oxide, 103–104
Metal oxide nanosheets, 149–151
  structure of, 151f
Metal-sulfur batteries, coordination
    materials for
  applied materials in multivalent, 318t
  cathode material in, 314t
  challenges of, 291
  description of, 287–289
  5 lithium–sulfur batteries, 293–308
    carbonaceous materials, 293–297
    metal-organic frameworks (MOFs),
      297–302
    MXenes, 302–305
    polymeric materials, 305–308
  mechanism of, 289–290
  multivalent metal-sulfur batteries,
    315–319
  potassium-sulfur batteries, 312–315
  requirements for host materials,
    291–293
  sodium-sulfur batteries, 310–312

Methanol oxidation reaction (MOR), 147–148
MFCs. *See* Microbial fuel cells (MFCs)
Microbial electrode interface microenvironment, 168–169
Microbial fuel cells (MFCs)
  anodes, 170, 177–180, 184, 185*f*, 186
  anodic modification materials, 177–191
  biomass-derived anodes, 178–179
  biomimetic anodes, 189–191
  carbon nanotubes-based anodes, 179–180
  commercial anodes, 177–178
  graphene and reduced graphene oxides-based anodes, 181
  heteroatom-doped carbonaceous anodes, 181–184
  metal- and metal compound-derived anodes, 184–186
  MXene-based anodes, 188–189
  polymer-based anodes, 186–188
  application of, 159–160
  basic principles of, 161–163
  charge transfer resistance of, 173
  density of, 182–183
  description of, 159–160
  EET pathways in microorganisms, 191–197
    internalize/biomineralize nanomaterials, 192–197
    polymer coated microorganisms, 191–192
  for electricity generation and wastewater treatment, 161
  electroactive microorganisms in, 163
  exoelectrogens and metal compounds, 167–169
  extracellular electron transfer between microorganisms and anodes, 163–167
  factors influencing anodic nanoelectrochemistry, 170–177
  hierarchical porous structures, 173–176
  surface micro-/nanostructures, 176–177
  surface roughness, 171–172
  surface wettability, 172–173

history of, 161
performance of, 170–171, 173–174, 176–180
with photoresponsive electricity production, 196
power generation
  and decontamination capabilities of, 170–171
  performance and operational stability of, 184–186
production capacity, 163
production capacity of, 172
publications on, 160*f*
structures of, 161
wastewater treatment performance of, 194–195
Microliter volume samples, 345
Microorganisms, 166, 169*f*
  extracellular electron transfer, 163–167
Micropollutants, 413
Microscopic radiation, 400
Mineralization, 347–348
Mixed metal oxides (MMOs), 361, 376–377, 380–381
  network homogeneity of, 381
  preparation of electrocatalyst from, 381
MOFs. *See* Metal–organic frameworks (MOFs)
Molecular recognition element (MRE), 346–347
Molecular self-assembled ether-based polyrotaxne, 235
Molten carbonate fuel cells (MCFCs), 136–137
Multichannel carbon fibers (MCCFs), 311–312
Multivalent metal-sulfur batteries, 315–319
Multiwalled CNTs (MWCNTs), 144–146, 150*f*
Municipal wastewater treatment, 349–350
Mushrooms, conductive electrodes using, 174
MXenes, 302–305
  based anodes, 188–189
  electrical conductivity of, 304–305
  electrochemical properties of, 306*t*

N

Nanoantenna, resonance properties of, 403–405
Nanobiosensors, 337–338
  types of, 338f
Nanocatalysts, 336–337
Nanocomposite materials, 9–10
Nanocomposite separators, 239–241
Nanoelectrochemistry
  alloying materials with, 228f
  conversion type materials with, 231f
  in electrolytes
    inorganic nanoparticle incorporated composite polymer electrolytes, 236–237
    localized high-concentration electrolytes, 232–233
    polymer electrolytes, 233–235
  in separators, 237–243
    general characteristics of LIB separator, 238–239
    nanocomposite separators, 239–241
    nanofibrous separators, 241–243
Nanoelectrodes, 442–443
  novel materials for, 144–151
    carbon nanotubes (CNTs), 144–146
    graphene and graphene oxide, 146–148
    metal-organic frameworks (MOFs), 148–149
    metal oxide nanosheets, 149–151
Nanofabrication, 337–338
Nanofertilizers, 337–338
Nanofibrous separators, 241–243
Nanofilms, 68
Nanofiltration, 349–350
Nanomaterials, 4, 90–94, 438
  applications of, 434–436, 448f
  assembly and integration of
    inkjet printing, 122
    layer-by-layer assembly, 122
    self-assembly, 122
  bandgap engineering with, 126
  as catalysts, 126
  categories of, 4, 431–432
  characteristic properties of, 428–431

  chemical structures of, 434f
  definition of, 428–429
  description of, 1–4
  dimensional structures of, 432–434
  for efficient charge separation and reduced recombination, 126
  electrochemistry and photoelectrochemistry of, 441
  electrochromic based on organic nanomaterials, 25–28
  electrochromic nanomaterials
    based on Prussian blue, 14–24
    based on transition metal oxide, 7–14
  energy state of, 435
  for enhanced light absorption, 126
  evolution of, 97–98
  features of, 431–432
  for flexible photo-electrochemical solar cells, 127
  forms of, 70
    nanofilms, nanosheets, or nanoplates, 68
    nanospheres, 68–69
    nanotubes, 67–68
    nanowires, 66–67
    Porous 3D nanostructures networks, 70
  for improved interfacial contacts, 127
  for increased stability and durability, 127
  manufacturing process of, 429–430
  molecules, 347
  morphology of, 440
  parameters for conventional electrochromic device performance, 4–6
  photoelectrochemistry of, 436–441, 447–448
  properties of, 436
  radical dimensions of, 438
  role of, 94–95
  technologies, 442
  types of, 196, 337–338, 429, 437f
  use of, 446–447
Nanomixed metal oxides, 336–337

464 Index

Nanoparticles, 12, 341−342
  encapsulation of, 337−338
  sputtering, 176−177
Nano/photoelectrochemistry, 334
Nanoplates, 68
Nanoroughness, 176−177
Nanoscale, 433
  material, 60f
  thermometry, 409−411, 415−416
Nanoscience in electrochemistry,
    333−334
Nanoscopic scale, 440−441
Nanosheets, 68, 432−433
Nanosized semiconductors, 77
Nanospheres, 68−69, 69f
Nanostructured cathode
  materials, 220−221
Nanostructures, 401−402
  cathodes, 214−215, 217−218
  dimensions on, 429
  fabrication processes of, 226−227
  films and layers, 97
  materials, 220
  morphology and size of, 214
Nanotechnology, 60, 64, 211, 413
  advances in, 436
  applications of, 65, 395−396
  challenges and limitations of,
    80−81
  features, 64−65
  materials in, 66−70
  and nanomaterials, 64−65
  studies, 431
Nanothermometers, gold nanoparticles as,
    414f
Nanotubes, 67−68
  anode, 382
Nanowires, 66−67, 168, 438
  interesting characteristic of, 66−67
  solar cells, 67
Natural mussels, 191−192
N-doped carbon nanotubes, 184−186
N doping carbon anode, 183f
Near-infrared (NIR) radiation, 405−408
Nernst−Donnan equation, 40
Nernst equation
  for anode potential, 162
  for cathode potential, 162−163

Next generation lithium batteries
  description of, 211
  nanoelectrochemistry in anode materials,
    222−232
  alloy-type materials and
    nanoelectrochemistry, 223−230
    silicon, 223−228
  conversion materials and
    nanoelectrochemistry, 230−232
  nanoelectrochemistry in cathode
    materials
    challenges and future perspectives,
      220−222
    fundamental properties of, 212−213
    nanoelectrochemical mechanisms in,
      216−219
    nanoengineering for high-
      performance cathodes, 219−220
    synthesis of nanostructured, 213−215
    types of advanced, 215−216
  nanoelectrochemistry in electrolytes
    inorganic nanoparticle incorporated
      composite polymer electrolytes,
      236−237
    localized high-concentration
      electrolytes, 232−233
    polymer electrolytes, 233−235
  nanoelectrochemistry in separators,
    237−243
    general characteristics of LIB
      separator, 238−239
    nanocomposite separators, 239−241
    nanofibrous separators, 241−243
Nitrogen-doped graphene, 294f
Noble metal
  based nanoelectrocatalysts, 371−383
  free nanomaterials, 383−384
  nanostructures, 398−400, 415−416
Nonradiative properties, 398−400
Nonrenewable energy, 287−289
Nonsilicon alloying anodes, 228−229
Novel device architectures, 95
Novel materials for nanoelectrodes in,
    144−151
N-type
  layer, 62−63
  semiconductor, 339−340
  silicon layers, 62

Index 465

O

OER. *See* Oxygen evolution reaction (OER)
1-D nanostructured materials, 432–433
One-dimensional nanomaterials (1D), 90f, 92
One-dimensional nanostructures, 97
Optical characterization
  Fourier transform infrared spectroscopy (FTIR), 125
  photoluminescence spectroscopy (PL), 124
  UV-Vis spectroscopy, 124
Optical manipulation, functionality of, 411–413
Optical transmittance, 23f
Organic electrochromic nanomaterials, 25
Organic-inorganic hybrid perovskites, 117–118
Organic polymers, 271
Organic solar cell (OSC), 69f
  buffer layer of, 69f
  polymer-based, 68–69
Organic synthesis operations, 334
OSC. *See* Organic solar cell (OSC)
Oxidation-reduction reaction, 2
Oxidized graphite, 295
Oxygen-containing groups, 146–147
Oxygen evolution reaction (OER), 149, 364–370
  phosphates and sulfates on, 369
  reactions, 366–368
    pathway of, 370
    rates, 367f
  selectivity, 368t
Oxygen reduction reaction (ORR)
  durability and, 143
  performance in, 149–151

P

Pathogen detection, 49–50
PCO
  material, 268–269
  nanosheet arrays electrode, 270f
PEC solar cells. *See* Photoelectrochemical (PEC) solar cells

Perovskite nanomaterials, 116–120
  all-inorganic perovskites, 118–119
  organic-inorganic hybrid perovskites, 117–118
Perovskite quantum dots, 113
Peroxytungstate precursor, 10–11
Phosphate-based cathodes, 216
Phosphoric acid fuel cells (PAFCs), 135–136
Photoanode, 96, 101
Photocatalysts, 347–348, 349t
Photochemical enzyme models, 349–350
Photo-driven semiconductors, 443
Photo electrocatalytic water splitting, 341f
Photoelectrochemical cell
  for hydrogen production, 439f
  schematic of, 446f
Photoelectrochemical (PEC) solar cells, 96, 98, 101, 109
  advanced functionalities, 129
  assembly and integration of nanomaterials
    inkjet printing, 122
    layer-by-layer assembly, 122
    self-assembly, 122
  assembly of, 90
  background and importance of, 89–90
  bandgap engineering with nanomaterials, 126
  bottom-up methods, 120–121
    chemical vapor deposition (CVD), 121
    electrospinning, 121
    hydrothermal/solvothermal synthesis, 121
  carbon-based nanomaterials, 106–109
    carbon nanotubes (CNTs), 107–108
    fullerenes, 108–109
    graphene, 107
  cells, 90
  charge
    collection and current generation, 99–101
    separation, 99
    transport, 99, 101
  components of, 101
  compositional characterization

Photoelectrochemical (PEC) solar cells
(*Continued*)
  energy-dispersive X-ray spectroscopy
(EDS), 124
  inductively coupled plasma mass
spectrometry (ICP-Ms), 124
  X-ray photoelectron spectroscopy
(XPS), 124
  conducting polymer nanomaterials,
113–116
   poly(3-hexylthiophene) (P3HT),
114–115
   polyaniline, 114
  cost-effective nanomaterials for solar
cells, 127
  counter electrode, 102
  description of, 89–99
  doping and surface functionalization of
nanomaterials, 122–123
   atomic layer deposition (ALD),
122–123
   ion implantation, 122
   plasma treatment, 123
  dye or quantum dots, 102
  economic viability, 128
  efficiency and performance, 128
  efficiency enhancements, 129
  electrical characterization
   current-voltage (I-V) measurements,
125
   electrochemical impedance
spectroscopy (EIS), 125
   quantum efficiency measurements,
125
  electrolyte, 102
  energy conversion efficiency, 100
  flexibility and integration, 129
  injection of excited electrons, 100
  light absorption and exciton generation,
99–100
  metal oxide nanomaterials, 102–106
   iron(III) oxide, 104–105
   titanium dioxide, 103
   tungsten trioxide, 105–106
   zinc oxide, 103–104
  nanomaterials, 90–94
   as catalysts, 126

   for efficient charge separation and
reduced recombination, 126
   for enhanced light absorption, 126
   evolution of, 97–98
   for flexible photo-electrochemical
solar cells, 127
   for improved interfacial contacts, 127
   for increased stability and durability,
127
   role of, 94–95
  optical characterization
   Fourier transform infrared
spectroscopy (FTIR), 125
   photoluminescence spectroscopy (PL),
124
   UV-Vis spectroscopy, 124
  overview of, 96
  perovskite nanomaterials, 116–120
   all-inorganic perovskites, 118–119
   organic-inorganic hybrid perovskites,
117–118
  photoanode, 101
  photovoltaic (PV) effect, 99
  quantum dots (QDs), 109–113
   cadmium telluride (CdTe) quantum
dots, 110–111
   lead selenide (PbSe) quantum dots,
111–112
  regeneration of dye or quantum dots,
101
  stability and durability, 128
  structural characterization, 123–124
   atomic force microscopy (AFM), 124
   scanning electron microscopy (SEM),
123
   transmission electron microscopy
(TEM), 123
   X-ray diffraction (XRD), 124
  surface characterization
   contact angle measurements, 125
   surface profilometry, 125–126
   X-ray photoelectron spectroscopy
(XPS), 125
  sustainability, 129
  synthesis and fabrication, 127–128
  top-down methods, 121–122
   lithography, 121

Index 467

physical vapor deposition (PVD), 121–122
sputtering, 122
Photoelectrochemistry, 334, 443, 445–446
applications of, 447
concept of, 438–439
of nanomaterials, 436–441
Photoelectrochromic smart windows, 2–3
Photoluminescence spectroscopy (PL), 124
Photolysis, 349–350
Photon wavelengths, 438
Photoredox catalysis, 334
Photo-sensitizer, 89–90
Photothermal
heat, formation of, 396–397
heating, 415–416
membrane, 413
motors, 415–416
nanoparticles, suspension of, 403–405
nanostructures, 407–408
Photovoltaic devices, power conversion efficiency in, 70
Photovoltaics (PV), 61, 99
Physical vapor deposition (PVD), 121–122
Plant heteroatoms, 178
Plasmonic heat energy, 408–409
Plasmonic heating
synergistic integrations of, 411–413
theory of, 400–402
Plasmonic metal nanostructures, 405–406
photothermal effects in, 408–409
Plasmonics, 395–396
nanoheaters, 398
nanoparticles, 403–405
nanostructures, 395–400
photothermal applications, 414–415
photothermal nanostructures, 409–411
polymers, 403–405
Platinum catalysts, 149
Poisson equation, 402
Pollutant degradation, 343–344
Poly3,4-ethylenedioxythiophene (PEDOT), 115–116, 187, 308
schematic illustration of, 310f
Poly(3-hexylthiophene) (P3HT), 114–115
Poly 3-octylthiophene (P3OT) polymer composite, 77–78

Polyacrylonitrile (PAN) nanofibers, 257–258
Polyaniline, 114
Polyaniline-doped mesoporous carbon nanocomposite, 175–176
Polycrystalline semiconductors, 442–443
Polydopamine (PDA), 173, 187–188, 236
Polyethylene, 25–26
Poly(ethylene glycol) (PEG), 235
Poly(ethylene glycol) methyl ether acrylate (PEGA), 234
Polymer
based anodes, 186–188
coated microorganisms, 191–192
electrolytes, 233–235
Nafion, 341–342
polyvinyl chloride (PVC), 359
Polymeric material is polyaniline (PANI), 307–308
schematic illustration of, 310f
Polymeric materials, 305–308
Polymeric nanoparticles, 438
Polymerization-induced microphase separation (PIMS)
design of, 234f
technique, 234–235
Polymerization process, 308
Polymers, 192, 308
conducting, 263–264
Polyolefin separators, 237–238
Poly phase photocatalysts, 339–340
Polypyrrole (PPy), 115, 307
schematic illustration of, 310f
Polyrotaxane solid polymer electrolyte, 235f
Polysulfides, 289–291
Polysulfide shuttle, 291, 312–315
suppression of, 292–293
Polythiophenes, 116
Polyvinylchloride incineration, dioxin emission for, 336–337
Polyvinyl pyrrolidone (PVP), 295
Porous 3D nanostructures networks, 70
Porous carbon, 183
materials, 178, 293
Potassium, abundance of, 312

468 Index

Potassium-sulfur batteries, 312−315
  cyclic voltammogram, 315f
  performance parameters of applied
    materials in, 316t
Potentiometric biosensors, 35−36, 39−41
Potentiometric sensors, 344−345
Potentiometry techniques, 345−346, 345f
Pourbaix diagram, 364−366, 366f
Power conversion efficiency (PCE), 66−69
Prostate-specific antigen (PSA), 45−46
Proton exchange membrane fuel cell
    (PEMFCs), 137−138, 142−143
  advantages of, 138
  nanoparticles for, 143
  schematic diagram for, 139f
  utilization of nanomaterials in, 142−143
Proton exchange membranes (PEM),
    338−339
  fuel cells, 147
Prussian blue (PB), 27−28
  applications, 17
  color of, 16−17
  cyclic voltammetry (CV) of, 24f
  electrochromic nanomaterials based on,
    14−24
  electrochromic properties of, 14−15
  films in electrolytes, 23−24
  nanoparticles and polyaniline, 24
  SEM and cross-section images of, 21
  SEM image for, 20f
  stability of, 19
  thin films, 18−19, 21
  unit cell, 15f
Prussian blue analog compounds (PBA), 16
Prussian blue unit cell, 15f
Pseudocapacitance, 258−259
Pseudocapacitive materials, 252
Pseudocapacitors (PCs), 253, 258−262
  electrode materials, 261−262
P-type semiconductor, 339−340
P-type silicon layers, 62

Q

Quantum dots (QDs), 64, 77, 102,
    109−113
  cadmium telluride (CdTe) quantum
    dots, 110−111

lead selenide (PbSe) quantum dots,
    111−112
  regeneration of, 101

R

Recrystallization, 27−28
Redox reactions, 37, 338−339
Reduced graphene oxide (rGO), 148−149
Reference electrode (RE), 35−36, 39−40
Remote area power supply (RAPS), 140
Reverse electrolysis process, 133−134
Rich redox reactions, 277−278
RuO$_2$, 378f
Ruthenium-based single-atom catalyst, 380

S

Sacrificial agents, 279−281
SAED patterns, 374f
Scanning electrochemical microscopy
    (SECM), 376
Scanning electron microscopy (SEM), 123
Scanning tunneling microscope (STM),
    427
Self-assembly, 122
Self-cleaning, 78
Self-monitoring of blood glucose (SMBG),
    46−47
Semiconductor, 439−440, 442−443
  electrochemistry, 442−443
  materials, inner engineering of, 66−67
  nanomaterials, 436
  nanowires (NWs), 66−67
  n-type, 339−340
  photochemistry, 445
  photo-driven, 443
  p-type, 339−340
Sensitivity, 52
Silicon (Si), 223−228
  based anode materials, 225f
  based solar cells, 89−90
  nanoparticles, 76−77
  nanostructured, 226−227
  nanowires, 226
  quantum dots, 112
Silver nanotubes, 65
Si microparticles (SiMPs), 227−228

Single-atom catalysts, rapid development of, 377–380
Single fuel cell, schematic of, 133–134, 134f
Single-walled CNTs (SWCNTs), 144–146, 150f
Smart materials, 1–2
Smonic photothermal nanostructures, 397–398
Sn-PC60 nanocomposite, 229–230
Sodium acetate, 162
Sodium-sulfur batteries, 310–312
Solar cells
  advantages of nanoscale, 70–80
    anti-reflective coating, 79–80
    carbon nanotubes, 77–78
    cost reduction, 80
    light absorption, 76–77
    quantum dots, 77
    self-cleaning, 78
  applications of nanotechnology in, 65
  asymmetry in, 62
  challenges and limitations of nanotechnology in, 80–81
  classification stems, 63–64
  components and mechanism of, 93f
  conventional, 76–77
  cost-effective nanomaterials for, 127
  description of, 59–60
  forms of nanomaterials, 70
    nanofilms, nanosheets, or nanoplates, 68
    nanospheres, 68–69
    nanotubes, 67–68
    nanowires, 66–67
    Porous 3D nanostructures networks, 70
  generations, types, components, and structures of, 92f
  history of, 60–61
  invention of, 60–61
  nanomaterials in, 71t
  nanotechnology, 64
    materials in, 66–70
    and nanomaterials, 64–65
  p-n junction of, 63f
  technology, 60–61

types of, 63–64
workings of, 62–63
Solar hydrogen production, 339
  photoelectrochemical for, 339
Solar panel, dust accumulation on, 79f
Solar-to-chemical processes, 440
Solid electrolyte interphase (SEI) layer, 231–232
Solid electrolyte nanoparticles, 241
Solid oxide fuel cells (SOFCs), 136, 141
  operating temperatures for, 142t
Solid polymer electrolytes (SPE), 233–234
Solid-state batteries, 221
Spherical plasmonic nanoparticle, 396f
Sputtering, 122
Stainless steel electrodes, 171–172
State-of-the-art nanomaterials, 140–144
  DAFCs, 144
  PEMFCs, 142–143
  SOFCs, 141
Strong-Metal-Support Interaction (SMSI), 149–151
Structural characterization, 123–124
  atomic force microscopy (AFM), 124
  scanning electron microscopy (SEM), 123
  transmission electron microscopy (TEM), 123
  X-ray diffraction (XRD), 124
Sulfur-based electrode, 300f
Sulfur cathodes, 317
  host materials for, 292f
  noncarbonaceous materials in, 305–307
Sulfur utilization
  hydrothermal reaction for, 295
Supercapacitors (SCs), 251–255
  charge storage mechanism in, 253
  classification of, 255
  cost-efficient coordination materials for electrochemical supercapacitors, 263–265
  effectiveness of, 255
  electrodes, 259, 272
  energy storage capability of electrodes, 265–282
    defect engineering, 278–282
    elemental doping, 265–268

Supercapacitors (SCs) (*Continued*)
  heterointerface engineering, 274−278
  hybridization with carbons, 272−274
  surface functionalization, 268−271
  overview of electrochemical
    supercapacitors, 254−263
  hybrid supercapacitors (HSCs),
    262−263
  pseudocapacitors, 258−262
  representative diagram of, 254f
Surface characterization
  contact angle measurements, 125
  surface profilometry, 125−126
  X-ray photoelectron spectroscopy
    (XPS), 125
Surface charges, accumulation of, 398−400
Surface functionalization, 268−271
Surface micro-/nanostructures, 176−177
Surface plasmon oscillation, 403−405
Surface plasmon resonances, 68−69,
  414−415
Surface profilometry, 125−126
Surface quasi-capacitive mechanism,
  12−14
Surface roughness, 21, 171−172
Surface wettability effect, 172−173
  on bacterial adhesion, 173f
Sustainability, 129
Sustainability Development Goals (SDGs),
  333−334
  perspective of, 159−160
Synthetic electrochemistry, 334
Synthetic strategies, 397−398
Synthetic techniques, 438
Systematic Evolution of Ligands by
  Exponential Enrichment
  (SELEX), 346−347

T

Temperature distribution, 402
Thermal energy, 400
Thermal stability, 148
Thermodynamics, 395−396
Thermoionic conversion, 409−411
Thermoplasmonics, 395−397
3-D structured nanomaterials
  features of, 433−434

Three-dimensional nanomaterials (3D),
  90f, 92
Tin (Sn), 229−230
TiO_2 catalyst on pollutant degradation,
  343−344
Titanium dioxide, 52, 103
Titanium dioxide nanotube arrays (TNAs),
  67−68
Titanium oxide film, 348−349
Toluene, degradation of, 413f
Top-down methods, 121−122
  lithography, 121
  physical vapor deposition (PVD),
    121−122
  sputtering, 122
Transition metal
  sulfides, 272−274
Transition metals, 382−383
  based electrodes, 259
  chalcogenides, 264−265
  ions, 12
  nitrides, 264−265
  oxides/hydroxides, 7, 263−264
  phosphides, 263−264
Transmission electron microscopy (TEM),
  123
Transparent flexible sheets, 3−4
Tungsten oxide, 8
  boundaries of, 10−11
  SEM images of, 11f
  two-dimensional, 9
Tungsten trioxide, 105−106
2-D nanostructured materials, 433
Two-dimensional metal oxides (TMOs), 9
Two-dimensional nanomaterials (2D), 90f, 92

U

Ultrathin nickel oxide nanosheet, 149−151
Urea detection, 40−41, 41f
Uric acid, 53
UV-Vis spectroscopy, 124

V

Vacancy-ordered perovskites, 120
Vanadium pentoxide, 12
Viologens, 26−28

Volmer—Heyrovsky pathway, 363—364
Volmer—Tafel pathway, 363
Voltammetry techniques, 345—346, 345f
Volumetric energy density, 288f

W
Wastewater
 photochemical enzymatic systems in,
  349—351
 treatment, 159—160
Water
 bodies, 343
 dissociation, 380
 pharmaceutical containments from,
  348—349
 sources, 287—289
Water splitting, 338—341, 441
 basic fundaments of water splitting,
  339—340

for H_2 production, 341
photoelectrochemical water splitting for
 solar hydrogen production, 339
reactions, 340f
Wide-bandgap nanomaterials, 66
Working electrode (WE), 35—36, 39—40

X
X-ray diffraction (XRD), 124
X-ray photoelectron spectroscopy (XPS),
 124—125

Z
Zeolitic imidazolate frameworks (ZIL),
 148—149
Zero-dimensional nanomaterials (0D), 92
 schematic diagram of, 90f
Zinc oxide, 103—104